DATE DUE

GAYLORD			PRINTED IN U.S.A

Manual of Physical Status and Performance in Childhood

Volume 1B: Physical Status

Manual of Physical Status and Performance in Childhood

Volume 1B: Physical Status

Alex F. Roche
Wright State University School of Medicine
Yellow Springs, Ohio

and

Robert M. Malina
University of Texas
Austin, Texas

PLENUM PRESS • NEW YORK AND LONDON

Library of Congress Cataloging in Publication Data

Roche, Alex F., 1921–
  Manual of physical status and performance in childhood.

  Author's names in reverse order in v. 2.
  Includes bibliographies and indexes.
  Contents: v. 1. Physical status (2 v.)—v. 2. Physical performance.
  1. Children—North America—Growth. 2. Children—North America—Anthropometry. I. Malina, Robert M. II. Title. (DNLM: 1. Child Development—Tables. 2. Motor skills—In infancy and childhood—Tables. WS16 R637m)
R637m)

RJ131.R592 1982   612'.65'0973   82-16515
ISBN 0-306-41136-9 (v. 1)
ISBN 0-306-41137-7 (v. 2)

© 1983 Plenum Press, New York
A Division of Plenum Publishing Corporation
233 Spring Street, New York, N.Y. 10013

Printed in the United States of America

CONTENTS

VOLUME 1A

INTRODUCTION.. 1

STATURE AND RECUMBENT LENGTH.. 9

WEIGHT... 170

HEIGHTS.. 395

CRANIOFACIAL MEASUREMENTS.. 441

VOLUME 1B

TRUNK MEASUREMENTS... 873

LOWER LIMB MEASURMENTS... 997

UPPER LIMB MEASURMENTS.. 1053

PHYSIQUE... 1115

SKELETAL MEASUREMENTS.. 1125

MATURATION... 1211

STATURE PREDICTION... 1361

SUPPLEMENTAL INFORMATION

    Published sources of raw data... 1371

    Relevant sources of data not included................................... 1383

ANNOTATED BIBLIOGRAPHY... 1387

INDEX.. 1425

TRUNK MEASUREMENTS

TABLE 811

SOME BODY LENGTHS (in)
IN MICHIGAN CHILDREN

Age (years)	Sex	Shoulder * to trochanter		Trochanter to knee	
		Mean	S.D.	Mean	S.D.
5	Boys	14.2	.86	8.6	.67
	Girls	13.1	1.64	8.4	.81
6	Boys	14.3	.81	8.8	.78
	Girls	13.9	.69	8.9	.76
7	Boys	15.3	.97	9.5	.89
	Girls	14.7	1.03	9.7	.97
8	Boys	15.8	.99	10.1	.91
	Girls	15.5	1.03	10.2	1.03
9	Boys	16.6	1.04	10.9	.76
	Girls	16.0	1.04	10.8	.82
10	Boys	17.0	1.42	11.4	.98
	Girls	17.0	1.18	11.5	.72
11	Boys	17.6	1.35	12.2	.93
	Girls	17.8	1.45	12.4	1.01
12	Boys	18.2	1.37	12.4	1.21
	Girls	19.0	1.35	12.5	.91
13	Boys	19.2	1.59	13.2	1.18
	Girls	19.8	1.13	13.1	.94
14	Boys	20.4	1.69	13.8	1.22
	Girls	21.6	1.19	13.5	.84
15	Boys	21.6	1.69	14.3	1.12
	Girls	21.1	1.15	13.4	.96
16	Boys	22.2	1.64	14.4	1.24
	Girls	21.2	1.32	13.3	1.46

TABLE 811   (continued)

SOME BODY LENGTHS (in)
IN MICHIGAN CHILDREN

Age (years)	Sex	Shoulder * to trochanter		Trochanter to knee	
		Mean	S.D.	Mean	S.D.
17	Boys	22.9	1.74	14.4	1.45
	Girls	21.6	1.38	13.1	1.19
18	Boys	23.3	1.41	14.3	1.24
	Girls	21.6	1.67	13.4	1.08

*="point of arm rotation"

(data from Martin, 1955)

TABLE 812

CERVICALE HEIGHT IN U.S. CHILDREN

Age						Percentiles			
(years)	Mean	S.D.	5	10	25	50	75	90	95

BOYS

Age	Mean	S.D.	5	10	25	50	75	90	95
12	128.1	7.51	115.4	118.1	122.8	128.2	132.8	137.3	140.4
13	135.0	8.38	121.4	124.1	129.5	134.6	140.6	145.8	148.9
14	141.1	7.92	127.7	130.4	135.8	141.9	146.6	150.5	153.1
15	145.4	6.70	133.3	136.7	141.0	145.8	149.5	153.7	155.9
16	147.7	6.59	136.8	139.4	143.6	147.8	152.2	155.7	158.0
17	148.9	6.54	137.6	140.4	144.7	149.2	153.3	157.2	160.1

GIRLS

Age	Mean	S.D.	5	10	25	50	75	90	95
12	131.1	6.72	119.3	121.9	127.0	131.2	135.3	139.4	141.7
13	134.4	6.43	123.2	126.2	129.9	134.8	138.9	142.4	144.0
14	136.8	5.89	127.4	129.3	133.1	136.5	140.6	143.9	146.4
15	137.6	6.44	127.4	129.2	133.3	137.6	142.1	145.5	148.3
16	138.0	6.16	127.6	129.9	133.9	138.0	141.9	145.8	148.5
17	138.0	6.07	128.0	130.5	133.6	138.3	142.4	145.6	147.4

CERVICALE HEIGHT (cm) IN U.S. WHITE CHILDREN

Age						Percentiles			
(years)	Mean	S.D.	5	10	25	50	75	90	95

BOYS

Age	Mean	S.D.	5	10	25	50	75	90	95
12	128.0	7.69	115.2	117.8	122.7	128.2	132.8	137.5	140.6
13	134.9	8.37	121.4	124.1	129.4	134.4	140.5	145.8	148.7
14	141.2	7.91	127.6	130.3	136.2	142.2	146.6	150.3	153.2
15	145.4	6.64	133.3	136.6	141.1	145.8	149.4	153.5	155.9
16	147.8	6.51	137.2	139.6	143.5	147.6	152.1	155.8	158.3
17	149.0	6.52	137.5	140.5	144.8	149.3	153.4	157.1	160.1

GIRLS

Age	Mean	S.D.	5	10	25	50	75	90	95
12	130.9	6.76	119.0	121.4	126.8	131.0	135.2	139.4	141.6
13	134.3	6.44	123.0	126.2	129.7	134.6	138.8	142.4	144.4
14	136.7	5.94	127.3	129.1	133.0	136.4	140.4	143.8	146.6
15	137.6	6.47	127.3	129.1	133 3	137.6	142.2	145.6	148.2
16	138.0	6.15	127.6	129.9	134.1	138.1	141.8	146.0	148.3
17	138.0	5.89	128.3	131.0	133.7	138.2	142.1	145.4	147.3

(data from Cycle III of the Health Examination Survey by the National Center for Health Statistics 1966-1970)

TABLE   813

CERVICALE HEIGHT (cm) IN U.S. NEGRO CHILDREN

Age					Percentiles				
(years)	Mean	S.D.	5	10	25	50	75	90	95

BOYS

12	128.3	6.31	117.0	121.1	123.3	128.4	133.1	136.8	138.6
13	135.4	8.54	122.0	124.2	129.6	136.3	141.3	147.3	149.1
14	140.7	8.00	129.0	131.2	133.7	141.0	145.8	151.3	152.6
15	145.5	7.19	133.6	136.9	140.7	145.3	150.7	154.7	157.4
16	147.9	7.10	133.8	137.6	144.8	149.3	152.6	155.4	157.0
17	148.8	6.70	137.6	139.8	144.9	148.2	152.6	157.6	161.8

GIRLS

12	132.7	6.34	122.3	125.2	129.3	133.1	136.3	139.7	144.8
13	135.4	6.28	124.6	126.2	131.0	136.2	139.5	143.1	145.7
14	137.4	5.49	128.2	130.1	133.7	138.0	141.4	144.5	145.8
15	137.9	6.05	127.5	131.0	133.9	137.5	141.8	144.9	149.7
16	138.0	6.14	128.2	130.0	133.4	137.8	142.5	144.8	148.0
17	138.4	7.01	127.2	128.5	133.4	139.1	143.7	146.5	148.4

(data from Cycle III of the Health Examination Survey by the National Center for Health Statistics, 1966-1970)

TABLE 814

SUPRASTERNALE HEIGHT (mm) IN CLEVELAND CHILDREN

Age	BOYS		GIRLS	
	Mean	S.D.	Mean	S.D.
18 mos.	617.54	25.51	607.90	24.90
2 yrs.	658.34	28.75	651.97	25.93
2½ "	698.57	28.40	694.16	29.28
3 "	735.78	28.26	730.34	30.74
3½ "	767.71	29.27	765.08	31.60
4 "	801.71	31.15	798.39	32.79
4½ "	832.08	35.03	832.56	36.63
5 "	861.91	34.06	862.78	37.88
6 "	922.77	37.56	925.69	43.10
7 "	979.83	40.32	981.09	46.32
8 "	1035.90	41.96	1036.37	48.36
9 "	1084.81	45.71	1086.23	49.22
10 "	1134.03	51.13	1138.96	50.77
11 "	1181.79	55.01	1195.74	56.74
12 "	1228.53	59.86	1254.14	61.22
13 "	1280.59	69.13	1302.75	52.22
14 "	1344.74	71.12	1333.17	46.94
15 "	1404.11	62.12	1344.50	44.60
16 "	1426.36	52.59	1345.79	47.70
17 "	1441.83	49.83	1346.64	42.96

(data from Simmons, 1944)

TABLE 815

SUPRASTERNALE HEIGHT (cm) IN PHILADELPHIA CHILDREN

Age (Years)	Boys White		Girls White		Boys Negro		Girls Negro	
	Mean	S.D.	Mean	S.D.	Mean	S.D.	Mean	S.D.
7	96.7	4.3	97.2	4.2	97.5	4.8	97.9	5.3
8	100.4	4.6	101.5	4.9	102.2	5.6	102.6	5.7
9	106.6	4.9	105.6	5.1	106.9	5.7	108.3	5.6
10	112.2	5.2	111.1	6.2	111.7	8.0	113.5	5.6
11	116.0	4.6	112.8	5.9	115.1	6.9	118.4	6.7
12	120.2	6.2	121.1	5.3	119.5	7.5	121.3	7.4
13	124.1	6.4	124.8	5.2	129.4	7.9	127.0	5.9
14	128.4	5.7	128.4	5.1	134.2	9.3	130.1	5.6

(data from Krogman, 1970)

TABLE 816

ANTERIOR TRUNK HEIGHT (cm) IN PHILADELPHIA CHILDREN

(Years)	Boys White Mean	S.D.	Girls White Mean	S.D.	Boys Negro Mean	S.D.	Girls Negro Mean	S.D.
7	36.2	1.5	36.0	2.1	35.7	2.2	35.4	2.3
8	37.4	2.4	37.1	2.3	37.2	2.5	37.3	4.1
9	39.1	2.3	38.2	2.4	38.5	2.8	38.6	2.4
10	40.5	2.3	39.7	2.5	40.1	3.5	39.6	2.6
11	41.1	2.3	40.9	2.2	39.9	2.6	40.7	2.6
12	42.5	2.5	42.7	2.4	40.6	2.6	41.7	2.9
13	44.6	2.8	44.2	2.1	44.3	3.2	42.9	3.0
14	47.8	2.6	44.5	2.1	46.2	4.3	44.8	2.6

(data from Krogman, 1970)

TABLE 817

LENGTHS OF UPPER AND LOWER SEGMENTS (cm) IN PHILADELPHIA CHILDREN

Age (Years)	Boys White LS	US	Girls White LS	US	Boys Negro LS	US	Girls Negro LS	US
7	60.4	61.4	60.7	60.7	61.8	60.4	62.5	59.3
8	64.3	62.5	64.3	60.9	65.1	62.3	65.5	61.2
9	67.7	63.7	67.8	62.6	68.3	64.0	69.9	62.2
10	71.6	66.7	70.7	66.0	71.5	66.0	73.9	65.3
11	74.7	66.1	74.3	67.3	75.2	66.6	77.6	68.3
12	78.1	69.7	78.4	68.8	78.7	67.9	80.1	69.5
13	83.0	75.4	80.0	72.4	85.2	69.7	84.1	69.9
14	84.8	78.6	83.5	72.9	83.8	72.9	86.0	73.8

Upper segment (US) = stature less symphyseal height; lower segment (LS) = symphyseal height.

(data from Krogman, 1970)

TABLE 818

BIACROMIAL DIAMETER (cm) IN U.S. CHILDREN

Age (years)	Mean	S.D.	Percentiles						
			5	10	25	50	75	90	95
BOYS									
12	32.8	2.00	29.3	30.1	31.5	32.8	34.2	35.4	36.1
13	34.6	2.45	30.6	31.4	33.0	34.6	36.4	37.9	38.9
14	36.4	2.52	32.2	33.2	34.9	36.5	38.2	39.6	40.4
15	37.8	2.32	33.5	34.6	36.3	38.0	39.4	40.6	41.4
16	38.8	2.08	35.3	36.1	37.4	38.9	40.3	41.4	42.1
17	39.5	1.99	36.1	36.9	38.3	39.5	40.8	42.2	42.7
GIRLS									
12	33.2	2.03	29.9	30.6	31.9	33.2	34.6	35.6	36.3
13	34.0	2.24	30.7	31.6	32.8	34.1	35.3	36.5	37.1
14	34.8	1.80	31.8	32.4	33.6	34.8	35.9	37.2	37.8
15	35.0	1.91	32.2	32.6	33.7	35.1	36.4	37.4	38.1
16	35.4	1.71	32.7	33.3	34.3	35.4	36.5	37.6	38.4
17	35.5	1.84	32.6	33.3	34.4	35.5	36.6	37.7	38.6

(data from Cycle III of the Health Examination Survey 1966-1970 by the National Center for Health Statistics).

TABLE   819

BIACROMIAL DIAMETER (cm) IN U.S. WHITE CHILDREN

						Percentiles			
Age (years)	Mean	S.D.	5	10	25	50	75	90	95

BOYS

Age	Mean	S.D.	5	10	25	50	75	90	95
6	26.0	1.59	23.4	24.0	25.1	26.1	27.2	28.1	28.8
7	27.3	1.57	24.7	25.3	26.3	27.3	28.4	29.5	30.1
8	28.4	1.76	25.4	26.2	27.3	28.4	29.6	30.7	31.4
9	29.6	1.92	27.0	27.4	28.4	29.5	30.8	32.2	32.9
10	30.5	1.97	27.3	28.2	29.4	30.5	31.7	32.7	33.6
11	31.8	2.06	28.8	29.4	30.5	31.8	33.1	34.5	35.3
12	32.8	2.03	29.2	30.1	31.5	32.9	34.2	35.4	36.2
13	34.6	2.40	30.6	31.4	32.9	34.5	36.3	37.7	38.7
14	36.5	2.53	32.1	33.2	35.0	36.6	38.3	39.6	40.4
15	37.9	2.29	33.6	34.8	36.4	38.1	39.5	40.6	41.4
16	38.8	2.05	35.3	36.1	37.4	38.8	40.2	41.4	42.0
17	39.6	2.00	36.1	36.7	38.3	39.6	40.9	42.2	42.7

GIRLS

Age	Mean	S.D.	5	10	25	50	75	90	95
6	25.8	1.52	23.3	24.0	24.7	25.7	26.7	27.7	28.5
7	27.0	1.54	24.5	25.1	26.0	27.0	27.8	28.8	29.6
8	28.1	1.62	25.4	26.2	27.1	28.1	29.4	30.5	30.8
9	29.2	1.87	26.3	27.1	28.1	29.3	30.5	31.6	32.3
10	30.4	1.93	27.4	28.1	29.1	30.4	31.7	33.1	33.7
11	31.8	2.22	28.6	29.3	30.5	31.7	33.3	34.6	35.5
12	33.1	2.03	29.7	30.5	31.8	33.1	34.5	35.6	36.2
13	33.9	2.25	30.4	31.5	32.8	34.0	35.3	36.4	36.9
14	34.7	1.77	31.9	32.4	33.5	34.7	35.9	37.2	37.7
15	35.0	1.92	32.2	32.6	33.7	35.1	36.4	37.4	38.2
16	35.4	1.59	32.7	33.3	34.3	35.3	36.5	37.5	38.2
17	35.4	1.80	32.6	33.3	34.4	35.5	36.6	37.6	38.3

(data from Cycles II and III of the Health Examination Survey by the National
Center for Health Statistics, 1963-1965 and 1966-1970)

TABLE 820

BIACROMIAL DIAMETER (cm) IN U.S. NEGRO CHILDREN

| | | | | | | Percentiles | | | |
Age (years)	Mean	S.D.	5	10	25	50	75	90	95
BOYS									
6	26.3	1.95	23.7	24.2	25.0	26.2	27.4	29.0	29.5
7	27.4	1.58	24.5	25.3	26.4	27.3	28.4	29.5	30.0
8	28.9	1.61	25.7	26.9	27.7	29.1	30.1	30.9	31.6
9	29.6	2.00	26.3	27.0	28.3	29.5	31.1	31.9	33.0
10	30.8	1.57	28.4	29.1	29.6	30.8	32.1	33.1	33.6
11	32.0	1.90	28.9	30.0	30.8	32.1	33.0	34.3	35.6
12	32.8	1.82	29.4	30.8	31.8	32.7	33.7	35.3	35.9
13	35.1	2.09	30.7	31.4	33.3	34.7	37.3	38.8	39.5
14	36.4	2.45	33.0	33.4	34.6	36.4	37.8	39.8	41.1
15	37.2	2.51	32.9	34.1	35.6	36.9	39.2	40.4	41.1
16	37.1	2.28	35.5	36.2	37.6	39.5	40.6	41.6	42.4
17	37.3	1.96	35.9	36.7	37.9	39.1	40.7	42.1	42.7
GIRLS									
6	26.0	1.73	23.2	23.7	24.8	25.7	27.3	28.5	28.9
7	27.2	1.52	25.1	25.3	26.1	27.2	28.1	29.1	30.1
8	28.1	2.01	25.5	26.1	26.7	27.8	29.6	30.7	31.5
9	29.8	2.00	26.6	27.3	28.3	29.6	31.2	32.5	33.4
10	31.1	2.14	27.1	28.4	29.7	31.2	32.6	33.6	34.1
11	32.3	2.47	28.8	29.4	30.5	32.2	34.4	35.6	35.9
12	33.9	1.89	30.6	31.3	32.6	34.1	35.2	35.9	36.8
13	34.4	2.11	31.4	31.9	32.8	34.3	35.5	37.1	37.8
14	35.1	1.97	31.6	32.5	33.9	35.3	36.3	37.5	38.2
15	35.1	1.94	32.1	32.7	34.2	35.3	36.3	37.2	38.0
16	35.5	2.31	32.4	33.2	34.2	35.4	36.8	38.4	39.3
17	35.9	1.98	32.6	33.2	34.6	35.8	37.3	39.1	39.4

(data from Cycles II and III of the Health Examination Survey by the National Center for Health Statistics, 1963-1965 and 1966-1970)

TABLE 821

MAXIMUM SHOULDER WIDTH (cm)
IN CHILDREN OF 8 STATES

Age (Months)	Boys		Girls	
	Mean	S.D.	Mean	S.D.
0 – 3	15.8	1.6	15.7	1.6
4 – 6	19.1	1.1	18.3	1.4
7 – 9	20.5	0.9	19.8	1.1
10 – 12	20.7	1.1	20.3	1.1
13 – 18	22.2	1.2	21.7	1.2
19 – 24	22.7	1.3	22.7	1.4
25 – 30	23.5	1.2	23.8	1.1
31 – 36	24.5	1.5	23.9	1.2
37 – 42	24.9	1.1	24.6	1.3
43 – 48	25.3	1.3	24.9	1.3
49 – 54	25.9	1.4	25.5	1.4
55 – 60	26.5	1.4	26.0	1.5
61 – 66	27.2	1.5	26.7	1.7
67 – 72	27.8	1.6	27.3	1.5
73 – 78	28.2	1.5	28.1	1.6
79 – 84	29.2	1.6	28.4	1.7
85 – 96	30.3	2.1	29.3	1.9
97 – 108	31.2	2.1	30.9	2.1
109 – 120	32.1	1.8	31.9	2.2
121 – 132	34.0	2.2	33.7	2.7
133 – 144	35.5	2.0	34.9	2.5
145 – 156	35.8	2.3	37.4	2.5

(data from Snyder et al., 1975)

TABLE 822

BIACROMIAL DIAMETER (mm) IN CLEVELAND CHILDREN

Age	BOYS		GIRLS	
	Mean	S.D.	Mean	S.D.
3 mos.	140.77	10.16	136.26	9.91
6 "	155.92	11.27	151.07	11.71
9 "	167.47	13.46	160.75	13.69
12 "	175.07	13.62	168.36	12.75
18 "	190.91	13.57	183.52	12.61
2 yrs.	204.69	11.31	197.90	14.07
2½ "	212.22	9.87	208.88	11.14
3 "	218.76	9.71	215.92	10.77
3½ "	226.09	11.13	223.71	10.11
4 "	232.38	11.54	229.31	11.39
4½ "	236.75	10.98	236.03	11.08
5 "	243.63	11.64	241.08	11.84
6 "	254.67	12.62	253.90	11.98
7 "	266.86	12.18	266.20	12.52
8 "	278.45	12.92	277.08	13.24
9 "	289.08	13.22	287.61	14.31
10 "	301.68	14.29	299.70	15.60
11 "	310.68	15.54	313.40	17.51
12 "	318.90	15.90	325.81	18.10
13 "	330.90	18.80	337.71	17.34
14 "	347.26	21.74	345.41	16.62
15 "	365.19	21.62	351.51	15.79
16 "	373.36	21.05	358.22	16.52
17 "	383.25	17.14	359.07	17.34

(data from Simmons, 1944)

TABLE 823

BIACROMIAL DIAMETER (cm)
IN DENVER CHILDREN

Age Yr.- Mo.	Boys		Girls	
	Mean	S.D.	Mean	S.D.
Supine				
Birth	11.2	1.06	11.1	0.81
0 - 1	12.4	1.31	12.2	1.19
0 - 2	13.8	1.09	13.5	1.05
0 - 3	14.5	1.12	13.8	0.87
0 - 4	15.3	0.97	14.6	0.95
0 - 5	15.8	0.92	15.2	0.97
0 - 6	15.7	1.50	15.3	1.10
0 - 9	17.1	1.31	16.5	1.04
1 - 0	17.6	1.44	17.1	1.25
1 - 6	18.0	1.55	17.8	1.38
2 - 0	18.6	1.20	18.5	1.21
Erect				
2 - 0	20.4	1.14	20.2	1.00
2 - 6	21.2	1.13	21.0	1.00
3 - 0	22.0	1.04	21.5	1.10
3 - 6	22.5	1.01	22.3	1.17
4 - 0	22.8	1.43	22.8	1.32
4 - 6	23.4	1.39	23.3	1.29
5 - 0	24.0	1.40	23.8	1.42
5 - 6	24.8	1.37	24.5	1.34
6 - 0	25.4	1.32	25.2	1.30
6 - 6	26.0	1.24	25.9	1.28
7 - 0	26.7	1.29	26.5	1.41
7 - 6	27.4	1.30	27.1	1.51
8 - 0	28.0	1.38	27.6	1.55
8 - 6	28.3	1.46	28.2	1.56
9 - 0	29.0	1.52	28.7	1.59
9 - 6	29.5	1.56	29.3	1.73
10 - 0	30.1	1.66	29.9	1.73
10 - 6	30.7	1.64	30.3	1.84
11 - 0	31.2	1.71	31.0	1.97
11 - 6	31.7	1.91	31.7	1.76
12 - 0	32.1	2.02	32.5	2.07
12 - 6	32.6	1.87	33.1	1.92
13 - 0	33.5	2.04	33.5	1.79
13 - 6	34.3	2.03	34.1	1.64

TABLE 823 (continued)

BIACROMIAL DIAMETER (cm)
IN DENVER CHILDREN

| Age | Boys | | Girls | |
Yr.- Mo.	Mean	S.D.	Mean	S.D.
14 - 0	35.0	2.07	34.9	1.78
14 - 6	35.8	2.02	35.1	1.56
15 - 0	36.7	2.23	35.4	1.47
15 - 6	37.5	2.06	35.4	1.78
16 - 0	38.0	2.08	35.8	1.52
16 - 6	38.9	1.76	--	--
17 - 0	39.1	1.71	36.1	1.75
17 - 6	39.4	1.46	--	--
18 - 0	39.4	1.45	36.7	1.24

(data from McCammon, Human Growth and Development, 1970.
Courtesy of Charles C Thomas, Publisher, Springfield, Illinois)

TABLE 824

BIACROMIAL DIAMETER (in)
IN MICHIGAN CHILDREN

Age (years)	Sex	Mean	S.D.
5	Boys	9.9	0.6
	Girls	9.9	.8
6	Boys	10.6	.8
	Girls	10.4	.8
7	Boys	11.0	.9
	Girls	10.9	.9
8	Boys	11.5	1.0
	Girls	11.4	.9
9	Boys	11.9	1.0
	Girls	11.6	.8
10	Boys	12.5	1.1
	Girls	12.2	.8
11	Boys	12.8	1.0
	Girls	12.5	.9
12	Boys	13.1	1.1
	Girls	13.3	1.0
13	Boys	13.6	1.0
	Girls	13.6	.9
14	Boys	13.8	1.0
	Girls	13.7	.9
15	Boys	14.5	1.0
	Girls	13.9	.8
16	Boys	14.7	.8
	Girls	13.8	.7
17	Boys	15.2	.7
	Girls	13.8	.7

(data from Martin, 1955)

TABLE 825

BIACROMIAL DIAMETER (cm) IN PHILADELPHIA CHILDREN

Age (Years)	Boys White		Girls White		Boys Negro		Girls Negro	
	Mean	S.D.	Mean	S.D.	Mean	S.D.	Mean	S.D.
7	26.8	1.5	26.6	1.6	27.3	1.7	27.1	1.8
8	30.1	1.6	27.6	1.8	28.3	1.7	28.0	2.2
9	29.1	1.5	28.5	1.7	29.4	2.4	29.2	2.4
10	30.2	1.7	29.8	1.6	30.3	5.0	30.5	2.5
11	31.2	1.7	30.9	1.8	31.7	3.8	31.9	3.3
12	32.1	1.9	32.1	1.9	33.3	3.2	33.2	4.1
13	37.4	1.9	33.4	1.8	34.7	5.9	34.7	4.0
14	35.8	2.3	34.2	2.1	35.9	6.2	35.4	4.3
15	37.8	2.2	35.6	2.1	37.3	7.3	36.3	5.2
16	38.4	1.9	35.4	2.0	--	--	--	--
17	39.1	2.7	35.7	2.6	--	--	--	--

(data from Krogman, 1970)

TABLE 826

BIACROMIAL DIAMETER (mm) FOR PHILADELPHIA CHILDREN

Age (years)	WHITE BOYS		BLACK BOYS	
	Mean	S.D.	Mean	S.D.
6	261.49	13.45	261.96	13.64
7	272.25	12.86	273.32	14.39
8	283.70	13.74	285.82	15.19
9	294.61	14.52	299.44	16.38
10	310.01	16.85	308.61	19.42
11	317.79	15.90	318.57	19.01
12	325.00	19.05	325.21	20.48
13	--	--	337.04	19.29

Age (years)	WHITE GIRLS		BLACK GIRLS	
	Mean	S.D.	Mean	S.D.
6	254.87	13.77	262.18	15.05
7	269.07	15.76	273.97	15.13
8	277.42	15.23	284.08	14.27
9	287.96	15.35	293.17	16.69
10	300.13	15.20	307.72	17.81
11	311.76	17.04	319.08	18.87
12	--	--	328.29	22.63
13	--	--	335.14	21.41

Age is for completed years.

(data from Malina, unpublished)

TABLE 827

BIACROMIAL DIAMETER (mm) OF PHILADELPHIA WHITE CHILDREN IN
RELATION TO CHRONOLOGICAL AND SKELETAL AGE (GREULICH-PYLE)

Age (years)	Chronological		Skeletal	
	Mean	S.D.	Mean	S.D.
BOYS				
7	267.36	14.42	268.73	14.20
8	281.21	14.84	279.52	13.79
9	291.74	15.97	285.22	17.74
10	299.87	14.26	296.61	14.41
11	311.38	15.09	302.61	15.31
12	323.16	19.99	313.89	13.77
13	338.02	22.23	328.20	15.50
14	357.19	22.93	352.54	17.49
15	376.74	19.67	366.42	18.94
16	387.08	21.01	382.94	18.66
17	--	--	393.00	20.21
GIRLS				
7	264.43	13.70	264.79	16.82
8	274.03	12.47	280.43	15.70
9	284.40	15.70	290.06	16.89
10	297.40	18.54	297.50	16.24
11	311.43	17.55	312.94	16.95
12	327.90	15.07	319.63	18.36
13	336.64	17.71	329.55	14.95
14	345.53	17.58	345.23	13.92
15	354.58	18.36	349.12	17.51
16	356.00	16.64	350.17	17.46
17	359.75	12.90	353.45	12.97

(data from Johnston, 1962)

TABLE 828

BIACROMIAL DIAMETER (mm) IN RELATION TO
AGE AT REACHING SKELETAL MATURITY
IN CALIFORNIA CHILDREN

BOYS

C.A. (yr.-mo.)	Early-Maturing		Average-Maturing		Late-Maturing	
	Mean	S.D.	Mean	S.D.	Mean	S.D.
10.7	311	14	299	17	296	17
11.2	312	16	303	12	304	16
11.7	318	15	307	12	304	18
12.2	336	18	313	12	310	14
12.7	338	19	319	15	314	17
13.2	347	20	326	16	317	17
13.7	361	21	337	16	321	20
14.2	371	20	346	17	331	22
14.7	378	16	357	17	337	23
15.2	382	21	364	15	349	23
15.7	392	21	375	15	360	22
16.2	398	22	381	13	372	23
16.7	398	15	383	15	376	23
17.2	404	23	386	17	380	22
17.7	407	21	389	18	388	19

TABLE 828 (continued)

BIACROMIAL DIAMETER (mm) IN RELATION TO
AGE AT REACHING SKELETAL MATURITY
IN CALIFORNIA CHILDREN

GIRLS

C.A. (yr.-mo.)	Early-Maturing		Average-Maturing		Late-Maturing	
	Mean	S.D.	Mean	S.D.	Mean	S.D.
10.7	314	13	305	17	323	10
11.2	319	18	309	14	314	18
11.7	326	16	316	14	314	19
12.2	331	18	323	16	321	18
12.7	335	17	331	14	328	19
13.2	342	18	338	13	343	19
13.7	344	17	342	15	340	22
14.2	350	18	345	12	345	19
14.7	349	17	346	13	350	17
15.2	347	22	352	14	352	17
15.7	351	18	354	11	358	17
16.2	351	16	353	13	362	19
16.7	353	16	356	12	360	20
17.2	353	15	359	10	366	19
17.7	353	17	358	12	369	20

C.A. = chronological age.

(data from Bayley, 1943b)

TABLE 829

BIACROMIAL AND BI-ILIAC
DIAMETERS (cm) IN CALIFORNIA CHILDREN

Age (years)	Boys		Girls	
	Mean (cm)	S.D. (cm)	Mean (cm)	S.D. (cm)
Biacromial Diameter				
9	29.36	1.57	29.31	1.72
10	30.56	1.45	30.60	1.84
11	31.68	1.48	31.93	1.87
12	32.88	1.72	33.33	1.99
13	34.41	2.04	34.37	2.04
14	36.14	2.21	35.11	1.92
15	37.84	2.14	35.73	1.73
16	39.37	1.92	36.14	1.66
17	40.33	1.80	36.34	1.62
18	40.90	1.72	36.55	1.57
Biiliac Diameter				
9	21.74	1.39	21.88	1.47
10	22.53	1.50	22.75	1.58
11	23.22	1.49	23.77	1.69
12	24.02	1.60	25.05	1.76
13	25.06	1.89	26.56	1.99
14	26.13	2.00	27.65	2.02
15	27.01	1.81	28.16	2.01
16	27.68	1.67	28.35	1.77
17	28.18	1.72	28.40	1.64
18	28.42	1.76	28.44	1.62

(data from Tuddenham and Snyder, 1954)

TABLE 830

BIACROMIAL DIAMETER (cm) OF BOYS IN MONTREAL

| Age | | | Percentiles | | |
(years)	Mean	S.D.	10	50	90
6.0	24.89	1.35	23.2	24.8	26.5
6.5	25.67	1.20	24.0	25.5	27.0
7.0	26.45	1.57	24.4	26.2	28.1
7.5	26.93	1.32	25.4	26.9	28.4
8.0	27.13	1.47	25.2	27.0	29.0
8.5	27.88	1.43	26.0	27.9	29.6
9.0	28.48	1.46	26.9	28.4	30.4
9.5	28.86	1.53	27.1	28.9	31.0
10.0	29.15	1.43	27.2	29.1	31.0
10.5	29.70	1.55	27.9	29.8	31.7
11.0	30.17	1.53	28.4	30.1	32.0
11.5	30.50	1.64	28.3	30.5	32.6
12.0	31.30	1.80	29.2	31.4	33.6
12.5	31.78	1.77	29.5	32.0	34.1
13.0	33.03	1.99	31.1	33.3	35.6
13.5	33.94	2.29	31.3	34.1	37.0
14.0	34.39	2.37	31.5	34.0	37.9
14.5	35.19	2.22	32.0	35.1	38.2
15.0	36.30	2.22	33.5	36.5	39.6
15.5	37.12	2.25	34.2	37.2	40.0
16.0	37.33	2.10	35.1	37.5	40.1
16.5	37.95	1.91	35.2	38.2	40.4
17.0	38.07	1.80	36.3	38.2	40.5

(data from Demirjian and Jeniček, 1972)

TABLE 831

BIACROMIAL DIAMETER (cm) OF GIRLS IN MONTREAL

Age (years)	Mean	S.D.	Percentiles		
			10	50	90
6.0	24.66	1.37	23.2	24.4	26.0
6.5	25.05	1.26	23.3	24.9	26.5
7.0	26.00	1.30	24.6	25.9	27.6
7.5	26.27	1.47	24.5	26.2	28.0
8.0	26.91	1.95	25.0	26.5	28.6
8.5	27.43	1.51	25.5	27.4	29.1
9.0	27.86	1.52	26.3	27.7	30.0
9.5	28.53	1.72	26.9	28.4	31.0
10.0	28.95	1.58	27.1	29.1	31.0
10.5	29.48	1.85	27.0	29.3	32.0
11.0	30.24	1.84	28.0	30.3	33.1
11.5	30.90	1.62	28.7	30.9	33.0
12.0	31.72	1.94	29.1	32.0	34.0
12.5	32.26	1.79	29.7	32.4	34.2
13.0	33.06	1.72	30.4	33.3	35.6
13.5	33.88	1.71	32.0	33.9	35.7
14.0	34.04	1.74	31.5	33.9	36.2
14.5	33.95	1.64	31.9	34.0	36.5
15.0	34.82	1.51	33.0	35.0	36.9
15.5	34.70	1.40	33.0	34.6	36.0
16.0	34.12	1.90	32.0	34.5	36.5

(data from Demirjian and Jeniček, 1972)

TABLE   832

BIACROMIAL DIAMETER (cm) OF MONTREAL BOYS
IN RELATION TO STATURE (cm)

| Stature | Mean | S.D. | Percentiles | | |
			10	50	90
114.5 - 115.4	25.18	0.99	24.4	25.1	26.4
119.5 - 120.4	26.43	0.84	25.3	26.4	27.6
124.5 - 125.4	27.57	0.78	26.6	27.6	28.5
129.5 - 130.4	28.32	1.11	27.0	28.3	29.7
134.5 - 135.4	29.27	1.10	28.0	29.3	30.6
139.5 - 140.4	30.34	1.30	28.4	30.4	32.0
144.5 - 145.4	30.85	1.08	29.8	31.0	32.1
149.5 - 150.4	32.38	1.34	30.6	32.3	33.6
154.5 - 155.4	33.51	1.21	32.0	33.3	35.2
159.5 - 160.4	34.92	1.60	33.3	35.0	37.0
164.5 - 165.4	36.03	1.44	34.5	36.2	37.5
169.5 - 170.4	37.24	1.65	35.0	37.3	39.2
174.5 - 175.4	38.68	1.63	36.6	38.7	41.3

(data from Demirjian and Jenicek, 1972)

TABLE   833

BIACROMIAL DIAMETER (cm) OF MONTREAL GIRLS
IN RELATION TO STATURE (cm)

| Stature | Mean | S.D. | Percentiles | | |
			10	50	90
109.5 - 110.4	24.12	1.13	22.7	24.3	25.5
114.5 - 115.4	24.71	0.83	23.5	24.9	25.6
119.5 - 120.4	26.01	1.00	25.0	26.0	27.3
124.5 - 125.4	26.87	0.97	25.5	27.0	28.1
129.5 - 130.4	28.04	1.17	26.5	28.0	29.3
134.5 - 135.4	29.29	1.13	27.7	29.1	30.6
139.5 - 140.4	29.69	1.16	27.8	29.8	31.3
144.5 - 145.4	31.19	1.17	29.6	31.1	33.0
149.5 - 150.4	32.45	1.23	31.1	32.4	33.9
154.5 - 155.4	33.42	1.25	32.0	33.2	35.1
159.5 - 160.4	33.94	1.13	32.6	34.2	35.4
164.5 - 165.4	34.58	2.29	32.4	35.0	36.8

(data from Demirjian and Jeniček, 1972)

TABLE   834

BIACROMIAL DIAMETER/STATURE IN MONTREAL BOYS

Age (years)	Mean	S.D.	Percentiles		
			10	50	90
6.0	21.89	0.76	20.9	22.0	22.7
6.5	21.78	0.72	20.8	21.9	22.6
7.0	22.00	0.69	21.0	22.0	22.9
7.5	21.97	0.84	20.9	22.0	23.0
8.0	21.58	0.73	20.7	21.5	22.5
8.5	21.69	0.73	20.8	21.6	22.8
9.0	21.84	0.71	20.9	21.9	22.7
9.5	21.86	0.85	20.8	21.8	22.8
10.0	21.45	0.76	20.5	21.4	22.6
10.5	21.54	0.77	20.7	21.5	22.5
11.0	21.51	0.80	20.6	21.6	22.4
11.5	21.55	0.81	20.7	21.5	22.5
12.0	21.47	0.84	20.3	21.5	22.5
12.5	21.40	0.84	20.4	21.4	22.5
13.0	21.56	0.75	20.7	21.5	22.7
13.5	21.52	0.84	20.3	21.5	22.6
14.0	21.64	0.90	20.4	21.7	22.8
14.5	21.58	0.89	20.4	21.6	22.6
15.0	21.84	0.90	20.6	21.8	23.0
15.5	21.91	1.02	20.7	22.0	23.2
16.0	21.95	0.96	20.9	22.0	23.0
16.5	22.19	1.00	21.0	22.2	23.6
17.0	22.24	0.96	21.1	22.1	23.5

(data from Demirjian and Jeniček, 1972)

TABLE   835

BIACROMIAL DIAMETER/STATURE IN MONTREAL GIRLS

Age (years)	Mean	S.D.	Percentiles		
			10	50	90
6.0	21.83	0.08	20.9	21.9	22.8
6.5	21.63	0.80	20.7	21.6	22.7
7.0	21.86	0.68	21.0	21.8	22.7
7.5	21.64	0.68	20.8	21.6	22.5
8.0	21.51	1.02	20.4	21.5	22.5
8.5	21.47	0.76	20.5	21.5	22.4
9.0	21.54	0.78	20.5	21.5	22.6
9.5	21.53	0.77	20.6	21.4	22.6
10.0	21.35	0.69	20.5	21.4	22.2
10.5	21.33	0.78	20.3	21.4	22.3
11.0	21.37	0.82	20.4	21.4	22.2
11.5	21.45	0.73	20.5	21.5	22.4
12.0	21.43	0.72	20.6	21.4	22.3
12.5	21.35	0.84	20.3	21.5	22.4
13.0	21.49	0.71	20.6	21.5	22.2
13.5	21.66	0.84	20.5	21.7	22.7
14.0	21.45	0.90	20.3	21.5	22.6
14.5	21.42	0.93	20.2	21.5	22.5
15.0	21.64	0.90	20.5	21.6	22.9
15.5	21.74	0.81	20.7	21.7	22.7
16.0	21.37	1.15	20.1	21.5	22.5

(data from Demirjian and Jeniček, 1972)

TABLE   836

MEAN SHOULDER WIDTH/HIP WIDTH RATIO
OF CALIFORNIA GIRLS AT DEVELOPMENTAL POINTS

Developmental Point	Mean	S.D.
Prepuberal	1.35	.10
Puberal onset	1.36	.07
Puberal end	1.32	.08
Postpuberal	1.30	.07

(data from Faust, 1977)

TABLE   837

TRUNK DIAMETERS (cm) FOR BLACK BOYS IN TEXAS

Age (years)	Biacromial diameter				Biiliac diameter			
	Lower Income		Middle Income		Lower Income		Middle Income	
	Mean	S.D.	Mean	S.D.	Mean	S.D.	Mean	S.D.
10.1 - 11	30.5	1.28	30.4	1.55	20.3	0.96	19.7	1.05
11.1 - 12	30.9	1.42	31.8	1.48	20.7	0.64	21.2	1.33
12.1 - 13	31.9	1.98	33.5	2.14	21.2	1.68	22.2	1.71
13.1 - 14	33.8	2.63	33.2	2.33	22.3	1.67	22.1	1.74
14.1 - 15	33.9	2.62	34.7	2.26	22.3	1.97	23.6	1.51
15.1 - 16	36.5	2.25	37.6	2.20	24.2	1.39	24.2	1.93
16.1 - 17	38.5	2.12	38.7	1.27	25.4	1.75	25.3	1.23
17.1 - 18	38.6	1.43	39.2	1.88	25.8	1.70	25.4	1.47

(data from Schutte, 1979)

TABLE 838

BODY DIAMETERS OF MEXICAN-
AMERICAN BOYS IN AUSTIN (TX)

Body Breadth	Age (years)	Mean	S.D.
Biacromial (cm)	9	29.1	1.4
	10	28.9	3.3
	11	30.6	2.0
	12	32.2	1.2
	13	33.1	1.8
	14	34.5	2.0
Bicristal (cm)	9	19.8	1.2
	10	20.6	1.7
	11	21.3	1.3
	12	22.3	1.3
	13	22.9	1.0
	14	23.7	1.8
Bicondylar (cm)	9	8.5	.73
	10	8.7	.70
	11	9.1	.80
	12	9.0	.57
	13	9.3	.56
	14	9.5	.51
Biepicondylar (cm)	9	5.3	.34
	10	5.5	.36
	11	5.8	.29
	12	6.1	.34
	13	6.3	.35
	14	6.4	.31

(data from Zavaleta, 1976)

TABLE 839

PERCENTILES FOR CHEST WIDTH (cm)
IN U.S. CHILDREN

Age (years)	5	10	25	50	75	90	95
			Boys				
6	16.4	17.0	17.6	18.4	19.2	19.9	20.5
7	17.1	17.5	18.2	19.1	19.7	20.6	21.1
8	18.0	18.3	19.0	19.7	20.7	21.5	22.0
9	18.3	18.7	19.5	20.5	21.5	22.5	23.2
10	19.0	19.3	20.1	20.9	21.9	22.8	23.5
11	19.8	20.2	21.1	21.8	23.0	24.3	24.9
			Girls				
6	16.1	16.3	17.1	17.8	18.6	19.4	19.8
7	16.5	17.1	17.7	18.5	19.3	19.9	20.7
8	17.3	17.6	18.3	19.2	20.1	21.0	21.6
9	18.0	18.2	18.9	19.8	20.7	21.9	22.8
10	18.3	18.7	19.6	20.7	21.8	23.1	24.1
11	19.1	19.6	20.5	21.6	22.8	24.2	25.3

(data from Malina et al., 1973)

TABLE 840

CHEST WIDTH (cm) IN U.S. CHILDREN
OF 8 STATES

Age (Months)	Boys		Girls	
	Mean	S.D.	Mean	S.D.
0 – 3	11.6	1.3	11.5	1.1
4 – 6	14.0	0.8	13.5	0.9
7 – 9	15.1	0.7	14.6	0.7
10 – 12	15.5	0.8	15.1	0.9
13 – 18	16.1	0.7	16.0	0.8
19 – 24	16.5	0.9	16.3	1.6
25 – 30	17.1	0.9	16.8	0.7
31 – 36	17.5	1.0	17.0	0.9
37 – 42	17.6	0.8	17.2	0.8
43 – 48	17.9	0.8	17.5	0.8
49 – 54	18.2	1.0	17.8	0.8
55 – 60	18.5	1.0	18.1	0.9
61 – 66	18.9	1.0	18.4	1.0
67 – 72	19.3	1.1	18.7	0.9
73 – 78	19.6	1.0	19.1	1.1
79 – 84	20.2	1.3	19.4	1.3
85 – 96	20.7	1.4	19.8	1.3
97 – 108	21.4	1.4	20.9	1.5
109 – 120	21.9	1.4	21.4	1.5
121 – 132	23.0	1.3	22.2	1.9
133 – 144	23.9	1.6	23.1	1.7
145 – 156	24.0	1.6	24.8	2.0

Measured at the level of the nipples.

(data from Snyder et al., 1975)

TABLE 841

CHEST WIDTH AT XIPHOID (recumbent, cm)
IN BOSTON CHILDREN

Age		Percentiles						
Yr.	Mo.	10	25	50	75	90	Mean	S.D.

Boys

Yr.	Mo.	10	25	50	75	90	Mean	S.D.
0	0	8.8	9.2	9.7	10.2	10.6	9.7	0.7
0	3	11.9	12.4	12.8	13.3	13.8	12.8	0.7
0	6	13.3	13.7	14.3	14.7	15.4	14.3	0.8
0	9	14.1	14.6	15.2	15.8	16.1	15.1	0.8
1	0	14.6	15.0	15.6	16.1	16.7	15.6	0.8
1	6	15.4	15.9	16.4	17.0	17.4	16.4	0.8
2	0	15.8	16.3	16.9	17.3	17.9	16.9	0.8
2	6	16.3	16.8	17.3	17.9	18.5	17.3	0.8
3	0	16.7	17.3	17.6	18.1	19.0	17.6	0.8
3	6	16.9	17.4	17.9	18.4	19.2	17.9	0.9
4	0	17.1	17.5	18.0	18.6	19.3	18.1	0.9
4	6	17.2	17.6	18.3	18.9	19.6	18.4	1.0
5	0	17.3	17.9	18.4	19.2	19.8	18.5	1.0
7	0	18.3	18.8	19.4	20.2	21.1	19.6	1.1
9	0	18.9	19.5	20.4	21.3	22.2	20.5	1.2

Girls

Yr.	Mo.	10	25	50	75	90	Mean	S.D.
0	0	8.9	9.2	9.6	10.0	10.5	9.6	0.6
0	3	11.6	12.0	12.6	13.1	13.5	12.6	0.8
0	6	12.9	13.3	13.9	14.3	14.6	13.9	0.7
0	9	13.7	14.3	14.8	15.1	15.6	14.7	0.7
1	0	14.3	14.8	15.2	15.9	16.3	15.3	0.8
1	6	14.9	15.4	15.9	16.4	16.9	15.9	0.8
2	0	15.3	15.8	16.4	17.1	17.6	16.4	0.8
2	6	16.0	16.4	17.0	17.6	17.9	17.0	0.8
3	0	16.2	16.7	17.3	17.8	18.3	17.3	0.8
3	6	16.3	17.0	17.4	18.0	18.7	17.5	0.9
4	0	16.5	17.1	17.7	18.5	19.0	17.8	0.9
4	6	16.8	17.3	17.9	18.6	19.1	17.9	0.9
5	0	16.9	17.4	18.0	18.6	19.0	18.1	0.9
7	0	18.0	18.4	19.1	19.8	20.4	19.1	0.9
9	0	18.6	19.1	20.0	21.1	21.9	20.2	1.3

(data from Vickers and Stuart, 1943, The Journal of Pediatrics 22:
155-170)

TABLE 842

CHEST WIDTH*(mm) IN CLEVELAND CHILDREN

	BOYS		GIRLS	
Age	Mean	S.D.	Mean	S.D.
3 mos.	128.11	7.64	123.22	8.19
6 "	140.78	8.65	136.74	7.65
9 "	146.01	8.33	143.36	7.36
12 "	151.98	9.12	147.75	8.60
18 "	159.10	7.80	155.59	8.49
2 yrs.	162.10	8.65	158.49	8.15
2½ "	165.83	8.77	162.32	8.64
3 "	167.59	8.47	164.61	8.09
3½ "	170.86	8.69	166.62	8.07
4 "	174.22	8.61	169.41	9.04
4½ "	175.64	9.49	171.36	8.86
5 "	178.21	9.66	174.17	8.92
6 "	183.01	9.44	178.68	10.57
7 "	189.87	10.96	185.21	10.93
8 "	196.56	10.74	192.42	11.43
9 "	203.42	11.30	199.72	12.62
10 "	211.89	12.31	207.96	12.89
11 "	218.48	15.41	216.59	16.34
12 "	225.54	15.80	227.64	15.66
13 "	235.15	16.28	236.98	15.46
14 "	247.15	18.48	245.38	15.22
15 "	259.29	16.25	249.94	13.44
16 "	266.52	17.00	252.86	13.72
17 "	274.33	15.52	257.11	14.94

*Level of nipples; mean of values at expiration and inspiration

(data from Simmons, 1944)

TABLE 843

CHEST WIDTH (cm) AT LEVEL OF XIPHI-STERNUM IN DENVER CHILDREN

Age (yr.-mo.)	Boys		Girls	
	Mean	S.D.	Mean	S.D.
Supine				
Birth	9.7	0.73	9.7	0.56
0 - 1	11.2	0.78	10.8	0.73
0 - 2	12.4	0.80	11.9	0.66
0 - 3	13.1	0.82	12.6	0.74
0 - 4	13.9	0.77	13.4	0.79
0 - 5	14.4	0.77	13.9	0.76
0 - 6	14.7	0.79	14.2	0.77
0 - 9	15.4	0.73	14.8	0.86
1 - 0	15.6	0.81	15.1	0.84
1 - 6	16.2	0.73	15.7	0.86
2 - 0	16.7	0.66	16.0	0.86
Erect				
2 - 0	16.6	0.89	16.3	0.83
2 - 6	17.1	0.83	16.6	0.77
3 - 0	17.5	0.71	16.9	0.84
3 - 6	17.7	0.77	17.1	0.75
4 - 0	17.8	0.84	17.4	0.81
4 - 6	18.0	0.90	17.5	0.78
5 - 0	18.3	0.89	17.8	0.91
5 - 6	18.6	0.95	18.1	0.92
6 - 0	18.9	1.00	18.4	0.87
6 - 6	19.2	1.04	18.7	0.93
7 - 0	19.6	1.06	19.0	0.97
7 - 6	19.9	1.10	19.4	1.04
8 - 0	20.3	1.22	19.7	1.20
8 - 6	20.6	1.24	20.0	1.27
9 - 0	21.2	1.32	20.4	1.25
9 - 6	21.5	1.38	20.8	1.29
10 - 0	22.0	1.51	21.2	1.58
10 - 6	22.3	1.42	21.6	1.74
11 - 0	22.9	1.62	22.1	1.82
11 - 6	23.3	1.64	22.4	1.81
12 - 0	23.7	1.88	23.1	1.98
12 - 6	24.0	1.81	23.4	1.79
13 - 0	24.8	1.94	23.8	1.68
13 - 6	25.2	2.05	24.2	1.49

TABLE 843 (continued)

CHEST WIDTH (cm) AT LEVEL OF XIPHI-STERNUM
IN DENVER CHILDREN

Age (yr.-mo.)	Boys		Girls	
	Mean	S.D.	Mean	S.D.
14 - 0	25.8	2.09	24.9	1.71
14 - 6	26.4	1.97	25.0	1.46
15 - 0	27.2	1.99	25.0	1.44
15 - 6	27.9	1.96	24.7	1.39
16 - 0	28.2	2.22	25.6	1.80
16 - 6	28.8	1.51	--	--
17 - 0	29.3	1.98	25.7	1.88
17 - 6	29.9	2.38	--	--
18 - 0	30.2	2.74	26.0	1.67

(data from McCammon, Human Growth and Development, 1970.
Courtesy of Charles C Thomas, Publisher, Springfield, Illinois)

TABLE   844

INTERNAL CHEST DIAMETER (cm)
IN DENVER CHILDREN

Age	Boys		Girls	
Yr. Mo.	Mean	S.D.	Mean	S.D.
Supine				
0 - 1	10.78	.65	10.55	.64
0 - 2	11.85	.95	11.68	.61
0 - 3	12.70	.73	12.33	.67
0 - 6	14.33	.83	13.76	.79
0 - 9	15.22	.89	14.76	.90
1 - 0	15.78	.86	15.28	.75
1 - 6	16.54	.84	16.10	.89
2 - 0	17.06	.78	16.70	.84
2 - 6	17.80	.86	17.38	.87
3 - 0	18.09	.79	17.66	.84
3 - 6	18.48	.84	17.93	.87
4 - 0	18.91	.72	17.91	1.10
Erect				
1 - 0	15.02	.80	14.82	.88
1 - 6	16.08	.82	15.36	1.00
2 - 0	16.49	.84	15.92	.82
2 - 6	16.86	.77	16.58	.80
3 - 0	17.50	1.01	17.00	.88
3 - 6	18.05	.88	17.49	.79
4 - 0	18.49	.88	17.79	.84
4 - 6	18.79	.86	18.05	.83
5 - 0	19.09	.88	18.27	.87
5 - 6	19.52	.91	18.61	.91
6 - 0	19.74	.93	18.92	.94
6 - 6	20.14	.99	19.19	.97
7 - 0	20.48	.93	19.54	.95
7 - 6	20.87	1.02	19.84	.96
8 - 0	21.20	1.02	20.20	1.05
8 - 6	21.53	1.11	20.54	1.09
9 - 0	21.93	1.16	20.91	1.08
9 - 6	22.34	1.15	21.31	1.16
10 - 0	22.65	1.16	21.43	1.24
10 - 6	22.87	1.28	21.89	1.36
11 - 0	23.25	1.35	22.45	1.44
11 - 6	23.61	1.25	22.75	1.49
12 - 0	23.99	1.42	23.20	1.58
12 - 6	24.34	1.47	23.56	1.61

TABLE 844 (continued)

INTERNAL CHEST DIAMETER (cm)
IN DENVER CHILDREN

| Age | Boys | | Girls | |
Yr. Mo.	Mean	S.D.	Mean	S.D.
13 - 0	24.86	1.59	24.03	1.56
13 - 6	25.37	1.70	24.49	1.54
14 - 0	26.06	1.79	24.87	1.54
14 - 6	26.53	1.77	24.97	1.52
15 - 0	27.17	1.67	25.12	1.37
16 - 0	28.13	1.80	25.30	1.35
17 - 0	28.62	1.85	25.43	1.33
18 - 0	29.16	1.83	25.74	1.40

(data from McCammon, Human Growth and Development,
1970. Courtesy of Charles C Thomas, Publisher, Springfield,
Illinois)

TABLE 845

CHEST WIDTH (cm) IN PHILADELPHIA CHILDREN

Age (Years)	Boys White		Girls White		Boys Negro		Girls Negro	
	Mean	S.D.	Mean	S.D.	Mean	S.D.	Mean	S.D.
7	19.4	1.2	18.8	1.3	18.6	1.0	18.3	1.3
8	20.3	1.5	19.3	1.4	19.1	1.4	18.9	1.8
9	21.0	1.7	19.8	1.5	19.9	1.6	19.8	1.4
10	22.1	1.8	20.6	1.7	20.9	1.7	20.5	1.6
11	22.6	1.7	21.2	1.7	21.6	1.7	21.6	1.9
12	23.4	1.9	22.0	1.8	22.6	1.5	22.8	2.1
13	24.6	1.5	23.1	1.6	24.3	2.0	23.6	1.4
14	26.2	2.0	24.4	2.3	24.4	2.7	24.7	1.3
15	27.7	2.1	25.3	1.8	25.0	2.3	25.3	1.4
16	28.6	1.9	25.5	1.6	--	--	--	--
17	29.9	2.1	25.3	1.8	--	--	--	--

Measured at the level of the fourth intercostal space.

(data from Krogman, 1970)

TABLE 846

MEASUREMENTS RELATIVE TO RECUMBENT LENGTH AND STATURE IN
COLORADO BOYS

Age (yr. mo.)	Xiphisternal Width Percentiles			Xiphisternal Depth Percentiles		
	10	50	90	10	50	90
Recumbent Length						
Birth	18.20	19.46	20.84	16.19	17.63	19.26
0 - 1	19.43	20.92	22.15	16.32	17.50	18.74
0 - 2	20.08	21.48	22.97	15.52	16.89	18.14
0 - 3	20.22	21.72	23.11	15.16	16.33	17.38
0 - 4	20.50	22.21	23.13	14.77	15.76	16.76
0 - 5	20.80	22.01	23.51	14.46	15.70	17.02
0 - 6	19.78	21.82	23.08	14.46	15.59	17.36
0 - 9	20.21	21.30	22.77	14.27	15.53	16.91
1 - 0	19.28	20.44	22.06	14.06	15.30	16.69
1 - 6	18.54	19.54	21.14	13.88	14.82	15.72
2 - 0	17.99	18.77	20.06	13.15	13.72	14.69
Stature						
2 - 0	18.25	19.30	20.47	12.32	13.61	15.12
2 - 6	17.89	18.84	19.89	12.28	13.35	14.58
3 - 0	17.36	18.34	19.48	12.17	12.98	13.82
3 - 6	16.93	18.03	19.03	11.97	12.74	13.73
4 - 0	16.30	17.60	18.54	11.65	12.60	13.70
4 - 6	15.80	16.97	18.05	11.35	12.21	13.32
5 - 0	15.64	16.77	17.89	11.05	12.00	13.08
5 - 6	15.26	16.36	17.81	10.86	11.82	12.85
6 - 0	15.02	16.23	17.48	10.81	11.71	12.75
6 - 6	14.92	16.07	17.28	10.60	11.58	12.56
7 - 0	14.94	15.83	17.17	10.41	11.54	12.66
7 - 6	14.62	15.88	16.96	10.23	11.45	12.63
8 - 0	14.57	15.73	17.06	10.26	11.39	12.29
8 - 6	14.50	15.71	16.90	10.12	11.26	12.38
9 - 0	14.42	15.78	17.04	10.06	11.34	12.38
9 - 6	14.48	15.68	16.99	9.99	11.31	12.53
10 - 0	14.54	15.78	17.17	10.10	11.33	12.48
10 - 6	14.67	15.73	17.05	10.00	11.12	12.66
11 - 0	14.50	15.74	17.25	9.86	11.16	12.38
11 - 6	14.59	15.77	17.33	9.71	11.11	12.61

TABLE 846 (continued)

MEASUREMENTS RELATIVE TO RECUMBENT LENGTH AND STATURE IN
COLORADO BOYS

Age	Xiphisternal Width Percentiles			Xiphisternal Depth Percentiles		
(yr. mo.)	10	50	90	10	50	90
12 - 0	14.56	15.64	17.33	9.90	11.21	12.56
12 - 6	14.42	15.74	17.08	9.84	11.24	12.51
13 - 0	14.50	15.88	17.33	9.91	11.16	12.66
13 - 6	14.45	15.68	17.35	9.97	11.12	12.79
14 - 0	14.35	15.80	17.21	9.93	11.21	12.58
14 - 6	14.57	15.82	17.38	9.91	11.21	12.72
15 - 0	14.75	16.11	17.48	10.00	11.25	12.76
15 - 6	14.97	16.08	17.60	10.24	11.28	12.73
16 - 0	14.60	16.06	17.66	10.05	11.32	12.56
16 - 6	15.28	16.22	17.70	10.49	11.38	12.86
17 - 0	15.13	16.46	18.32	10.24	11.33	13.06
17 - 6	15.40	16.82	18.63	10.48	11.52	13.75
18 - 0	15.01	16.75	19.22	10.27	11.52	13.38

(data from McCammon, Human Growth and Development, 1970. Courtesy of Charles C Thomas,
Publisher, Springfield, Illinois)

TABLE 847

MEASUREMENTS RELATIVE TO RECUMBENT LENGTH AND STATURE IN
COLORADO GIRLS

Age (yr. mo.)	Xiphisternal Width Percentiles			Xiphisternal Depth Percentiles		
	10	50	90	10	50	90
Recumbent Length						
Birth	18.69	19.74	20.96	16.11	17.41	18.87
0 - 1	18.99	20.68	22.02	16.04	17.40	18.89
0 - 2	19.66	21.15	22.69	16.02	16.81	18.12
0 - 3	19.69	21.14	22.93	15.44	16.27	17.72
0 - 4	20.27	21.80	23.60	15.12	16.24	17.45
0 - 5	20.55	21.86	23.40	14.80	16.00	17.21
0 - 6	20.47	21.65	23.41	14.69	16.00	17.22
0 - 9	19.59	21.22	22.68	14.48	15.51	16.75
1 - 0	18.96	20.40	21.85	14.40	15.30	16.54
1 - 6	18.40	19.53	21.01	13.34	14.67	15.72
2 - 0	17.22	18.51	19.44	12.37	13.42	14.70
Stature						
2 - 0	17.86	19.11	20.33	12.51	13.70	14.77
2 - 6	17.46	18.62	19.51	11.95	13.06	14.17
3 - 0	16.80	18.04	19.19	11.87	12.68	13.79
3 - 6	16.58	17.47	18.69	11.82	12.66	13.60
4 - 0	15.96	17.21	18.01	11.46	12.37	13.20
4 - 6	15.56	16.72	17.59	11.15	12.02	12.95
5 - 0	15.10	16.45	17.39	10.90	11.86	12.95
5 - 6	15.03	16.12	17.26	10.79	11.70	12.58
6 - 0	14.87	15.93	16.95	10.58	11.46	12.45
6 - 6	14.62	15.78	16.85	10.45	11.44	12.27
7 - 0	14.52	15.64	16.53	10.29	11.30	12.17
7 - 6	14.53	15.61	16.64	10.08	11.23	12.17
8 - 0	14.31	15.40	16.65	10.06	11.21	12.20
8 - 6	14.28	15.41	16.53	10.17	11.07	12.18
9 - 0	14.48	15.21	16.72	10.05	11.02	12.30
9 - 6	14.32	15.31	16.39	10.23	11.03	12.07
10 - 0	14.19	15.18	16.55	10.11	10.98	12.38
10 - 6	14.01	15.20	16.78	9.92	10.89	12.29
11 - 0	14.11	15.23	16.62	9.90	10.96	12.30
11 - 6	14.00	14.99	16.31	10.00	11.00	12.46

TABLE 847   (continued)

MEASUREMENTS RELATIVE TO RECUMBENT LENGTH AND STATURE IN
COLORADO GIRLS

Age	Xiphisternal Width Percentiles			Xiphisternal Depth Percentiles		
(yr. mo.)	10	50	90	10	50	90
12 - 0	14.00	15.07	16.50	9.70	10.84	12.22
12 - 6	13.96	14.98	16.14	9.63	10.79	11.90
13 - 0	13.97	15.04	16.13	9.64	10.62	12.10
13 - 6	13.97	15.09	16.32	9.64	10.76	12.08
14 - 0	14.19	15.31	16.67	9.42	10.69	11.95
14 - 6	14.41	15.22	16.27	9.56	10.78	12.26
15 - 0	14.07	15.00	16.30	9.51	10.69	12.08
15 - 6	13.80	14.67	16.50	9.68	10.76	11.98
16 - 0	14.15	15.29	16.92	9.74	10.78	12.49
16 - 6	--	--	--	--	--	--
17 - 0	13.91	15.24	17.02	9.79	10.72	12.09
17 - 6	--	--	--	--	--	--
18 - 0	14.37	15.42	17.05	10.06	10.89	12.75

(data from McCammon, Human Growth and Development, 1970. Courtesy of Charles C Thomas, Publisher, Springfield, Illinois)

TABLE 848

THORACIC DIAMETERS (cm) AND INDEX ON THE
FIRST, FOURTH, AND EIGHTH POSTNATAL DAYS
IN IOWA INFANTS

Age	Mean Chest Width	Mean Chest Depth	Mean Thoracic Index*
First day (8.1 hours)	10.50	8.94	85.2
Fourth day	10.34	8.53	82.6
Eighth day	10.38	8.67	83.6

*The thoracic index was obtained from the formula $\dfrac{\text{depth of chest} \times 100}{\text{width of chest}}$

(data from Goodman, 1942, American Journal of Diseases of Children 64:674-679. Copyright 1942, American Medical Association)

TABLE 849

MEAN CHEST MEASUREMENTS (cm) IN AUSTRALIAN CHILDREN

Age (years)	Sex	Chest Width	Internipple distance	Nipple Width	Sternal length	Stature
2	M	16.3	11.5	1.3	11.1	91.9
	F	16.1	11.1	1.4	10.9	91.0
4	M	17.7	12.5	1.6	12.5	107.0
	F	17.1	12.0	1.6	11.6	106.1
6	M	18.7	13.0	1.7	12.9	121.5
	F	18.5	12.8	1.8	13.4	119.8
8	M	20.5	13.8	1.7	13.8	130.4
	F	19.3	13.4	1.8	13.4	128.2
10	M	21.7	15.2	1.9	14.6	140.1
	F	20.3	14.1	2.0	13.7	138.6

(data from Collins, 1973, The Journal of Pediatrics 83:557-561)

TABLE 850

LOWER TORSO WIDTH (cm) IN CHILDREN OF 8 STATES

Age (Months)	Boys		Girls	
	Mean	S.D.	Mean	S.D.
0 - 3	11.6	1.5	11.6	1.5
4 - 6	14.4	1.1	14.1	1.2
7 - 9	14.9	1.1	14.6	1.1
10 - 12	15.2	1.0	14.6	1.1
13 - 18	15.8	1.0	15.7	1.1
19 - 24	16.8	1.1	16.3	1.3
25 - 30	17.1	1.1	17.0	1.1
31 - 36	17.3	1.1	17.3	1.2
37 - 42	17.7	1.1	17.6	1.2
43 - 48	17.8	1.0	17.9	1.1
49 - 54	18.1	1.1	18.3	1.1
55 - 60	18.6	1.2	18.7	1.2
61 - 66	19.1	1.3	19.3	1.4
67 - 72	19.6	1.5	19.8	1.3
73 - 78	19.9	1.3	20.2	1.4
79 - 84	20.7	1.4	20.6	1.2
85 - 96	21.3	1.8	21.4	1.6
97 - 108	22.1	1.5	22.8	1.8
109 - 120	22.8	1.4	23.6	1.6
121 - 132	24.5	1.9	25.1	2.1
133 - 144	25.5	1.6	26.3	2.2
145 - 156	26.2	2.0	28.7	1.9

Measured at the level of the upper thigh.

(data from Snyder et al., 1975)

TABLE 851

CHEST WIDTH (cm) IN RELATION TO TYPE OF ARTIFICIAL FEEDING
IN PHILADELPHIA CHILDREN

Group	Race	3 Months		1 Year		2 Years	
		Mean	S.D.	Mean	S.D.	Mean	S.D.
All	All White	12.7	+0.72	15.4	+0.68	16.8	+0.65
	All Negro	12.4	+0.77	15.0	+0.71	16.4	+0.74
Group I (irradiated evaporated milk)	White	12.6	+0.80	15.5	+0.62	16.8	+0.66
	Negro	12.3	+0.81	15.1	+0.66	16.3	+0.73
Group II (nonirradiated evaporated milk plus cod-liver oil)	White	12.6	+0.60	15.3	+0.58	16.7	+0.64
	Negro	12.3	+0.73	14.9	+0.75	16.3	+0.70
Group III (irradiated evaporated milk plus carotene)	White	12.8	+0.66	15.6	+0.57	16.8	+0.51
	Negro	12.4	+0.77	15.0	+0.73	16.4	+0.97
Group IV (irradiated evaporated milk plus carotene and yeast)	White	12.8	+0.76	15.6	+0.85	16.9	+0.75
	Negro	12.7	+0.70	15.2	+0.71	16.7	+0.61

Group	Race	3 Years		4 Years	
		Mean	S.D.	Mean	S.D.
All	All White	17.6	+0.67	18.2	+0.77
	All Negro	17.2	+0.87	17.9	+0.88
Group I (irradiated evaporated milk)	White	17.6	+0.74	18.0	+0.79
	Negro	17.1	+0.94	17.8	+0.86
Group II (nonirradiated evaporated milk plus cod-liver oil)	White	17.5	+0.71	18.1	+0.69
	Negro	17.2	+0.88	17.9	+0.77
Group III (irradiated evaporated milk plus carotene)	White	17.7	+0.54	18.6	+0.59
	Negro	17.1	+1.07	17.8	+1.18
Group IV (irradiated evaporated milk plus carotene and yeast)	White	17.8	+0.64	18.3	+1.04
	Negro	17.3	+0.59	18.0	+0.90

(data from Rhoads, Rapaport, Kennedy, et al., 1945, The Journal of Pediatrics 26:415-454)

TABLE 852

INTERNIPPLE DISTANCES (cm) IN MICHIGAN CHILDREN

Age Group	Mean	S.D.
Age < 3 d	8.133	0.868
3 d < Age < 2 m	8.418	1.009
2 m < Age < 4 m	9.933	1.048
4 m < Age < 6 m	10.182	0.893
6 m < Age < 1 yr	10.850	0.968
1 yr < Age < 2 yr	11.030	1.088
2 yr < Age < 3 yr	11.610	1.085
3 yr < Age < 4 yr	12.015	1.064
4 yr < Age < 5 yr	12.708	1.300
5 yr < Age < 6 yr	13.147	1.133
6 yr < Age < 7 yr	14.058	0.808
7 yr < Age < 8 yr	14.246	0.766
8 yr < Age < 10 yr	14.290	1.191
10 yr < Age < 12 yr	16.095	2.136
Age > 12 yr	17.200	2.484

d = days; m = months; yr = years.

(data from Chen et al., 1974)

TABLE 853

VALUES OF RATIOS INVOLVING INTERNIPPLE DISTANCE (ID) IN MICHIGAN CHILDREN

Age (years)	Ratio	Mean	S.D.
3.1	ID/CC	0.24	0.02
4.6	ID/CW	0.77	0.10
4.7	ID/AP	1.01	0.11

CC = chest circumference; CW = chest width;
AP = chest depth.

(data from Chen et al., 1974)

TABLE 854

PERCENTAGE RATIOS OF INTERNIPPLE DISTANCE (IND) TO CHEST WIDTH (CW) IN AUSTRALIAN CHILDREN

Age (years)	BOYS Mean IND/CW (%)	2 S.D.	GIRLS Mean IND/CW (%)	2 S.D.
2	70.4	8.9	68.4	6.9
4	70.5	8.1	69.4	5.9
6	69.6	9.3	69.1	8.5
8	67.4	8.3	69.4	9.8
10	70.1	8.5	69.6	7.9

(data from Collins, 1973, The Journal of Pediatrics 83:557-561)

TABLE 855

WAIST WIDTH (cm) IN CHILDREN OF 8 STATES

Age	(Months)	Boys		Girls	
		Mean	S.D.	Mean	S.D.
0 -	3	10.4	1.1	10.4	1.2
4 -	6	12.6	0.9	12.2	1.0
7 -	9	13.4	0.8	12.9	1.0
10 -	12	13.4	0.7	13.1	1.1
13 -	18	14.0	0.9	13.8	0.9
19 -	24	14.8	0.9	14.3	1.5
25 -	30	15.4	1.0	15.0	0.8
31 -	36	15.4	1.1	14.8	1.1
37 -	42	15.5	1.3	15.0	1.4
43 -	48	15.5	1.3	15.0	1.4
49 -	54	15.9	1.4	15.2	1.5
55 -	60	16.1	1.5	15.5	1.5
61 -	66	16.4	1.4	15.7	1.7
67 -	72	16.6	1.7	16.1	1.8
73 -	78	16.2	1.9	16.5	1.5
79 -	84	17.5	1.7	16.5	1.6
85 -	96	17.8	2.0	16.8	2.0
97 -	108	18.3	1.9	17.7	1.9
109 -	120	18.8	1.9	18.2	2.0
121 -	132	19.8	1.9	19.3	1.9
133 -	144	20.3	2.1	19.7	1.9
145 -	156	20.6	2.1	20.4	2.4

(data from Snyder et al., 1975)

TABLE 856

BICRISTAL DIAMETER (cm) IN U.S. CHILDREN

Age (years)	Mean	S.D.	Percentiles						
			5	10	25	50	75	90	95
BOYS									
12	21.7	2.04	18.6	19.3	20.3	21.6	23.2	24.5	25.1
13	22.8	2.33	19.3	20.0	21.2	22.7	24.2	25.7	26.7
14	23.9	2.49	20.2	20.8	22.2	23.7	25.4	26.8	28.0
15	24.6	2.23	21.2	21.8	23.2	24.5	25.8	27.6	28.6
16	25.1	2.34	21.8	22.4	23.5	25.0	26.5	27.8	28.7
17	25.6	2.28	21.7	22.6	24.1	25.5	27.1	28.6	29.5
GIRLS									
12	23.3	2.26	20.0	20.5	21.8	23.3	24.6	26.0	27.1
13	24.1	2.69	20.2	21.1	22.5	23.8	25.7	27.3	28.5
14	24.9	2.38	21.0	22.1	23.3	24.8	26.4	27.8	28.8
15	25.3	2.45	21.9	22.4	23.6	25.2	26.6	28.5	29.8
16	25.5	2.59	21.6	22.4	23.8	25.4	27.0	28.6	29.9
17	25.4	2.40	21.7	22.6	23.8	25.3	26.8	28.6	29.5

(data from Cycle III of the Health Examination Survey, 1966-1970 of the National Center for Health Statistics).

TABLE   857

BICRISTAL DIAMETER (cm) IN U.S. WHITE CHILDREN

Age (years)	Mean	S.D.	Percentiles						
			5	10	25	50	75	90	95
Boys									
6	18.2	1.28	16.1	16.5	17.3	18.2	19.2	19.8	20.4
7	19.0	1.36	16.8	17.3	18.1	19.1	19.8	20.7	21.2
8	19.8	1.61	17.3	17.9	18.7	19.7	20.6	21.6	22.4
9	20.7	1.90	18.2	18.7	19.6	20.6	21.6	22.7	23.9
10	21.3	2.02	18.4	19.2	20.2	21.2	22.3	23.4	24.4
11	22.3	2.16	19.5	20.2	21.1	22.2	23.4	24.8	25.9
12	21.9	2.05	18.8	19.4	20.4	21.8	23.4	24.6	25.3
13	23.0	2.29	19.4	20.2	21.4	23.0	24.4	25.8	26.7
14	24.1	2.45	20.4	21.1	22.5	24.0	25.6	27.0	28.3
15	24.8	2.21	21.4	22.2	23.4	24.7	26.1	27.7	28.7
16	25.3	2.29	22.1	22.6	23.8	25.2	26.6	27.9	28.8
17	25.8	2.22	22.1	23.0	24.3	25.1	27.3	28.7	29.6
Girls									
6	18.2	1.44	16.1	16.5	17.3	18.2	19.2	20.1	20.7
7	19.0	1.47	16.5	17.1	18.1	19.1	20.0	20.9	21.6
8	20.0	1.86	17.2	18.0	18.8	20.1	21.1	22.4	23.0
9	21.0	2.01	18.1	18.7	19.7	20.8	22.3	23.8	24.8
10	21.9	2.13	18.7	19.4	20.4	21.7	23.1	25.0	26.2
11	23.3	2.46	20.0	20.5	21.6	23.2	24.7	26.6	27.7
12	23.4	2.22	20.1	20.6	22.0	23.4	24.7	25.9	27.1
13	24.4	2.64	20.6	21.5	22.7	24.2	25.8	27.4	28.6
14	25.1	2.32	21.3	22.2	23.5	25.1	26.5	27.9	28.8
15	25.5	2.42 ·	22.1	22.6	23.9	25.4	26.7	28.6	29.9
16	25.6	2.56	21.8	22.6	24.1	25.5	27.1	28.6	30.1
17	25.6	2.36	23.1	23.0	24.1	25.4	26.8	28.5	30.0

(data from Cycles II and III of the Health Examination Survey by the National Center for Health Statistics 1963-1965 and 1966-1970)

TABLE 858

BICRISTAL DIAMETER (cm) IN U.S. NEGRO CHILDREN

Age (years)	Mean	S.D.	Percentiles						
			5	10	25	50	75	90	95
BOYS									
6	17.0	1.05	15.3	15.6	16.3	17.0	17.7	18.5	19.1
7	17.8	1.07	16.1	16.3	17.0	17.7	18.5	19.3	20.0
8	18.8	1.21	16.5	17.0	18.0	19.0	19.6	20.2	20.7
9	19.1	1.60	16.4	17.1	18.1	18.8	20.3	21.3	21.9
10	19.6	1.66	17.2	17.6	18.5	19.5	20.6	21.8	22.7
11	20.9	1.56	18.6	19.2	20.1	20.7	21.8	22.8	23.5
12	20.7	1.63	18.1	18.0	19.5	20.0	21.8	22.7	23.4
13	21.7	2.32	18.0	19.1	20.2	21.4	23.0	24.2	24.7
14	22.2	2.06	19.2	19.6	20.7	22.2	23.5	24.8	25.6
15	23.0	1.71	20.4	21.1	21.8	22.9	24.3	25.3	26.2
16	23.5	1.98	21.0	21.4	22.3	23.2	24.6	25.7	26.4
17	24.1	2.10	20.7	21.2	22.6	24.1	25.2	26.7	28.5
GIRLS									
6	17.0	1.27	14.9	15.4	16.2	17.0	17.8	18.7	19.1
7	17.7	1.43	15.7	16.2	17.0	17.7	18.6	19.5	20.3
8	18.6	1.81	16.2	16.6	17.3	18.3	19.7	21.3	22.6
9	19.6	1.78	17.1	17.5	18.4	19.5	20.8	21.9	22.8
10	21.3	2.39	17.8	18.6	19.5	21.1	22.8	24.5	25.1
11	22.4	2.65	19.0	19.4	20.5	21.7	23.9	26.3	27.6
12	22.7	2.39	19.2	19.8	20.8	22.7	24.0	26.3	27.2
13	22.6	2.49	19.4	19.8	20.8	22.6	23.8	25.6	27.0
14	23.9	2.48	20.2	20.8	22.5	23.7	25.4	26.6	28.2
15	24.1	2.26	21.1	21.8	22.6	23.7	25.4	27.0	28.4
16	24.5	2.58	21.1	21.7	22.8	24.1	26.2	28.3	28.8
17	24.6	2.57	20.9	21.4	22.8	24.3	26.3	27.0	29.5

(data from Cycle III of the Health Examination Survey by the National Center of Health Statistics 1966-1970)

TABLE   859

BICRISTAL DIAMETER (cm)
RECUMBENT USING SLIDING-ARM CALIPERS
IN BOSTON CHILDREN

Age		Percentiles					Mean	S.D.
Yr.	Mo.	10	25	50	75	90		
				Boys				
0	0	7.4	7.7	8.1	8.4	8.7	8.1	0.5
0	3	10.0	10.2	10.6	11.2	11.5	10.6	0.6
0	6	10.8	11.3	11.8	12.1	12.6	11.8	0.7
0	9	11.6	12.1	12.4	12.8	13.1	12.4	0.6
1	0	12.1	12.4	12.8	13.2	13.7	12.8	0.7
1	6	12.7	13.2	12.7	14.3	14.7	13.7	0.8
2	0	13.6	14.0	14.5	15.2	15.6	14.6	0.8
2	6	14.2	14.7	15.1	15.7	16.2	15.2	0.8
3	0	14.8	15.2	15.9	16.5	17.1	15.9	0.9
3	6	15.4	15.7	16.4	17.0	17.4	16.4	0.8
4	0	15.9	16.2	17.0	17.5	18.0	16.9	0.8
4	6	16.2	16.6	17.3	18.0	18.4	17.3	0.9
5	0	16.7	17.1	17.8	18.5	19.1	17.8	0.9
5	6	17.0	17.4	18.1	18.9	19.6	18.2	1.0
6	0	17.7	18.0	18.6	19.4	20.0	18.7	1.0
				Girls				
0	0	7.2	7.4	7.7	8.2	8.5	7.8	0.5
0	3	9.6	9.9	10.4	10.9	11.4	10.5	0.7
0	6	10.7	11.0	11.5	12.0	12.6	11.6	0.7
0	9	11.3	11.6	12.0	12.5	13.2	12.1	0.8
1	0	11.8	12.0	12.5	13.0	13.6	12.6	0.8
1	6	12.4	12.8	13.2	13.8	14.4	13.4	0.8
2	0	13.0	13.6	14.2	14.9	15.4	14.2	0.9
2	6	13.7	14.4	14.9	15.4	16.2	14.9	0.9
3	0	14.5	14.9	15.5	16.2	16.8	15.6	1.0
3	6	15.0	15.5	16.0	16.8	17.5	16.2	1.0
4	0	15.4	15.8	16.5	17.4	18.0	16.6	1.0
4	6	15.8	16.2	17.0	17.6	18.3	17.0	1.0
5	0	16.4	16.9	17.6	18.2	19.0	17.6	1.0
5	6	16.9	17.4	18.0	18.7	19.5	18.1	1.0
6	0	17.0	17.7	18.3	18.9	19.5	18.3	0.9

(data from Vickers and Stuart, 1943, The Journal of Pediatrics 22: 155-170)

TABLE   860

BICRISTAL DIAMETER (cm)
STANDING, USING SLIDING-ARM CALIPERS
IN BOSTON CHILDREN

Age Yr. Mo.		Percentiles					Mean	S.D.
		10	25	50	75	90		
Boys								
3	0	15.3	16.1	16.4	17.1	17.7	16.5	0.9
3	6	16.0	16.4	17.1	17.6	18.2	17.1	1.0
4	0	16.4	17.0	17.5	18.2	18.6	17.6	0.9
4	6	16.8	17.2	17.9	18.4	18.9	17.8	0.9
5	0	17.1	17.5	18.1	18.7	19.4	18.1	0.9
5	6	17.5	17.8	18.6	19.2	19.9	18.6	1.0
6	0	17.8	18.3	19.0	19.6	20.4	19.0	1.0
6	6	18.2	18.5	19.2	20.0	20.7	19.3	1.0
7	0	18.4	18.8	19.5	20.2	21.0	19.6	1.0
7	6	18.7	19.2	19.8	20.6	21.4	20.0	1.0
8	0	18.9	19.6	20.2	21.2	22.1	20.4	1.2
8	6	19.2	19.7	20.5	21.4	22.5	20.7	1.2
9	0	19.6	20.1	20.9	21.8	22.5	21.0	1.2
9	6	19.8	20.4	21.4	22.6	23.4	21.6	1.4
10	0	20.1	20.6	21.2	22.6	24.2	21.7	1.4
Girls								
3	0	15.4	15.9	16.3	17.0	17.5	16.4	0.9
3	6	15.6	16.2	17.0	17.6	17.9	16.9	1.0
4	0	16.0	16.4	17.4	18.2	18.7	17.4	1.1
4	6	16.4	17.1	17.7	18.5	19.1	17.8	1.0
5	0	17.0	17.5	18.1	18.9	19.8	18.2	1.0
5	6	17.2	17.8	18.4	19.3	20.0	18.6	1.1
6	0	17.6	18.1	18.7	19.5	20.1	18.8	0.9
6	6	17.9	18.5	19.2	19.8	20.5	19.2	1.0
7	0	18.2	18.8	19.6	20.2	20.9	19.5	1.0
7	6	18.6	19.3	19.9	20.7	21.4	20.1	1.0
8	0	19.0	19.4	20.3	21.1	21.7	20.4	1.1
8	6	19.6	20.2	21.0	21.6	22.7	21.0	1.2
9	0	19.9	20.5	21.3	22.0	22.9	21.4	1.1
9	6	20.1	20.9	21.6	22.6	23.7	21.8	1.5
10	0	20.7	21.6	22.3	23.2	23.9	22.5	1.5

(data from Vickers and Stuart, 1943, The Journal of Pediatrics 22:
155-170)

TABLE 861

BICRISTAL DIAMETER (mm) IN CLEVELAND CHILDREN

		BOYS		GIRLS	
Age		Mean	S.D.	Mean	S.D.
3	mos.	112.50	7.08	107.88	6.86
6	"	123.43	7.69	119.88	7.99
9	"	129.74	8.54	124.71	9.54
12	"	133.92	9.24	129.28	9.56
18	"	143.54	9.39	138.06	10.77
2	yrs.	154.01	10.07	150.49	11.19
2½	"	162.52	9.96	159.81	11.00
3	"	167.71	9.11	166.25	11.76
3½	"	172.68	9.25	171.14	10.35
4	"	178.05	9.61	175.50	10.77
4½	"	182.46	10.46	179.91	10.93
5	"	185.64	9.75	183.29	11.91
6	"	193.28	11.72	190.08	12.76
7	"	201.31	11.53	198.81	13.49
8	"	208.72	11.99	208.71	13.54
9	"	217.33	12.44	216.54	14.20
10	"	225.01	13.51	227.81	15.93
11	"	233.25	16.27	239.29	18.06
12	"	241.52	16.84	251.22	17.73
13	"	251.36	16.46	263.96	17.15
14	"	262.27	16.35	274.81	17.00
15	"	275.11	15.10	279.76	18.57
16	"	280.14	15.50	285.64	16.18
17	"	285.00	14.06	286.89	18.97

(data from Simmons, 1944)

TABLE 862

BICRISTAL DIAMETER (cm) IN PHILADELPHIA CHILDREN

Age (years)	White Boys		White Girls		Negro Boys		Negro Girls	
	Mean	S.D.	Mean	S.D.	Mean	S.D.	Mean	S.D.
7	20.2	1.1	20.0	1.5	19.2	1.3	19.1	1.7
8	20.9	1.5	20.8	1.6	19.7	1.3	20.0	2.0
9	21.8	1.7	21.8	1.7	20.7	1.7	20.6	1.7
10	22.8	1.8	22.8	2.2	21.3	1.9	21.8	1.6
11	23.9	1.8	23.7	2.0	22.2	2.1	23.3	1.7
12	24.3	1.9	25.5	1.9	23.6	1.9	23.9	2.5
13	25.5	1.8	26.0	1.8	25.4	2.3	24.9	2.0
14	27.3	1.9	27.0	1.6	25.7	2.4	25.5	1.8
15	28.1	2.1	27.7	1.5	25.4	1.5	27.1	2.2
16	28.5	1.8	27.4	1.8	--	--	--	--
17	29.4	2.1	28.2	1.2	--	--	--	--

(data from Krogman, 1970)

TABLE 863

BICRISTAL DIAMETER (mm) FOR PHILADELPHIA CHILDREN

Age (years)	WHITE BOYS Mean	S.D.	BLACK BOYS Mean	S.D.
6	192.56	12.35	176.86	10.96
7	199.47	12.24	184.27	11.50
8	206.96	11.86	191.81	11.12
9	214.52	13.21	199.52	13.51
10	224.69	16.33	209.24	20.62
11	231.25	16.89	216.14	20.77
12	234.31	15.83	219.51	17.05
13	--	--	225.23	17.23

Age (years)	WHITE GIRLS Mean	S.D.	BLACK GIRLS Mean	S.D.
6	186.87	10.75	179.81	12.95
7	197.92	14.35	186.37	14.29
8	205.18	13.91	193.11	13.59
9	214.69	15.18	200.05	12.34
10	225.02	15.37	210.98	15.34
11	230.69	16.57	220.91	17.01
12	--	--	230.06	19.10
13	--	--	238.14	20.34

Age is for completed years.

(data from Malina, unpublished)

TABLE 864

BICRISTAL DIAMETER (cm) IN NEGRO INFANTS
IN WASHINGTON, D.C.

Age	Boys		Girls	
(weeks)	Mean	S.D.	Mean	S.D.
4	8.75	0.65	8.76	0.76
8	9.32	0.60	9.24	0.68
12	9.88	0.66	9.80	0.72
16	10.34	0.71	10.32	0.76
20	10.69	0.73	10.63	0.80
24	10.98	0.77	10.87	0.89
28	11.24	0.81	11.09	0.95
32	11.51	0.88	11.28	0.96
36	11.74	0.96	11.47	0.94
40	11.96	1.00	11.61	0.93
44	12.17	0.93	11.75	0.89
48	12.37	0.89	11.89	0.88
52	12.56	0.85	12.01	0.87

(data from Scott, Hiatt, Clark, et al., 1962, Pediatrics 29:65-81. Copyright American
Academy of Pediatrics 1962)

TABLE 865

BICRISTAL DIAMETER (mm) OF PHILADELPHIA WHITE CHILDREN IN
RELATION TO CHRONOLOGICAL AND SKELETAL AGES (GREULICH-PYLE)

Age (years)	Chronological Mean	Chronological S.D.	Skeletal Mean	Skeletal S.D.
BOYS				
7	200.86	14.29	199.47	16.80
8	209.04	14.87	207.63	12.18
9	216.29	13.99	211.65	13.65
10	223.98	12.99	220.94	13.73
11	234.72	14.95	227.58	13.59
12	242.38	15.63	235.31	12.32
13	255.24	15.55	247.83	14.43
14	265.24	17.81	263.91	12.23
15	278.74	16.82	274.38	13.27
16	285.33	18.46	281.00	17.56
17	--	--	290.00	13.92
GIRLS				
7	196.18	11.31	198.00	15.70
8	205.79	12.90	210.19	11.64
9	217.15	21.26	216.76	11.19
10	229.98	21.36	227.69	12.66
11	237.94	17.25	239.28	12.08
12	256.69	18.01	248.00	18.17
13	262.53	15.70	255.98	13.02
14	272.38	20.42	267.06	12.90
15	274.21	17.60	272.49	17.10
16	277.52	14.30	275.24	17.14
17	283.58	11.87	275.15	12.21

(data from Johnston, 1962)

TABLE 866

BICRISTAL DIAMETER (cm) OF BOYS IN MONTREAL

Age (years)	Mean	S.D.	Percentiles		
			10	50	90
6.0	17.96	1.08	16.6	17.8	19.3
6.5	18.24	1.03	16.9	18.2	19.4
7.0	18.69	1.17	17.0	18.6	20.2
7.5	19.19	1.22	17.6	19.1	20.5
8.0	19.16	1.27	17.7	19.1	20.5
8.5	19.67	1.23	18.0	19.8	21.2
9.0	20.08	1.25	18.5	19.9	21.7
9.5	20.46	1.84	18.6	20.3	22.0
10.0	20.56	1.19	19.0	20.4	22.2
10.5	20.94	1.38	19.1	21.0	22.4
11.0	21.36	1.22	20.0	21.4	23.0
11.5	21.56	1.39	20.0	21.4	23.5
12.0	22.18	1.49	20.5	22.2	24.0
12.5	22.33	1.42	20.7	22.2	24.3
13.0	23.36	1.67	21.4	23.7	25.9
13.5	24.08	1.83	22.0	24.0	26.5
14.0	24.29	1.97	22.3	24.0	27.3
14.5	24.85	1.87	22.6	24.5	27.3
15.0	25.41	1.67	23.3	25.5	27.6
15.5	26.00	1.75	24.0	26.0	28.1
16.0	26.25	1.70	24.1	26.2	28.3
16.5	26.32	1.46	24.5	26.4	28.1
17.0	26.30	1.42	24.1	26.4	28.8

(data from Demirjian and Jeniček, 1972)

TABLE   867

BICRISTAL DIAMETER (cm) OF GIRLS IN MONTREAL

Age (years)	Mean	S.D.	Percentiles		
			10	50	90
6.0	17.60	1.02	16.2	17.5	18.6
6.5	18.05	1.12	16.5	17.9	19.0
7.0	18.39	1.06	17.0	18.5	19.9
7.5	18.77	1.49	17.2	18.5	20.4
8.0	19.07	1.27	17.5	18.8	20.6
8.5	19.59	1.27	18.1	19.5	21.2
9.0	19.76	1.26	18.2	19.6	21.5
9.5	20.27	1.24	18.8	20.1	21.7
10.0	20.64	1.53	19.0	20.5	22.6
10.5	21.17	1.43	19.3	21.1	23.1
11.0	21.87	1.75	19.6	21.7	24.1
11.5	22.46	1.54	20.6	22.5	24.5
12.0	23.06	1.87	20.9	22.9	25.4
12.5	23.54	1.63	21.5	23.7	25.5
13.0	24.59	1.33	23.0	24.5	26.5
13.5	25.23	1.76	23.0	25.1	27.3
14.0	25.65	1.71	23.5	25.5	28.0
14.5	25.31	1.50	23.5	25.3	27.2
15.0	26.02	1.41	24.0	26.0	27.9
15.5	26.18	1.87	24.0	26.0	28.0
16.0	26.24	1.59	24.2	26.4	28.1

(data from Demirjian and Jeniček, 1972)

TABLE 868

BICRISTAL DIAMETER (cm) OF MONTREAL BOYS
IN RELATION TO STATURE (cm)

Stature	Mean	S.D.	Percentiles		
			10	50	90
114.5 - 115.4	18.04	0.91	17.3	17.9	19.4
119.5 - 120.4	18.82	0.73	18.0	18.7	19.9
124.5 - 125.4	19.20	0.77	18.3	19.2	20.2
129.5 - 130.4	20.08	2.19	18.6	19.7	21.6
134.5 - 135.4	20.74	1.01	19.7	20.7	21.8
139.5 - 140.4	20.93	0.82	20.0	21.0	21.7
144.5 - 145.5	22.01	1.09	20.6	22.2	23.1
149.5 - 150.4	22.88	1.21	21.2	23.1	24.0
154.5 - 155.4	23.65	1.27	22.2	23.7	25.4
159.5 - 160.4	24.54	0.94	23.5	24.3	25.8
164.5 - 165.4	25.48	1.25	24.2	25.5	27.5
169.5 - 170.4	26.28	1.21	24.7	26.3	28.2
174.5 - 175.4	26.65	1.41	24.3	27.0	28.0

(data from Demirjian and Jeniček, 1972)

TABLE 869

BICRISTAL DIAMETER (cm) OF MONTREAL GIRLS
IN RELATION TO STATURE (cm)

Stature	Mean	S.D.	Percentiles		
			10	50	90
109.5 - 110.4	17.00	0.95	16.0	17.0	18.1
114.5 - 115.4	18.05	1.74	17.0	17.7	19.1
119.5 - 120.4	18.33	0.58	17.6	18.3	19.1
124.5 - 125.4	19.12	0.92	18.0	19.2	20.4
129.5 - 130.4	19.76	0.91	18.6	19.8	20.8
134.5 - 135.4	20.67	1.00	19.6	20.9	21.7
139.5 - 140.4	21.31	0.91	19.9	21.6	22.5
144.5 - 145.4	22.87	1.20	21.2	23.0	24.3
149.5 - 150.4	24.31	1.40	22.6	24.1	26.0
154.5 - 155.4	24.62	1.34	23.0	24.5	26.0
159.5 - 160.4	25.79	2.07	23.5	25.7	27.5
164.5 - 165.4	26.64	1.37	24.9	26.6	28.2

(data from Demirjian and Jeniček, 1972)

TABLE 870

BICRISTAL DIAMETER/STATURE IN MONTREAL BOYS

Age (years)	Mean	S.D.	Percentiles		
			10	50	90
6.0	15.71	0.69	14.8	15.7	16.6
6.5	15.56	0.73	14.8	15.6	16.5
7.0	15.57	0.68	14.9	15.5	16.5
7.5	15.53	0.67	14.7	15.6	16.4
8.0	15.26	0.70	14.5	15.2	16.3
8.5	15.32	0.72	14.4	15.3	16.1
9.0	15.38	0.82	14.5	15.3	16.5
9.5	15.43	1.32	14.2	15.4	16.2
10.0	15.10	0.70	14.1	15.1	15.9
10.5	15.18	0.77	14.3	15.1	16.1
11.0	15.22	0.62	14.5	15.2	16.0
11.5	15.21	0.74	14.3	15.2	16.2
12.0	15.22	0.77	14.2	15.2	16.2
12.5	15.06	0.72	14.1	15.1	15.9
13.0	15.20	0.68	14.3	15.3	16.1
13.5	15.25	0.65	14.5	15.3	16.0
14.0	15.31	0.73	14.5	15.2	16.3
14.5	15.23	0.72	14.3	15.2	16.1
15.0	15.24	0.68	14.4	15.2	16.1
15.5	15.39	0.81	14.5	15.4	16.3
16.0	15.41	0.82	14.5	15.4	16.4
16.5	15.37	0.72	14.5	15.4	16.3
17.0	15.28	0.65	14.3	15.3	16.2

(data from Demirjian and Jeniček, 1972)

TABLE 871

BICRISTAL DIAMETER/STATURE IN MONTREAL GIRLS

Age (years)	Mean	S.D.	Percentiles		
			10	50	90
6.0	15.53	0.61	14.8	15.5	16.4
6.5	15.56	0.83	14.6	15.5	16.6
7.0	15.48	0.70	14.6	15.5	16.4
7.5	15.41	1.06	14.5	15.4	16.2
8.0	15.28	0.69	14.5	15.3	16.1
8.5	15.32	0.70	14.6	15.2	16.2
9.0	15.23	0.67	14.5	15.3	16.1
9.5	15.29	0.62	14.5	15.3	16.0
10.0	15.20	0.74	14.1	15.2	16.0
10.5	15.32	0.68	14.5	15.3	16.2
11.0	15.40	0.71	14.4	15.4	16.3
11.5	15.58	0.72	14.7	15.7	16.4
12.0	15.54	0.80	14.6	15.5	16.7
12.5	15.55	0.78	14.7	15.5	16.6
13.0	15.86	0.73	15.0	15.9	16.8
13.5	16.06	0.85	14.9	16.1	17.1
14.0	16.17	0.92	15.1	16.0	17.4
14.5	15.91	0.85	14.8	15.8	16.9
15.0	16.14	0.75	14.9	16.1	17.2
15.5	16.43	1.10	15.3	16.4	17.5
16.0	16.36	0.89	15.3	16.4	17.3

(data from Demirjian and Jeniček, 1972)

TABLE 872

BIILIAC DIAMETER (cm) IN COLORADO CHILDREN

Age (yr.-mo.)	BOYS		GIRLS	
	Mean	S.D.	Mean	S.D.
Supine				
Birth	8.2	0.70	8.0	0.77
0 - 1	9.0	0.67	8.6	0.58
0 - 2	9.9	0.69	9.5	0.71
0 - 3	10.4	0.68	10.0	0.68
0 - 4	11.0	0.77	10.5	0.72
0 - 5	11.4	0.80	10.9	0.87
0 - 6	11.7	0.80	11.2	0.76
0 - 9	12.3	1.04	11.8	1.02
1 - 0	12.7	0.95	12.3	0.87
1 - 6	13.6	0.96	13.0	0.95
2 - 0	14.3	0.85	13.8	0.85
Erect				
2 - 0	14.9	0.94	14.7	0.85
2 - 6	15.5	1.02	15.2	0.92
3 - 0	16.2	1.06	16.0	0.94
3 - 6	16.7	0.91	16.6	0.87
4 - 0	17.0	0.90	17.0	0.95
4 - 6	17.4	0.86	17.4	1.01
5 - 0	17.8	0.96	17.8	1.09
5 - 6	18.3	0.98	18.2	1.16
6 - 0	18.7	0.98	18.6	1.13
6 - 6	19.0	0.96	19.0	1.12
7 - 0	19.5	1.00	19.5	1.32
7 - 6	19.8	1.04	19.9	1.31
8 - 0	20.2	1.00	20.3	1.39
8 - 6	20.5	1.09	20.7	1.44
9 - 0	21.1	1.16	21.2	1.42
9 - 6	21.5	1.20	21.7	1.41
10 - 0	21.9	1.36	22.2	1.56
10 - 6	22.2	1.51	22.7	1.84
11 - 0	22.8	1.64	23.3	1.85
11 - 6	23.3	1.70	23.8	1.69
12 - 0	23.5	1.93	24.5	2.02
12 - 6	23.9	1.94	25.0	1.96
13 - 0	24.6	1.91	25.4	1.97
13 - 6	25.1	2.15	25.9	1.78

TABLE   872   (continued)

BIILIAC DIAMETER (cm) IN COLORADO CHILDREN

| | BOYS | | GIRLS | |
Age (yr.-mo.)	Mean	S.D.	Mean	S.D.
14 - 0	25.5	1.94	26.8	1.70
14 - 6	26.1	1.98	27.0	1.70
15 - 0	27.0	2.08	27.5	1.83
15 - 6	27.5	2.00	27.4	1.53
16 - 0	27.8	1.84	27.8	1.58
16 - 6	28.1	1.70	--	--
17 - 0	28.3	1.73	28.2	1.66
17 - 6	28.9	1.96	--	--
18 - 0	28.9	1.68	28.8	2.22

(data from McCammon, Human Growth and Development, 1970.
Courtesy of Charles C Thomas, Publisher, Springfield, Illinois)

TABLE 873

BIILIAC DIAMETER (mm) IN RELATION TO
AGE AT REACHING SKELETAL MATURITY
IN CALIFORNIA CHILDREN

BOYS

C.A. (yr.-mo.)	Early-Maturing		Average-Maturing		Late-Maturing	
	Mean	S.D.	Mean	S.D.	Mean	S.D.
10.7	240	13	225	14	218	11
11.2	237	14	224	14	226	9
11.7	242	13	230	12	224	10
12.2	252	15	234	13	229	10
12.7	257	15	238	14	233	13
13.2	264	15	244	15	237	12
13.7	271	14	250	14	240	11
14.2	278	14	257	14	246	12
14.7	279	10	263	14	250	12
15.2	283	15	268	14	259	13
15.7	288	15	273	14	264	11
16.2	290	13	275	13	272	12
16.7	288	10	276	15	274	11
17.2	291	14	276	14	275	10
17.7	293	13	278	14	277	10

TABLE 873 (continued)

BIILIAC DIAMETER (mm) IN RELATION TO
AGE AT REACHING SKELETAL MATURITY
IN CALIFORNIA CHILDREN

GIRLS

C.A. (yr.-mo.)	Early-Maturing		Average-Maturing		Late-Maturing	
	Mean	S.D.	Mean	S.D.	Mean	S.D.
10.7	241	16	228	10	229	9
11.2	246	14	235	14	232	14
11.7	248	14	242	14	234	13
12.2	257	15	245	15	240	14
12.7	261	14	253	14	245	12
13.2	262	13	260	13	253	13
13.7	266	12	264	14	257	14
14.2	272	13	268	14	262	14
14.7	276	15	271	14	266	14
15.2	278	16	273	15	270	16
15.7	278	14	279	9	274	15
16.2	280	16	280	14	280	15
16.7	280	14	282	13	279	16
17.2	281	15	283	14	283	15
17.7	283	17	280	14	284	13

C.A. = chronological age.

(data from Bayley, 1943b)

TABLE   874

BIILIAC DIAMETER/STATURE IN RELATION TO
AGE AT REACHING SKELETAL MATURITY

BOYS

C.A. (years)	Early-Maturing		Average-Maturing		Late-Maturing	
	Mean	S.D.	Mean	S.D.	Mean	S.D.
10.7	16.1	.93	15.8	.88	15.4	.63
11.2	15.9	.42	15.6	.85	15.7	.44
11.7	16.0	.66	15.8	.71	15.4	.47
12.2	16.0	.56	15.7	.72	15.6	.48
12.7	16.2	.54	15.7	.74	15.5	.54
13.2	16.0	.60	15.7	.78	15.6	.45
13.7	16.1	.58	15.7	.69	15.5	.47
14.2	16.0	.60	15.7	.72	15.5	.43
14.7	16.1	.60	15.7	.71	15.4	.50
15.2	16.0	.67	15.7	.73	15.5	.49
15.7	16.2	.57	15.8	.70	15.4	.42
16.2	16.1	.54	15.7	.70	15.6	.54
16.7	16.0	.63	15.7	.72	15.6	.46
17.2	16.0	.58	15.7	.70	15.6	.50
17.7	16.1	.58	15.8	.66	15.6	.57

TABLE 874 (continued)

BIILIAC DIAMETER/STATURE IN RELATION TO
AGE AT REACHING SKELETAL MATURITY

GIRLS

C.A. (years)	Early-Maturing		Average-Maturing		Late-Maturing	
	Mean	S.D.	Mean	S.D.	Mean	S.D.
10.7	16.7	.94	16.2	.63	16.0	.55
11.2	16.5	.84	16.3	.74	16.6	.56
11.7	16.4	.73	16.4	.75	16.4	.59
12.2	16.6	.72	16.4	.80	16.2	.55
12.7	16.7	.78	16.4	.80	16.2	.53
13.2	16.6	.76	16.6	.86	16.2	.55
13.7	16.7	.75	16.6	.88	16.3	.60
14.2	16.9	.78	16.7	.87	16.5	.61
14.7	17.2	.81	16.9	.86	16.6	.62
15.2	17.2	.86	16.9	.94	16.7	.69
15.7	17.2	.80	17.2	.71	16.8	.64
16.2	17.2	.85	17.2	.84	17.1	.70
16.7	17.3	.79	17.3	.80	17.1	.65
17.2	17.3	.90	17.3	.88	17.2	.68
17.7	17.5	.81	17.3	.82	17.2	.57

C.A = chronological age.

(data from Bayley, 1943b)

TABLE 875

MEASUREMENTS RELATIVE TO RECUMBENT LENGTH AND STATURE IN
COLORADO BOYS

Age (yr. mo.)	Iliac Diameter Percentiles			Biacromial Diameter Percentiles		
	10	50	90	10	50	90
Recumbent Length						
Birth	15.54	16.45	17.57	20.30	22.49	24.82
0 - 1	15.70	16.84	17.69	19.80	22.90	25.71
0 - 2	15.94	17.29	18.32	21.91	23.69	26.38
0 - 3	15.92	17.34	18.29	21.92	23.91	26.04
0 - 4	16.08	17.36	18.96	22.45	24.14	26.20
0 - 5	15.92	17.49	18.50	22.53	24.05	26.18
0 - 6	15.97	17.29	18.66	20.57	23.35	25.72
0 - 9	15.58	16.98	18.29	21.46	23.75	26.15
1 - 0	15.55	16.73	18.80	20.64	23.24	25.67
1 - 6	15.36	16.38	17.73	19.68	21.56	24.69
2 - 0	15.24	16.22	17.30	19.64	21.05	22.54
Stature						
2 - 0	16.07	17.31	18.41	22.49	23.68	25.08
2 - 6	15.98	17.00	18.22	22.17	23.41	24.52
3 - 0	15.89	17.08	18.20	22.08	23.04	24.34
3 - 6	15.97	16.89	17.98	21.78	22.70	24.16
4 - 0	15.61	16.71	17.70	20.68	22.49	23.70
4 - 6	15.46	16.35	17.39	19.91	22.17	23.51
5 - 0	15.23	16.22	17.41	20.14	22.19	23.16
5 - 6	15.12	16.08	17.36	20.38	22.01	23.31
6 - 0	15.13	16.12	17.22	20.61	21.85	23.19
6 - 6	14.87	16.00	16.89	20.58	21.84	22.95
7 - 0	14.87	15.84	16.87	20.59	21.79	22.95
7 - 6	14.89	15.72	16.91	20.50	21.81	22.94
8 - 0	14.79	15.65	16.73	20.40	21.81	22.71
8 - 6	14.82	15.58	16.76	20.34	21.69	22.67
9 - 0	14.62	15.71	16.84	20.41	21.69	22.68
9 - 6	14.64	15.68	16.89	20.38	21.67	22.86
10 - 0	14.66	15.63	17.13	20.58	21.77	22.87
10 - 6	14.55	15.57	16.94	20.56	21.68	22.99
11 - 0	14.58	15.62	17.26	20.28	21.72	22.97
11 - 6	14.68	15.67	17.55	20.29	21.60	22.85

TABLE 875 (continued)

MEASUREMENTS RELATIVE TO RECUMBENT LENGTH AND STATURE IN
COLORADO BOYS

Age	Iliac Diameter Percentiles			Biacromial Diameter Percentiles		
(yr. mo.)	10	50	90	10	50	90
12 - 0	14.53	15.50	17.58	20.31	21.54	22.75
12 - 6	14.48	15.63	17.16	20.42	21.52	22.69
13 - 0	14.70	15.65	17.40	20.42	21.46	22.57
13 - 6	14.53	15.56	17.04	20.54	21.44	22.51
14 - 0	14.58	15.55	16.86	20.15	21.52	22.61
14 - 6	14.63	15.54	17.21	20.42	21.48	22.78
15 - 0	14.78	15.73	17.31	20.24	21.68	23.06
15 - 6	14.73	15.82	17.61	20.46	21.79	23.39
16 - 0	14.90	15.72	17.04	20.32	21.87	23.17
16 - 6	14.79	16.04	17.26	21.02	22.08	23.33
17 - 0	14.88	15.88	17.39	20.76	21.99	23.54
17 - 6	14.89	16.42	18.06	21.00	22.21	23.64
18 - 0	14.99	16.05	17.72	21.18	22.13	23.24

(data from McCammon, Human Growth and Development, 1970. Courtesy of Charles C Thomas,
Publisher, Springfield, Illinois)

TABLE 876

MEASUREMENTS RELATIVE TO RECUMBENT LENGTH AND STATURE IN
COLORADO GIRLS

Age (yr. mo.)	Iliac Diameter Percentiles			Biacromial Diameter Percentiles		
	10	50	90	10	50	90
Recumbent Length						
Birth	14.73	16.11	18.47	20.77	22.58	25.20
0 - 1	15.17	16.37	17.35	20.95	23.22	25.64
0 - 2	15.60	16.80	18.56	21.64	24.16	26.11
0 - 3	15.58	16.84	18.25	21.56	23.26	24.95
0 - 4	15.78	17.08	18.51	21.93	23.69	25.61
0 - 5	15.76	16.96	18.92	22.15	23.77	25.92
0 - 6	16.16	17.07	18.58	21.35	23.42	25.42
0 - 9	15.60	16.76	18.56	21.75	23.64	25.35
1 - 0	15.34	16.53	18.18	21.41	23.34	24.71
1 - 6	14.91	16.16	17.43	19.89	21.95	24.19
2 - 0	14.52	15.85	17.02	19.33	21.02	23.51
Stature						
2 - 0	16.12	17.40	18.42	22.54	23.64	25.12
2 - 6	16.01	17.06	18.34	22.42	23.29	24.95
3 - 0	15.92	17.04	18.18	21.82	22.82	24.17
3 - 6	16.06	16.83	18.16	21.83	22.71	24.07
4 - 0	15.88	16.62	17.77	21.15	22.54	23.68
4 - 6	15.36	16.52	17.52	20.32	22.10	23.39
5 - 0	15.09	16.34	17.48	20.14	21.98	23.01
5 - 6	15.02	16.16	17.39	20.43	21.85	23.14
6 - 0	15.08	16.07	17.22	20.51	21.92	22.96
6 - 6	14.94	15.96	17.06	20.77	21.90	22.90
7 - 0	14.95	15.88	17.23	20.70	21.81	22.91
7 - 6	14.95	15.92	16.99	20.55	21.75	22.84
8 - 0	14.80	15.82	16.98	20.42	21.70	22.84
8 - 6	14.81	15.91	17.01	20.61	21.80	22.78
9 - 0	14.88	15.89	17.25	20.50	21.71	22.83
9 - 6	15.07	15.99	16.97	20.58	21.57	22.74
10 - 0	15.00	15.96	17.02	20.42	21.71	22.69
10 - 6	14.93	15.96	17.48	20.36	21.46	22.62
11 - 0	14.94	16.00	17.46	20.32	21.47	22.62
11 - 6	15.03	15.94	17.22	20.27	21.43	22.36

TABLE 876 (continued)

MEASUREMENTS RELATIVE TO RECUMBENT LENGTH AND STATURE IN
COLORADO GIRLS

Age	Iliac Diameter Percentiles			Biacromial Diameter Percentiles		
(yr. mo.)	10	50	90	10	50	90
12 - 0	14.85	15.98	17.36	20.24	21.40	22.42
12 - 6	14.98	16.04	17.31	20.57	21.32	22.55
13 - 0	14.97	16.00	17.49	20.22	21.26	22.45
13 - 6	15.11	16.06	17.39	20.20	21.39	22.47
14 - 0	15.44	16.37	17.88	20.47	21.48	22.71
14 - 6	15.41	16.52	17.79	20.61	21.63	22.19
15 - 0	15.65	16.55	17.97	20.48	21.54	22.40
15 - 6	15.54	16.54	17.77	20.18	21.54	22.77
16 - 0	15.82	16.64	18.09	20.39	21.69	22.56
16 - 6	--	--	--	--	--	--
17 - 0	15.92	16.79	18.24	20.43	21.82	22.64
17 - 6	--	--	--	--	--	--
18 - 0	15.55	17.13	18.96	21.01	22.19	22.98

(data from McCammon, Human Growth and Development, 1970. Courtesy of Charles C Thomas, Publisher, Springfield, Illinois)

TABLE 877

BIILIAC/BIACROMIAL DIAMETER IN RELATION
TO AGE AT REACHING SKELETAL MATURITY
IN CALIFORNIA CHILDREN

BOYS

C.A. (yr.-mo.)	Early-Maturing		Average-Maturing		Late-Maturing	
	Mean	S.D.	Mean	S.D.	Mean	S.D.
10.7	76.9	2.0	75.7	4.4	74.0	4.3
11.2	76.1	2.5	73.8	4.4	74.7	2.2
11.7	76.1	2.8	75.1	3.1	74.0	3.2
12.2	74.7	2.6	74.9	3.4	74.0	2.6
12.7	76.1	2.5	74.8	3.7	74.2	3.3
13.2	76.0	3.0	74.8	3.1	74.8	2.9
13.7	75.3	2.9	74.3	2.8	74.8	3.4
14.2	75.0	2.8	74.3	3.6	74.6	3.8
14.7	74.0	2.3	73.8	3.4	74.6	4.4
15.2	74.0	2.0	73.5	3.3	74.5	4.2
15.7	73.4	2.5	72.8	3.2	73.7	3.8
16.2	72.7	2.3	72.2	3.0	73.4	4.5
16.7	72.3	2.8	72.2	3.2	73.0	4.1
17.2	71.9	2.7	71.6	2.9	72.7	4.1
17.7	71.8	2.5	71.3	2.9	71.5	3.2

TABLE 877 (continued)

BIILIAC/BIACROMIAL DIAMETER IN RELATION
TO AGE AT REACHING SKELETAL MATURITY
IN CALIFORNIA CHILDREN

GIRLS

C.A. (yr.-mo.)	Early-Maturing		Average-Maturing		Late-Maturing	
	Mean	S.D.	Mean	S.D.	Mean	S.D.
10.7	77.2	4.0	75.0	3.3	71.8	1.4
11.2	77.4	4.6	75.9	5.0	74.7	2.4
11.7	76.7	4.4	76.5	3.8	74.8	3.4
12.2	77.5	3.8	75.8	3.8	74.7	2.8
12.7	78.4	4.4	76.3	3.8	74.7	3.7
13.2	77.2	3.8	76.9	4.1	74.1	3.2
13.7	77.6	4.0	77.5	4.4	75.8	3.7
14.2	77.8	4.1	77.8	4.6	75.9	3.3
14.7	79.2	4.3	78.3	4.5	76.0	3.5
15.2	79.9	5.4	77.5	3.8	76.8	3.8
15.7	79.4	4.8	79.1	3.4	76.6	3.1
16.2	79.8	4.2	79.2	3.6	77.4	3.0
16.7	79.4	3.9	79.1	3.6	77.6	3.5
17.2	80.0	4.3	78.9	3.6	77.4	3.9
17.7	80.2	4.9	78.4	3.4	76.9	3.7

C.A. = chronlogical age.

(data from Bayley, 1943b)

TABLE 878

BITROCHANTERIC DIAMETER (cm) IN U.S. WHITE CHILDREN

Age (years)	Mean	S.D.	Percentiles						
			5	10	25	50	75	90	95
BOYS									
12	26.0	2.11	23.0	23.5	24.5	25.8	27.4	28.8	29.6
13	27.6	2.42	24.1	24.6	26.0	27.4	29.3	31.0	31.8
14	29.4	2.37	25.4	26.2	27.8	29.5	30.8	32.4	33.4
15	30.5	2.00	27.2	27.8	29.1	30.5	31.8	33.1	33.8
16	31.1	1.92	28.0	28.7	30.1	31.2	32.4	33.7	34.4
17	31.5	1.91	28.3	29.2	30.3	31.6	32.7	34.1	35.0
GIRLS									
12	27.8	2.21	24.2	25.0	26.3	27.9	29.3	30.6	31.5
13	29.0	2.09	25.5	26.4	27.6	29.1	30.5	31.6	32.4
14	30.0	1.90	27.1	27.7	28.7	29.8	31.3	32.5	33.1
15	30.5	2.09	27.3	27.9	29.1	30.4	31.7	33.3	34.4
16	30.8	1.85	28.1	28.5	29.6	30.7	32.0	33.0	33.8
17	30.9	1.94	28.1	28.6	29.6	30.8	32.0	33.3	33.9

BITROCHANTERIC DIAMETER (cm) IN U.S. NEGRO CHILDREN

Age (years)	Mean	S.D.	Percentiles						
			5	10	25	50	75	90	95
BOYS									
12	25.2	1.78	22.6	23.2	24.1	25.1	26.3	27.5	28.4
13	27.2	2.60	22.5	23.5	25.7	27.3	28.8	30.3	31.6
14	28.1	2.24	24.8	25.2	26.2	28.7	29.7	30.8	31.6
15	29.1	2.04	26.1	26.6	27.5	29.0	30.4	31.7	32.4
16	29.9	1.87	26.6	27.3	28.7	29.8	30.8	31.8	33.0
17	34.2	1.86	27.4	27.9	29.0	30.2	31.6	32.6	33.7
GIRLS									
12	27.8	2.35	24.2	24.7	26.4	27.6	29.5	31.2	31.8
13	28.3	2.35	25.2	25.6	26.8	28.1	29.5	31.4	32.6
14	29.5	1.90	26.2	27.0	28.4	29.6	30.7	32.0	32.7
15	29.7	2.08	26.8	27.3	28.3	29.5	30.8	32.9	33.5
16	30.2	2.50	27.0	27.4	28.4	29.7	31.4	34.1	35.5
17	30.2	2.35	26.6	27.7	28.6	30.2	32.0	33.2	33.9

(data from Cycle III of the Health Examination Survey by the National Center of Health Statistics 1966-1970)

TABLE   879

BITROCHANTERIC DIAMETER (mm) IN CLEVELAND CHILDREN

	BOYS		GIRLS	
Age	Mean	S.D.	Mean	S.D.
2 yrs.	165.90	10.03	162.37	13.36
2½ "	172.38	9.87	170.49	8.96
3 "	177.76	9.15	175.65	9.61
3½ "	182.42	10.18	182.24	9.63
4 "	186.75	10.57	187.41	10.84
4½ "	190.96	11.11	191.73	11.61
5 "	195.13	10.68	195.87	12.39
6 "	203.04	13.22	204.59	13.17
7 "	211.89	12.72	215.30	13.76
8 "	222.30	13.26	226.14	14.25
9 "	232.06	14.26	236.00	15.90
10 "	242.78	15.44	247.40	16.63
11 "	252.92	18.08	261.39	19.08
12 "	263.53	20.09	278.59	21.33
13 "	275.66	20.53	295.49	19.33
14 "	292.75	21.49	308.92	17.07
15 "	309.11	19.04	314.41	16.54
16 "	315.96	15.71	320.57	18.19
17 "	320.33	14.86	322.22	18.72

(data from Simmons, 1944)

TABLE 880

BITROCHANTERIC DIAMETER (cm) IN PHILADELPHIA CHILDREN

Age (years)	White Boys		White Girls		Negro Boys		Negro Girls	
	Mean	S.D.	Mean	S.D.	Mean	S.D.	Mean	S.D.
7	21.5	1.8	21.9	1.7	20.7	1.2	21.0	1.8
8	22.7	1.8	22.8	1.7	21.5	1.6	21.8	2.3
9	23.7	1.9	23.8	1.8	22.6	1.8	22.7	1.7
10	25.1	2.5	25.2	2.9	23.7	2.5	24.2	1.7
11	26.0	2.6	26.3	2.4	24.5	2.2	25.2	2.8
12	27.0	2.3	27.9	2.2	25.6	2.2	26.7	3.0
13	28.3	1.9	29.2	1.7	28.3	2.3	28.2	1.9
14	30.4	1.5	30.3	1.4	29.2	2.8	30.0	2.2
15	32.0	1.6	31.7	1.3	28.4	1.9	31.1	1.9
16	33.3	1.5	31.2	1.5	--	--	--	--
17	34.2	2.8	31.8	1.1	--	--	--	--

(data from Krogman, 1970)

TABLE 881

BITROCHANTERIC DIAMETER (mm) FOR PHILADELPHIA CHILDREN

Age (years)	WHITE BOYS		BLACK BOYS	
	Mean	S.D.	Mean	S.D.
6	203.41	12.57	198.28	13.01
7	212.77	13.94	207.47	12.76
8	223.37	13.86	218.28	12.41
9	233.63	15.68	230.74	14.31
10	247.30	18.50	240.42	22.03
11	255.06	19.51	247.33	21.49
12	260.37	18.38	253.49	17.90
13	--	--	263.92	21.24

Age (years)	WHITE GIRLS		BLACK GIRLS	
	Mean	S.D.	Mean	S.D.
6	201.36	10.67	200.75	13.09
7	214.76	14.20	211.01	15.09
8	223.52	14.45	219.42	16.19
9	236.45	15.65	226.46	16.59
10	247.13	16.42	240.49	20.38
11	256.64	19.56	253.80	22.24
12	--	--	267.44	22.92
13	--	--	280.07	20.38

Age is for completed years.

(data from Malina, unpublished)

TABLE 882

BITROCHANTERIC DIAMETER (mm) OF PHILADELPHIA WHITE CHILDREN IN
RELATION TO CHRONOLOGICAL AND SKELETAL AGES (Greulich-Pyle)

Age (years)	Chronological		Skeletal	
	Mean	S.D.	Mean	S.D.
		BOYS		
7	218.25	16.50	220.69	21.38
8	228.48	16.66	227.34	15.98
9	238.40	16.47	321.10	17.53
10	248.84	16.16	253.57	16.57
11	258.57	18.27	250.31	17.16
12	269.64	19.47	260.19	16.80
13	286.74	20.02	278.39	19.06
14	302.86	22.72	296.52	16.93
15	313.74	21.30	315.85	12.41
16	327.00	25.60	320.46	23.16
17	--	--	338.55	21.18
		GIRLS		
7	212.40	11.73	215.00	12.62
8	225.03	15.97	230.00	13.81
9	235.97	16.36	239.48	13.19
10	259.89	19.99	251.93	13.55
11	262.80	19.52	263.93	15.26
12	285.66	19.50	274.22	18.49
13	292.45	16.67	286.00	15.00
14	304.69	14.18	300.77	12.35
15	312.50	14.84	307.00	12.24
16	316.45	12.91	312.37	10.04
17	--	--	315.29	11.91

(data from Johnston, 1962)

TABLE   883

PERCENTILES FOR CHEST DEPTH (cm) IN UNITED STATES CHILDREN

Age (years)	5	10	25	Boys 50	75	90	95
6	11.5	12.0	12.5	13.3	14.1	14.8	15.4
7	11.9	12.2	13.0	13.6	14.5	15.3	15.9
8	12.2	12.5	13.3	14.2	14.8	15.8	16.7
9	12.5	13.1	13.7	14.6	15.6	16.6	17.5
10	13.0	13.3	14.1	14.8	15.7	16.7	17.5
11	13.4	14.0	14.7	15.6	16.6	17.7	18.6
				Girls			
6	11.2	11.5	12.2	12.8	13.6	14.4	14.8
7	11.4	11.8	12.4	13.2	14.0	14.8	15.5
8	11.8	12.2	12.7	13.6	14.6	15.5	16.3
9	12.1	12.4	13.2	14.0	15.1	16.5	17.3
10	12.3	12.8	13.5	14.5	15.7	16.9	17.9
11	13.0	13.3	14.1	15.2	16.6	17.8	18.7

(data from Malina et al., 1973)

TABLE   884

CHEST DEPTH (cm) IN CHILDREN OF 8 STATES

Age (Months)	Boys		Girls	
	Mean	S.D.	Mean	S.D.
0 - 3	9.3	0.9	9.0	0.9
4 - 6	9.9	0.9	9.9	0.9
7 - 9	10.4	0.9	10.5	0.7
10 - 12	11.0	0.6	10.4	1.1
13 - 18	10.9	1.0	10.9	0.8
19 - 24	11.6	1.0	11.3	1.0
25 - 30	11.7	0.7	11.6	0.9
31 - 36	12.0	1.2	11.8	0.8
37 - 42	12.4	0.9	12.1	0.9
43 - 48	12.5	0.9	12.2	0.8
49 - 54	12.8	0.9	12.5	0.9
55 - 60	13.0	1.0	12.7	1.1
61 - 66	13.2	0.8	12.8	0.9
67 - 72	13.3	1.1	13.2	1.0
73 - 78	13.6	0.9	13.4	0.9
79 - 84	14.1	1.1	13.5	1.0
85 - 96	14.3	1.3	13.7	1.4
97 - 108	14.8	1.3	14.4	1.4
109 - 120	15.2	1.3	14.7	1.5
121 - 132	16.2	1.6	15.7	2.0
133 - 144	16.8	1.6	16.2	1.7
145 - 156	17.2	1.7	17.9	2.2

Measured at the level of the nipples.

(data from Snyder et al., 1975)

TABLE   885

CHEST DEPTH (cm) AT THE XIPHI-STERNUM IN COLORADO CHILDREN

Age (yr.-mo.)	Boys		Girls	
	Mean	S.D.	Mean	S.D.
Supine				
Birth	8.8	0.65	8.6	0.59
0 - 1	9.4	0.60	9.2	0.58
0 - 2	9.7	0.55	9.5	0.48
0 - 3	9.9	0.54	9.7	0.54
0 - 4	9.9	0.51	10.0	0.51
0 - 5	10.3	0.62	10.2	0.56
0 - 6	10.6	0.74	10.5	0.64
0 - 9	11.2	0.72	10.9	0.60
1 - 0	11.6	0.78	11.4	0.57
1 - 6	12.2	0.58	11.7	0.72
2 - 0	12.2	0.51	11.7	0.65
Erect				
2 - 0	11.8	0.93	11.6	0.80
2 - 6	12.2	0.90	11.6	0.78
3 - 0	12.4	0.72	12.0	0.74
3 - 6	12.6	0.68	12.4	0.75
4 - 0	12.9	0.80	12.6	0.77
4 - 6	13.0	0.81	12.7	0.74
5 - 0	13.2	0.77	12.9	0.75
5 - 6	13.4	0.85	13.1	0.76
6 - 0	13.6	0.91	13.3	0.78
6 - 6	13.9	1.00	13.5	0.84
7 - 0	14.2	1.11	13.7	0.90
7 - 6	14.4	1.13	14.0	1.01
8 - 0	14.6	1.09	14.3	1.13
8 - 6	14.8	1.23	14.4	1.08
9 - 0	15.1	1.24	14.7	1.18
9 - 6	15.4	1.26	15.0	1.15
10 - 0	15.6	1.36	15.3	1.45
10 - 6	15.9	1.36	15.6	1.53
11 - 0	16.1	1.44	16.0	1.66
11 - 6	16.4	1.52	16.4	1.50
12 - 0	16.8	1.55	16.7	1.82
12 - 6	17.0	1.58	16.7	1.56
13 - 0	17.4	1.71	16.9	1.62
13 - 6	17.9	1.83	17.3	1.47

TABLE 885 (continued)

CHEST DEPTH (cm) AT THE XIPHI-STERNUM
IN COLORADO CHILDREN

Age (yr.-mo.)	Boys		Girls	
	Mean	S.D.	Mean	S.D.
14 - 0	18.3	1.74	17.5	1.62
14 - 6	18.6	1.94	17.7	1.71
15 - 0	19.1	1.87	17.9	1.73
15 - 6	19.5	1.78	17.8	1.18
16 - 0	19.8	1.88	18.1	1.74
16 - 6	20.2	1.84	--	--
17 - 0	20.3	2.06	18.0	1.47
17 - 6	20.9	2.18	--	--
18 - 0	21.1	2.04	18.6	1.71

(data from McCammon, Human Growth and Development, 1970.
Courtesy of Charles C Thomas, Publisher, Springfield, Illinois)

TABLE 886

CHEST DEPTH (in)
IN OKLAHOMA CHILDREN

Age (years)	BOYS		GIRLS	
	Mean	S.D.	Mean	S.D.
5	5.53	0.33	5.38	0.31
6	5.71	0.38	5.55	0.33
7	5.93	0.35	5.77	0.41
8	6.21	0.49	5.95	0.38
9	6.50	0.56	6.31	0.54
10	6.83	0.65	6.48	0.65
11	6.90	0.54	6.89	0.77
12	7.28	0.77	7.14	0.68
13	7.56	0.81	7.30	0.59
14	7.95	0.68	6.99	0.56
15	8.26	0.55	7.23	0.56
16	8.63	0.74	7.29	0.64
17	8.93	0.93	7.45	0.59
18	9.03	0.57	7.56	0.61

Maximum horizontal distance from the sternum to the back in the midline with the child standing.

(data from Swearingen and Young, 1965)

TABLE   887

CHEST DEPTH (cm) IN PHILADELPHIA CHILDREN

Age (Years)	White Boys		White Girls		Negro Boys		Negro Girls	
	Mean	S.D.	Mean	S.D.	Mean	S.D.	Mean	S.D.
7	14.2	1.2	13.5	1.3	14.1	1.1	13.7	1.4
8	14.8	1.4	14.2	1.2	14.4	1.2	14.3	1.5
9	15.6	1.8	14.6	1.3	14.9	1.3	14.9	1.4
10	16.2	1.7	15.1	1.7	15.4	1.6	15.3	1.6
11	16.7	1.7	15.6	1.6	15.9	1.6	16.0	1.9
12	17.2	1.9	16.4	1.6	16.9	1.8	17.3	2.1
13	18.0	1.9	16.8	1.6	19.4	1.8	17.5	1.6
14	18.7	2.0	17.5	1.4	18.2	2.2	18.1	2.2
15	19.9	2.0	19.4	2.1	18.2	1.1	19.8	2.5
16	20.6	1.9	18.7	1.5	--	--	--	--
17	21.4	2.1	18.8	2.0	--	--	--	--

Measured at the maximum depth.

(data from Krogman, 1970)

TABLE 888

WAIST DEPTH (in)
IN OKLAHOMA CHILDREN

Age (years)	BOYS		GIRLS	
	Mean	S.D.	Mean	S.D.
5	5.26	0.41	5.25	0.38
6	5.46	0.50	5.29	0.37
7	5.65	0.46	5.47	0.46
8	5.87	0.62	5.59	0.43
9	6.08	0.77	6.00	0.71
10	6.23	0.79	6.13	0.66
11	6.36	0.59	6.44	1.01
12	6.77	0.88	6.67	0.87
13	6.98	0.90	6.77	0.79
14	7.01	0.69	6.58	0.80
15	7.24	0.66	6.44	0.72
16	7.56	0.83	6.53	0.65
17	7.75	1.02	6.68	0.60
18	7.74	0.68	6.77	0.67

(data from Swearingen and Young, 1965)

TABLE 889

CHEST CIRCUMFERENCE (cm) IN U.S. CHILDREN

Age (years)	Mean	S.D.	Percentiles						
			5	10	25	50	75	90	95

BOYS

Age (years)	Mean	S.D.	5	10	25	50	75	90	95
6	58.0	3.22	53.4	54.7	55.5	57.8	60.3	62.6	63.7
7	59.8	2.75	55.9	56.5	57.9	59.7	61.6	63.5	64.9
8	62.4	2.89	58.0	59.2	60.4	62.2	64.2	65.7	67.5
9	64.1	4.48	57.5	58.7	60.4	64.0	61.2	70.8	71.7
10	66.0	4.17	60.8	61.4	63.2	65.5	68.2	72.4	79.1
11	68.3	4.77	61.4	63.0	65.7	67.5	70.1	73.7	79.4
12	71.2	5.86	63.6	65.7	68.1	69.9	74.1	78.6	80.5
13	76.7	8.10	66.1	67.1	71.3	75.5	82.2	86.4	92.6
14	79.2	7.26	69.3	70.4	74.1	78.3	84.6	87.8	91.1
15	80.6	6.27	71.2	72.4	75.8	80.7	83.7	88.9	93.0
16	84.7	7.13	74.8	76.5	78.7	84.1	87.9	94.0	95.7
17	87.3	7.03	76.4	80.0	84.0	85.8	89.8	96.6	101.3

GIRLS

Age (years)	Mean	S.D.	5	10	25	50	75	90	95
6	55.9	2.91	51.2	52.2	54.1	55.8	58.3	59.6	61.3
7	57.8	3.88	53.2	53.8	55.0	57.1	60.2	62.3	63.7
8	60.2	5.01	53.8	54.9	56.6	59.2	63.4	66.9	68.6
9	63.2	5.61	56.5	57.3	58.9	63.1	64.9	70.9	75.6
10	66.2	66.4	56.6	60.2	61.8	65.4	69.0	73.2	81.0
11	69.7	7.09	60.5	61.6	64.3	69.0	74.2	80.7	83.8
12	75.7	6.25	66.4	68.6	71.6	75.2	80.0	84.1	84.9
13	76.5	6.36	68.3	69.7	72.2	76.0	79.1	84.4	89.5
14	78.9	5.86	70.4	72.5	74.9	78.3	81.8	87.7	91.1
15	79.6	5.93	72.1	72.6	75.9	79.3	82.5	87.4	90.4
16	80.6	6.01	71.7	73.7	76.4	80.3	83.3	86.8	93.5
17	81.5	6.42	73.1	75.0	77.6	80.6	84.6	90.3	94.6

(data from Cycles II and III of the Health Examination Survey by the National Center for Health Statistics 1963-1965 and 1966-1970)

TABLE 890

CHEST CIRCUMFERENCE (cm) IN U.S. WHITE CHILDREN

Age					Percentiles				
(years)	Mean	S.D.	5	10	25	50	75	90	95

					BOYS				
6	58.7	3.50	54.2	54.9	56.4	58.4	60.6	63.1	69.4
7	60.9	4.08	55.5	56.5	58.3	60.5	62.8	65.8	68.2
8	63.3	4.90	57.2	58.4	60.1	62.5	65.6	68.8	71.8
9	66.3	5.92	59.2	63.4	62.5	65.4	68.6	72.8	78.1
10	67.6	5.33	60.4	61.8	64.3	67.2	70.1	73.9	76.8
11	71.3	6.23	63.8	65.2	67.3	70.1	74.2	70.3	83.3
12	73.5	6.75	64.3	65.8	68.8	72.6	77.3	83.0	87.2
13	77.6	8.03	67.0	68.5	71.4	76.6	82.3	88.7	92.8
14	82.1	7.81	70.8	73.0	76.6	81.4	86.6	92.4	97.3
15	85.8	7.51	74.8	77.3	81.0	85.2	89.8	96.2	99.6
16	87.6	7.04	78.0	79.5	83.2	87.0	91.2	97.0	99.8
17	89.8	7.10	80.4	82.0	85.3	89.0	93.8	99.2	101.2

					GIRLS				
6	57.0	3.94	51.8	52.8	54.5	56.1	59.1	61.4	63.8
7	59.0	4.21	53.3	54.4	56.3	58.4	61.4	64.6	67.3
8	61.9	5.16	55.5	56.5	58.3	61.4	64.4	68.1	71.8
9	64.6	5.93	57.3	58.3	60.6	63.5	68.0	72.7	76.8
10	67.2	6.67	58.6	60.0	62.4	66.4	71.0	76.6	80.0
11	70.6	7.12	60.4	62.6	65.6	69.5	75.5	80.3	83.2
12	74.7	6.36	64.8	66.7	70.6	74.2	78.6	82.4	85.7
13	77.3	6.53	68.1	69.7	73.5	77.2	80.8	84.9	88.3
14	79.1	5.91	71.2	73.1	75.5	78.4	82.1	85.9	89.3
15	80.6	6.42	72.3	74.0	76.3	79.6	83.6	89.8	92.5
16	81.1	5.83	73.4	75.1	77.2	80.1	84.0	88.1	93.8
17	80.8	6.28	72.7	74.1	77.2	80.3	83.8	87.8	91.6

(data from Cycles II and III of the Health Examination Survey by the National Center for Health Statistics, 1963-1965 and 1966-1970)

TABLE 891

CHEST CIRCUMFERENCE (cm) IN CHILDREN OF 8 STATES

Age (Months)	Boys		Girls	
	Mean	S.D.	Mean	S.D.
0 - 3	35.9	3.4	35.9	2.5
4 - 6	42.0	2.1	40.5	2.4
7 - 9	45.2	2.0	43.5	1.6
10 - 12	46.2	2.3	44.2	2.2
13 - 18	47.2	2.4	44.7	1.8
19 - 24	48.6	2.2	47.6	2.4
25 - 30	49.2	2.0	48.8	2.5
31 - 36	51.0	2.3	49.1	2.4
37 - 42	51.4	2.0	50.4	2.4
43 - 48	52.2	2.1	51.3	2.3
49 - 54	53.6	2.6	52.1	2.2
55 - 60	54.7	2.6	53.1	2.7
61 - 66	55.5	2.7	54.0	3.4
67 - 72	56.2	3.2	54.8	3.1
73 - 78	57.1	2.7	55.8	3.4
79 - 84	59.2	3.4	56.9	2.9
85 - 96	60.5	4.3	58.5	4.3
97 - 108	62.4	4.0	61.3	4.8
109 - 120	64.1	4.3	63.7	5.1
121 - 132	67.6	4.6	67.2	6.5
133 - 144	70.6	4.4	70.2	6.5
145 - 156	71.6	4.9	78.3	7.4

Measured at the level of the nipples.

(data from Snyder et al., 1975)

TABLE 892

CHEST CIRCUMFERENCE AT XIPHOID (recumbent, cm)
IN BOSTON CHILDREN

Age Yr. Mo.	Percentiles					Mean	S.D.
	10	25	50	75	90		

Boys

0 0	30.6	31.8	33.2	34.4	35.7	33.2	1.8
0 3	38.3	39.3	40.6	41.6	42.9	40.6	1.7
0 6	42.0	42.8	43.9	45.2	46.5	44.0	1.8
0 9	44.0	45.0	46.3	48.0	49.4	46.5	2.2
1 0	45.2	46.5	47.8	49.4	50.9	47.9	2.1
1 6	47.3	48.4	49.8	51.3	52.9	49.9	2.1
2 0	48.6	49.6	50.9	52.1	54.0	51.0	2.0
2 6	49.3	50.4	51.8	53.2	54.9	51.9	2.0
3 0	50.0	51.1	52.4	54.3	55.9	52.7	2.2
3 6	50.5	51.6	53.1	54.7	56.6	53.4	2.4
4 0	51.1	52.2	53.8	55.6	57.2	54.0	2.3
4 6	51.7	52.9	54.2	56.3	57.9	54.5	2.4
5 0	52.2	53.5	55.1	57.0	58.8	55.4	2.4
5 6	53.0	54.1	55.8	57.6	59.6	56.0	2.6
6 0	53.8	54.6	56.1	58.4	60.2	56.5	2.5
7 0	55.3	56.4	57.8	60.3	61.9	58.3	2.6
8 0	55.7	57.8	59.3	62.0	65.1	59.9	3.2
9 0	58.2	59.0	60.9	63.6	68.0	61.7	3.4
10 0	60.2	61.4	62.9	66.1	70.3	64.2	4.2

Girls

0 0	30.8	31.8	32.9	34.0	35.0	32.9	1.6
0 3	37.6	38.8	39.8	40.9	42.0	39.8	1.7
0 6	40.8	42.0	43.2	44.6	45.7	43.2	1.9
0 9	43.0	44.3	45.6	46.8	48.3	45.6	1.9
1 0	44.4	45.8	47.4	48.4	49.6	47.2	2.0
1 6	46.2	47.4	48.8	50.4	51.4	48.9	2.0
2 0	47.5	48.7	50.1	51.8	53.3	50.3	2.2
2 6	48.5	49.8	51.5	53.0	54.4	51.5	2.2
3 0	49.2	50.6	51.9	53.5	55.1	52.1	2.4
3 6	50.3	51.2	52.4	54.0	55.8	52.8	2.3
4 0	50.8	51.8	53.3	54.8	57.6	53.5	2.6
4 6	51.3	52.2	53.5	55.3	57.2	53.9	2.4
5 0	51.7	52.8	54.4	56.2	58.0	54.6	2.5
5 6	62.0	53.3	54.7	56.5	59.1	55.1	2.6
6 0	52.8	54.2	56.0	57.4	59.1	56.0	2.6
7 0	54.8	56.0	57.0	59.1	61.8	57.7	2.9
8 0	55.6	57.1	59.0	61.3	63.4	59.1	3.0
9 0	57.7	58.8	60.3	63.2	69.6	61.5	3.9
10 0	59.4	59.8	63.0	65.5	74.3	64.4	5.3

(data from Vickers and Stuart, 1943, The Journal of Pediatrics
22:155-170)

TABLE 893

CHEST CIRCUMFERENCE (expiration, mm) IN CLEVELAND CHILDREN*

Age (years)	BOYS		GIRLS	
	Mean	S.D.	Mean	S.D.
4 years	501.06	20.97	491.75	18.98
4½ years	508.85	22.15	503.28	20.91
5 years	516.48	23.49	504.43	22.71
6 years	533.65	28.47	521.60	31.17
7 years	552.18	26.91	540.37	34.53
8 years	573.94	34.91	559.21	38.90
9 years	592.71	37.30	579.83	41.53
10 years	618.60	39.70	604.69	46.47
11 years	638.03	50.41	621.99	46.13
12 years	658.29	51.68	644.69	45.50
13 years	679.65	50.63	664.94	47.62
14 years	713.19	54.86	678.51	42.92
15 years	739.28	51.75	686.76	44.64
16 years	757.69	45.98	693.87	50.55
17 years	770.17	44.70	697.28	63.11

* level of xiphoid

(data from Simmons, 1944)

TABLE 894

CHEST CIRCUMFERENCE (inspiration, mm) IN CLEVELAND CHILDREN*

Age (years)	BOYS		GIRLS	
	Mean	S.D.	Mean	S.D.
4 years	542.19	30.23	541.21	20.41
4½ years	556.94	23.97	550.28	26.68
5 years	569.10	25.83	553.64	28.57
6 years	590.94	31.71	577.67	31.46
7 years	611.14	32.32	600.89	37.02
8 years	636.98	33.74	623.03	38.95
9 years	658.39	40.24	649.51	40.15
10 years	689.39	42.46	678.99	45.46
11 years	713.73	51.62	695.38	46.84
12 years	738.45	52.72	721.17	44.36
13 years	765.21	53.71	744.03	47.05
14 years	803.00	56.89	760.46	41.72
15 years	832.80	51.69	769.19	47.09
16 years	857.43	48.04	775.25	50.42
17 years	876.17	46.90	783.89	60.22

* level of xiphoid.

(data from Simmons, 1944)

TABLE 895

CHEST CIRCUMFERENCE (cm) AT LEVEL OF XIPHI-STERNUM
IN COLORADO CHILDREN

Age	Boys		Girls	
(yr.-mo.)	Mean	S.D.	Mean	S.D.
Supine				
Birth	32.2	1.97	32.3	1.53
0 - 1	35.3	1.75	34.5	1.59
0 - 2	38.1	1.90	36.9	1.38
0 - 3	39.8	1.73	38.7	1.67
0 - 4	41.4	1.66	40.2	1.74
0 - 5	42.5	1.85	41.3	1.65
0 - 6	43.5	1.80	42.6	1.87
0 - 9	45.7	1.95	44.4	1.88
1 - 0	46.9	1.87	45.7	1.78
1 - 6	48.5	2.18	47.1	1.76
2 - 0	49.3	1.68	48.0	1.51
Erect				
2 - 0	48.9	2.67	47.7	1.90
2 - 6	50.0	2.18	48.5	1.74
3 - 0	50.8	2.06	49.2	1.89
3 - 6	51.6	2.16	50.2	1.74
4 - 0	52.2	2.13	50.9	1.70
4 - 6	52.9	2.06	51.8	1.94
5 - 0	53.9	1.87	52.5	2.17
5 - 6	54.7	2.08	53.4	2.26
6 - 0	55.8	2.30	54.3	2.41
6 - 6	56.9	2.59	55.0	2.35
7 - 0	57.8	2.50	56.2	2.94
7 - 6	58.9	2.75	57.3	3.12
8 - 0	60.0	3.04	58.6	3.34
8 - 6	61.0	3.19	59.5	3.66
9 - 0	62.3	3.42	60.3	3.91
9 - 6	63.3	3.34	61.5	3.83
10 - 0	64.7	4.09	62.8	4.28
10 - 6	65.5	3.77	63.8	4.98
11 - 0	66.7	4.22	65.0	5.34
11 - 6	67.7	4.42	66.1	4.90
12 - 0	69.2	4.73	67.4	5.39
12 - 6	70.4	4.92	67.7	4.77
13 - 0	72.4	5.32	68.9	4.78
13 - 6	73.6	5.86	69.6	4.48

TABLE 895 (continued)

CHEST CIRCUMFERENCE (cm) AT LEVEL OF XIPHI-STERNUM
IN COLORADO CHILDREN

Age (yr.-mo.)	Boys		Girls	
	Mean	S.D.	Mean	S.D.
14 - 0	75.6	5.88	71.3	4.91
15 - 6	76.6	5.90	71.1	4.34
15 - 0	79.0	5.76	71.6	4.60
15 - 6	80.1	5.68	71.2	3.20
16 - 0	81.9	6.33	73.0	5.04
16 - 6	82.7	5.45	--	--
17 - 0	84.6	6.59	73.3	5.35
17 - 6	86.4	6.99	--	--
18 - 0	87.1	6.49	74.0	5.00

(data from McCammon, Human Growth and Development, 1970.
Courtesy of Charles C Thomas, Publisher, Springfield, Illinois)

TABLE 896

CHEST CIRCUMFERENCES IN ILLINOIS BOYS

Per-centiles	Age (years)								
	6	7	8	9	10	11	12	13	14
Chest expansion (in)									
95	3.0	4.2	4.1	4.6	4.3	5.1	5.3	6.6	5.3
75	2.7	3.5	3.4	3.8	3.7	4.2	4.4	5.2	4.3
50	2.2	2.5	2.5	2.8	2.8	3.0	3.2	3.4	2.9
25	1.8	1.6	1.6	1.8	2.0	1.9	2.1	1.7	1.6
5	1.4	0.8	0.9	1.0	1.3	0.9	1.1	0.2	0.5
Expanded chest circumference minus abdominal circumference (in)									
95	5.3	7.3	6.9	8.4	8.5	9.3	8.6	10.2	7.9
75	4.6	5.9	5.7	6.6	6.7	7.1	6.8	7.7	6.1
50	3.7	4.2	4.2	4.3	4.3	4.3	4.6	4.6	3.8
25	2.8	2.5	2.7	2.1	2.0	1.6	2.3	1.5	1.5
5	2.1	1.1	1.5	0.3	0.1	-0.4	0.5	-1.0	-0.4

(data from Cureton and Barry, 1964)

TABLE   897

CHEST CIRCUMFERENCE (cm)* IN NEGRO INFANTS
IN WASHINGTON (D.C.)

Age (weeks)	Boys		Girls	
	Mean	S.D.	Mean	S.D.
0	32.06	1.79	32.68	1.46
4	34.48	1.77	34.35	1.14
8	36.40	1.97	36.06	1.26
12	38.21	2.22	37.79	1.66
16	39.69	2.22	39.17	1.93
20	40.88	2.15	40.41	1.84
24	41.90	2.05	41.48	2.00
28	42.78	2.01	42.26	2.03
32	43.52	1.92	42.88	1.90
36	44.17	1.83	43.45	1.77
40	44.76	1.79	44.01	1.67
44	45.32	1.76	44.50	1.57
48	45.89	1.78	44.98	1.50
52	46.40	1.81	45.45	1.50

* = in mid-respiration at level of xiphoid

(data from Scott, Hiatt, Clark, et al., 1962, Pediatrics 29:65-81.
Copyright American Academy of Pediatrics 1962)

TABLE 898

CHEST CIRCUMFERENCE*(cm) IN UNDERPRIVILEGED NEGRO CHILDREN
IN NEW ORLEANS (LA)

Age (months)	BOYS		GIRLS	
	Mean	S.D.	Mean	S.D.
Birth	31.5	2.5	31.5	2.2
3	39.7	2.3	39.7	2.3
6	43.7	2.4	43.7	2.9
18	48.8	2.2	48.0	2.4
24	49.7	2.3	49.0	2.4
42	51.9	2.1	51.0	2.1
60	54.3	2.3	53.2	2.6

*level of nipple

(data from Cherry, 1968)

TABLE 899

TRUNK CIRCUMFERENCES (cm) FOR BLACK BOYS IN TEXAS

Age (years)	Minimum chest				Maximum chest				Buttocks			
	Lower Income		Middle Income		Lower Income		Middle Income		Lower Income		Middle Income	
	Mean	S.D.	Mean	S.D.	Mean	S.D.	Mean	S.D.	Mean	S.D.	Mean	S.D.
10.1-11	64.9	4.35	63.4	2.99	68.4	4.33	67.6	2.93	72.1	6.59	69.9	2.96
11.1-12	65.7	2.58	68.0	4.16	69.5	2.81	71.0	3.56	74.6	5.25	76.0	5.67
12.1-13	70.6	8.78	71.2	6.88	73.9	8.26	74.9	6.43	76.5	8.74	79.1	7.24
13.1-14	71.8	4.62	74.8	8.34	76.2	5.86	76.7	7.23	79.1	6.16	81.7	7.39
14.1-15	74.4	7.20	77.0	6.03	77.6	7.28	80.3	5.41	80.4	7.16	84.0	5.26
15.1-16	77.3	5.08	80.3	4.22	85.8	3.12	86.7	4.57	84.4	6.26	87.6	4.83
16.1-17	82.2	3.31	83.4	4.75	81.1	5.37	83.9	4.34	90.0	4.66	92.0	4.99
17.1-18	84.8	4.20	82.7	4.23	88.0	4.77	86.6	4.57	92.0	6.67	89.4	4.57

(data from Schutte, 1979)

TABLE 900

CHEST CIRCUMFERENCE (cm) IN RELATION TO TYPES OF ARTIFICIAL FEEDING
IN PHILADELPHIA CHILDREN

Group	Race	3 Months		1 Year		2 Years	
		Mean	S.D.	Mean	S.D.	Mean	S.D.
All	All White	39.9	+1.69	47.2	+1.56	50.4	+1.72
	All Negro	39.4	+1.84	46.5	+2.03	49.7	+1.97
Group I (irradiated evaporated milk)	White	39.6	+1.81	47.2	+1.76	50.1	+1.77
	Negro	39.2	+1.91	46.5	+1.90	49.4	+1.94
Group II (nonirradiated evaporated milk plus cod-liver oil)	White	40.0	+1.68	47.0	+1.54	50.4	+1.60
	Negro	39.4	+2.07	46.6	+2.37	49.8	+1.97
Group III (irradiated evaporated milk plus carotene)	White	40.2	+1.28	47.5	+1.69	50.7	+1.82
	Negro	39.5	+1.92	46.7	+2.04	49.8	+2.09
Group IV (irradiated evaporated milk plus carotene and yeast)	White	40.0	+1.84	47.0	+1.54	50.7	+1.77
	Negro	39.5	+1.22	46.4	+1.84	50.2	+2.02

Group	Race	3 Years		4 Years	
		Mean	S.D.	Mean	S.D.
All	All White	52.8	+1.92	54.7	+1.83
	All Negro	52.0	+1.99	53.9	+2.31
Group I (irradiated evaporated milk)	White	52.8	+1.86	54.5	+1.88
	Negro	51.7	+1.89	53.6	+2.32
Group II (nonirradiated evaporated milk plus cod-liver oil)	White	52.4	+1.77	54.5	+1.36
	Negro	52.2	+2.23	54.1	+2.33
Group III (irradiated evaporated milk plus carotene)	White	53.3	+2.29	55.3	+2.34
	Negro	51.9	+2.15	53.7	+2.57
Group IV (irradiated evaporated milk plus carotene and yeast)	White	52.9	+1.92	54.8	+1.73
	Negro	52.0	+1.70	53.9	+2.22

(data from Rhoads, Rapaport, Kennedy, et al., 1945, The Journal of Pediatrics 26:415-454)

TABLE 901

CHEST CIRCUMFERENCE (mm) IN JAPANESE-AMERICAN
CHILDREN IN LOS ANGELES

Age	Boys		Girls	
(years)	Mean	S.D.	Mean	S.D.
6	578.1	47.4	541.1	28.0
7	584.4	48.9	562.2	45.7
8	612.4	45.1	576.7	47.8
9	607.2	45.7	593.5	57.5
10	644.5	53.3	630.3	55.0
11	668.7	58.5	656.1	68.1
12	702.6	70.6	698.5	61.2
13	726.2	64.0	736.4	75.5
14	767.0	77.6	758.2	58.9
15	788.3	77.7	763.4	54.1
16	817.1	37.6	--	--
18	--	--	788.6	79.2

(data from Kondo and Eto, 1975, pp. 13-45 in Comparative
Studies on Human Adaptability of Japanese, Caucasians
and Japanese Americans, S. Horvath et al., eds. University
of Tokyo Press)

TABLE 902

MEANS FOR CHEST CIRCUMFERENCE (cm) AT SUPRAMAMMILLARY AND XIPHOID LEVELS
IN NEW JERSEY CHILDREN

Age	Boys		Girls	
	Supramammillary	Xiphoid	Supramammillary	Xiphoid
Newborn	30.5	31.6	30.0	31.2
3 mo.	39.0	39.2	37.4	38.1
6 mo.	42.6	43.0	41.2	41.7
9 mo.	45.8	45.8	44.2	44.4
1 yr.	47.1	47.2	46.0	46.0
1½ yr.	48.9	48.4	47.6	47.2
2 yr.	49.4	48.8	48.2	47.7
2½ yr.	50.5	49.4	49.5	48.5
3 yr.	51.9	50.8	50.3	49.2
3½ yr.	52.6	51.0	51.4	50.2
4 yr.	53.6	52.3	52.4	51.2
4½ yr.	54.8	53.4	53.5	52.0
5 yr.	55.7	53.9	55.1	53.1
5½ yr.	57.2	55.4	55.4	53.5
6 yr.	58.4	56.6	56.6	54.5

(data from Kornfeld, 1953, The Journal of Pediatrics 42:715-720)

TABLE   903

DISTRIBUTIONS OF THE DIFFERENCES (cm) BETWEEN CHEST CIRCUMFERENCES AT SUPRAMAMMILLARY
AND XIPHOID LEVELS IN NEW JERSEY CHILDREN (boys and girls combined)

Difference	Newborn	3 Mo.	6 Mo.	9 Mo.	1 Yr.	1½ Yr.	2 Yr.	2½ Yr.	3 Yr.	3½ Yr.	4 Yr.
+ 4.5	--	--	--	--	--	--	--	--	--	--	1
+ 4.0	--	--	--	--	--	--	--	--	--	2	--
+ 3.5	--	--	--	--	--	--	--	--	--	2	5
+ 3.0	--	--	--	--	--	--	3	3	6	6	11
+ 2.5	--	--	1	1	--	3	1	3	8	7	20
+ 2.0	--	1	1	3	2	2	10	22	29	28	46
+ 1.5	--	5	2	1	2	20	17	19	28	27	40
+ 1.0	2	12	11	13	31	31	44	48	61	46	52
+ 0.5	1	7	10	14	17	27	24	19	33	11	18
± 0.0	12	52	69	50	66	27	32	18	21	8	17
- 0.5	22	39	24	22	22	16	12	2	2	1	2
- 1.0	87	45	66	24	25	13	6	3	3	--	--
- 1.5	36	21	14	6	10	3	1	1	1	1	--
- 2.0	27	7	12	3	3	--	--	--	--	--	--
- 2.5	7	--	--	--	--	--	--	--	--	--	--
- 3.0	--	--	1	--	--	--	--	--	--	--	--

(data from Kornfeld, 1953, The Journal of Pediatrics 42:715-720)

TABLE   904

NECK CIRCUMFERENCE (cm) IN CHILDREN OF 8 STATES

Age	(Months)	Boys		Girls	
		Mean	S.D.	Mean	S.D.
0 –	3	30.7	3.6	31.0	3.3
4 –	6	37.0	2.5	36.3	2.7
7 –	9	39.7	2.2	38.8	2.8
10 –	12	39.5	3.1	39.0	2.7
13 –	18	42.0	2.6	39.9	2.4
19 –	24	44.7	3.4	43.9	3.4
25 –	30	46.4	1.8	46.4	3.2
31 –	36	48.2	2.7	46.7	3.1
37 –	42	48.7	2.8	47.8	2.6
43 –	48	49.3	2.5	48.1	2.5
49 –	54	49.8	2.7	48.5	2.4
55 –	60	50.7	2.9	49.4	3.1
61 –	66	51.2	3.2	49.9	3.4
67 –	72	51.4	3.8	49.9	3.8
73 –	78	52.1	2.4	50.8	3.8
79 –	84	53.6	3.8	51.7	3.1
85 –	96	54.5	4.6	52.8	4.4
97 –	108	56.0	4.4	54.6	5.2
109 –	120	57.5	4.7	56.1	5.5
121 –	132	60.3	5.1	58.7	5.7
133 –	144	61.9	4.6	60.3	5.8
145 –	156	63.0	5.8	64.0	7.0

(data from Snyder et al., 1975)

TABLE 905

PERCENTILES FOR WAIST CIRCUMFERENCE (cm) IN U.S. CHILDREN*

Sex and age				Percentiles			
	5	10	25	50	75	90	95
Boys							
6 years.........47.4	48.4	50.2	52.4	54.9	58.2	60.3
7 years.........47.9	49.5	51.8	54.1	56.5	59.5	61.3
8 years.........50.0	50.6	53.0	55.7	58.6	62.3	65.5
9 years.........51.1	52.1	54.3	57.1	60.3	65.8	70.8
10 years........52.3	53.3	55.3	58.2	61.8	66.8	69.8
11 years........54.1	55.4	57.6	60.5	64.8	70.9	76.8
Girls							
6 years.........45.5	46.8	49.2	51.7	53.9	56.7	58.8
7 years.........47.2	48.3	50.1	52.5	55.4	58.9	61.5
8 years.........47.8	59.4	51.5	54.0	57.5	61.7	65.8
9 years.........50.1	51.1	53.4	56.1	59.5	65.1	68.3
10 years........50.4	51.6	54.3	57.3	61.4	65.8	71.6
11 years........52.1	53.2	55.5	59.2	63.5	68.9	72.7

*level of waist.

(data from Malina et al., 1973)

TABLE 906

WAIST CIRCUMFERENCE (cm) IN U.S. CHILDREN

Age (years)	Mean	S.D.				Percentiles			
			5	10	25	50	75	90	95
					BOYS				
12	64.5	7.26	55.3	57.0	60.1	63.2	67.4	74.6	79.7
13	67.6	8.56	57.5	59.2	62.0	65.7	70.8	79.2	85.5
14	70.4	8.43	61.1	62.6	65.1	68.5	73.5	81.6	85.5
15	72.3	8.02	62.7	64.4	67.5	70.7	75.0	81.4	89.5
16	73.4	7.64	64.1	65.6	68.4	72.2	76.8	82.6	86.7
17	75.5	8.04	66.1	67.6	70.2	73.6	79.4	86.5	91.8
					GIRLS				
12	62.9	7.09	54.2	55.4	58.0	61.8	65.8	72.6	77.4
13	64.0	7.11	55.1	56.6	59.3	62.7	66.8	73.2	79.3
14	65.1	7.24	56.6	58.0	60.3	63.8	68.3	73.9	80.0
15	66.3	8.05	56.7	58.0	61.1	64.7	69.4	78.6	82.2
16	67.0	8.07	58.1	59.2	61.8	64.9	69.7	77.5	84.5
17	66.7	7.80	57.6	59.1	61.7	65.3	69.8	75.6	83.2

(data from Cycle III of the Health Examination Survey 1966-1970 conducted by the National Center for Health Statistics).

TABLE 907

WAIST CIRCUMFERENCE (cm) IN U.S. WHITE CHILDREN

Age (years)	Mean	S.D.	Percentiles						
			5	10	25	50	75	90	95
BOYS									
6	53.0	4.20	47.4	48.4	50.3	52.5	54.9	58.3	60.2
7	54.5	4.64	47.9	49.6	51.7	54.1	56.7	59.7	61.7
8	56.4	5.64	49.5	50.7	53.0	55.8	58.7	62.5	66.1
9	58.3	6.33	51.3	52.3	54.4	57.3	60.5	66.3	71.5
10	59.5	6.08	52.3	51.4	55.3	58.4	62.3	67.2	72.3
11	62.4	7.27	54.2	55.5	57.7	60.6	65.5	71.6	77.5
12	64.7	7.38	55.2	57.0	60.3	63.4	67.5	74.9	80.1
13	67.7	8.60	57.6	59.3	62.0	65.7	70.8	79.5	84.9
14	76.8	8.58	61.3	62.9	65.4	68.8	74.1	82.0	86.4
15	72.8	8.04	63.2	64.8	68.0	71.2	75.6	82.5	90.6
16	73.7	7.70	64.3	65.8	69.0	72.4	77.1	83.2	87.4
17	76.0	7.99	66.4	67.9	70.4	74.3	79.9	86.7	92.4
GIRLS									
6	51.8	4.58	45.4	46.8	49.1	51.6	53.9	56.9	59.2
7	53.1	4.54	47.2	48.3	50.2	52.5	55.5	59.0	61.5
8	54.9	5.34	47.7	49.4	51.6	54.0	57.4	61.7	66.1
9	57.1	5.87	50.1	50.9	53.3	56.1	59.9	65.3	68.4
10	58.2	6.38	50.4	51.5	54.1	57.3	61.3	65.8	71.4
11	60.1	6.45	52.1	53.1	55.5	59.3	63.6	69.0	72.5
12	62.6	7.06	54.1	55.3	57.8	61.6	65.4	72.4	77.1
13	64.0	6.98	55.2	56.7	59.4	62.8	66.8	73.2	78.9
14	65.0	7.35	56.6	58.0	60.2	63.6	68.1	73.9	80.2
15	66.4	8.23	56.7	57.8	61.1	64.8	69.5	78.8	82.4
16	66.8	7.89	58.2	59.2	61.8	64.8	69.4	72.1	84.3
17	66.5	7.65	57.6	59.0	61.6	64.8	69.7	75.8	83.0

(data from Cyles II and III of the Health Examination Survey by the National Center for Health Statistics 1963-1965 and 1966-1970)

TABLE 908

WAIST CIRCUMFERENCE (cm) IN U.S. NEGRO CHILDREN

| Age | | | Percentiles | | | | | | |
(years)	Mean	S.D.	5	10	25	50	75	90	95
				BOYS					
6	52.4	4.28	46.9	48.3	49.7	51.7	54.7	57.8	60.6
7	53.4	3.15	47.9	49.4	51.8	53.6	55.2	57.4	59.0
8	55.3	3.85	49.7	50.6	52.9	54.9	57.8	61.1	63.0
9	56.2	4.71	49.3	50.8	53.1	57.8	58.8	61.6	64.3
10	57.7	4.29	52.3	52.6	55.2	57.4	59.6	63.9	67.4
11	60.0	5.25	53.0	55.1	57.1	59.5	62.3	65.3	72.5
12	62.8	6.00	55.6	56.8	59.1	61.7	65.2	69.3	71.5
13	66.7	8.36	57.1	57.8	62.0	65.5	70.2	75.2	87.1
14	67.8	6.88	57.0	61.2	63.5	66.6	71.2	76.2	78.7
15	68.2	5.75	61.1	62.2	64.8	67.8	70.6	74.9	77.6
16	71.5	7.06	63.0	64.2	67.1	69.8	74.6	80.4	82.6
17	72.6	7.76	63.2	66.2	68.5	72.0	73.8	79.4	87.7
				GIRLS					
6	51.5	3.28	45.5	46.7	49.6	51.8	53.9	55.2	56.4
7	52.6	4.11	47.5	48.4	49.6	52.2	54.6	57.6	61.7
8	54.8	5.47	47.8	49.1	50.8	53.7	58.3	61.7	63.9
9	56.8	5.27	51.0	51.6	53.4	56.7	58.5	64.2	67.8
10	58.3	5.57	51.1	52.7	54.7	56.8	60.9	64.7	71.5
11	60.9	8.30	53.0	53.5	56.2	58.8	63.1	68.0	78.4
12	64.2	7.22	54.4	56.7	59.4	62.9	68.1	75.0	79.6
13	63.8	7.89	53.8	55.8	58.7	62.2	67.1	76.2	80.3
14	65.7	6.47	56.5	57.7	61.1	65.4	69.6	76.2	78.9
15	65.9	6.78	58.1	58.7	61.3	64.5	69.6	74.9	81.2
16	68.1	9.08	56.8	58.2	61.8	65.6	72.0	82.3	88.4
17	68.2	8.56	57.7	59.3	64.1	67.4	71.6	74.7	91.6

(data from Cycles II and III of the Health Examination Survey by the National Center for Health Statistics 1963-1965 and 1966-1970)

TABLE 909

PERCENTILES FOR HIP CIRCUMFERENCE (cm) IN U.S. CHILDREN*

Sex and age	Percentiles						
	5	10	25	50	75	90	95
Boys							
6 years..........51.3	52.3	54.8	57.6	60.7	64.3	66.3
7 years..........53.5	55.1	57.5	60.2	63.3	67.3	69.4
8 years..........56.3	57.6	60.3	63.5	67.1	71.2	74.5
9 years..........58.0	59.4	62.5	65.8	70.2	75.8	80.5
10 years........60.1	62.2	64.8	68.3	72.8	78.1	80.8
11 years........62.6	65.2	68.2	72.2	77.1	82.7	87.7
Girls							
6 years..........50.7	52.7	55.4	58.5	62.1	65.3	67.4
7 years..........54.2	55.7	57.9	61.2	65.5	69.4	72.3
8 years..........55.9	57.6	61.3	65.2	69.2	73.5	76.9
9 years..........58.8	60.7	63.5	67.8	73.5	79.0	81.5
10 years........61.1	63.1	66.8	71.8	76.9	82.9	86.5
11 years........64.4	66.4	70.8	75.6	82.1	87.8	92.4

* level of greater trochanter.

(data from Malina et al., 1973)

TABLE 910

HIP CIRCUMFERENCE (cm) IN U.S. CHILDREN

Age (years)	Mean	S.D.	Percentiles						
			5	10	25	50	75	90	95
				BOYS					
12	78.2	7.50	68.0	70.0	73.2	76.9	82.9	89.2	92.0
13	82.7	8.37	70.8	73.3	77.1	81.8	87.4	93.1	97.6
14	86.6	8.53	74.5	76.8	81.2	85.9	91.2	96.8	101.4
15	89.3	7.48	79.4	81.1	84.6	88.6	92.8	98.3	104.6
16	90.7	7.37	81.4	83.3	86.6	89.9	94.6	99.4	103.2
17	92.6	7.32	83.4	84.8	87.6	91.6	95.8	102.3	107.5
				GIRLS					
12	84.3	8.24	72.5	74.2	78.3	84.1	89.1	95.4	99.4
13	87.4	8.02	76.0	78.2	81.8	87.1	92.3	97.8	100.8
14	90.3	7.73	79.1	81.6	85.7	89.7	94.4	99.7	103.1
15	92.3	8.33	81.7	83.3	86.8	91.4	96.2	103.2	108.1
16	93.4	8.12	84.0	85.5	88.1	92.1	97.4	102.2	108.5
17	93.5	8.20	83.0	85.0	88.6	92.6	96.6	102.3	109.1

(data from Cycle III of the Health Examination Survey 1966-1970 conducted by the National Center for Health Statistics).

TABLE 911

HIP CIRCUMFERENCE (cm) IN U.S. WHITE CHILDREN

Age (years)	Mean	S.D.	Percentiles						
			5	10	25	50	75	90	95

BOYS

Age (years)	Mean	S.D.	5	10	25	50	75	90	95
6	58.3	4.82	51.4	52.5	55.1	53.0	61.1	64.5	66.6
7	61.1	5.24	53.7	55.4	57.8	60.5	63.8	67.7	69.8
8	64.2	5.93	56.5	57.7	60.4	63.7	67.2	71.6	75.1
9	67.3	7.23	58.3	59.6	62.6	66.1	70.6	76.2	81.7
10	69.4	6.54	60.6	62.4	65.0	68.5	73.4	78.4	80.9
11	73.5	7.49	63.0	65.6	68.3	72.4	77.5	83.8	88.1
12	78.4	7.52	68.0	69.8	73.3	77.4	83.1	89.4	92.2
13	82.7	8.23	71.3	73.5	77.1	81.8	87.3	92.8	97.8
14	86.9	8.50	75.1	79.1	81.4	86.2	91.4	97.1	101.6
15	89.7	7.49	80.0	81.5	85.1	89.1	93.3	99.0	105.2
16	91.2	6.94	82.3	83.8	86.9	90.1	95.0	99.6	103.4
17	92.8	7.32	83.3	84.8	87.7	91.9	96.3	102.8	107.8

GIRLS

Age (years)	Mean	S.D.	5	10	25	50	75	90	95
6	59.1	5.21	51.1	53.1	55.6	58.8	62.3	65.6	67.6
7	62.3	5.55	54.6	56.1	58.3	61.6	65.8	69.6	72.5
8	65.7	6.14	56.2	51.7	61.7	65.4	69.3	73.5	76.7
9	68.9	7.12	59.1	61.0	63.7	68.0	73.7	79.2	81.6
10	72.4	7.89	61.2	63.2	66.7	71.8	76.9	82.7	86.3
11	76.6	8.25	64.6	66.5	71.0	75.7	82.2	87.4	92.0
12	84.0	8.01	72.5	74.2	78.1	84.0	88.7	94.1	98.5
13	87.5	7.73	75.9	78.4	82.4	87.5	92.4	97.7	100.3
14	90.2	7.62	79.3	81.8	85.8	89.6	94.2	99.2	102.2
15	92.5	8.40	81.6	83.4	87.2	91.5	96.4	103.4	108.2
16	93.4	7.80	84.6	85.7	88.2	92.3	97.4	101.4	107.6
17	93.4	7.88	83.0	85.2	88.8	92.4	96.6	102.2	108.9

(data from Cycles II and III of the Health Examination Survey by the National Center for Health Statistics 1963-1965 and 1966-1970)

TABLE 912

HIP CIRCUMFERENCE (cm) IN U.S. NEGRO CHILDREN

Age						Percentiles			
(years)	Mean	S.D.	5	10	25	50	75	90	95

BOYS

6	56.2	3.86	51.1	51.5	53.4	56.3	58.4	61.3	63.1
7	58.4	3.34	52.8	54.1	56.2	58.5	61.4	62.5	63.1
8	62.8	4.74	55.2	56.5	60.1	62.8	65.8	68.6	71.0
9	64.9	6.14	57.3	58.1	60.3	64.7	67.6	72.5	77.9
10	67.4	5.79	58.8	59.9	64.3	66.7	70.1	75.3	78.9
11	70.9	6.32	61.1	62.0	66.8	70.8	74.7	78.0	80.2
12	79.0	7.05	67.9	70.2	72.4	76.0	80.0	86.4	89.5
13	82.6	9.39	68.7	70.7	76.2	81.5	88.4	95.2	96.6
14	84.9	8.62	72.5	76.1	78.7	83.8	88.7	95.1	98.2
15	86.1	6.75	74.8	79.1	82.2	85.6	90.1	93.3	98.2
16	87.6	9.41	61.8	78.3	83.1	88.8	93.3	97.0	101.1
17	91.3	7.31	83.6	84.6	87.1	90.0	93.0	97.7	106.5

GIRLS

6	57.2	4.44	49.7	51.1	54.4	57.3	60.3	63.5	64.5
7	60.0	5.67	53.3	54.2	56.4	59.4	62.4	66.6	70.5
8	64.3	6.90	54.5	55.9	59.3	63.6	69.0	73.4	76.9
9	67.8	7.53	58.1	59.6	62.6	66.7	71.4	77.6	79.8
10	72.4	8.55	60.1	61.9	66.9	71.5	76.8	83.8	86.7
11	77.1	10.56	63.4	64.9	70.4	74.9	81.8	91.5	100.7
12	86.2	9.47	72.6	74.3	79.7	84.5	94.3	99.4	102.5
13	86.5	9.61	76.1	76.9	80.2	83.7	90.4	99.6	108.2
14	90.8	8.35	76.6	80.4	85.2	91.5	96.5	101.9	103.7
15	91.4	7.81	82.1	83.1	85.9	90.5	95.1	100.1	107.9
16	93.3	9.96	81.2	82.5	86.8	91.4	96.3	110.8	115.7
17	94.4	10.02	83.0	83.6	87.6	93.9	97.1	164.8	114.2

(data from Cycles II and III of the Health Examination Survey by the National Center for Health Statistics 1963-1965 and 1966-1970)

TABLE   913

DISTRIBUTION STATISTICS FOR
ILIAC CIRCUMFERENCE (cm)
IN DENVER CHILDREN

| Age | Boys | | Girls | |
Yr.- Mo.	Mean	S.D.	Mean	S.D.
Supine				
Birth	26.6	2.24	26.8	2.17
0 - 1	30.3	2.14	29.9	2.58
0 - 2	33.6	2.86	32.8	2.12
0 - 3	36.2	2.79	34.8	2.52
0 - 4	37.9	2.84	36.6	2.38
0 - 5	38.9	3.01	38.1	2.68
0 - 6	40.0	3.00	39.4	2.86
0 - 9	42.2	3.13	41.2	2.87
1 - 0	42.6	2.97	42.0	2.78
1 - 6	43.5	2.96	43.2	2.71
2 - 0	44.0	2.27	43.0	2.49
Erect				
2 - 0	47.0	2.96	46.7	3.02
2 - 6	48.5	2.76	48.0	2.64
3 - 0	49.4	2.62	49.0	2.65
3 - 6	50.3	2.61	50.2	2.47
4 - 0	51.2	2.84	51.0	2.52
4 - 6	51.7	2.46	51.8	2.69
5 - 0	52.9	2.66	52.7	2.80
5 - 6	53.9	2.75	53.8	3.18
6 - 0	54.7	2.84	54.7	3.52
6 - 6	55.8	3.38	55.5	3.56
7 - 0	56.4	3.31	57.1	4.29
7 - 6	57.5	3.41	58.1	4.52
8 - 0	58.8	3.75	59.2	4.81
8 - 6	59.6	4.06	60.2	5.12
9 - 0	60.9	4.47	61.2	5.42
9 - 6	62.1	4.62	62.8	5.68
10 - 0	63.4	5.28	63.9	6.55
10 - 6	64.0	5.37	65.3	6.91
11 - 0	65.5	5.62	66.8	7.19
11 - 6	66.4	5.86	68.1	6.98
12 - 0	67.2	6.59	70.2	7.37
12 - 6	68.0	6.39	71.0	6.71
13 - 0	70.0	6.71	72.1	6.55
13 - 6	71.4	7.05	72.8	6.24

TABLE 913 (continued)

DISTRIBUTION STATISTICS FOR
ILIAC CIRCUMFERENCE (cm)
IN DENVER CHILDREN

Age Yr. - Mo.	Boys Mean	S.D.	Girls Mean	S.D.
14 - 0	72.8	6.69	75.9	6.40
14 - 6	73.4	6.50	76.2	6.40
15 - 0	75.5	6.80	77.6	6.29
15 - 6	76.6	6.53	76.2	4.60
16 - 0	77.5	7.34	79.3	6.47
16 - 6	78.7	6.98	--	--
17 - 0	79.3	5.94	79.4	6.12
17 - 6	80.9	7.70	--	--
18 - 0	80.3	7.59	80.2	5.79

(data from McCammon, Human Growth and Development, 1970.
Courtesy of Charles C Thomas, Publisher, Springfield, Illinois)

TABLE 914

ANTHROPOMETRIC DATA FOR BOYS IN CALIFORNIA
(ninth grade; mean age 14.49 years)

	Caucasian		Negro		Oriental	
	Mean	S.D.	Mean	S.D.	Mean	S.D.
Circumference (cm)						
Abdomen	70.96	7.814	67.69	6.539	66.18	7.728
Ankle	21.80	1.571	21.12	1.546	20.57	1.583
Biceps	27.10	2.972	26.95	3.998	25.82	3.275
Buttocks	85.70	7.318	82.05	7.315	81.29	7.907
Calf	33.74	2.838	32.85	2.967	33.08	3.003
Chest	83.34	7.215	80.04	7.076	78.20	7.896
Forearm	24.13	1.849	23.92	2.062	22.83	2.198
Knee	34.71	2.472	34.37	2.544	33.25	2.628
Shoulder	100.23	7.691	97.71	7.533	96.09	9.273
Thigh	50.52	5.738	49.49	5.882	47.95	6.146
Wrist	15.88	0.860	15.89	0.966	15.06	0.981
Diameters (cm)						
Ankle	13.86	0.706	13.83	0.731	13.23	0.681
Biacromial	36.78	2.418	36.25	2.763	35.94	2.682
Bi-iliac	26.25	1.834	24.45	1.804	25.15	1.926
Bitrochanteric	30.25	2.180	28.73	2.220	28.79	2.298
Chest	26.22	2.081	25.21	2.139	25.14	2.336
Wrist	10.76	0.653	10.74	0.720	10.20	0.714
Accessory Information						
Stature (dm)	16.84	0.837	16.41	0.807	15.96	0.736
Lean body wt. (kg)	50.58	7.305	46.41	6.971	45.07	7.668
Weight (kg)	57.96	12.487	53.31	10.779	50.42	12.338
% body fat	11.50	7.990	11.96	7.222	9.33	6.575

(data from Huenemann, Hampton, Behnke, et al., Teenage Nutrition and Physique, 1974.
Courtesy of Charles C Thomas, Publisher, Springfield, Illinois)

TABLE 915

ANTHROPOMETRIC DATA FOR BOYS IN CALIFORNIA
(tenth grade; mean age 15.32 years)

	Caucasian		Negro		Oriental	
	Mean	S.D.	Mean	S.D.	Mean	S.D.
Circumferences (cm)						
Abdomen	72.20	7.228	70.00	6.775	67.28	5.588
Ankle	22.34	1.502	21.66	1.415	21.09	1.304
Biceps	28.50	2.847	29.19	2.953	26.99	2.731
Buttocks	88.25	6.406	86.17	7.129	83.40	6.193
Calf	34.53	2.639	34.12	2.825	33.88	2.755
Chest	86.01	6.451	85.15	7.055	80.53	6.544
Forearm	25.07	1.692	25.32	1.949	23.61	1.706
Knee	35.46	2.232	35.12	2.400	33.77	2.112
Shoulder	104.41	7.041	104.23	7.275	100.42	7.751
Thigh	52.52	5.349	52.48	6.008	49.70	5.071
Wrist	16.23	0.749	16.29	0.793	15.34	0.791
Diameters (cm)						
Ankle	14.24	0.646	14.24	0.681	13.57	0.739
Biacromial	37.77	2.239	38.16	2.380	37.05	2.367
Bi-iliac	27.07	1.687	25.57	1.705	25.88	1.694
Bitrochanteric	31.25	1.865	30.19	1.973	29.81	2.078
Chest	27.06	1.853	26.47	1.878	25.80	1.882
Wrist	11.05	0.577	11.16	0.601	10.50	0.669
Accessory Information						
Stature (dm)	17.25	0.733	17.08	0.736	16.37	0.656
Lean body wt. (kg)	54.48	6.486	52.24	6.729	48.54	6.550
Weight (kg)	62.22	10.642	61.03	11.233	53.99	9.350
% body fat	11.58	7.374	13.37	7.662	9.38	6.017

(data from Huenemann, Hampton, Behnke, et al., Teenage Nutrition and Physique, 1974.
Courtesy of Charles C Thomas, Publisher, Springfield, Illinois)

TABLE 916

ANTHROPOMETRIC DATA FOR BOYS IN CALIFORNIA
(eleventh grade; mean age 16.36 years)

	Caucasian		Negro		Oriental	
	Mean	S.D.	Mean	S.D.	Mean	S.D.
Circumferences (cm)						
Abdomen	73.89	7.533	71.20	6.014	68.99	6.737
Ankle	22.54	1.507	21.82	1.402	21.33	1.250
Biceps	29.59	2.792	30.31	2.545	28.24	2.680
Buttocks	90.13	6.292	87.27	6.053	85.25	5.649
Calf	35.35	2.669	34.70	2.598	34.61	2.712
Chest	88.87	6.429	87.39	6.075	83.73	6.611
Forearm	25.85	1.667	25.95	1.714	24.32	1.606
Knee	35.84	2.186	35.38	2.245	34.19	2.050
Shoulder	107.38	6.668	107.40	6.257	103.66	7.177
Thigh	53.44	5.506	52.96	5.097	50.54	5.114
Wrist	16.48	0.748	16.44	0.764	15.51	0.680
Diameters (cm)						
Ankle	14.25	0.697	14.20	0.705	13.58	0.645
Biacromial	38.80	2.205	38.96	2.084	38.12	2.079
Bi-iliac	27.79	1.720	25.97	1.523	26.42	1.645
Bitrochanteric	32.12	1.751	30.71	1.643	30.44	1.488
Chest	27.76	1.898	26.87	1.687	26.44	2.157
Wrist	11.29	0.599	11.33	0.590	10.66	0.553
Accessory Information						
Stature (dm)	17.63	0.679	17.34	0.736	16.66	0.501
Lean body wt. (kg)	58.01	6.420	54.37	6.016	51.20	5.891
Weight (kg)	66.52	11.306	64.02	10.042	57.47	9.531
% body fat	11.87	7.451	14.31	6.970	10.05	7.033

(data from Huenemann, Hampton, Behnke, et al., Teenage Nutrition and Physique, 1974.
Courtesy of Charles C Thomas, Publisher, Springfield, Illinois)

TABLE   917

ANTHROPOMETRIC DATA FOR BOYS IN CALIFORNIA
(twelfth grade; mean age 17.29 years)

	Caucasian		Negro		Oriental	
	Mean	S.D.	Mean	S.D.	Mean	S.D.
Circumferences (cm)						
Abdomen	75.50	7.906	72.93	6.586	71.09	6.423
Ankle	22.37	1.451	21.83	1.250	21.29	1.310
Biceps	30.46	2.774	31.58	2.788	29.20	2.614
Buttocks	91.56	6.577	89.11	6.353	87.06	5.540
Calf	35.55	2.550	35.05	2.596	35.09	2.644
Chest	90.39	6.504	89.00	6.484	84.92	6.118
Forearm	26.36	1.588	26.77	1.790	24.94	1.627
Knee	35.77	2.162	35.51	2.100	34.37	1.954
Shoulder	109.34	6.384	110.26	6.210	106.26	6.737
Thigh	53.99	5.253	53.68	5,300	51.61	4.938
Wrist	16.61	0.758	16.71	0.758	15.67	0.717
Diameters (cm)						
Ankle	14.36	0.727	14.38	0.695	13.77	0.789
Biacromial	39.67	1.958	40.35	1.951	39.27	1.780
Bi-iliac	28.11	1.754	26.36	1.473	26.91	1.715
Bitrochanteric	32.63	1.693	31.43	1.638	31.06	1.604
Chest	28.30	1.725	27.68	1.695	26.89	2.111
Wrist	11.37	0.583	11.49	0.573	10.77	0.619
Accessory Information						
Stature (dm)	17.81	0.658	17.62	0.712	16.87	0.587
Lean body wt. (kg)	60.25	6.278	57.68	5.891	53.80	6.384
Weight (kg)	69.33	11.620	67.55	10.465	60.77	9.770
% body fat	12.16	7.463	13.75	7.317	10.76	6.137

(data from Huenemann, Hampton, Behnke, et al., Teenage Nutrition and Physique, 1974.
Courtesy of Charles C Thomas, Publisher, Springfield, Illinois)

TABLE   918

ANTHROPOMETRIC DATA FOR GIRLS IN CALIFORNIA
(ninth grade; mean age 14.51 years)

	Caucasian		Negro		Oriental	
	Mean	S.D.	Mean	S.D.	Mean	S.D.
Circumferences (cm)						
Abdomen	71.69	6.165	72.30	7.912	68.75	5.588
Ankle	21.41	1.338	21.18	1.362	20.33	1.395
Biceps	26.18	2.428	26.59	2.770	24.64	2.579
Buttocks	92.77	6.260	92.99	7.723	87.44	6.675
Calf	34.43	2.583	34.25	2.724	33.01	2.748
Chest	82.79	4.942	83.51	6.180	79.32	5.549
Forearm	23.02	1.490	23.44	1.731	22.06	1.694
Knee	34.87	2.322	35.71	2.563	33.74	2.346
Shoulder	96.30	5.428	97.89	6.507	93.78	6.165
Thigh	56.71	5.097	57.57	6.455	53.37	5.033
Wrist	15.15	0.700	15.56	0.760	14.59	0.654
Diameters (cm)						
Ankle	12.27	0.579	12.33	0.665	11.88	0.663
Biacromial	35.17	1.695	35.70	1.864	34.47	1.780
Bi-iliac	26.57	1.680	25.75	1.921	25.67	1.724
Bitrochanteric	31.02	1.772	30.63	2.107	29.75	1.662
Chest	25.35	1.509	25.17	1.703	24.62	1.604
Wrist	9.76	0.542	9.91	0.591	9.50	0.502
Accessory Information						
Stature (dm)	16.26	0.619	16.17	0.547	15.56	0.448
Lean body wt. (kg)	45.98	5.085	45.33	5.552	41.37	4.613
Weight (kg)	55.44	8.507	56.56	10.380	48.35	7.871
% body fat	16.33	7.157	18.84	7.482	13.42	8.638

(data from Huenemann, Hampton, Behnke, et al., Teenage Nutrition and Physique, 1974.
Courtesy of Charles C Thomas, Publisher, Springfield, Illinois)

TABLE   919

ANTHROPOMETRIC DATA FOR GIRLS IN CALIFORNIA
(tenth grade; mean age 15.24 years)

	Caucasian		Negro		Oriental	
	Mean	S.D.	Mean	S.D.	Mean	S.D.
Circumferences (cm)						
Abdomen	73.06	6.377	73.09	7.814	69.68	5.074
Ankle	21.32	1.356	20.98	1.366	20.06	1.518
Biceps	26.16	2.514	26.52	2.640	24.65	2.762
Buttocks	94.17	6.418	94.26	7.300	88.67	6.062
Calf	34.56	2.623	34.21	2.648	32.93	2.859
Chest	84.44	4.997	84.14	5.661	80.64	5.339
Forearm	22.62	1.598	23.26	1.609	21.65	1.998
Knee	36.19	2.485	36.42	2.728	34.47	2.328
Shoulder	98.97	5.572	99.47	6.535	95.18	6.290
Thigh	57.15	4.885	57.62	5.696	53.80	5.064
Wrist	14.97	0.776	15.39	0.764	14.35	0.826
Diameters (cm)						
Ankle	12.28	0.562	12.43	0.571	11.90	0.606
Biacromial	35.74	1.757	36.24	1.811	34.83	1.692
Bi-iliac	27.29	1.761	26.14	1.932	26.14	1.187
Bitrochanteric	31.27	1.802	30.68	1.804	29.81	1.301
Chest	25.49	1.522	25.54	1.686	24.62	1.523
Wrist	9.91	0.538	10.22	0.544	9.51	0.533
Accessory Information						
Stature (dm)	16.37	0.611	16.31	0.550	15.56	0.505
Lean body wt. (kg)	47.51	5.052	46.86	5.343	41.84	4.170
Weight (kg)	56.83	8.724	58.03	10.459	49.23	7.938
% body fat	15.66	7.397	18.26	7.317	13.85	9.648

(data from Huenemann, Hampton, Behnke, et al., Teenage Nutrition and Physique, 1974.
Courtesy of Charles C Thomas, Publisher, Springfield, Illinois)

TABLE   920

ANTHROPOMETRIC DATA FOR GIRLS IN CALIFORNIA
(eleventh grade; mean age 16.23 years)

	Caucasian		Negro		Oriental	
	Mean	S.D.	Mean	S.D.	Mean	S.D.
Circumferences (cm)						
Abdomen	69.91	6.042	70.35	7.388	66.81	5.223
Ankle	21.50	1.317	21.08	1.358	20.42	1.361
Biceps	26.92	2.452	27.01	2.753	25.56	2.617
Buttocks	94.34	5.907	93.51	7.236	88.65	5.773
Calf	35.05	2.420	34.49	2.627	33.51	2.785
Chest	83.94	4.776	83.55	5.686	80.59	5.457
Forearm	22.95	1.484	23.46	1.625	22.16	1.689
Knee	36.17	2.367	36.31	2.544	34.54	2.396
Shoulder	97.98	5.763	99.12	5.820	95.76	6.105
Thigh	57.76	4.420	57.53	6.631	54.34	4.668
Wrist	15.22	0.716	15.51	0.727	14.63	0.637
Diameters (cm)						
Ankle	12.14	0.562	12.35	0.633	11.91	0.584
Biacromial	35.70	1.626	36.25	1.756	34.74	1.598
Bi-iliac	27.43	1.540	25.90	1.713	26.67	1.415
Bitrochanteric	31.30	1.623	30.68	1.838	30.14	1.285
Chest	24.63	1.381	24.30	1.455	24.10	1.441
Wrist	9.58	0.510	9.83	0.509	9.39	0.409
Accessory Information						
Stature (dm)	16.45	0.617	16.34	0.538	15.71	0.509
Lean body wt. (kg)	46.92	4.998	45.64	4.996	42.34	3.931
Weight (kg)	57.65	8.552	57.98	10.108	50.07	7.496
% body fat	17.93	7.032	20.33	7.303	14.34	9.707

(data from Huenemann, Hampton, Behnke, et al., Teenage Nutrition and Physique, 1974.
Courtesy of Charles C Thomas, Publisher, Springfield, Illinois)

TABLE 921

ANTHROPOMETRIC DATA FOR GIRLS IN CALIFORNIA
(twelfth grade; mean age 17.23 years)

	Caucasian		Negro		Oriental	
	Mean	S.D.	Mean	S.D.	Mean	S.D.
Circumferences (cm)						
Abdomen	72.92	6.686	73.25	8.133	70.22	6.723
Ankle	21.57	1.379	21.20	1.384	20.57	1.565
Biceps	26.88	2.733	27.11	2.788	25.78	2.883
Buttocks	94.95	6.701	95.09	7.994	89.99	6.274
Calf	35.19	2.677	34.85	2.792	33.81	2.894
Chest	84.91	5.264	84.98	5.847	81.99	5.637
Forearm	23.53	1.592	24.03	1.615	22.72	1.797
Knee	36.34	2.703	36.86	2.920	34.94	2.586
Shoulder	98.98	5.863	99.83	6.142	96.52	6.132
Thigh	56.39	5.121	57.13	6.067	52.77	5.422
Wrist	15.19	0.774	15.51	0.814	14.57	0.610
Diameters (cm)						
Ankle	12.54	0.587	12.72	0.628	12.16	0.518
Biacromial	35.78	1.751	36.15	1.952	35.07	1.630
Bi-iliac	28.15	1.718	26.82	1.937	27.26	1.790
Bitrochanteric	31.67	1.827	31.24	2.025	30.54	1.622
Chest	24.71	1.529	24.45	1.513	24.06	1.539
Wrist	10.02	0.528	10.22	0.545	9.62	0.359
Accessory Information						
Stature (dm)	16.50	0.623	16.39	0.561	15.70	0.499
Lean body wt. (kg)	48.51	5.632	47.32	5.675	43.44	4.441
Weight (kg)	58.21	9.542	59.60	11.020	51.20	8.489
% body fat	15.86	7.626	19.63	7.378	14.02	9.258

(data from Huenemann, Hampton, Behnke, et al., Teenage Nutrition and Physique, 1974.
Courtesy of Charles C Thomas, Publisher, Springfield, Illinois)

TABLE 922

HEIGHTS AND AREAS OF THORACIC AND LUMBAR SEGMENTS OF THE TRUNK FOR
ENGLISH BOYS

Age (years)	Thorax height (cm)		Lumbar height (cm)		Thorax area (cm^2)		Lumbar area (cm^2)	
	Mean	S.D.	Mean	S.D.	Mean	S.D.	Mean	S.D.
5	26.7	1.2	17.3	0.9	525.2	33.6	354.4	26.7
6	28.0	1.2	18.0	1.0	557.5	35.9	376.3	30.2
7	29.1	1.2	18.7	1.0	599.0	40.8	401.8	33.0
8	30.3	1.3	19.3	1.0	642.7	44.3	430.6	37.2
9	31.5	1.3	20.1	1.2	694.1	50.6	468.1	48.7
10	32.4	1.4	20.9	1.3	740.9	52.0	503.4	54.1
11	33.4	1.4	21.6	1.3	788.5	57.1	541.1	62.1
12	34.4	1.5	22.3	1.3	836.6	64.4	578.9	69.7
13	35.8	2.0	23.2	1.4	909.5	88.0	630.2	78.8
14	37.4	2.2	24.1	1.5	994.6	112.9	685.3	88.3
15	39.5	2.8	25.2	1.7	1100.4	129.3	745.4	89.9
16	41.3	2.3	26.2	1.4	1193.8	114.3	804.4	78.1
17	42.4	1.9	26.6	1.3	1255.7	100.6	833.3	75.2
18	43.0	1.9	27.2	1.4	1300.8	108.5	864.7	86.0

(data from Walker, 1979, Annals of Human Biology 6:315-336)

TABLE 923

TRUNK INDEX IN CONNECTICUT BOYS

Age (years)	Mean	S.D.
5	1.49	0.08
6	1.49	0.08
7	1.49	0.08
8	1.50	0.09
9	1.49	0.10
10	1.48	0.11
11	1.47	0.11
12	1.45	0.10
13	1.46	0.10
14	1.46	0.10
15	1.48	0.10
16	1.49	0.09
17	1.51	0.09
18	1.51	0.10

Trunk index = $\dfrac{\text{area of thoracic trunk}}{\text{area of lumbar trunk}}$
on somatotype photographs averaging
the values for frontal and rear
views.

(data from Walker, 1979, Annals of
Human Biology 6:315-336)

TABLE   924

TRUNK AREAS (ANTERIOR AND POSTERIOR VIEWS) IN ENGLISH BOYS

Age (years)	Thorax (area cm^2)		Abdomen (area cm^2)		Trunk index	
	Mean	S.D.	Mean	S.D.	Mean	S.D.
8.5	652.6	60.9	483.9	60.7	136.2	16.5
9.0	670.0	60.4	501.7	58.7	134.6	14.8
9.5	709.1	56.4	514.9	67.3	139.2	16.0
10.0	716.7	52.5	549.9	64.4	131.2	10.6
10.5	741.4	63.9	552.4	62.7	135.4	14.1
11.0	756.0	63.5	562.6	58.7	135.1	12.1
11.5	780.2	62.5	578.0	67.7	136.2	15.1
12.0	797.7	60.7	600.5	74.0	134.0	12.8
12.5	822.1	66.9	629.9	73.4	131.7	15.6
13.0	843.4	66.1	656.1	78.8	129.6	12.4
13.5	890.7	79.2	689.3	79.0	130.1	12.6
14.0	938.4	90.8	734.9	98.4	129.1	16.2
14.5	999.5	85.6	771.7	81.5	130.1	9.4
15.0	1054.6	90.6	815.8	88.9	129.9	9.9
15.5	1084.4	59.8	864.7	76.1	125.8	6.4
16.0	1144.5	88.0	894.8	85.8	128.5	10.5
17.0	1198.3	106.0	921.6	98.5	130.8	12.6
18.0	1242.1	107.9	941.6	113.7	132.7	8.8

(data from Singh, 1976, Annals of Human Biology 3:181-186)

LOWER LIMB MEASUREMENTS

TABLE 925

SUBISCHIAL LENGTH (cm) FOR U.S. CHILDREN

Age						Percentiles			
(years)	Mean	S.D.	5	10	25	50	75	90	95

BOYS

Age	Mean	S.D.	5	10	25	50	75	90	95
12	74.4	4.97	63.9	68.0	71.2	74.6	77.7	80.4	82.4
13	78.1	5.14	69.9	71.8	74.9	78.0	81.6	84.8	86.9
14	81.4	4.76	73.9	75.4	78.0	81.5	84.8	87.4	88.9
15	83.2	4.58	75.4	77.5	80.3	83.2	85.9	88.9	90.8
16	84.3	4.44	77.3	78.6	81.5	84.1	87.4	90.4	91.7
17	84.5	4.66	76.8	78.4	81.4	84.5	87.8	90.4	91.9

GIRLS

Age	Mean	S.D.	5	10	25	50	75	90	95
12	74.8	4.56	66.9	69.1	72.0	74.7	77.6	80.8	82.1
13	76.1	4.49	68.8	70.5	72.9	76.0	79.3	81.8	83.5
14	77.0	4.23	70.4	71.6	74.0	77.0	79.6	82.4	84.5
15	77.1	4.65	69.8	71.1	74.0	76.9	80.4	83.2	84.6
16	76.9	4.47	69.7	71.3	74.1	76.9	79.8	82.7	84.3
17	76.9	4.45	70.4	71.3	73.8	76.8	80.0	82.4	84.7

(data from Cycle III of The Health Examination Survey by the National Center for Health Statistics, 1966-1970)

TABLE 926

SUBISCHIAL LENGTH (cm) FOR U.S. WHITE CHILDREN

Age (years)	Mean	S.D.	Percentiles						
			5	10	25	50	75	90	95
			BOYS						
12	74.1	4.97	63.5	67.7	70.7	74.3	77.4	79.9	81.8
13	77.9	4.97	69.8	71.6	74.7	77.9	81.4	84.4	86.5
14	81.2	4.74	73.8	75.3	77.8	81.3	84.7	87.2	88.4
15	82.9	4.52	75.2	77.2	78.0	82.9	85.6	88.3	90.2
16	84.0	4.35	77.2	78.4	81.3	83.9	86.5	89.9	91.2
17	84.3	4.61	76.4	78.3	81.2	84.4	87.7	90.0	91.6
			GIRLS						
12	74.3	4.50	66.2	68.4	71.6	74.4	77.1	80.0	81.7
13	75.7	4.40	68.4	70.3	72.7	75.6	78.8	81.4	82.9
14	76.7	4.12	70.3	71.4	73.8	76.6	79.3	81.8	83.5
15	76.9	4.57	69.8	71.0	73.9	76.6	80.0	82.9	84.5
16	76.6	4.42	69.4	71.0	73.8	76.6	79.4	82.4	84.0
17	76.6	4.31	70.3	71.3	73.7	76.4	79.8	81.8	83.8

SUBISCHIAL LENGTH (cm) FOR U.S. NEGRO CHILDREN

Age (years)	Mean	S.D.	Percentiles						
			5	10	25	50	75	90	95
			BOYS						
12	76.6	4.42	68.8	70.6	73.8	76.4	79.5	82.8	83.8
13	79.4	6.08	70.9	73.8	76.4	79.5	83.4	86.3	87.8
14	83.0	4.64	76.4	77.4	79.8	82.7	86.4	89.6	90.4
15	85.5	4.37	78.1	79.4	82.8	84.9	88.2	92.0	94.4
16	87.0	4.04	79.2	81.7	84.3	87.8	89.9	92.2	93.2
17	86.2	4.65	78.0	80.3	82.8	86.4	89.4	92.6	95.0
			GIRLS						
12	77.6	4.00	71.8	72.3	74.7	77.3	79.9	82.5	84.1
13	78.6	4.21	71.4	72.8	75.4	78.8	81.5	84.0	85.7
14	79.2	4.27	72.3	73.5	75.9	78.8	82.2	84.8	86.3
15	79.0	4.65	70.4	73.7	75.5	79.0	82.8	84.4	86.4
16	79.3	4.07	72.8	74.2	76.3	79.1	82.7	84.0	85.8
17	79.0	4.63	70.6	72.9	76.2	78.6	82.3	85.3	86.1

(data from Cycle III of The Health Examination Survey by the National Center for Health Statistics, 1966-1970)

TABLE 927

RUMP - SOLE LENGTH (cm)
FOR CHILDREN IN 8 STATES

Age (months)	Boys Mean	S.D.	Girls Mean	S.D.
0 - 3	23.5	1.9	23.3	1.9
4 - 6	28.6	2.0	28.0	2.0
7 - 9	31.7	1.6	31.0	1.5
10 - 12	35.0	1.7	34.1	1.5
13 - 18	37.3	2.4	37.2	2.7
19 - 24	42.5	2.5	42.4	3.0
25 - 30	45.0	2.6	46.0	2.8
31 - 36	48.9	3.5	49.4	3.4
37 - 42	51.7	2.5	51.6	2.8
43 - 48	53.4	2.9	53.8	2.8
49 - 54	56.0	2.9	56.1	2.9
55 - 60	58.2	3.3	58.0	3.2
61 - 66	60.3	3.6	60.8	3.4
67 - 72	62.5	3.2	63.0	3.5
73 - 78	64.4	3.2	64.6	3.7
79 - 84	67.2	3.4	67.0	3.5
85 - 96	70.5	4.3	69.9	4.0
97 - 108	73.7	4.2	75.4	4.4
109 - 120	77.3	4.9	77.8	4.4
121 - 132	82.3	4.3	83.0	5.2
133 - 144	86.1	4.9	85.6	5.0
145 - 156	86.9	4.6	92.2	4.9

(data from Snyder et al., 1975)

TABLE   928

LEG LENGTH (cm) IN PHILADELPHIA CHILDREN

Age (Years)	White Boys		White Girls		Negro Boys		Negro Girls	
	Mean	S.D.	Mean	S.D.	Mean	S.D.	Mean	S.D.
7	63.6	3.2	63.6	3.4	65.4	3.7	66.2	4.1
8	67.0	3.5	67.0	3.8	69.1	4.4	69.4	4.4
9	70.9	4.0	71.0	4.2	72.3	4.3	73.9	4.1
10	75.0	3.9	73.7	5.1	75.7	5.2	77.6	4.4
11	78.9	4.2	77.5	4.7	78.8	5.7	81.1	5.0
12	81.3	4.9	81.8	3.7	82.3	5.6	83.9	5.3
13	85.9	4.9	82.6	4.5	89.5	5.7	88.0	4.6
14	90.4	4.6	86.9	4.1	91.9	6.2	89.8	5.0
15	--	--	--	--	92.7	6.1	92.2	5.8

Leg length is measured to a level midway between the most superior point on the pubic symphysis and the anterior superior iliac spine.

(data from Krogman, 1970)

TABLE   929

SYMPHYSEAL HEIGHT (cm) IN PHILADELPHIA CHILDREN

Age (Years)	White Boys		White Girls		Negro Boys		Negro Girls	
	Mean	S.D.	Mean	S.D.	Mean	S.D.	Mean	S.D.
7	60.4	3.0	60.7	3.4	61.8	3.5	62.5	4.0
8	64.3	3.3	64.3	3.8	65.1	4.1	65.5	4.0
9	67.7	3.8	67.3	3.9	68.3	4.1	69.9	4.0
10	71.6	3.6	70.7	4.7	71.5	5.3	73.9	4.1
11	74.7	4.0	74.3	4.4	75.2	5.1	77.6	4.9
12	78.1	4.6	78.4	3.8	78.7	5.5	80.1	5.1
13	83.0	4.7	80.0	4.8	85.2	5.5	84.1	4.5
14	84.8	4.4	83.5	2.6	87.8	6.0	86.0	4.9

(data from Krogman, 1970)

TABLE 930

LOWER LEG LENGTH (SPHYRION FIBULARE HEIGHT
TO TIBIALE HEIGHT: cm) IN PHILADELPHIA CHILDREN

Age (years)	White Boys		White Girls		Negro Boys		Negro Girls	
	Mean	S.D.	Mean	S.D.	Mean	S.D.	Mean	S.D.
7	28.9	1.5	27.6	1.7	29.3	3.0	29.4	2.7
8	28.9	1.8	29.0	1.8	30.2	2.5	30.2	3.2
9	30.8	2.0	31.9	2.4	31.5	2.7	32.3	3.8
10	32.3	2.2	32.9	2.4	33.0	3.2	34.1	3.7
11	34.4	2.1	34.1	2.4	34.4	2.9	35.0	3.2
12	35.5	2.3	35.6	1.9	36.5	3.3	36.2	5.0
13	38.0	2.3	36.5	1.6	39.8	3.3	37.4	3.9
14	39.5	2.2	37.3	2.1	40.7	3.3	38.7	6.9
15	--	--	--	--	40.7	3.4	38.9	10.0

(data from Krogman, 1970)

TABLE 931

LOWER LIMB MEASUREMENTS (cm) FOR WHITE BOYS IN IOWA

Age (years)	Mean	S.D.	Percentiles			
			5	25	75	95

Lower limb length

5	48.5	2.32	44.7	46.6	50.3	52.2
6	52.5	2.57	48.4	50.4	54.5	56.6
7	56.5	2.70	52.1	54.3	58.7	61.0
8	60.2	3.01	55.5	57.9	62.5	65.0
9	63.6	3.19	58.5	61.3	65.9	68.6
10	66.8	3.43	61.3	64.5	69.2	72.2
11	69.9	3.61	63.9	67.5	72.3	75.6

Calf circumference

5	23.1	1.62	20.6	21.9	24.2	26.3
6	24.2	1.87	21.4	22.9	25.3	27.8
7	25.3	2.14	22.2	23.9	26.4	29.3
8	26.5	2.35	23.1	24.9	27.6	30.8
9	27.7	2.56	24.1	26.0	28.8	32.3
10	28.9	2.88	25.0	27.0	30.2	34.2
11	30.2	3.04	26.1	28.1	31.7	36.2

Lower limb index

5	47.7	2.88	43.7	45.7	49.4	53.2
6	46.0	3.09	41.8	43.9	47.8	51.5
7	44.8	3.25	40.6	42.6	46.7	50.5
8	44.1	3.47	39.5	41.6	46.0	50.0
9	43.6	3.59	38.8	41.0	45.8	49.7
10	43.4	3.77	38.4	40.7	45.6	49.8
11	43.2	3.89	38.3	40.6	45.5	50.0

The index is circumference/length.

(data from Meredith, 1955)

TABLE 932

ANNUAL INCREMENTS (cm/year) FOR LOWER LIMB
MEASUREMENTS OF WHITE BOYS IN IOWA

Age Interval (years)	S.D.	Percentiles			
		5	25	75	95

Lower limb length

5- 6	0.46	3.3	3.8	4.4	4.9
6- 7	0.39	3.2	3.7	4.2	4.7
7- 8	0.37	3.0	3.5	4.0	4.5
8- 9	0.36	2.8	3.2	3.7	4.2
9-10	0.37	2.6	3.0	3.5	4.0
10-11	0.40	2.3	2.8	3.3	3.8

Calf circumference

5- 6	0.41	0.4	0.7	1.2	1.8
6- 7	0.36	0.6	0.8	1.3	1.9
7- 8	0.34	0.7	0.9	1.4	2.0
8- 9	0.40	0.7	0.9	1.4	2.1
9-10	0.43	0.7	1.0	1.5	2.1
10-11	0.41	0.8	1.0	1.5	2.1

Lower limb index

5- 6	0.72	-2.6	-2.2	-1.3	-0.4
6- 7	0.52	-2.0	-1.6	-0.8	-0.1
7- 8	0.48	-1.5	-1.0	-0.5	+0.1
8- 9	0.49	-1.2	-0.8	-0.2	+0.6
9-10	0.52	-1.0	-0.6	0.0	+0.8
10-11	0.48	-0.8	-0.5	+0.2	+0.8

The index is circumference/length.

(data from Meredith, 1955)

TABLE 933

RELATIVE LEG LENGTH FOR CANADIAN BOYS

| Age (years) | Mean | S.D. | Percentiles | | | | | | |
			5	10	25	50	75	90	95
3	53.1	4.0	47.0	48.3	50.7	53.6	55.7	57.2	58.4
4	53.2	2.7	49.5	50.3	51.2	53.2	55.1	56.4	57.9
5	53.9	2.8	49.7	51.1	52.1	53.2	55.9	57.4	58.2
6	55.1	3.4	51.7	52.5	53.5	54.7	56.9	58.1	59.6
7	55.1	1.8	52.9	53.3	53.8	55.0	56.1	58.1	58.6
8	56.2	1.9	53.7	54.1	55.2	55.7	57.1	58.8	59.1
9	57.1	2.3	53.1	54.9	55.4	57.3	59.1	59.3	60.4
10	57.6	2.3	54.0	55.0	55.9	57.3	59.7	60.5	61.4
11	58.4	2.4	55.3	55.6	56.7	57.8	60.2	62.2	62.4
12	58.5	2.6	54.6	55.7	56.7	58.1	60.6	62.3	62.9
13	58.5	2.8	54.0	55.3	56.8	58.1	60.3	62.2	62.5
14	59.0	1.8	56.6	56.9	57.6	58.9	60.1	61.8	62.2
15	58.5	1.7	55.8	56.6	57.2	58.8	59.7	60.6	61.9
16	58.9	2.4	54.8	55.9	57.1	58.6	61.1	62.0	62.2
17	58.0	2.0	54.8	56.0	56.7	57.7	59.3	60.9	61.6
18	57.8	1.9	55.0	55.8	56.1	57.8	59.2	60.9	60.9

Data from Demirjian, 1980, Anthropometry Report: Height, Weight and Body Dimensions. Minister of National Health and Welfare, Ottawa, Canada. Reproduced by permission of the Minister of Supply and Services Canada.

TABLE 934

RELATIVE LEG LENGTH FOR CANADIAN GIRLS

Age (years)	Mean	S.D.	5	10	Percentiles 25	50	75	90	95
3	52.2	2.3	49.4	49.8	50.4	51.8	54.1	55.3	55.8
4	53.6	2.1	50.9	51.3	52.0	52.9	54.7	56.5	58.0
5	54.2	2.6	50.4	51.4	52.6	54.5	56.0	58.2	59.2
6	56.3	2.3	52.7	52.8	54.4	56.5	57.7	59.5	60.5
7	56.2	2.3	52.2	52.7	54.7	56.1	58.3	60.3	60.6
8	56.8	2.8	53.2	53.7	54.6	56.1	58.4	60.9	62.3
9	58.2	1.9	55.1	55.4	56.5	58.5	59.6	60.5	60.9
10	58.1	2.2	55.1	55.2	56.6	57.6	59.8	61.0	61.7
11	59.0	2.3	55.8	56.1	57.1	58.9	60.7	62.4	62.4
12	58.0	3.2	55.7	55.9	57.1	58.3	59.5	60.9	61.9
13	58.0	2.1	54.5	54.9	56.5	58.1	59.6	60.9	61.7
14	57.7	1.7	54.8	55.4	56.7	57.5	58.9	60.0	60.4
15	58.6	2.4	55.1	55.4	56.7	57.8	61.2	62.3	62.3
16	57.1	1.9	54.5	55.2	56.0	56.8	58.2	59.1	60.4
17	56.9	1.7	54.2	55.1	55.9	56.5	57.7	59.5	60.4
18	57.5	2.5	54.8	54.9	55.2	56.8	58.7	61.7	62.1

Data from Demirjian, 1980, Anthropometry Report: Height, Weight and Body Dimensions. Minister of National Health and Welfare, Ottawa, Canada. Reproduced by permission of the Minister of Supply and Services Canada.

TABLE 935

SKELIC INDEX IN BLACK AND WHITE CHILDREN AND YOUTHS OF SOUTH CAROLINA

Ethnic Group	Mean	S.D.	Percentiles		
			25	75	95
Girls age 9 years					
White	86.2	4.1	83.3	89.2	93.2
Black	91.2	4.3	88.2	94.0	98.2
Girls age 13 years					
White	88.0	4.6	84.7	90.7	95.9
Black	93.2	4.8	89.9	96.7	100.7
Boys age 11 years					
White	88.7	3.9	86.1	91.5	94.7
Black	93.4	5.0	90.1	97.1	100.1
Boys age 15 years					
White	91.1	5.2	87.3	94.4	98.7
Black	97.5	5.1	94.3	101.0	105.1

Skelic index = leg length/sitting height.

(data from Meredith and Spurgeon, 1976a)

TABLE 936

SELECTED ANTHROPOMETRIC INDICES IN
BLACK CHILDREN OF SOUTH CAROLINA

Index	Mean	S.D.
Girls age 6 years:		
Skelic index	85.4	3.97
Hip/lower limb	32.2	1.87
Calf/lower limb	43.9	3.59
Girls age 9 years:		
Skelic index	91.1	4.42
Hip/lower limb	30.7	2.21
Calf/lower limb	42.2	4.35
Boys age 6 years:		
Skelic index	84.0	3.73
Hip/lower limb	32.4	1.90
Calf/lower limb	44.0	3.34
Boys age 9 years:		
Skelic index	89.5	3.88
Hip/lower limb	30.5	1.69
Calf/lower limb	42.4	3.65
Boys age 11 years:		
Skelic index	93.9	4.59
Hip/lower limb	29.6	1.71
Calf/lower limb	41.3	3.64

Skelic index = lower limb length x 100/sitting height;
hip and calf values are circumferences.

(data from Spurgeon, Meredith, and Meredith, 1978,
Annals of Human Biology 5:229-246)

TABLE   937

THIGH LENGTH (cm) IN U.S. WHITE CHILDREN

Age						Percentiles			
(years)	Mean	S.D.	5	10	25	50	75	90	95

BOYS

Age	Mean	S.D.	5	10	25	50	75	90	95
12	38.4	2.96	33.1	34.8	36.3	38.3	40.4	42.0	43.2
13	40.5	2.94	35.9	36.8	38.6	40.4	42.4	44.4	45.3
14	42.1	3.10	36.8	38.1	40.1	42.2	44.2	46.1	47.0
15	43.5	2.97	38.7	37.8	41.7	43.5	45.3	46.8	48.0
16	43.9	2.76	39.6	40.6	42.0	43.8	45.7	47.4	48.6
17	44.1	2.86	39.5	40.4	42.3	44.1	46.2	47.7	48.6

GIRLS

Age	Mean	S.D.	5	10	25	50	75	90	95
12	39.8	2.89	35.1	36.2	38.1	39.6	41.7	43.5	44.4
13	40.8	2.82	35.9	37.0	39.1	40.8	42.8	44.4	45.3
14	41.5	2.95	36.6	37.8	39.8	41.5	43.6	45.0	46.2
15	41.6	3.00	36.8	37.9	39.5	41.7	43.5	45.5	46.4
16	41.8	2.95	36.8	38.1	40.0	43.0	43.9	45.9	46.3
17	41.7	2.89	37.2	38.1	39.8	41.5	43.0	45.6	46.4

THIGH LENGTH (cm) IN U.S. NEGRO CHILDREN

Age						Percentiles			
(years)	Mean	S.D.	5	10	25	50	75	90	95

BOYS

Age	Mean	S.D.	5	10	25	50	75	90	95
12	39.2	2.74	34.2	35.1	37.2	39.5	41.5	42.9	43.0
13	40.8	3.08	34.3	36.9	38.7	41.4	42.9	44.4	45.3
14	43.1	3.35	37.5	39.3	40.8	43.2	44.8	47.4	48.9
15	44.3	2.63	39.9	41.2	42.4	44.2	45.7	47.4	49.0
16	45.6	2.49	41.2	42.5	44.0	45.6	47.3	48.9	50.0
17	45.2	3.24	39.5	40.5	43.5	45.6	47.5	48.9	50.0

GIRLS

Age	Mean	S.D.	5	10	25	50	75	90	95
12	40.5	2.29	36.3	37.7	38.8	40.8	42.0	43.5	44.1
13	41.4	2.89	36.0	37.1	39.6	41.4	43.3	45.3	45.9
14	41.8	2.74	37.4	38.4	40.2	41.6	43.8	45.8	46.3
15	42.3	2.85	37.4	39.3	40.2	42.9	44.1	46.1	46.7
16	42.6	3.03	37.4	38.4	40.6	42.9	44.2	46.2	46.8
17	42.0	3.73	35.7	36.7	39.5	41.8	45.0	47.2	48.1

(data from Cycle III of Health Examination Survey 1966-1970 by National Center for Health Statistics)

TABLE   938

THIGH LENGTH (cm) IN U.S. CHILDREN

Age (years)	Mean	S.D.	Percentiles						
			5	10	25	50	75	90	95
BOYS									
12	38.5	2.94	33.4	34.9	36.4	38.5	40.5	42.3	43.2
13	40.6	2.95	35.8	36.8	38.6	40.5	42.6	44.4	45.3
14	42.3	3.15	36.9	38.3	40.2	42.3	44.4	46.2	47.3
15	43.6	2.94	38.8	39.9	41.8	43.7	45.4	47.0	48.0
16	44.1	2.80	39.6	40.7	42.2	44.0	45.9	47.7	48.9
17	44.3	2.93	39.5	40.4	42.4	44.3	46.3	47.9	48.7
GIRLS									
12	39.9	2.88	35.3	36.3	38.1	39.8	41.7	43.5	44.4
13	40.9	2.84	36.0	37.0	39.2	40.8	42.8	44.4	45.4
14	41.6	2.93	36.8	37.8	39.8	41.5	43.4	45.0	46.3
15	41.7	3.00	36.8	37.9	39.6	41.7	43.6	45.5	46.4
16	41.9	2.97	37.0	38.2	40.1	42.0	43.9	45.5	46.3
17	41.7	3.01	37.0	38.0	39.8	41.6	43.7	45.8	47.0

(data from Cycle III of Health Examination Survey 1966-1970 by National Center for Health Statistics)

TABLE   939

THIGH LENGTH (INGUINALE-TIBIALE; cm) IN PHILADELPHIA CHILDREN

Age (years)	White Boys		White Girls		Negro Boys		Negro Girls	
	Mean	S.D.	Mean	S.D.	Mean	S.D.	Mean	S.D.
7	31.2	1.8	30.9	1.8	31.5	2.3	32.7	2.8
8	32.6	1.8	32.8	2.3	34.0	2.8	35.4	2.8
9	34.2	2.3	34.8	2.5	35.8	3.1	37.0	2.6
10	36.1	2.1	36.2	2.9	37.6	3.3	38.6	2.8
11	38.4	3.2	37.9	2.5	39.2	3.7	40.8	3.8
12	39.7	2.6	40.2	2.5	39.6	3.9	42.2	3.3
13	41.6	3.1	41.2	2.1	43.8	3.7	44.9	3.4
14	44.7	2.6	43.9	1.7	45.8	3.7	45.3	3.1
15	--	--	--	--	46.1	3.1	46.9	3.5

THIGH LENGTH (ILIOSPINALE-TIBIALE; cm) IN PHILADELPHIA CHILDREN

Age (years)	White Boys		White Girls		Negro Boys		Negro Girls	
	Mean	S.D.	Mean	S.D.	Mean	S.D.	Mean	S.D.
7	33.8	2.3	33.9	2.0	34.9	2.7	36.4	3.2
8	35.6	1.8	35.4	2.0	37.8	3.1	39.0	4.0
9	37.6	2.5	37.7	2.1	40.0	3.2	40.9	2.9
10	39.9	2.4	39.4	2.8	41.6	3.4	42.5	3.1
11	41.7	2.7	41.1	2.8	43.1	4.1	44.8	3.8
12	42.9	2.8	43.1	3.0	44.2	3.3	46.0	3.7
13	45.2	2.8	44.7	2.4	48.1	4.0	48.3	3.3
14	48.3	2.9	47.0	3.0	49.4	3.8	49.0	3.2
15	--	--	--	--	50.4	3.3	50.9	3.6

(data from Krogman, 1970)

TABLE 940

CALF LENGTH (cm) IN U.S. CHILDREN

Age						Percentiles			
(years)	Mean	S.D.	5	10	25	50	75	90	95

BOYS

Age	Mean	S.D.	5	10	25	50	75	90	95
12	37.5	2.61	32.8	34.1	35.8	37.5	39.4	40.6	41.7
13	39.4	2.82	34.9	36.0	37.5	39.4	41.2	42.8	44.4
14	40.9	2.59	36.8	37.6	39.1	40.9	42.7	44.0	45.1
15	41.5	2.65	37.6	38.4	39.9	41.5	43.2	44.7	45.9
16	42.0	2.42	37.9	39.0	40.4	42.0	43.6	45.3	46.0
17	42.1	2.69	37.9	38.5	40.3	42.1	43.9	45.5	46.6

GIRLS

Age	Mean	S.D.	5	10	25	50	75	90	95
12	37.8	2.44	33.8	34.9	36.2	37.7	39.5	40.8	41.8
13	38.4	2.44	34.7	35.4	36.7	38.4	40.0	41.7	42.8
14	38.7	2.20	35.2	35.8	37.3	38.8	40.2	41.5	42.2
15	38.7	2.43	34.9	35.6	37.0	38.6	40.3	41.8	43.0
16	38.7	2.42	34.9	35.7	37.0	38.6	40.2	41.9	42.7
17	38.6	2.40	34.9	35.7	37.0	38.6	40.1	41.8	42.6

(data from Cycle III of the Health Examination Survey by the National Center of Health Statistics 1966-1970)

TABLE 941

CALF LENGTH (cm) IN U.S. WHITE CHILDREN

Age (years)	Mean	S.D.	Percentiles						
			5	10	25	50	75	90	95
				BOYS					
12	37.3	2.60	32.3	34.0	35.8	37.3	39.1	40.5	41.3
13	39.2	2.78	34.9	35.9	37.3	39.1	40.9	42.7	44.2
14	40.7	2.53	36.8	37.6	39.1	40.8	42.4	43.7	44.9
15	41.2	2.60	37.5	38.2	39.7	41.3	42.8	44.2	45.2
16	41.8	2.35	37.8	38.9	40.3	41.8	43.3	44.9	45.6
17	42.0	2.64	37.8	38.5	40.3	42.0	43.8	45.4	46.1
				GIRLS					
12	37.6	2.41	33.7	34.7	36.1	37.6	39.4	40.6	41.5
13	38.2	2.32	34.6	35.2	36.5	38.2	39.7	41.0	42.1
14	38.5	2.15	35.9	35.7	37.0	38.5	40.0	41.2	42.0
15	38.6	2.36	34.9	35.6	37.0	38.5	40.0	41.6	42.8
16	38.5	2.35	34.8	35.5	36.9	38.4	40.0	41.5	42.6
17	38.4	2.28	35.0	35.7	36.9	38.4	40.0	41.4	42.1

CALF LENGTH (cm) IN U.S. NEGRO CHILDREN

Age (years)	Mean	S.D.	Percentiles						
			5	10	25	50	75	90	95
				BOYS					
12	38.8	2.32	35.2	36.3	37.3	38.9	40.3	41.8	42.5
13	40.6	2.85	35.7	37.2	38.8	40.6	42.2	43.9	46.3
14	41.8	2.74	37.5	38.7	40.3	41.8	43.7	45.6	46.1
15	43.2	2.45	39.3	39.9	41.2	43.1	45.6	46.3	47.2
16	43.7	2.22	39.7	40.3	41.8	44.0	45.4	46.3	47.0
17	43.0	2.75	39.0	39.4	41.2	42.5	45.0	47.0	48.3
				GIRLS					
12	39.2	2.13	36.0	36.2	37.4	39.7	40.6	41.8	42.7
13	40.0	2.54	35.8	36.4	37.5	40.1	41.9	43.4	44.0
14	40.2	1.98	37.0	37.5	39.0	40.3	41.6	42.6	43.3
15	39.8	2.60	35.2	36.4	37.9	39.8	41.6	42.9	43.4
16	39.9	2.48	36.5	37.1	38.3	39.9	41.7	42.7	43.1
17	40.0	2.62	35.5	36.2	38.5	40.0	41.9	43.6	44.0

(data from Cycle III of the Health Examination Survey by the National Center of Health Statistics 1966-1970)

TABLE   942

KNEE – SOLE LENGTH (cm)
IN CHILDREN OF 8 STATES

Age (months)	Boys		Girls	
	Mean	S.D.	Mean	S.D.
0 – 3	14.4	1.4	14.0	1.1
4 – 6	17.1	1.0	16.4	1.0
7 – 9	18.6	1.0	18.1	0.8
10 – 12	20.1	0.8	19.3	1.2
13 – 18	21.7	1.3	21.0	1.2
19 – 24	23.8	1.6	23.4	1.5
25 – 30	25.4	1.6	25.5	1.3
31 – 36	27.2	1.6	26.7	1.5
37 – 42	28.3	1.3	28.2	1.5
43 – 48	29.3	1.4	29.1	1.5
49 – 54	30.8	1.5	30.5	1.6
55 – 60	31.9	1.7	31.6	1.7
61 – 66	33.1	1.8	32.9	1.7
67 – 72	34.1	1.8	34.0	1.9
73 – 78	35.3	1.8	35.0	1.9
79 – 84	36.5	2.0	36.0	1.8
85 – 96	38.3	2.3	37.8	2.1
97 – 108	40.2	2.2	40.8	2.5
109 – 120	42.2	2.5	41.9	2.3
121 – 132	44.6	2.1	44.4	2.7
133 – 144	46.8	2.7	45.7	2.6
145 – 156	47.5	3.0	49.0	2.1

(data from Snyder et al., 1975)

TABLE   943

FOOT LENGTH (cm) IN U.S. CHILDREN

Age (years)	5	10	25	50	75	90	95
			BOYS				
6	16.2	16.6	17.3	18.0	18.7	19.5	19.9
7	17.1	17.4	18.2	18.8	19.7	20.5	20.8
8	17.8	18.2	18.8	19.7	20.6	21.4	21.8
9	18.5	19.1	20.0	20.7	21.6	22.5	22.8
10	19.3	19.8	20.5	21.5	22.5	23.3	23.8
11	20.1	20.5	21.4	22.4	23.3	24.1	24.7
			GIRLS				
6	15.8	16.3	17.1	17.7	18.6	19.4	19.8
7	16.7	17.2	17.8	18.6	19.5	20.1	20.6
8	17.5	18.1	18.6	19.5	20.4	21.2	21.7
9	18.2	18.7	19.6	20.4	21.3	22.3	22.8
10	19.1	19.5	20.4	21.3	22.3	22.9	23.6
11	20.0	20.3	21.2	22.1	23.1	23.8	24.5

(data from Malina et al., 1973)

TABLE   944

FOOT LENGTH (cm) IN U.S. CHILDREN

Age (years)	Mean	S.D.	Percentiles						
			5	10	25	50	75	90	95
				BOYS					
12	23.6	1.48	21.1	21.6	22.5	23.6	24.6	25.7	26.3
13	24.6	1.56	22.0	22.5	23.5	24.6	25.7	26.7	27.3
14	25.4	1.57	23.1	23.6	24.4	25.4	26.4	27.3	27.7
15	25.8	1.33	23.4	24.1	25.0	25.8	26.7	27.6	28.1
16	25.9	1.31	23.6	24.2	25.1	25.9	26.8	27.6	28.0
17	26.1	1.32	24.0	24.4	25.2	26.1	26.9	27.8	28.4
				GIRLS					
12	23.1	1.24	21.2	21.5	22.3	23.1	23.8	24.7	25.3
13	23.3	1.40	21.2	21.6	22.4	23.3	24.2	24.8	25.5
14	23.4	1.20	21.4	22.0	22.5	23.4	24.4	25.2	25.6
15	23.4	1.29	21.3	21.7	22.4	23.4	24.4	25.2	25.7
16	23.5	1.32	21.3	21.8	22.6	23.5	24.4	25.3	25.7
17	23.5	1.22	21.3	21.8	22.6	23.5	24.4	25.3	25.7

(data from Cycle III of the Health Examination Survey, 1966-1970, conducted by the National Center for Health Statistics).

TABLE   945

FOOT LENGTH (cm) IN U.S. WHITE CHILDREN

| | | | \multicolumn{7}{c}{Percentiles} | | | | | | |
Age (years)	Mean	S.D.	5	10	25	50	75	90	95
\multicolumn{10}{c}{BOYS}									
6	17.9	1.01	16.2	16.5	17.2	17.8	18.6	19.4	19.8
7	18.8	1.05	17.1	17.4	18.1	18.8	19.6	20.4	20.8
8	19.6	1.20	17.7	18.1	18.7	19.6	20.5	21.3	21.7
9	20.7	1.23	18.5	19.1	19.9	20.6	21.5	22.4	22.8
10	21.4	1.30	19.2	19.7	20.4	21.4	22.4	23.2	23.7
11	22.3	1.34	20.1	20.4	21.3	22.3	23.2	24.1	24.7
12	23.5	1.51	21.0	21.4	22.4	23.6	24.6	25.6	26.2
13	24.5	1.52	22.0	22.4	23.4	24.5	25.6	26.6	27.0
14	25.4	1.59	23.0	23.5	24.4	25.4	26.4	27.1	27.6
15	25.7	1.28	23.4	24.1	24.8	25.7	26.6	27.4	27.8
16	25.9	1.26	23.6	24.2	25.1	25.8	26.7	27.6	28.0
17	26.0	1.29	24.0	24.3	25.2	25.9	26.8	27.6	28.0
\multicolumn{10}{c}{GIRLS}									
6	17.7	1.07	15.7	16.2	17.1	17.7	18.5	19.2	19.6
7	18.5	1.06	16.6	17.1	17.7	18.5	19.3	19.9	20.5
8	19.4	1.14	17.4	18.1	18.6	19.5	20.3	21.0	21.5
9	20.3	1.20	18.2	18.6	19.5	20.4	21.1	21.9	22.5
10	21.1	1.33	19.1	19.4	20.3	21.2	22.1	22.8	23.3
11	22.0	1.27	20.0	20.3	21.1	21.9	22.9	23.7	24.2
12	23.0	1.20	21.1	21.4	22.2	22.9	23.7	24.5	24.8
13	23.1	1.16	21.2	21.5	22.3	23.2	23.9	24.7	25.1
14	23.3	1.18	21.4	21.8	22.4	23.3	24.2	24.9	25.5
15	23.3	1.26	21.3	21.6	22.4	23.3	24.2	25.0	25.6
16	23.4	1.31	21.2	21.7	22.5	23.4	24.3	25.1	25.6
17	23.4	1.16	21.3	21.8	22.5	23.4	24.2	24.9	25.5

(data from Cycles II and III of the Health Examination Survey by the National
Center for Health Statistics, 1963-1965 and 1966-1970)

TABLE 946

FOOT LENGTH (cm) IN U.S. NEGRO CHILDREN

Age (years)	Mean	S.D.	Percentiles						
			5	10	25	50	75	90	95
BOYS									
6	18.5	1.07	16.7	17.1	17.7	18.5	19.2	18.9	20.5
7	19.5	1.02	17.5	18.0	18.7	19.5	20.2	20.8	21.3
8	20.4	1.06	18.8	19.1	19.6	20.4	21.2	22.0	22.4
9	21.2	1.30	18.9	19.4	20.3	21.2	21.9	22.8	24.1
10	22.0	1.22	20.1	20.3	20.9	22.1	22.9	23.6	23.9
11	22.9	1.20	20.7	21.2	22.1	23.1	23.6	24.3	24.8
12	24.0	1.23	22.1	22.4	23.2	23.8	24.8	26.0	26.3
13	25.2	1.71	22.2	22.9	24.0	25.3	26.6	27.4	27.7
14	25.9	1.37	23.6	24.1	24.7	26.1	27.0	27.7	28.2
15	26.5	1.44	23.4	24.4	25.7	26.7	27.6	28.4	28.8
16	26.3	1.59	23.4	23.9	25.2	26.6	27.4	27.8	28.3
17	26.7	1.35	24.1	25.0	25.8	26.7	27.7	28.7	29.0
GIRLS									
6	18.4	1.10	16.4	17.1	17.6	18.4	19.3	19.8	20.2
7	19.3	.97	17.6	17.8	18.6	19.4	20.1	20.7	21.0
8	20.0	1.26	18.2	18.4	19.0	20.1	21.0	21.8	22.4
9	21.2	1.40	18.8	19.2	20.2	21.0	22.4	23.2	23.7
10	22.2	1.20	20.1	20.5	21.3	22.2	23.1	23.8	24.4
11	22.7	1.52	20.1	20.6	21.5	22.7	23.8	24.9	25.3
12	23.9	1.14	22.0	22.3	23.0	23.7	25.0	25.6	25.8
13	24.2	2.16	22.1	22.3	23.1	24.2	25.1	26.1	26.6
14	24.2	1.02	22.4	22.9	23.5	24.3	25.1	25.7	25.9
15	24.1	1.21	22.0	22.5	23.3	24.2	25.0	25.8	26.2
16	24.2	1.16	22.2	22.7	23.4	24.2	25.2	25.8	26.2
17	24.3	1.30	21.8	22.4	23.4	24.4	25.4	26.1	26.3

(data from Cycles II and III of the Health Examination Survey by the National Center for Health Statistics, 1963-1965 and 1966-1970)

TABLE 947

FOOT LENGTH (cm)
IN CHILDREN OF 8 STATES

Age (months)	Boys Mean	Boys S.D.	Girls Mean	Girls S.D.
0 - 3	8.2	0.6	8.1	0.6
4 - 6	9.7	0.8	9.2	0.6
7 - 9	10.5	0.7	10.3	0.4
10 - 12	11.1	0.8	10.8	0.8
13 - 18	12.0	0.9	11.4	0.7
19 - 24	13.0	0.8	12.6	0.8
25 - 30	13.9	0.8	13.5	0.6
31 - 36	14.4	0.9	14.0	0.9
37 - 42	14.9	0.8	14.6	0.8
43 - 48	15.2	1.0	15.1	0.8
49 - 54	15.9	0.8	15.5	0.8
55 - 60	16.3	1.0	16.0	1.0
61 - 66	16.8	1.1	16.6	1.0
67 - 72	17.3	1.0	17.0	1.0
73 - 78	17.7	0.9	17.5	1.0
79 - 84	18.3	1.0	17.9	0.9
85 - 96	19.0	1.1	18.6	1.1
97 - 108	19.7	1.1	19.8	1.1
109 - 120	20.4	1.2	20.1	1.1
121 - 132	21.5	1.1	21.0	1.2
133 - 144	22.2	1.4	21.7	1.1
145 - 156	22.6	1.2	22.5	1.0

(data from Snyder et al., 1975)

TABLE   948

LENGTH OF THE NORMAL FOOT (cm)
IN WEIGHT-BEARING POSITION
IN BOSTON CHILDREN

Age	GIRLS Percentiles					BOYS Percentiles				
(years)	3	25	50	75	97	3	25	50	75	97
1	10.5	11.4	12.0	12.3	12.6	10.9	11.6	12.0	12.2	13.1
2	11.6	13.0	13.6	14.0	14.7	11.8	12.8	13.6	14.1	15.1
3	13.2	14.3	14.8	15.4	16.9	13.2	14.4	14.9	15.8	16.8
4	14.0	15.4	16.0	16.4	17.8	14.5	15.7	16.2	17.0	17.8
5	15.0	16.5	17.2	17.6	18.9	15.4	16.8	17.2	17.9	19.2
6	16.1	17.8	18.3	18.9	20.4	16.4	17.6	18.2	18.9	20.1
7	16.8	18.6	19.2	20.0	21.4	17.3	18.5	19.2	19.9	21.3
8	17.3	19.2	20.0	20.7	22.4	18.6	19.7	20.2	20.7	22.8
9	18.3	20.3	20.8	21.5	23.1	19.2	20.4	21.1	21.6	23.5
10	18.9	20.9	21.7	22.4	24.2	19.9	21.2	21.9	22.4	24.0
11	19.9	21.6	22.5	23.4	25.0	20.4	21.8	22.6	23.3	24.8
12	20.6	22.3	23.2	23.9	25.7	21.2	22.8	23.5	24.2	25.9
13	20.9	22.7	23.6	24.3	26.5	21.8	23.3	24.2	25.1	27.0
14	21.4	22.8	23.8	24.5	26.4	22.6	24.0	25.1	26.0	27.8
15	21.5	22.8	23.8	24.7	26.4	23.3	24.7	25.7	26.7	28.3
16	21.4	22.8	23.8	24.7	26.7	23.7	25.2	25.9	26.9	28.3
17	21.1	22.8	23.9	24.7	26.8	23.9	25.2	26.1	27.0	28.3
18	20.8	22.8	24.0	24.7	26.7	23.8	25.2	26.2	27.1	28.4

(data from Blais et al., 1956)

TABLE 949

LENGTH (cm) OF THE NORMAL WEIGHT-BEARING FOOT
FROM HEEL TO TIP OF GREAT TOE
IN BOSTON CHILDREN

Age (years)	GIRLS			BOYS		
	Mean	S.D.	Median	Mean	S.D.	Median
1	11.87	0.635	12.0	11.90	0.579	12.0
2	13.47	0.679	13.6	13.50	0.776	13.6
3	14.86	0.843	14.8	15.07	0.938	14.9
4	15.93	0.855	16.0	16.29	0.915	16.2
5	17.07	0.956	17.2	17.27	0.920	17.2
6	18.25	0.998	18.3	18.19	0.981	18.2
7	19.13	1.064	19.2	19.23	0.969	19.2
8	19.91	1.148	20.0	20.16	0.965	20.2
9	20.86	1.174	20.8	21.08	1.026	21.1
10	21.65	1.287	21.7	21.89	1.017	21.9
11	22.44	1.334	22.5	22.58	1.115	22.6
12	23.15	1.264	23.2	23.51	1.205	23.5
13	23.57	1.323	23.6	24.22	1.266	24.2
14	23.77	1.347	23.8	25.06	1.415	25.1
15	23.84	1.323	23.8	25.71	1.331	25.7
16	23.82	1.310	23.8	26.04	1.276	25.9
17	23.84	1.325	23.9	26.11	1.249	26.1
18	23.87	1.294	24.0	26.14	1.272	26.2

(data from Anderson et al., 1956)

TABLE 950

FOOT LENGTH (mm) OF MONTREAL INFANTS AT
BIRTH IN RELATION TO GESTATIONAL AGE

Gestation (weeks)	Mean	S.D.
24-26	50.1	3.33
27-28	55.4	3.08
29-30	57.7	2.12
31-32	62.6	3.32
33	67.4	4.78
34	69.6	3.76
35	71.4	3.76
36	72.8	4.12
37	78.0	3.91
38	78.2	4.11
39	77.3	3.56
40	78.4	3.45
41	79.8	3.49
42	78.9	4.60
43	80.0	2.80
44	78.0	3.83

(data from Usher and McLean, 1969, The Journal
of Pediatrics 74:901-910)

TABLE   951

FOOT MEASUREMENTS (in)
IN MICHIGAN CHILDREN

Age (Years)	Sex	Ankle height with shoe		Foot length without shoe		Heel to ankle without shoe		Foot width without shoe	
		Mean	S.D.	Mean	S.D.	Mean	S.D.	Mean	S.D.
5	Boys	3.0	0.33	7.0	0.28	1.0	0.10	2.6	0.17
	Girls	2.7	0.20	6.6	0.40	0.9	0.14	2.4	0.20
6	Boys	2.8	0.30	7.0	0.37	1.0	0.10	2.6	0.17
	Girls	2.7	0.26	6.9	0.32	1.0	0.10	2.5	0.22
7	Boys	3.1	0.26	7.4	0.53	1.2	0.17	2.8	0.17
	Girls	2.9	0.26	7.3	0.47	1.1	0.10	2.6	0.22
8	Boys	3.2	0.30	7.8	0.51	1.2	0.14	2.9	0.24
	Girls	3.0	0.24	7.8	0.51	1.3	0.14	2.8	0.17
9	Boys	3.4	0.32	8.2	0.53	1.3	0.17	3.0	0.14
	Girls	3.0	0.30	8.0	0.40	1.3	0.14	2.8	0.17
10	Boys	3.4	0.36	8.4	0.97	1.4	0.17	3.0	0.20
	Girls	3.2	0.28	8.5	0.46	1.4	0.10	3.0	0.22
11	Boys	3.6	0.26	8.8	0.57	1.5	0.14	3.2	0.26
	Girls	3.4	0.35	8.8	0.57	1.5	0.14	3.1	0.22
12	Boys	3.6	0.36	9.0	0.64	1.6	0.24	3.2	0.28
	Girls	3.4	0.33	9.1	0.45	1.6	0.20	3.2	0.24
13	Boys	3.6	0.40	9.5	0.64	1.8	0.17	3.4	0.14
	Girls	3.3	0.30	9.2	0.55	1.7	0.22	3.3	0.22
14	Boys	3.7	0.32	9.9	0.60	1.8	0.20	3.6	0.22
	Girls	3.3	0.32	9.3	0.51	1.8	0.22	3.3	0.20
15	Boys	3.9	0.26	10.3	0.47	1.9	0.17	3.7	0.28
	Girls	3.3	0.33	9.4	0.32	1.8	0.24	3.3	0.24
16	Boys	3.8	0.36	10.4	0.49	1.9	0.17	3.7	0.26
	Girls	3.3	0.20	9.4	0.37	1.8	0.14	3.3	0.30

TABLE 951 (continued).

FOOT MEASUREMENTS (in)
IN MICHIGAN CHILDREN

Age (Years)	Sex	Ankle height with shoe		Foot length without shoe		Heel to ankle without shoe		Foot width without shoe	
		Mean	S.D.	Mean	S.D.	Mean	S.D.	Mean	S.D.
17	Boys	3.8	0.25	10.5	0.56	1.9	0.20	3.7	0.17
	Girls	3.2	0.22	9.4	0.49	1.7	0.14	3.3	0.22
18	Boys	3.9	0.30	10.4	0.53	1.9	0.22	3.7	0.10
	Girls	3.3	0.30	9.5	0.55	1.7	0.10	3.3	0.17

(data from Martin, 1955)

TABLE 952

FOOT WIDTH (cm)
IN CHILDREN OF 8 STATES

Age (months)	Boys		Girls	
	Mean	S.D.	Mean	S.D.
0 - 3	3.5	0.4	3.4	0.4
4 - 6	4.2	0.5	3.9	0.4
7 - 9	4.4	0.4	4.4	0.4
10 - 12	4.7	0.4	4.4	0.4
13 - 18	4.9	0.4	4.7	0.3
19 - 24	5.4	0.4	5.0	0.4
25 - 30	5.4	0.4	5.4	0.3
31 - 36	5.6	0.3	5.4	0.5
37 - 42	5.7	0.4	5.6	0.4
43 - 48	5.8	0.4	5.6	0.4
49 - 54	6.0	0.4	5.8	0.4
55 - 60	6.1	0.4	5.9	0.4
61 - 66	6.3	0.5	6.2	0.5
67 - 72	6.4	0.5	6.2	0.4
73 - 78	6.6	0.5	6.4	0.4
79 - 84	6.9	0.4	6.6	0.4
85 - 96	7.0	0.5	6.8	0.4
97 - 108	7.2	0.5	7.1	0.5
109 - 120	7.4	0.5	7.3	0.5
121 - 132	7.9	0.5	7.6	0.5
133 - 144	8.0	0.5	8.0	0.5
145 - 156	8.4	0.5	8.3	0.5

(data from Snyder et al., 1975)

TABLE   953

FOOT LENGTH (mm) AT BIRTH IN RELATION TO BIRTH WEIGHT
IN MONTREAL CHILDREN

Weight group (Gm)	Mean	S.D.
501-1,000	49.7	3.82
1,001-1,500	57.3	2.30
1,501-2,000	65.9	4.00
2,001-2,500	71.5	3.90
2,501-3,000	75.4	2.86
3,001-3,500	78.2	4.38
3,501-4,000	80.4	3.18
4,001-4,500	81.7	3.35

(data from Usher and McLean, 1969, The Journal
of Pediatrics 74:901-910)

TABLE 954

MEAN PROPORTIONS OF THE HUMAN FOOT
("HEEL," MID-TARSUS AND METATARSUS RELATED TO TOTAL FOOT LENGTH
AS MEASURED FROM LATERAL ROENTGENOGRAMS) IN BOSTON CHILDREN

AGE (years)	MEANS, BOYS				MEANS, GIRLS			
	Calcified Os Calcis	"Heel"[1]	Cuboid[2]	Meta-tarsus[3]	Calcified Os Calcis	"Heel"[1]	Cuboid[2]	Meta-tarsus[3]
	%	%	%	%	%	%	%	%
1	21.5	32.2	13.9	15.3	22.6	32.1	13.7	16.5
2	23.9	33.1	14.3	16.1	24.3	32.7	14.0	16.7
3	25.2	33.4	14.2	15.5	25.4	32.6	13.8	16.3
4	26.4	33.7	14.2	15.9	26.0	32.6	13.8	16.3
5	27.1	34.0	14.3	15.8	26.1	32.4	13.6	17.0
6	27.2	33.6	14.2	16.0	26.3	32.3	13.6	16.9
7	27.3	33.3	14.1	16.0	26.6	32.2	13.6	16.7
8	27.4	33.1	14.0	15.8	27.3	32.4	13.4	16.6
9	27.8	33.0	13.7	15.6	27.6	32.3	13.1	16.6
10	28.2	33.1	13.5	15.4	28.2	32.4	13.0	16.8
11	28.5	33.2	13.3	15.3	28.5	32.3	12.4	16.8
12	28.9	33.2	13.1	15.2	28.6	32.5	12.3	17.1
13	29.3	33.3	12.9	15.3	28.7	32.6	12.1	17.0
14	29.7	33.7	12.7	15.6	28.9	32.6	12.1	17.2
15	30.1	33.8	12.5	16.3	28.9	32.7	12.1	17.2
16	30.1	33.5	12.5	16.6	29.0	32.8	12.1	17.2
17	30.3	33.4	12.5	16.6	28.9	32.8	12.1	17.2
18	30.2	33.4	12.5	16.6	28.9	32.8	12.1	17.2

[1]Skin at back heel to mid-point between os calcis and cuboid.
[2]Mid-point between os calcis and cuboid to mid-point between cuboid and fourth metatarsal
[3]Mid-point between cuboid and fourth metatarsal to distal epiphyseal line of fifth metatarsal.

(data from Anderson et al., 1956)

TABLE 955

CHANGING RELATIONSHIPS OF FOOT LENGTH, LONG BONE LENGTHS, AND STATURE
IN BOSTON CHILDREN

Age (years)	FOOT: TIBIA		FEMUR + TIBIA: STATURE		FOOT: STATURE	
	Girls %	Boys %	Girls %	Boys %	Girls %	Boys %
1	102.9	102.5	34.4	34.6	15.9	15.9
2	96.6	95.6	37.9	37.2	16.1	15.8
3	93.0	91.5	39.2	39.3	16.1	15.9
4	89.8	87.6	40.7	40.7	16.0	15.7
5	86.1	83.8	42.7	42.6	16.1	15.7
6	84.0	82.6	43.5	43.5	16.0	15.7
7	82.0	81.0	44.5	44.5	16.0	15.7
8	80.2	79.4	45.2	45.4	15.9	15.8
9	78.8	78.1	46.1	46.2	16.0	15.8
10	76.9	77.2	46.9	46.8	15.9	15.9
11	76.2	76.4	47.4	47.6	15.9	15.9
12	75.1	75.6	47.8	48.2	15.8	16.1
13	73.9	74.4	48.1	48.6	15.5	16.0
14	72.6	72.6	47.9	48.8	15.3	15.6
15	72.5	71.3	47.8	48.6	15.2	15.3
16	72.4	71.1	47.7	48.3	15.1	15.1
17	72.4	71.2	47.5	48.0	15.1	15.0
18	72.4	71.2	47.4	47.7	15.1	14.9

(data from Anderson et al., 1956)

TABLE   956

MID-THIGH CIRCUMFERENCE (cm)
IN CHILDREN OF 8 STATES

Age	Boys		Girls	
(months)	Mean	S.D.	Mean	S.D.
0 - 3	17.5	3.0	18.0	2.7
4 - 6	21.8	2.3	21.4	2.2
7 - 9	23.0	2.6	22.6	1.8
10 - 12	23.3	1.9	23.6	2.6
13 - 18	23.6	2.7	23.5	2.3
19 - 24	25.6	2.1	24.8	2.0
25 - 30	25.4	1.6	26.5	2.0
31 - 36	26.9	2.0	26.6	2.1
37 - 42	27.4	1.9	27.7	2.6
43 - 48	27.9	2.2	28.2	2.0
49 - 54	28.6	2.1	29.0	2.1
55 - 60	29.3	2.2	29.9	2.3
61 - 66	30.3	2.2	30.5	2.6
67 - 72	30.4	2.7	30.7	2.7
73 - 78	31.1	2.5	31.6	2.9
79 - 84	32.8	2.8	32.2	2.5
85 - 96	33.3	3.6	33.8	3.4
97 - 108	35.2	3.5	35.8	3.7
109 - 120	36.0	2.9	37.2	3.8
121 - 132	39.1	3.9	39.6	4.4
133 - 144	40.9	2.9	41.1	3.5
145 - 156	42.0	3.5	45.2	4.2

(data from Snyder et al., 1975)

TABLE   957

CALF CIRCUMFERENCE (cm) IN U.S. CHILDREN

Age			Percentiles						
(years)	Mean	S.D.	5	10	25	50	75	90	95

BOYS

Age	Mean	S.D.	5	10	25	50	75	90	95
12	30.8	3.05	26.4	27.2	28.7	30.5	32.6	35.2	36.4
13	32.6	3.27	27.9	28.7	30.3	32.3	34.6	37.1	38.3
14	34.0	3.33	29.2	30.1	32.0	33.8	36.2	38.2	39.6
15	34.9	2.99	30.6	31.4	33.0	34.7	36.7	38.9	40.5
16	35.6	2.83	31.1	32.1	33.7	35.5	37.3	39.4	40.6
17	36.1	3.02	31.6	32.6	34.2	35.8	37.7	39.6	41.6

GIRLS

Age	Mean	S.D.	5	10	25	50	75	90	95
12	31.6	3.09	27.1	27.7	29.6	31.3	33.6	36.1	37.3
13	32.6	3.04	28.0	28.8	30.4	32.4	34.6	36.6	38.0
14	33.6	3.01	28.9	29.9	31.7	33.5	35.6	37.5	38.7
15	34.2	3.07	29.7	30.5	32.1	34.1	35.8	38.0	39.7
16	34.6	3.25	30.1	31.1	32.5	34.3	36.4	38.8	40.7
17	34.5	2.88	30.2	31.3	32.6	34.4	36.1	37.8	39.2

(data from Cycle III of the Health Examination Survey, 1966-1970,conducted by
the National Center for Health Statistics).

TABLE 958

CALF CIRCUMFERENCE (cm) IN U.S. WHITE CHILDREN

Age (years)	Mean	S.D.	Percentiles						
			5	10	25	50	75	90	95
			BOYS						
6	23.6	1.87	20.8	21.3	22.3	23.5	24.6	26.1	27.1
7	24.7	2.14	21.6	22.3	23.3	24.6	25.8	27.3	28.4
8	25.7	2.36	22.2	23.1	24.3	25.6	27.0	28.7	29.9
9	26.8	2.65	23.4	24.1	25.1	26.5	28.2	30.4	31.8
10	27.6	2.49	24.1	24.7	26.0	27.5	29.3	30.7	32.1
11	29.0	2.82	25.2	26.0	27.2	28.6	30.6	32.8	34.6
12	30.8	3.08	26.4	27.2	28.7	30.6	32.6	35.2	36.5
13	32.6	3.22	28.0	28.7	30.3	32.3	34.6	36.8	38.2
14	34.1	3.32	29.2	30.2	32.2	34.1	36.2	38.2	39.6
15	35.2	2.99	30.9	31.5	33.2	34.8	36.9	39.3	40.6
16	35.7	2.86	31.2	32.2	33.7	35.6	37.4	39.5	40.6
17	36.1	3.05	31.4	32.5	34.2	35.8	37.7	39.6	41.6
			GIRLS						
6	23.7	1.93	20.6	21.3	22.4	23.6	24.9	26.3	27.3
7	24.6	1.98	21.6	22.3	23.3	24.5	25.9	27.4	28.2
8	25.8	2.21	22.6	23.3	24.3	25.6	27.3	28.8	30.0
9	27.0	2.53	23.4	24.1	25.2	26.7	28.6	30.4	31.7
10	27.9	2.70	24.0	24.5	26.1	27.9	29.8	31.5	32.7
11	29.2	2.87	25.1	25.6	27.3	28.9	31.3	33.2	34.3
12	31.5	3.04	27.1	27.7	29.5	31.3	33.4	36.0	37.1
13	32.7	2.92	28.2	29.1	30.6	32.6	34.7	36.5	37.8
14	33.6	2.98	29.0	30.0	31.8	33.5	35.5	37.4	38.6
15	34.3	3.09	29.7	30.6	32.2	34.2	35.9	33.1	39.8
16	34.7	3.22	30.3	31.2	32.6	34.4	36.4	38.7	40.3
17	34.5	2.82	30.4	31.4	32.7	34.4	35.9	37.8	39.2

(data from Cycles II and III of the Health Examination Survey by the National
Center for Health Statistics, 1963-1965 and 1966-1970)

TABLE 959

CALF CIRCUMFERENCE (cm) IN U.S. NEGRO CHILDREN

Age (years)	Mean	S.D.	Percentiles						
			5	10	25	50	75	90	95
BOYS									
6	23.2	1.78	20.3	20.7	22.1	23.2	24.3	27.8	26.6
7	24.2	2.13	21.3	22.0	22.6	24.2	25.4	26.6	27.4
8	25.5	1.80	22.9	23.3	24.2	25.4	26.8	28.1	28.9
9	26.2	2.62	22.7	23.4	24.5	25.7	27.6	29.6	30.7
10	27.2	2.30	23.8	24.3	25.6	26.8	28.5	30.3	31.7
11	28.3	2.42	24.5	25.2	26.8	27.9	29.9	31.6	32.5
12	30.4	2.74	26.7	27.4	28.6	30.1	32.1	34.3	35.5
13	32.8	3.63	27.2	28.3	30.4	32.5	34.9	37.6	38.7
14	33.5	3.40	28.9	29.6	31.2	33.1	36.1	37.8	39.6
15	33.5	2.57	29.3	30.1	31.9	33.3	35.4	36.7	37.7
16	35.0	2.61	31.0	31.7	33.3	34.6	36.6	38.8	39.9
17	36.1	2.81	32.3	33.2	34.4	35.6	37.4	39.6	40.8
GIRLS									
6	23.2	1.49	21.0	21.3	22.2	23.1	24.3	25.3	26.1
7	24.2	1.86	21.5	22.1	22.8	24.3	25.2	26.3	26.7
8	25.4	2.71	21.5	22.2	23.5	24.9	27.1	29.3	30.4
9	26.4	2.54	22.5	23.2	24.8	26.2	28.1	30.4	31.2
10	27.8	3.24	23.6	24.2	25.4	27.6	30.1	31.8	33.9
11	29.1	3.55	24.3	24.9	26.5	28.6	31.3	34.8	35.9
12	32.1	3.42	26.6	28.1	29.9	31.5	34.6	36.6	38.5
13	32.0	3.66	27.0	27.8	29.3	31.4	34.1	36.8	39.5
14	33.5	3.22	27.7	29.5	31.4	33.5	35.7	38.1	39.1
15	33.7	2.92	29.2	30.2	31.6	33.7	35.5	37.5	38.8
16	34.4	3.44	29.4	30.2	32.2	34.1	35.6	41.0	41.6
17	34.2	3.25	29.6	30.4	31.6	34.2	36.4	37.7	39.4

(data from Cycles II and III of the Health Examination Survey by the National
Center for Health Statistics, 1963-1965 and 1966-1970)

TABLE   960

MAXIMUM CALF CIRCUMFERENCE (cm)
IN CHILDREN OF 8 STATES

Age (months)	Boys		Girls	
	Mean	S.D.	Mean	S.D.
0 – 3	12.5	1.8	12.8	1.8
4 – 6	16.1	1.3	15.6	1.3
7 – 9	17.3	1.2	17.0	1.3
10 – 12	18.0	1.1	17.5	1.4
13 – 18	18.5	1.5	17.9	1.3
19 – 24	19.7	1.3	18.7	1.2
25 – 30	19.8	1.2	19.8	1.2
31 – 36	20.3	1.2	19.9	1.7
37 – 42	20.5	1.1	20.5	1.4
43 – 48	20.8	1.3	20.7	1.2
49 – 54	21.3	1.4	21.2	1.3
55 – 60	21.6	1.5	21.6	1.5
61 – 66	22.2	1.4	21.9	1.7
67 – 72	22.5	1.6	22.3	1.8
73 – 78	22.9	1.5	22.9	1.8
79 – 84	23.9	1.7	23.2	1.4
85 – 96	24.4	2.2	24.1	1.9
97 – 108	25.6	2.2	25.3	2.1
109 – 120	26.5	2.1	26.2	2.2
121 – 132	28.2	2.8	27.8	2.6
133 – 144	29.1	2.1	29.0	2.5
145 – 156	29.6	2.3	31.4	2.6

(data from Snyder et al., 1975)

TABLE 961

CALF CIRCUMFERENCE (cm)
IN CALIFORNIA CHILDREN

Age (years)	Boys		Girls	
	Mean (cm)	S.D. (cm)	Mean (cm)	S.D. (cm)
9	27.48	2.46	27.84	2.43
10	28.56	2.51	28.93	2.65
11	29.58	2.67	30.02	2.84
12	30.97	2.80	31.30	2.76
13	32.50	3.06	32.48	2.84
14	33.88	3.18	33.47	2.69
15	34.96	2.79	34.41	2.64
16	35.76	2.63	35.22	2.60
17	36.09	2.52	35.53	2.58
18	36.28	2.49	35.40	2.57

(data from Tuddenham and Snyder, 1954)

TABLE 962

CALF CIRCUMFERENCE (mm) FOR PHILADELPHIA CHILDREN

Age (years)	WHITE BOYS		BLACK BOYS	
	Mean	S.D.	Mean	S.D.
6	242.72	19.69	235.45	21.43
7	254.27	25.05	245.83	21.14
8	264.69	22.47	257.39	19.96
9	277.39	26.38	272.35	23.26
10	293.01	32.17	280.65	29.55
11	300.34	26.69	291.31	31.45
12	312.22	30.13	299.22	26.15
13	--	--	317.04	30.45

Age (years)	WHITE GIRLS		BLACK GIRLS	
	Mean	S.D.	Mean	S.D.
6	242.20	13.87	238.91	20.98
7	256.62	19.74	253.22	22.75
8	262.00	22.93	263.77	24.88
9	273.54	25.33	273.30	26.88
10	286.52	25.21	288.99	29.30
11	295.83	23.78	300.29	32.91
12	--	--	313.44	36.57
13	--	--	318.14	28.07

Age is for completed years.

(data from Malina, unpublished)

TABLE   963

CALF CIRCUMFERENCE (cm) OF PUERTO RICAN PRIVATE SCHOOL CHILDREN AT SAN JUAN

Mean Age (years)	Mean	S.D.	Percentiles				
			10	30	50	70	90
BOYS							
7	25.1	2.2	22.6	23.7	25.1	25.9	27.9
8	26.5	2.4	23.7	25.1	26.2	27.2	30.1
9	27.6	2.8	24.6	26.0	26.9	28.6	32.0
10	29.0	3.1	25.3	27.2	28.8	30.4	33.0
11	30.3	3.1	26.3	28.6	30.0	31.6	35.1
12	31.3	3.2	27.7	29.1	30.5	33.1	36.5
13	32.6	3.5	28.6	30.5	32.4	34.1	36.7
14	33.8	3.6	29.6	32.3	33.4	35.3	38.6
15	34.3	3.6	29.9	32.5	34.0	36.1	39.2
16	35.4	2.9	31.8	33.9	35.3	36.8	39.3
17	35.9	3.5	31.5	34.1	35.3	37.0	40.4
GIRLS							
7	25.2	1.9	23.1	24.0	24.9	26.1	28.1
8	26.6	2.2	24.0	25.3	26.2	27.3	29.8
9	27.9	2.3	24.8	26.8	27.8	29.0	30.9
10	28.8	2.6	25.4	27.5	28.7	30.0	32.4
11	29.7	2.9	26.0	28.1	29.5	31.5	33.7
12	31.5	2.8	27.9	29.9	31.3	33.0	35.4
13	32.4	3.5	28.2	30.3	31.8	34.3	37.2
14	33.0	2.4	29.7	31.7	32.9	34.3	36.1
15	33.1	2.6	30.3	31.4	32.8	34.4	36.7
16	34.1	2.6	31.2	32.7	33.6	35.2	37.2
17	33.7	2.5	30.7	32.4	33.8	35.0	36.6

(data from Knott, 1963)

TABLE 964

CALF CIRCUMFERENCE (cm) AT EXAMINATIONS ONE YEAR APART
OF PHILADELPHIA CHILDREN

Ages (years)	Examination 1 Mean	S.D.	Examination 2 Mean	S.D.	Increase Mean	S.D.
Negro Males						
6- 7	22.8	1.7	24.0	2.2	1.2	0.6
7- 8	23.7	1.7	25.3	2.1	1.6	0.9
8- 9	25.3	2.2	26.7	2.5	1.4	0.7
9-10	26.0	1.7	27.7	2.1	1.7	0.7
10-11	27.3	2.7	29.0	3.2	1.7	0.9
11-12	28.4	2.4	30.2	2.9	1.8	0.8
12-13	28.6	1.9	30.6	1.9	1.9	0.7
White Males						
6- 7	23.5	1.6	25.0	1.7	1.5	0.5
7- 8	24.8	2.4	26.1	3.0	1.3	0.9
8- 9	25.4	1.7	27.0	2.1	1.6	0.7
9-10	26.7	2.4	28.2	2.6	1.5	0.7
10-11	28.6	3.0	30.6	3.5	2.0	0.7
11-12	28.8	2.3	30.7	2.7	1.9	0.7
Negro Females						
6- 7	23.0	1.6	24.5	1.8	1.5	0.7
7- 8	24.8	1.9	26.4	2.2	1.6	0.7
8- 9	25.4	2.0	27.1	2.1	1.7	0.7
9-10	26.7	3.1	28.5	3.4	1.8	0.9
10-11	28.5	3.2	30.3	3.5	1.8	0.7
11-12	28.4	2.7	30.1	2.9	1.7	0.7
12-13	30.5	3.3	32.0	3.5	1.5	0.7
White Females						
6- 7	23.6	1.1	25.1	1.3	1.5	0.7
7- 8	24.9	1.7	26.4	1.9	1.5	0.7
8- 9	25.1	2.1	26.4	2.3	1.3	0.6
9-10	26.3	2.4	27.7	2.7	1.4	0.6
10-11	27.2	1.9	28.7	2.0	1.5	0.6
11-12	28.8	2.5	30.0	2.7	1.2	0.8

(data from Malina, 1972)

TABLE   965

ANKLE CIRCUMFERENCE (cm)
IN CHILDREN OF 8 STATES

| Age | Boys | | Girls | |
(months)	Mean	S.D.	Mean	S.D.
0 – 3	9.9	1.3	9.8	1.2
4 – 6	12.2	1.0	11.5	1.1
7 – 9	12.7	1.0	12.0	1.1
10 – 12	12.9	0.7	12.7	0.9
13 – 18	13.2	0.7	13.1	1.0
18 – 24	14.2	0.9	13.4	1.2
25 – 30	14.1	0.9	13.9	1.0
31 – 36	14.1	1.1	13.7	1.0
37 – 42	14.2	0.9	14.1	1.0
43 – 48	14.3	1.0	14.2	0.9
49 – 54	14.6	0.9	14.4	0.9
55 – 60	14.8	1.1	14.7	1.1
61 – 66	15.1	1.0	14.7	1.1
67 – 72	15.3	1.1	15.0	1.2
73 – 78	15.4	1.0	15.3	1.2
79 – 84	15.9	1.2	15.4	0.9
85 – 96	16.2	1.4	16.1	1.2
97 – 108	16.8	1.3	16.7	1.4
109 – 120	17.5	1.4	17.2	1.4
121 – 132	18.5	1.5	18.2	1.6
133 – 144	19.1	1.3	18.8	1.6
145 – 156	19.5	1.4	20.1	1.4

(data from Snyder et al., 1975)

TABLE   966

LEG CIRCUMFERENCE IN COLORADO BOYS (cm)

Age (yr-mo)	Foot Percentiles			Ankle Percentiles			Calf Percentiles			Knee Percentiles		
	10	50	90	10	50	90	10	50	90	10	50	90
Supine												
Birth	7.8	8.6	9.3	7.8	8.6	9.6	9.7	11.0	12.3	11.1	12.6	14.3
0 - 1	8.8	9.3	10.4	8.3	9.2	10.1	10.8	11.8	13.4	12.5	13.4	15.6
0 - 2	9.4	10.3	11.1	9.2	10.2	11.2	12.0	13.5	15.0	13.3	15.4	17.5
0 - 3	9.8	10.6	11.6	9.8	10.9	12.0	12.8	14.6	16.2	14.0	16.3	18.5
0 - 4	10.4	11.1	12.0	10.5	11.4	12.5	14.1	15.7	17.2	15.8	17.7	19.5
0 - 5	10.6	11.5	12.3	11.0	11.8	13.2	15.0	16.4	17.9	16.2	18.5	20.5
0 - 6	10.8	11.7	12.7	11.2	12.2	13.7	15.0	17.0	19.2	16.4	18.7	21.0
0 - 9	11.5	12.3	13.8	11.7	13.0	14.4	16.5	18.4	20.0	17.6	20.4	23.1
1 - 0	11.9	12.9	14.0	12.1	13.4	14.7	17.1	18.6	20.4	18.0	20.4	22.5
1 - 6	12.4	13.2	14.4	12.8	13.3	14.8	17.4	19.0	20.7	18.8	20.6	22.6
2 - 0	12.8	13.7	14.5	13.0	13.5	14.5	18.4	19.4	20.8	19.2	20.6	22.3
Erect												
2 - 0	12.7	14.3	15.5	12.7	14.5	15.9	18.1	19.9	21.7	19.5	22.0	24.0
2 - 6	13.3	14.6	16.2	12.7	14.6	16.2	18.6	20.2	22.0	19.9	22.5	24.1
3 - 0	14.0	15.3	16.4	13.0	14.7	16.3	19.2	20.6	22.4	20.8	22.9	24.5
3 - 6	13.9	15.5	16.7	13.2	14.6	16.2	19.4	20.8	22.6	20.7	23.1	24.7
4 - 0	13.8	15.8	17.0	13.4	14.8	16.2	19.6	21.0	23.0	21.4	23.1	24.9
4 - 6	14.5	15.7	17.4	13.7	14.9	16.1	19.9	21.5	23.1	21.8	23.6	25.0
5 - 0	14.9	16.3	18.0	13.8	15.1	16.6	20.4	22.1	23.7	22.4	24.0	25.4
5 - 6	15.2	16.6	18.4	14.3	15.5	16.7	21.0	22.6	24.3	22.4	24.4	26.0
6 - 0	15.9	17.1	18.6	14.6	15.7	16.9	21.2	23.1	24.7	23.0	24.6	26.3
6 - 6	16.2	17.6	18.9	14.9	16.1	17.3	21.6	23.7	25.3	23.6	25.3	26.9
7 - 0	16.5	17.9	19.7	15.2	16.4	17.6	22.2	24.2	25.9	24.1	25.7	28.1
7 - 6	16.9	18.3	19.8	15.6	16.6	18.2	22.4	24.7	26.8	24.7	26.3	28.7
8 - 0	17.5	18.8	20.4	15.2	16.9	18.4	22.9	25.3	27.8	25.2	27.0	29.5
8 - 6	17.6	19.2	20.7	15.7	17.3	18.8	23.4	25.8	28.2	25.4	27.2	30.1
9 - 0	18.1	19.8	21.1	15.7	17.5	19.2	23.8	26.3	29.2	25.6	28.3	31.0
9 - 6	18.4	20.2	21.4	15.9	18.0	19.5	24.1	27.0	29.5	26.4	28.8	31.1
10 - 0	18.5	20.4	22.2	16.6	18.3	20.2	24.9	27.7	30.2	26.7	29.3	32.3
10 - 6	18.4	20.9	22.2	16.7	18.4	20.3	25.6	28.0	30.8	27.4	30.0	32.8
11 - 0	19.0	21.3	22.9	17.2	18.8	20.5	26.0	28.8	31.8	28.2	30.8	34.2
11 - 6	19.6	21.6	23.1	17.7	19.2	20.7	26.5	29.2	32.4	28.5	31.0	34.2
12 - 0	19.7	22.0	23.8	17.6	19.5	21.2	26.3	29.6	33.5	29.0	31.8	35.0
12 - 6	20.0	22.2	24.4	18.0	19.7	21.7	27.0	30.3	33.8	29.1	31.6	35.9
13 - 0	21.0	22.7	24.6	18.5	20.4	22.1	27.9	31.3	34.3	30.0	33.1	36.9
13 - 6	21.2	23.4	25.3	18.8	20.7	23.1	28.3	31.9	35.3	31.0	33.9	36.8

TABLE   966   (continued)

LEG CIRCUMFERENCE IN COLORADO BOYS (cm)

Age	Foot Percentiles			Ankle Percentiles			Calf Percentiles			Knee Percentiles		
(yr-mo)	10	50	90	10	50	90	10	50	90	10	50	90
14 - 0	21.5	23.8	25.6	19.5	21.3	23.1	29.2	32.8	35.4	32.0	35.0	36.7
14 - 6	22.0	23.9	26.0	19.3	21.5	23.2	28.9	32.9	36.7	31.9	35.4	37.6
15 - 0	22.2	24.4	26.0	20.3	21.6	23.5	30.2	33.3	37.0	32.5	35.5	38.2
15 - 6	22.5	24.7	26.1	20.4	22.1	23.3	30.5	33.8	37.2	33.9	36.0	38.7
16 - 0	23.4	24.8	26.5	20.9	22.3	24.0	31.5	34.4	37.3	34.6	36.4	38.1
16 - 6	23.5	24.7	26.8	21.1	21.9	23.7	31.8	34.6	37.8	34.4	36.5	38.9
17 - 0	23.3	24.7	26.7	20.7	22.7	24.5	32.1	35.6	39.0	34.0	36.8	38.7
17 - 6	23.1	25.2	26.7	20.5	22.9	24.2	32.1	36.0	38.3	34.0	37.3	39.4
18 - 0	23.3	25.3	27.0	20.8	22.2	23.9	32.5	34.9	38.1	34.0	37.2	39.0

(data from McCammon, Human Growth and Development, 1970. Courtesy of Charles C Thomas, Publisher, Springfield, Illinois)

TABLE 967

LEG CIRCUMFERENCE IN COLORADO GIRLS (cm)

Age (yr-mo)	Foot Percentiles			Ankle Percentiles			Calf Percentiles			Knee Percentiles		
	10th	50th	90th	10th	50th	90th	10th	50th	90th	10th	50th	90th
Supine												
Birth	7.5	8.4	9.0	7.9	8.7	9.4	9.7	11.0	12.2	11.5	12.5	14.0
0-1	8.0	8.9	9.6	8.0	8.9	9.8	10.0	11.8	12.7	11.7	13.1	14.5
0-2	8.6	9.6	10.2	9.1	9.7	10.7	11.9	13.2	14.3	13.4	15.0	16.7
0-3	9.0	10.0	10.5	9.7	10.5	11.6	13.0	14.3	15.5	14.1	15.8	17.8
0-4	9.9	10.4	11.1	10.3	10.9	12.0	14.0	15.3	16.3	15.2	16.8	18.3
0-5	10.0	10.7	11.7	10.8	11.5	12.4	14.6	16.0	17.8	16.1	18.1	19.6
0-6	10.3	11.0	11.9	11.0	11.9	12.9	14.9	16.7	18.2	16.3	18.4	21.2
0-9	10.8	12.0	12.6	11.9	12.8	13.5	16.4	18.1	19.7	17.8	19.4	22.5
1-0	11.4	12.2	13.2	12.0	12.9	13.8	17.2	18.5	20.3	17.9	19.8	23.4
1-6	11.9	12.7	13.7	12.3	13.2	14.3	17.5	19.0	20.7	18.2	20.2	22.8
2-0	12.1	12.8	13.8	12.0	13.2	13.9	17.7	19.4	20.7	18.6	19.8	21.1
Erect												
2-0	12.8	13.6	14.6	12.7	14.0	15.0	18.2	19.9	21.5	20.1	21.3	24.0
2-6	13.1	14.1	15.1	12.8	14.1	15.2	18.5	20.0	21.7	20.2	21.6	23.7
3-0	13.7	14.6	15.6	13.1	14.6	16.0	18.8	20.3	22.2	20.8	22.1	24.0
3-6	14.0	15.1	16.2	13.4	14.9	15.9	19.6	20.8	22.8	21.4	22.4	24.3
4-0	13.8	15.4	16.4	13.4	15.0	16.0	19.5	21.3	22.9	21.6	22.9	24.7
4-6	14.1	15.6	17.0	13.6	15.1	16.3	20.1	22.0	23.5	22.0	23.4	25.2
5-0	14.5	15.9	17.3	13.8	15.4	16.3	20.5	22.4	24.1	22.2	23.7	25.5
5-6	14.9	16.3	17.8	14.2	15.6	16.7	20.8	23.0	24.7	22.6	24.3	26.3
6-0	15.1	16.8	18.1	14.2	16.0	17.0	21.3	23.4	25.4	23.0	24.6	27.0
6-6	15.4	16.9	18.3	14.4	16.1	17.4	21.6	23.9	25.9	23.5	25.0	27.3
7-0	15.8	17.4	18.6	14.8	16.5	17.8	22.0	24.1	26.6	23.9	25.5	28.1
7-6	16.4	17.7	19.1	15.1	16.7	18.1	22.6	24.8	27.4	24.1	26.0	28.6
8-0	16.6	18.0	19.2	15.3	17.1	18.4	23.1	25.4	27.9	24.7	26.8	29.2
8-6	17.2	18.3	19.8	15.5	17.4	18.9	23.4	26.1	28.7	25.4	27.5	30.3
9-0	17.4	18.9	20.2	15.7	17.7	19.5	23.7	26.4	29.5	26.2	28.2	31.4
9-6	17.8	19.1	20.6	15.8	17.8	19.4	24.0	26.9	29.8	26.6	28.6	31.9
10-0	18.0	19.5	21.1	16.1	18.2	19.8	24.9	27.3	30.4	27.0	29.2	32.7
10-6	18.3	19.9	21.1	16.1	18.5	20.1	24.8	28.1	31.4	27.2	29.5	33.3
11-0	18.6	20.4	21.9	16.7	18.8	20.6	25.5	28.5	32.2	27.8	30.3	34.1
11-6	19.2	20.6	22.1	17.1	19.1	20.9	26.3	29.0	33.1	28.8	31.0	34.2
12-0	19.5	20.9	22.4	17.3	19.8	21.2	27.0	30.1	34.4	28.9	31.8	35.2
12-6	19.8	21.6	22.7	18.1	20.1	21.4	27.4	31.2	33.9	29.4	33.0	36.1
13-0	20.2	21.9	23.2	17.9	20.5	21.8	28.0	31.2	34.0	30.6	33.2	37.4
13-6	20.7	22.0	23.1	19.0	20.7	21.8	28.6	32.2	34.7	31.3	33.6	36.5

TABLE   967   (continued)

LEG CIRCUMFERENCE   IN COLORADO GIRLS (cm)

Age	Foot Percentiles			Ankle Percentiles			Calf Percentiles			Knee Percentiles		
(yr-mo)	10th	50th	90th	10th	50th	90th	10th	50th	90th	10th	50th	90th
14-0	21.4	22.2	23.8	19.6	21.1	22.2	30.0	32.6	35.5	32.4	34.8	37.4
14-6	21.4	22.1	23.7	19.0	21.1	22.4	29.7	32.7	35.9	32.8	35.5	38.2
15-0	21.6	22.7	23.8	19.2	21.4	22.2	30.1	33.3	36.1	33.1	35.6	38.6
15-6	21.7	22.8	23.8	19.4	21.4	22.6	29.0	33.0	35.6	32.3	36.6	38.2
16-0	21.4	22.6	23.5	19.6	21.1	22.5	30.4	33.6	36.8	33.2	36.1	39.0
17-0	21.3	22.6	24.5	19.6	21.4	22.6	31.0	34.2	36.8	33.3	36.3	40.1
18-0	21.8	22.7	24.1	19.4	21.2	22.4	31.4	34.7	36.7	32.9	36.4	40.8

(data from McCammon, Human Growth and Development, 1970. Courtesy of Charles C Thomas,
Publisher, Springfield, Illinois)

TABLE   968

LOWER LIMB CIRCUMFERENCES (cm) FOR BLACK BOYS IN TEXAS

Age	Thigh				Calf			
	Lower Income		Middle Income		Lower Income		Middle Income	
(years)	Mean	S.D.	Mean	S.D.	Mean	S.D.	Mean	S.D.
10.1 - 11	40.5	2.95	39.0	1.84	28.7	2.20	28.1	1.72
11.1 - 12	41.7	2.69	42.8	4.04	29.9	1.51	31.4	5.47
12.1 - 13	44.0	6.29	45.2	5.15	30.8	4.22	31.8	3.42
13.1 - 14	44.7	4.33	46.0	4.86	31.2	3.21	32.4	2.54
14.1 - 15	45.1	3.68	47.9	3.48	33.3	5.15	33.5	2.56
15.1 - 16	47.5	3.84	50.5	2.67	34.5	4.67	35.2	1.74
16.1 - 17	50.7	3.13	53.0	3.17	35.5	2.16	36.6	2.58
17.1 - 18	52.3	3.15	50.2	3.39	36.0	2.14	35.5	1.75

(data from Schutte, 1979)

TABLE 969

KNEE BREADTH (cm) IN U.S. CHILDREN

Age (years)	Mean	S.D.	5	10	25	50	75	90	95
						Percentiles			
					BOYS				
12	9.0	0.56	8.1	8.3	8.6	9.1	9.4	9.8	10.0
13	9.3	0.58	8.3	8.6	9.0	9.3	9.8	10.2	10.3
14	9.5	0.52	8.6	8.8	9.2	9.6	9.8	10.2	10.4
15	9.6	0.47	8.7	9.0	9.2	9.6	9.9	10.3	10.4
16	9.6	0.58	8.7	9.0	9.2	9.6	9.9	10.2	10.4
17	9.6	0.49	8.8	0.0	9.3	9.7	9.9	10.3	10.5
					GIRLS				
12	8.6	0.50	7.7	8.0	8.2	8.6	8.9	9.3	9.4
13	8.7	0.50	7.8	8.0	8.3	8.7	9.0	9.3	9.6
14	8.8	0.50	8.0	8.1	8.4	8.7	9.1	9.4	9.7
15	8.8	0.54	8.0	8.2	8.5	8.8	9.2	9.4	9.8
16	8.8	0.58	8.0	8.1	8.5	8.8	9.2	9.6	9.9
17	8.8	0.53	8.0	8.2	8.5	8.8	9.2	9.5	9.7

(data from Cycle III of the Health Examination Survey by the National Center of Health Statistics 1966-1970)

TABLE 970

KNEE BREADTH (cm) IN U.S. WHITE CHILDREN

Age (years)	Mean	S.D.	Percentiles						
			5	10	25	50	75	90	95

BOYS

Age (years)	Mean	S.D.	5	10	25	50	75	90	95
12	9.0	0.57	8.1	8.3	8.6	9.1	9.4	9.8	10.1
13	9.3	0.56	8.4	8.6	9.0	9.3	9.8	10.2	10.3
14	9.5	0.51	8.6	8.9	9.2	9.6	9.8	10.2	10.4
15	9.6	0.46	8.8	9.0	9.2	9.6	9.9	10.3	10.4
16	9.6	0.47	8.8	9.0	9.2	9.6	9.9	10.3	10.4
17	9.7	0.49	8.8	9.0	9.3	9.7	10.0	10.3	10.6

GIRLS

Age (years)	Mean	S.D.	5	10	25	50	75	90	95
12	8.6	0.48	7.7	8.0	8.2	8.6	8.8	9.2	9.4
13	8.7	0.48	7.8	8.1	8.3	8.7	9.1	9.3	9.6
14	8.8	0.49	8.0	8.1	8.4	8.7	9.1	9.4	9.7
15	8.8	0.55	8.0	8.2	8.5	8.8	9.2	9.4	9.8
16	8.9	0.56	8.0	8.2	8.5	8.8	9.2	9.6	9.8
17	8.8	0.51	8.1	8.2	8.6	8.8	9.2	9.4	9.7

KNEE BREADTH (cm) IN U.S. NEGRO CHILDREN

Age (years)	Mean	S.D.	Percentiles						
			5	10	25	50	75	90	95

BOYS

Age (years)	Mean	S.D.	5	10	25	50	75	90	95
12	8.9	0.44	8.0	8.2	8.6	8.9	9.2	9.4	9.7
13	9.3	0.68	8.1	8.4	8.9	9.3	9.7	10.2	10.7
14	9.4	0.58	8.6	8.7	9.1	9.3	9.8	10.2	10.3
15	9.4	0.45	8.6	8.7	9.1	9.3	9.7	10.1	10.3
16	9.3	1.02	5.9	8.7	9.1	9.6	9.8	10.2	10.3
17	9.5	0.47	8.7	9.0	9.2	9.6	9.9	10.2	10.3

GIRLS

Age (years)	Mean	S.D.	5	10	25	50	75	90	95
12	8.7	0.58	7.7	8.0	8.2	8.6	9.1	9.4	9.8
13	8.6	0.60	7.6	7.8	8.2	8.6	8.9	9.3	9.8
14	8.8	0.50	8.0	8.1	8.5	8.8	9.2	9.5	9.8
15	8.8	0.49	8.0	8.1	8.4	8.8	9.2	9.4	9.7
16	8.8	0.70	8.0	8.1	8.3	8.7	9.2	9.8	10.6
17	8.9	0.64	7.9	8.1	8.3	8.9	9.3	9.7	9.8

(data from Cycle III of the Health Examination Survey by the National Center of Health Statistics 1966-1970)

TABLE 971

BICONDYLAR DIAMETER OF THE FEMUR
(mm) FOR PHILADELPHIA CHILDREN

Age (years)	WHITE BOYS		BLACK BOYS	
	Mean	S.D.	Mean	S.D.
6	75.04	4.64	72.94	5.06
7	78.14	5.12	75.99	4.21
8	81.14	4.95	79.53	4.40
9	83.99	5.66	82.81	5.74
10	87.92	6.66	85.22	7.11
11	89.47	6.17	87.69	6.92
12	90.75	6.39	88.97	6.27
13	--	--	92.19	5.80

Age (years)	WHITE GIRLS		BLACK GIRLS	
	Mean	S.D.	Mean	S.D.
6	70.63	3.41	69.99	4.61
7	73.65	4.11	73.46	4.64
8	75.95	4.65	76.67	4.98
9	78.79	5.19	78.85	5.71
10	81.52	4.93	82.29	6.51
11	83.83	4.60	85.25	6.90
12	--	--	87.39	7.87
13	--	--	87.14	6.93

Age is for completed years.

(data from Malina, unpublished)

TABLE   972

LOWER LIMB DIAMETERS (cm) FOR BLACK BOYS IN TEXAS

	Knee				Ankle			
Age (years)	Lower Income		Middle Income		Lower Income		Middle Income	
	Mean	S.D.	Mean	S.D.	Mean	S.D.	Mean	S.D.
10.1 - 11	8.5	0.41	8.2	0.28	6.9	0.40	6.7	0.25
11.1 - 12	8.7	0.34	8.8	0.49	6.9	0.28	7.0	0.31
12.1 - 13	8.7	0.41	9.1	0.62	7.0	0.28	7.3	0.35
13.1 - 14	9.0	0.53	9.1	0.34	7.3	0.42	7.2	0.33
14.1 - 15	9.0	0.60	9.4	0.37	7.1	0.42	7.6	0.35
15.1 - 16	9.4	0.41	9.4	0.35	7.5	0.27	7.5	0.44
16.1 - 17	9.6	0.57	9.4	0.32	7.8	0.45	7.6	0.32
17.1 - 18	9.5	0.36	9.5	0.40	7.5	0.27	7.7	0.39

(data from Schutte, 1979)

TABLE   973

INCREASE (mm) IN CALF BREADTH AND TISSUE BREADTHS IN
OHIO GIRLS BETWEEN 7.5 AND 12.5 YEARS OF AGE

	Total Breadth of Calf	Breadth of Fat	Breadth of Muscle	Breadth of Bone
	Early-Maturing			
Gain (mm)	26.5	4.7	15.2	6.6
Gain (%)	32.1	32.6	35.6	26.0
	Late-Maturing			
Gain (mm)	11.9	1.7	5.7	4.5
Gain (%)	15.0	14.9	13.0	18.7

(data from Reynolds, 1946)

TABLE   974

CALF DIMENSIONS (mm) IN OHIO GIRLS

Age (years)	Total Breadth of Calf		Breadth of Fat		Breadth of Muscle Mass		Breadth of Bone	
	Mean	S.D.	Mean	S.D.	Mean	S.D.	Mean	S.D.
Early-Maturing								
7.5	82.6	6.4	14.5	2.6	42.8	6.1	25.4	1.3
8.5	87.3	6.3	15.0	3.0	46.0	5.9	26.3	1.4
9.5	91.0	7.6	15.6	3.3	48.0	6.6	27.4	1.5
10.5	97.2	7.3	16.0	2.8	52.0	5.8	29.2	1.9
11.5	103.9	10.2	18.1	2.6	54.5	7.4	31.3	2.0
12.5	109.1	11.0	19.2	2.4	58.0	8.6	32.0	2.0
Late-Maturing								
7.5	79.3	6.2	11.6	3.6	43.6	4.3	24.1	2.2
8.5	80.9	5.8	12.1	3.5	44.4	4.0	24.4	2.0
9.5	83.3	6.4	12.2	3.1	45.5	4.5	25.6	1.9
10.5	85.9	6.4	13.0	3.1	46.2	4.4	26.7	2.4
11.5	88.8	8.0	13.4	4.0	47.3	5.1	28.1	2.7
12.5	91.2	6.4	13.3	3.4	49.3	4.9	28.6	2.5

(data from Reynolds, 1946)

TABLE   975

TOTAL CALF WIDTH (cm)
IN BOSTON CHILDREN

Age (years)	Boys Percentiles			Girls Percentiles		
	10	50	90	10	50	90
6	7.04	7.87	8.71	7.24	8.20	8.72
7	7.21	8.09	9.22	7.52	8.46	9.04
8	7.62	8.43	9.46	7.73	8.60	9.54
9	7.82	8.66	9.88	8.02	8.70	10.06
10	7.83	8.79	10.17	7.80	8.75	10.80

(data from Stuart and Dwinell, 1942)

TABLE   976

ANKLE BREADTH (cm) IN U.S. CHILDREN

| | | | | | | Percentiles | | | |
Age (years)	Mean	S.D.	5	10	25	50	75	90	95
Boys									
12	6.7	0.40	6.1	6.2	6.5	6.7	7.1	7.3	7.4
13	7.0	0.43	6.2	6.4	6.7	7.0	7.3	7.6	7.8
14	7.1	0.41	6.4	6.6	6.8	7.2	7.4	7.7	7.8
15	7.2	0.40	6.5	6.6	6.9	7.2	7.4	7.8	7.9
16	7.2	0.41	6.5	6.6	6.9	7.2	7.6	7.8	7.9
17	7.2	0.41	6.6	6.6	7.0	7.2	7.5	7.8	7.9
Girls									
12	6.4	0.36	5.7	5.9	6.1	6.3	6.6	6.8	6.9
13	6.4	0.34	5.7	6.0	6.1	6.3	6.7	6.8	6.9
14	6.4	0.33	5.8	6.0	6.1	6.4	6.7	6.9	7.0
15	6.4	0.37	5.7	6.0	6.1	6.3	6.7	6.8	7.0
16	6.4	0.37	5.7	6.0	6.1	6.3	6.7	6.9	7.1
17	6.4	0.36	5.7	5.8	6.1	6.4	6.7	6.8	7.0

(data from Cycle III of the Health Examination Survey by the National Center for Health Statistics 1966-1970)

TABLE   977

ANKLE BREADTH (cm) IN WHITE U.S. CHILDREN

Age					Percentiles				
(years)	Mean	S.D.	5	10	25	50	75	90	95

Boys

12	6.7	0.41	6.1	6.2	6.5	6.7	7.1	7.3	7.4
13	7.0	0.43	6.2	6.4	6.6	7.0	7.3	7.6	7.8
14	7.1	0.42	6.4	6.6	6.8	7.2	7.4	7.7	7.8
15	7.2	0.39	6.6	6.6	6.9	7.2	7.4	7.8	7.9
16	7.2	0.41	6.5	6.6	6.9	7.2	7.6	7.8	7.9
17	7.2	0.41	6.6	6.6	7.0	7.2	7.6	7.8	7.9

Girls

12	6.3	0.36	5.7	5.9	6.1	6.3	6.6	6.8	6.9
13	6.4	0.34	5.7	6.0	6.1	6.3	6.7	6.8	6.9
14	6.4	0.33	5.8	6.0	6.1	6.3	6.7	6.9	7.1
15	6.4	0.38	5.7	6.0	6.1	6.3	6.7	6.9	7.1
16	6.4	0.38	5.7	6.0	6.1	6.3	6.7	6.9	7.1
17	6.4	0.36	5.7	5.8	6.1	6.4	6.7	6.8	6.9

TABLE   978

ANKLE BREADTH (cm) IN NEGRO U.S. CHILDREN

Age					Percentiles				
(years)	Mean	S.D.	5	10	25	50	75	90	95

Boys

12	6.7	0.33	6.1	6.2	6.6	6.7	6.9	7.2	7.3
13	7.0	0.46	6.3	6.5	6.7	7.1	7.4	7.7	7.8
14	7.1	0.57	6.5	6.6	6.8	7.1	7.3	7.6	7.8
15	7.1	0.42	6.4	6.5	6.8	7.1	7.4	7.8	7.8
16	7.2	0.41	6.6	6.6	6.8	7.2	7.4	7.8	7.9
17	7.2	0.42	6.5	6.7	7.0	7.2	7.4	7.8	8.0

Girls

12	6.4	0.35	5.7	6.0	6.2	6.4	6.7	6.9	7.0
13	6.4	0.38	5.7	6.0	6.1	6.3	6.7	6.9	7.1
14	6.4	0.31	6.0	6.0	6.2	6.4	6.7	6.8	6.9
15	6.4	0.31	5.7	6.0	6.1	6.3	6.6	6.8	6.9
16	6.4	0.31	5.8	6.0	6.2	6.4	6.7	6.8	6.9
17	6.4	0.35	5.7	6.0	6.1	6.4	6.7	6.9	7.0

(data from Cycle III of the Health Examination Survey by the National Center for
Health Statistics 1966-1970)

TABLE 979

FOOT WIDTH (cm) IN U.S. CHILDREN

Boys

Age (years)	5	10	25	50	75	90	95
6	5.7	6.1	6.3	6.6	7.2	7.7	7.8
7	6.1	6.1	6.4	7.0	7.5	7.8	7.9
8	6.1	6.3	6.7	7.3	7.6	7.9	8.3
9	6.3	6.6	7.1	7.5	7.8	8.5	8.7
10	6.5	7.0	7.3	7.6	8.2	8.6	8.8
11	7.0	7.1	7.4	7.9	8.5	8.8	9.2

Girls

Age (years)	5	10	25	50	75	90	95
6	5.3	5.7	6.2	6.5	6.8	7.4	7.6
7	6.0	6.1	6.3	6.7	7.2	7.7	7.8
8	6.1	6.2	6.5	7.1	7.5	7.8	7.9
9	6.1	6.3	6.8	7.3	7.7	8.2	8.6
10	6.3	6.6	7.2	7.5	7.9	8.6	8.8
11	6.6	7.1	7.3	7.8	8.4	8.8	8.9

(data from Malina et al., 1973)

TABLE 980

NORMAL RANGE OF FOOT ANGLES IN DEGREES WITH THE CHILDREN STANDING (Mean $\pm$ 1 Sigma);
DATA FROM OHIO CHILDREN

Age (years)	Angle			
	A	B	C	D
2½	121.5 to 133.5	14.5 to 20.5	137.5 to 149.5	20.0 to 28.0
3	121.5 to 135.5	15.5 to 22.5	137.0 to 149.0	20.0 to 28.0
3½	121.0 to 134.0	16.0 to 23.0	138.0 to 149.0	20.0 to 28.0
4	120.5 to 133.5	16.0 to 24.0	136.5 to 147.5	20.5 to 28.5
4½	120.5 to 133.5	15.5 to 21.5	133.5 to 146.5	21.0 to 30.0
5	121.0 to 133.0	16.0 to 22.0	135.5 to 146.5	21.5 to 29.5
5½	120.5 to 131.5	14.5 to 21.5	133.0 to 146.0	21.5 to 30.5
6	121.5 to 131.5	14.5 to 21.5	133.0 to 145.0	22.5 to 30.5
6½	121.5 to 131.5	14.0 to 22.0	133.5 to 144.5	22.5 to 29.5
7	121.0 to 133.0	13.5 to 20.5	133.5 to 144.5	22.0 to 30.0
7½	122.3 to 133.5	13.5 to 20.5	134.0 to 146.0	21.5 to 29.5
8	121.5 to 132.5	13.5 to 20.5	133.0 to 145.0	22.0 to 30.0
8½	121.5 to 132.5	13.0 to 19.0	134.0 to 147.0	21.5 to 29.5
9	121.0 to 133.0	12.5 to 18.5	135.0 to 145.0	22.0 to 29.0
9½	122.0 to 133.0	12.0 to 18.0	135.5 to 147.5	20.5 to 28.5
10	121.0 to 131.0	12.5 to 18.5	136.0 to 146.0	22.0 to 28.0
10½	121.5 to 132.5	11.5 to 17.5	135.5 to 146.5	22.0 to 29.0
11	120.0 to 132.0	11.0 to 18.0	138.0 to 148.0	21.0 to 27.0

A = height of medial aspect of longitudinal arch.
B = anterior medial part of longitudinal arch.
C = height of lateral aspect of longitudinal arch.
D = inclination of calcaneum with the horizontal.

(data from Robinow, Johnston and Anderson, 1943, The Journal of Pediatrics 23: 141-149)

UPPER LIMB MEASUREMENTS

TABLE 981

TOTAL ARM LENGTH   (mm) IN CLEVELAND CHILDREN

Age (years)	BOYS		GIRLS	
	Mean	S.D.	Mean	S.D.
2 years	346.28	14.54	344.94	15.59
2½ years	368.85	16.64	358.88	18.89
3 years	389.21	17.11	379.23	18.99
3½ years	404.26	17.56	396.10	19.60
4 years	422.76	19.22	414.32	18.98
4½ years	438.47	20.19	432.28	22.39
5 years	453.29	21.22	447.96	21.94
6 years	486.02	22.64	480.46	26.36
7 years	515.58	22.93	508.85	26.60
8 years	542.68	24.19	536.63	27.31
9 years	567.27	25.75	562.95	27.82
10 years	592.85	27.38	590.47	29.95
11 years	617.98	30.72	618.88	30.46
12 years	641.61	34.96	649.14	30.87
13 years	667.65	39.20	673.38	29.87
14 years	704.93	40.94	690.93	28.10
15 years	734.81	36.72	694.95	29.85
16 years	748.58	33.04	703.64	30.59
17 years	756.67	34.21	697.00	28.86

(data from Simmons, 1944)

TABLE 982

TOTAL ARM LENGTH (cm) IN PHILADELPHIA CHILDREN

Age (years)	White Boys		White Girls		Negro Boys		Negro Girls	
	Mean	S.D.	Mean	S.D.	Mean	S.D.	Mean	S.D.
7	52.0	2.6	50.3	4.3	53.4	3.3	53.4	3.3
8	55.0	2.5	53.8	3.3	56.2	4.1	55.0	3.4
9	57.8	2.9	56.9	3.1	58.1	3.6	59.0	3.6
10	60.4	3.0	58.8	3.7	60.7	4.8	61.4	3.5
11	62.4	3.8	61.5	3.5	63.2	4.2	64.0	4.5
12	64.7	3.6	64.2	4.1	65.3	4.0	66.6	4.3
13	68.7	3.9	66.4	3.4	71.4	4.3	69.6	3.5
14	70.4	4.3	67.2	3.1	73.0	5.8	71.4	4.0

UPPER ARM LENGTH (cm) IN PHILADELPHIA CHILDREN

Age (years)	White Boys		White Girls		Negro Boys		Negro Girls	
	Mean	S.D.	Mean	S.D.	Mean	S.D.	Mean	S.D.
7	21.9	1.4	21.7	1.5	21.8	1.6	21.9	1.6
8	23.0	1.4	22.6	1.6	22.7	2.5	22.5	1.6
9	24.3	1.5	23.7	1.6	24.0	1.7	24.2	1.7
10	25.6	1.7	24.9	1.8	25.0	2.3	25.5	1.9
11	26.5	2.0	26.0	1.8	26.3	2.2	27.1	2.3
12	27.5	2.0	27.3	1.5	27.3	1.7	28.2	2.1
13	28.9	2.0	28.3	1.6	29.6	2.3	30.0	1.9
14	30.6	1.7	28.6	1.8	31.1	2.5	30.0	2.1

(data from Krogman, 1970)

TABLE   983

UPPER ARM LENGTH (cm) IN U.S. CHILDREN

Boys

Age (years)	5	10	25	50	75	90	95
6	21.6	22.1	22.7	23.7	24.7	25.6	26.1
7	22.6	23.2	24.1	25.1	25.9	26.8	27.6
8	24.0	24.5	25.4	26.4	27.3	28.1	28.7
9	25.1	25.5	26.4	27.5	28.6	29.7	30.4
10	25.7	26.4	27.5	28.6	29.8	30.7	31.4
11	27.2	27.8	28.8	30.1	31.2	32.4	33.1

Girls

6	21.2	21.6	22.5	23.6	24.5	25.4	25.8
7	22.4	23.0	23.7	24.7	25.7	26.6	27.1
8	23.5	24.2	25.2	26.2	27.3	28.4	29.1
9	24.9	25.5	26.5	27.5	28.7	29.9	30.6
10	26.0	26.6	27.7	29.0	30.2	31.4	31.9
11	27.5	28.2	29.2	30.5	32.0	33.1	33.7

(data from Malina et al., 1973)

TABLE   984

UPPER ARM LENGTH (cm) IN U.S. CHILDREN

Age			Percentiles						
(years)	Mean	S.D.	5	10	25	50	75	90	95

BOYS

12	28.2	1.90	25.3	25.9	27.0	28.3	29.4	30.7	31.6
13	29.8	2.08	26.5	27.1	28.4	29.8	31.2	32.5	33.4
14	31.2	2.06	27.8	28.7	29.8	31.3	32.5	33.8	34.4
15	32.1	1.97	29.2	29.8	30.7	32.1	33.4	34.5	35.2
16	32.7	1.88	29.4	30.3	31.6	32.8	34.0	35.3	35.9
17	33.1	2.15	30.0	30.7	31.6	33.1	34.4	35.4	36.2

GIRLS

12	28.8	1.99	25.4	26.4	27.4	28.7	29.8	31.2	32.2
13	29.5	1.81	26.6	27.2	28.3	29.5	30.7	31.8	32.4
14	30.0	1.70	27.4	27.9	28.9	29.9	31.1	32.2	33.0
15	30.2	1.87	27.2	27.8	29.0	30.2	31.4	32.6	33.4
16	30.3	1.80	27.6	28.2	29.2	30.3	31.6	32.7	33.1
17	30.2	1.69	27.6	28.2	29.2	30.2	31.3	32.3	33.0

(data from Cycle III of the Health Examination Survey by the National Center for Health Statistics 1966-1970)

TABLE 985

UPPER ARM LENGTH (cm) IN U.S. WHITE CHILDREN

Age (years)	Mean	S.D.	Percentiles						
			5	10	25	50	75	90	95
BOYS									
12	28.2	1.89	25.3	25.9	26.9	28.2	29.4	30.6	31.5
13	29.8	2.11	26.4	27.0	28.3	29.7	31.2	32.5	33.6
14	31.2	2.05	27.7	28.7	29.8	31.3	32.5	33.8	34.3
15	32.1	2.01	29.1	29.8	30.8	32.1	33.4	34.6	35.2
16	32.7	1.88	29.5	30.4	31.6	32.8	33.9	35.3	36.0
17	33.1	2.21	29.8	30.7	31.6	33.1	34.4	35.6	36.2
GIRLS									
12	28.6	1.98	25.3	26.2	27.4	28.6	29.8	31.1	31.9
13	29.4	1.81	26.5	27.1	28.3	29.3	30.5	31.7	38.3
14	29.9	1.68	27.4	27.8	28.9	29.8	31.0	32.0	32.8
15	30.2	1.88	27.1	27.7	28.9	30.1	31.3	32.5	33.2
16	30.2	1.79	27.4	28.1	29.2	30.2	31.4	32.5	33.1
17	30.1	1.63	27.5	28.1	29.1	30.2	31.2	32.2	32.8

UPPER ARM LENGTH (cm) IN U.S. NEGRO CHILDREN

Age (years)	Mean	S.D.	Percentiles						
			5	10	25	50	75	90	95
BOYS									
12	28.4	2.01	25.3	25.7	27.3	28.6	29.6	31.1	31.6
13	30.1	1.84	27.1	27.8	28.7	30.1	31.2	32.5	32.8
14	31.3	2.16	28.0	28.7	29.5	31.3	33.1	34.5	34.8
15	32.2	1.80	29.3	30.1	30.7	32.1	33.4	34.4	34.9
16	33.0	1.88	29.2	30.3	32.2	33.1	34.4	35.1	35.9
17	33.1	1.67	30.6	31.0	31.8	33.0	34.4	35.0	36.3
GIRLS									
12	29.6	1.81	26.8	27.1	28.6	29.5	30.5	32.1	32.5
13	30.3	1.64	27.5	28.0	29.2	30.4	31.2	32.3	33.0
14	30.6	1.69	28.0	28.8	20.5	30.4	31.8	33.0	33.4
15	30.9	1.70	27.9	28.8	29.9	30.5	31.9	33.2	33.8
16	30.9	1.69	28.0	28.6	29.6	31.0	32.0	33.0	33.4
17	30.9	1.91	28.2	28.4	29.5	30.9	32.2	33.3	34.0

(data from Cycle III of Health Examination Survey 1966-1970 by National Center for Health Statistics)

TABLE   986

UPPER ARM LENGTH (cm)
FOR CHILDREN IN 8 STATES

| Age | Boys | | Girls | |
(months)	Mean	S.D.	Mean	S.D.
0 - 3	10.9	0.9	10.6	0.8
4 - 6	12.9	0.7	12.4	0.9
7 - 9	13.9	0.7	13.5	0.7
10 - 12	14.7	0.6	14.4	0.7
13 - 18	15.6	0.9	15.4	0.7
19 - 24	17.2	0.9	16.9	1.0
25 - 30	18.1	1.2	18.0	0.9
31 - 36	19.2	1.0	18.7	1.0
37 - 42	19.9	0.9	19.5	1.1
43 - 48	20.4	1.1	20.3	1.1
49 - 54	21.4	1.1	21.1	1.0
55 - 60	22.1	1.3	21.8	1.2
61 - 66	22.8	1.2	22.4	1.2
67 - 72	23.4	1.3	23.4	1.2
73 - 78	24.1	1.2	23.9	1.3
79 - 84	24.9	1.3	24.6	1.3
85 - 96	26.1	1.6	25.6	1.3
97 - 108	27.2	1.4	27.3	1.4
109 - 120	28.3	1.5	28.1	1.3
121 - 132	29.9	1.3	29.8	1.8
133 - 144	31.2	1.7	30.8	1.7
145 - 156	31.8	2.0	33.2	1.7

(data from Snyder et al., 1975)

TABLE 987

UPPER ARM LENGTH (mm) IN CLEVELAND CHILDREN

Age (years)	BOYS		GIRLS	
	Mean	S.D.	Mean	S.D.
2 years	130.84	7.94	129.51	10.03
2½ years	138.97	9.99	134.96	9.94
3 years	148.65	10.42	144.83	11.38
3½ years	155.15	10.60	151.40	10.66
4 years	163.05	10.56	161.94	11.30
4½ years	170.98	11.59	169.43	12.89
5 years	178.40	12.32	178.06	13.17
6 years	194.14	12.51	193.96	15.28
7 years	207.23	15.04	208.40	16.33
8 years	219.75	13.38	219.77	15.11
9 years	230.51	13.95	231.32	15.45
10 years	239.94	15.41	241.90	17.09
11 years	251.26	16.66	254.22	17.53
12 years	261.78	17.78	268.13	16.61
13 years	272.62	19.60	280.09	17.69
14 years	289.83	19.12	289.54	16.58
15 years	302.02	17.04	290.86	17.65
16 years	308.97	17.24	296.93	18.46
17 years	312.50	19.64	292.64	14.63

(data from Simmons, 1944)

TABLE 988

UPPER-ARM LENGTH (in)
IN OKLAHOMA CHILDREN

Age (years)	BOYS		GIRLS	
	Mean	S.D.	Mean	S.D.
5	8.29	0.49	8.04	0.51
6	8.74	0.79	8.49	0.71
7	9.38	0.68	9.21	0.67
8	10.00	0.51	9.78	0.67
9	10.36	0.57	10.52	0.52
10	11.21	0.76	10.96	0.71
11	11.50	0.68	11.70	0.79
12	11.89	0.96	11.83	1.05
13	11.98	0.84	11.96	0.74
14	12.22	0.83	11.86	0.68
15	13.04	0.77	12.07	0.51
16	13.05	0.76	12.16	0.69
17	13.29	0.65	12.56	0.72
18	13.82	0.88	12.44	0.47

(data from Swearingen and Young, 1965)

TABLE   989

UPPER ARM AND FOREARM LENGTH
(cm) IN BRITISH GIRLS

Age*	Upper arm		Forearm	
	Mean	S.D.	Mean	S.D.
5	19.12	1.13	19.72	0.89
6	20.39	1.28	15.73	0.98
7	21.58	1.36	16.55	1.02
8	22.80	1.34	17.45	1.10
9	23.98	1.38	18.27	1.10
10	25.07	1.53	19.27	1.14
11	26.28	1.68	20.01	1.22
12	27.66	1.69	21.07	1.25
13	28.83	1.64	21.90	1.22
14	29.62	1.55	22.52	1.19
15	30.31	1.64	22.71	1.23
16	30.19	1.65	22.69	1.13
17	30.29	1.40	22.70	1.03
18	30.83	1.55	22.92	1.30

* Age intervals are 4.76-5.24 years, etc. The measurements were made from
  photographs.

(data from Marshall and Ahmed, 1976, Annals of Human Biology 3:61-70)

TABLE 990

UPPER ARM AND FOREARM LENGTH (cm) IN RELATION
TO SITTING HEIGHT (cm) IN BRITISH GIRLS

Sitting height	Upper Arm		Forearm	
(cm)	Mean	S.D.	Mean	S.D.
60.0-62.4	19.48	0.88	14.95	0.78
62.5-64.9	20.22	1.10	15.67	0.86
65.0-67.4	21.58	1.18	16.56	1.00
67.5-69.9	22.77	1.24	17.49	1.12
70.0-72.4	24.04	1.26	18.42	1.02
72.5-74.9	25.23	1.31	19.21	1.05
75.0-77.4	26.57	1.26	20.18	1.02
77.5-79.9	27.57	1.20	21.03	0.94
80.0-82.4	28.65	1.30	21.82	1.03
82.5-84.9	29.34	1.39	22.21	1.06
85.0-87.4	30.07	1.22	22.62	0.97
87.5-89.9	30.84	1.36	23.22	1.01
90.0-92.4	31.87	1.16	23.58	1.04

(data from Marshall and Ahmed, 1976, Annals of Human Biology 3:61-70)

TABLE 991

FOREARM LENGTH (cm) IN U.S. CHILDREN

Age (years)	Mean	S.D.	Percentiles						
			5	10	25	50	75	90	95
BOYS									
12	22.1	1.66	19.4	20.0	21.1	22.1	23.2	24.2	24.8
13	23.2	1.81	20.4	20.9	21.8	23.2	24.4	25.5	26.3
14	24.3	1.79	21.4	22.1	23.2	24.4	25.4	26.5	27.1
15	25.1	1.97	22.3	22.9	24.0	25.0	26.2	27.2	28.0
16	25.4	1.63	22.6	23.2	24.3	25.4	26.5	27.5	28.0
17	25.7	2.15	22.7	23.5	24.4	25.6	26.8	27.8	28.7
GIRLS									
12	22.2	1.66	19.5	20.2	21.0	22.1	23.2	24.2	24.7
13	22.6	1.59	20.1	20.6	21.6	22.6	23.5	24.7	25.2
14	23.0	1.60	20.5	21.0	21.9	23.0	24.1	25.0	25.5
15	23.2	1.61	20.7	21.1	22.1	23.1	24.2	25.3	25.9
16	23.1	1.58	20.7	21.2	22.1	23.1	24.2	25.1	25.7
17	23.1	1.49	20.6	21.2	22.1	23.0	24.1	25.0	25.4

(data from Cycle III of Health Examination Survey 1966-1970 by National Center for Health Statistics)

TABLE 992

FOREARM LENGTH (cm) IN U.S. CHILDREN

Age (years)	Percentiles						
	5	10	25	50	75	90	95
BOYS							
6	16.6	17.1	17.5	18.3	18.9	19.7	20.2
7	17.4	18.0	18.5	19.3	20.1	20.8	21.4
8	18.2	18.6	19.4	20.2	20.9	21.8	22.3
9	19.1	19.5	20.3	21.2	22.0	22.8	23.5
10	19.8	20.3	21.1	22.0	22.8	23.7	24.2
11	20.6	21.2	22.1	23.0	23.9	24.8	25.4
GIRLS							
6	16.1	16.3	17.1	17.8	18.6	19.4	19.7
7	17.0	17.3	18.1	18.7	19.6	20.5	20.8
8	17.8	18.2	19.0	19.8	20.7	21.5	21.9
9	19.0	19.2	20.0	20.8	21.8	22.8	23.4
10	19.6	20.1	20.8	21.8	22.9	23.8	24.5
11	20.6	21.2	22.1	23.0	24.2	25.1	25.8

(data from Malina et al., 1973)

TABLE   993

FOREARM LENGTH (cm) IN U.S. WHITE CHILDREN

Age (years)	Mean	S.D.	Percentiles						
			5	10	25	50	75	90	95
					BOYS				
12	22.0	1.67	19.2	20.0	21.0	22.0	23.1	24.0	24.8
13	23.0	1.76	20.3	20.9	21.8	23.1	24.2	25.2	26.0
14	24.2	1.72	21.4	22.1	23.2	24.3	25.3	26.2	27.0
15	24.9	1.95	22.2	22.8	23.9	24.9	26.0	27.0	27.7
16	25.2	1.60	22.4	23.1	24.1	25.3	26.3	27.4	27.9
17	25.5	2.18	22.7	23.4	24.4	25.6	26.7	27.7	28.6
					GIRLS				
12	22.0	1.62	19.4	20.0	20.9	22.0	23.0	24.0	24.5
13	22.4	1.50	20.0	20.5	21.5	22.4	23.3	24.3	24.8
14	22.8	1.53	20.5	20.9	21.8	22.8	23.9	24.7	25.2
15	23.0	1.59	20.6	21.0	22.0	23.0	23.9	25.1	25.8
16	23.0	1.51	20.6	21.2	22.0	23.0	23.9	24.8	25.3
17	22.9	1.42	20.6	21.1	22.0	22.8	23.9	24.8	25.3

FOREARM LENGTH (cm) IN U.S. NEGRO CHILDREN

Age (years)	Mean	S.D.	Percentiles						
			5	10	25	50	75	90	95
					BOYS				
12	22.8	1.41	20.3	20.8	21.9	22.8	23.7	24.6	25.0
13	24.0	1.96	20.9	21.5	22.4	23.8	25.5	26.8	27.6
14	25.0	2.10	21.4	22.7	23.8	25.1	26.0	27.7	28.5
15	26.0	1.93	23.9	24.1	24.6	25.8	27.2	28.6	29.1
16	26.4	1.44	24.3	24.5	25.2	26.2	27.4	28.0	29.0
17	26.5	1.78	24.3	24.4	25.2	26.2	27.6	28.7	29.7
					GIRLS				
12	23.1	1.54	20.8	21.2	22.3	23.0	24.1	25.2	25.9
13	23.8	1.60	20.7	21.8	23.9	24.1	24.8	25.6	26.2
14	24.0	1.62	21.0	21.8	23.0	24.4	25.3	26.0	26.3
15	24.2	1.37	21.4	22.1	23.4	24.4	25.0	25.9	26.2
16	24.2	1.60	21.2	22.1	23.0	24.4	25.4	26.2	26.8
17	24.1	1.57	21.7	22.3	23.0	23.9	25.0	25.9	27.1

(data from Cycle III of the Health Examination Survey by the National Center of Health Statistics 1966-1970)

TABLE 994

FOREARM LENGTH (cm)
FOR CHILDREN OF 8 STATES

Age (months)	Boys		Girls	
	Mean	S.D.	Mean	S.D.
0 - 3	15.0	1.1	14.7	1.2
4 - 6	17.4	1.0	16.5	1.1
7 - 9	18.7	0.8	18.2	0.7
10 - 12	19.7	0.9	19.1	0.9
13 - 18	21.0	1.1	20.3	0.8
19 - 24	22.7	1.5	22.2	1.4
25 - 30	24.0	1.5	23.5	1.1
31 - 36	25.3	1.5	24.7	1.5
37 - 42	26.0	1.3	25.6	1.3
43 - 48	26.6	1.6	26.3	1.3
49 - 54	27.9	1.4	27.2	1.4
55 - 60	28.7	1.5	28.0	1.4
61 - 66	29.5	1.6	28.7	1.5
67 - 72	30.2	1.5	29.7	1.5
73 - 78	31.0	1.4	30.3	1.6
79 - 84	32.0	1.7	31.2	1.5
85 - 96	33.3	1.9	32.4	1.7
97 - 108	34.5	1.8	34.4	2.0
109 - 120	36.0	2.0	35.4	1.8
121 - 132	38.0	1.8	37.5	2.3
133 - 144	39.4	2.1	39.0	2.3
145 - 156	40.1	2.5	41.5	2.0

(data from Snyder et al., 1975)

TABLE 995

FOREARM LENGTH (mm) IN CLEVELAND CHILDREN

Age (years)	BOYS		GIRLS	
	Mean	S.D.	Mean	S.D.
2 years	117.57	6.18	116.30	6.58
2½ years	123.81	6.88	120.56	7.36
3 years	130.84	6.41	126.20	7.23
3½ years	135.63	7.30	132.89	7.98
4 years	141.27	8.32	137.35	7.32
4½ years	146.28	8.69	142.98	7.91
5 years	151.02	8.64	148.55	8.42
6 years	162.58	9.09	158.27	9.77
7 years	171.47	10.40	167.65	10.03
8 years	180.50	10.25	176.40	10.37
9 years	188.70	10.19	184.88	10.71
10 years	198.29	10.77	195.39	10.82
11 years	206.29	11.92	204.77	11.28
12 years	214.17	12.53	214.75	12.07
13 years	222.59	15.80	222.47	11.64
14 years	235.04	14.90	227.96	10.77
15 years	245.54	14.38	228.72	10.98
16 years	250.81	12.44	230.00	12.32
17 years	253.00	11.87	230.68	12.50

(data from Simmons, 1944)

TABLE 996

FOREARM LENGTH (in)
IN OKLAHOMA CHILDREN

Age (years)	BOYS		GIRLS	
	Mean	S.D.	Mean	S.D.
5	11.46	0.51	11.31	0.32
6	12.10	0.76	11.77	0.62
7	12.78	0.69	12.49	0.73
8	13.58	0.65	13.03	0.63
9	13.89	0.66	13.90	0.62
10	14.72	0.79	14.50	0.72
11	15.16	0.84	15.35	0.89
12	16.21	1.20	16.03	0.74
13	16.83	0.92	16.48	0.77
14	17.36	1.06	16.78	0.76
15	18.52	0.97	16.71	0.67
16	18.36	1.28	16.94	0.83
17	18.68	0.74	17.20	0.97
18	19.19	0.66	17.17	0.90

(data from Swearingen and Young, 1965)

TABLE 997

FOREARM LENGTH (cm) IN PHILADELPHIA CHILDREN

Age (years)	White Boys		White Girls		Negro Boys		Negro Girls	
	Mean	S.D.	Mean	S.D.	Mean	S.D.	Mean	S.D.
7	16.6	1.8	16.7	1.4	17.3	2.0	17.1	1.8
8	18.0	1.3	17.6	1.4	18.6	1.9	17.9	1.9
9	19.0	1.3	18.6	1.5	19.4	1.8	19.6	1.9
10	20.0	1.4	19.4	1.7	20.0	1.4	20.1	1.3
11	20.8	1.7	20.3	1.6	20.9	1.8	20.7	1.9
12	21.6	1.5	21.4	1.6	21.8	1.8	21.8	1.6
13	22.7	1.7	21.9	1.4	23.5	1.7	22.3	1.2
14	23.3	1.8	22.1	1.4	23.7	2.2	23.4	1.5

(data from Krogman, 1970)

TABLE 998

HAND LENGTH (mm) IN CLEVELAND CHILDREN

Age	BOYS		GIRLS	
	Mean	S.D.	Mean	S.D.
3 mos.	72.94	4.91	69.68	4.99
6 "	79.93	5.74	76.21	5.40
9 "	85.66	5.89	81.55	5.95
12 "	91.45	6.34	87.01	6.95
18 "	97.51	6.37	94.55	5.75
2 yrs.	103.58	6.19	100.95	5.64
2½ "	107.44	6.16	104.76	6.28
3 "	110.92	6.26	108.88	6.36
3½ "	114.78	6.03	112.77	5.72
4 "	118.63	6.47	116.18	6.06
4½ "	121.46	6.39	119.58	6.12
5 "	124.78	6.49	122.43	7.07
6 "	129.75	7.37	128.44	7.71
7 "	136.37	7.00	135.42	8.00
8 "	143.64	7.54	141.58	7.59
9 "	149.62	8.19	146.81	8.14
10 "	155.38	8.47	153.94	9.51
11 "	160.54	9.13	161.79	10.14
12 "	166.75	10.59	167.81	9.12
13 "	172.98	11.84	172.91	8.41
14 "	182.05	12.22	175.39	8.50
15 "	189.58	11.64	176.58	9.09
16 "	192.70	9.77	178.41	9.26
17 "	194.17	8.62	177.78	7.16

(data from Simmons, 1944)

TABLE 999

PERCENTILES FOR HAND LENGTH (cm) IN UNITED STATES CHILDREN

Age (years)	5	10	25	50	75	90	95
				Boys			
6	11.9	12.1	12.5	13.2	13.6	14.0	14.5
7	12.2	12.5	13.1	13.6	14.2	14.7	14.9
8	12.6	13.1	13.6	14.3	14.8	15.4	15.7
9	13.2	13.6	14.2	14.6	15.3	15.8	16.2
10	13.8	14.1	14.5	15.2	15.7	16.4	16.7
11	14.2	14.5	15.2	15.7	16.4	16.8	17.4
				Girls			
6	11.4	11.9	12.3	12.8	13.5	14.0	14.5
7	12.1	12.3	13.0	13.4	13.9	14.6	14.8
8	12.5	13.0	13.4	13.9	14.6	15.2	15.6
9	13.2	13.4	14.1	14.6	15.3	15.9	16.4
10	13.6	14.1	14.5	15.3	15.8	16.6	16.8
11	14.3	14.7	15.3	15.9	16.7	17.4	17.8

(data from Malina et al., 1973)

TABLE 1000

HAND LENGTH (cm)
IN CHILDREN OF 8 STATES

| Age | Boys | | Girls | |
(months)	Mean	S.D.	Mean	S.D.
0 - 3	6.8	0.6	6.7	0.6
4 - 6	7.8	0.6	7.5	0.6
7 - 9	8.5	0.5	8.2	0.5
10 - 12	8.8	0.4	8.8	0.5
13 - 18	9.3	0.6	9.1	0.6
19 - 24	9.9	0.5	9.7	0.7
25 - 30	10.3	0.7	10.1	0.7
31 - 36	10.8	0.7	10.5	0.7
37 - 42	11.0	0.6	10.9	0.6
43 - 48	11.3	0.7	11.1	0.6
49 - 54	11.7	0.6	11.4	0.6
55 - 60	12.0	0.7	11.8	0.6
61 - 66	12.3	0.7	12.1	0.7
67 - 72	12.6	0.6	12.4	0.7
73 - 78	12.9	0.6	12.7	0.8
79 - 84	13.2	0.7	13.0	0.7
85 - 96	13.8	0.8	13.6	0.8
97 - 108	14.2	0.8	14.3	0.8
109 - 120	14.7	0.8	14.5	0.8
121 - 132	15.4	0.8	15.3	0.9
133 - 144	15.7	1.0	15.8	0.9
145 - 156	15.9	1.0	16.8	0.9

(data from Snyder et al., 1975)

TABLE 1001

ARM SEGMENTS (in)
FOR CHILDREN IN MICHIGAN

Age (years)	Sex	Upper arm length		Elbow to wrist		Wrist to knuckle		Wrist to finger tip		Hand width	
		Mean	S.D.	Mean	S.D.	Mean	S.D.	Mean	S.D.	Mean	S.D.
5	Boys	8.2	0.44	6.7	0.28	2.7	0.30	5.5	0.26	2.4	0.10
	Girls	7.9	0.68	6.3	0.26	2.5	0.30	5.2	0.28	2.3	0.10
6	Boys	8.3	0.64	6.7	0.35	2.7	0.30	5.5	0.29	2.4	0.17
	Girls	8.3	0.73	6.6	0.25	2.6	0.30	5.4	0.30	2.3	0.17
7	Boys	8.9	0.85	7.2	0.36	2.8	0.30	5.8	0.28	2.6	0.54
	Girls	8.9	0.62	7.0	0.32	2.8	0.30	5.6	0.31	2.4	0.20
8	Boys	9.4	0.65	7.5	0.41	2.0	0.30	6.0	0.30	2.6	0.17
	Girls	9.4	0.69	7.4	0.45	2.9	0.30	5.9	0.30	2.6	0.10
9	Boys	10.0	0.60	7.9	0.35	3.1	0.30	6.3	0.32	2.7	0.17
	Girls	9.6	0.70	7.7	0.33	2.9	0.30	6.1	0.32	2.6	0.17
10	Boys	10.1	0.86	8.2	0.55	3.2	0.30	6.5	0.33	2.8	0.22
	Girls	10.2	1.39	8.1	0.47	3.1	0.40	6.4	0.37	2.7	0.22
11	Boys	10.7	0.84	8.6	0.55	3.3	0.40	6.7	0.36	2.9	0.10
	Girls	10.9	0.91	8.7	0.58	3.3	0.40	6.7	0.40	2.8	0.10
12	Boys	11.0	1.01	8.9	0.60	3.3	0.40	6.9	0.42	2.9	0.10
	Girls	11.4	0.78	9.0	0.16	3.4	0.40	7.1	0.36	2.9	0.10
13	Boys	11.6	1.04	9.4	0.62	3.5	0.50	7.2	0.47	3.1	0.26
	Girls	11.8	0.72	9.3	0.60	3.5	0.30	7.2	0.33	3.0	0.17
14	Boys	12.3	0.91	9.8	0.75	3.7	0.50	7.7	0.48	3.2	0.30
	Girls	12.2	0.66	9.6	0.57	3.6	0.30	7.4	0.33	3.0	0.14
15	Boys	13.1	0.99	10.4	0.59	3.9	0.50	8.0	0.46	3.4	0.10
	Girls	12.3	0.74	9.7	0.55	3.6	0.40	7.4	0.36	3.0	0.10
16	Boys	13.3	0.82	10.7	0.54	3.9	0.40	8.1	0.38	3.4	0.22
	Girls	12.2	0.71	9.7	0.49	3.6	0.40	7.4	0.36	3.0	0.22

TABLE 1001 (continued)

ARM SEGMENTS (in)
FOR CHILDREN IN MICHIGAN

Age (years)	Sex	Upper arm length		Elbow to wrist		Wrist to knuckle		Wrist to finger tip		Hand width	
		Mean	S.D.	Mean	S.D.	Mean	S.D.	Mean	S.D.	Mean	S.D.
17	Boys	13.6	0.82	10.8	0.65	4.0	0.30	8.2	0.34	3.5	0.20
	Girls	12.3	0.72	9.8	0.55	3.6	0.30	7.4	0.28	3.0	0.10
18	Boys	13.4	0.79	10.8	0.57	3.9	0.30	8.1	0.34	3.5	0.22
	Girls	12.3	0.69	9.8	0.50	3.6	0.30	7.4	0.34	3.0	0.22

(data from Martin, 1955)

TABLE 1002

HAND LENGTH (cm) IN PHILADELPHIA CHILDREN

Age (years)	Boys White		Girls White		Boys Negro		Girls Negro	
	Mean	S.D.	Mean	S.D.	Mean	S.D.	Mean	S.D.
7	13.5	1.0	13.3	1.1	14.3	1.4	14.2	1.5
8	14.0	1.2	13.6	1.4	15.3	1.3	14.6	1.5
9	14.4	1.2	14.0	1.2	15.0	1.8	15.2	1.2
10	15.2	1.1	14.8	1.6	15.8	2.0	15.8	1.3
11	15.4	1.4	15.4	1.2	16.0	1.4	16.3	1.6
12	15.9	1.1	16.2	1.3	16.1	1.2	16.6	1.3
13	17.3	1.4	16.6	1.3	18.3	1.2	17.5	1.3
14	18.2	2.1	17.8	1.3	18.2	1.8	18.0	2.8

(data from Krogman, 1970)

TABLE 1003

THIRD FINGER LENGTH (cm)
IN CHILDREN OF 8 STATES

Age	Boys		Girls	
(months)	Mean	S.D.	Mean	S.D.
0 - 3	3.0	0.4	2.9	0.4
4 - 6	3.4	0.4	3.2	0.4
7 - 9	3.7	0.5	3.5	0.3
10 - 12	3.6	0.4	3.6	0.3
13 - 18	3.9	0.3	3.8	0.3
19 - 24	4.2	0.4	4.2	0.4
25 - 30	4.3	0.4	4.3	0.3
31 - 36	4.5	0.3	4.5	0.4
37 - 42	4.6	0.3	4.6	0.3
43 - 48	4.7	0.4	4.7	0.3
49 - 54	4.9	0.3	4.8	0.3
55 - 60	5.0	0.4	5.0	0.3
61 - 66	5.1	0.4	5.2	0.4
67 - 72	5.2	0.4	5.3	0.4
73 - 78	5.4	0.4	5.4	0.4
79 - 84	5.6	0.4	5.6	0.3
85 - 96	5.8	0.4	5.8	0.4
97 - 108	6.0	0.4	6.1	0.4
109 - 120	6.1	0.4	6.2	0.4
121 - 132	6.5	0.3	6.5	0.4
133 - 144	6.6	0.5	6.8	0.4
145 - 156	6.9	0.4	7.2	0.5

(data from Snyder et al., 1975)

TABLE 1004

FIFTH FINGER LENGTH (cm)
IN CHILDREN OF 8 STATES

Age (months)	Boys		Girls	
	Mean	S.D.	Mean	S.D.
0 - 3	2.4	0.4	2.4	0.4
4 - 6	2.8	0.4	2.6	0.4
7 - 9	3.0	0.5	2.9	0.4
10 - 12	2.9	0.4	3.0	0.3
13 - 18	3.0	0.3	2.9	0.2
19 - 24	3.1	0.2	3.1	0.3
25 - 30	3.4	0.5	3.2	0.3
31 - 36	3.4	0.3	3.3	0.3
37 - 42	3.4	0.3	3.3	0.3
43 - 48	3.5	0.4	3.4	0.3
49 - 54	3.6	0.3	3.5	0.3
55 - 60	3.6	0.3	3.6	0.3
61 - 66	3.8	0.3	3.8	0.4
67 - 72	3.9	0.4	3.9	0.4
73 - 78	4.1	0.3	4.0	0.4
79 - 84	4.1	0.4	4.1	0.4
85 - 96	4.3	0.4	4.3	0.5
97 - 108	4.5	0.4	4.5	0.4
109 - 120	4.7	0.4	4.6	0.4
121 - 132	4.9	0.4	4.8	0.4
133 - 144	5.1	0.4	5.0	0.4
145 - 156	5.3	0.3	5.3	0.4

(data from Snyder et al., 1975)

TABLE 1005

UPPER ARM CIRCUMFERENCE (cm) IN U.S. CHILDREN

Age (years)	Mean	S.D.	Percentiles						
			5	10	25	50	75	90	95
			BOYS						
6	17.6	1.60	15.3	15.1	16.5	17.4	18.5	19.7	20.6
7	18.2	1.83	15.7	16.2	17.1	18.1	19.2	20.4	21.6
8	19.0	2.16	16.2	16.7	17.6	18.7	20.1	21.7	23.2
9	19.8	2.62	16.4	17.2	18.2	19.4	21.1	23.3	25.4
10	20.4	2.42	17.3	17.9	18.8	19.9	21.6	23.8	25.3
11	21.5	2.80	18.1	18.7	19.6	21.1	23.2	25.3	27.1
12	22.9	2.98	19.0	19.5	20.8	22.5	24.6	27.0	28.6
13	24.3	3.30	20.0	20.7	21.9	23.8	26.2	28.7	30.5
14	25.7	3.42	20.8	21.7	23.5	25.4	27.7	30.2	31.7
15	27.0	3.07	22.5	23.4	24.9	26.6	28.8	31.1	32.3
16	27.8	3.02	23.3	24.4	25.8	27.6	29.4	31.6	33.4
17	28.7	3.11	24.2	25.1	26.6	28.3	30.5	32.6	34.4
			GIRLS						
6	17.6	1.81	15.1	15.5	16.4	17.4	18.6	19.8	20.7
7	18.2	2.09	15.6	16.1	16.9	18.1	19.4	21.1	22.0
8	19.2	2.38	16.0	16.6	17.6	18.9	20.8	22.4	23.8
9	20.1	2.69	16.6	17.2	18.3	19.7	21.6	23.9	25.6
10	20.9	2.99	16.8	17.4	18.7	20.6	22.7	24.9	26.6
11	21.7	3.07	17.5	18.2	19.5	21.4	23.6	26.2	27.5
12	23.4	3.12	18.9	19.7	21.2	23.1	25.2	27.9	29.4
13	24.2	3.17	19.7	20.5	22.0	23.7	26.2	28.6	30.3
14	25.1	3.20	20.6	21.6	23.1	24.6	26.8	29.3	30.8
15	25.8	3.39	21.3	22.2	23.5	25.4	27.4	30.5	32.7
16	26.2	3.48	21.9	22.6	23.8	25.6	27.8	30.4	33.6
17	26.1	3.20	21.8	22.5	24.0	25.7	27.7	29.8	31.6

(data from Cycles II and III of the Health Examination Survey by the National
Center for Health Statistics, 1963-1965 and 1966-1970)

TABLE 1006

UPPER ARM CIRCUMFERENCE (cm) IN U.S. WHITE CHILDREN

Age (years)	Mean	S.D.	5	10	25	Percentiles		90	95
						50	75		
Boys									
6	17.6	1.61	15.3	15.8	16.5	17.5	18.5	19.7	20.6
7	18.3	1.90	15.7	16.2	17.1	18.2	19.3	20.7	21.7
8	19.0	2.24	16.2	16.7	17.6	18.7	20.3	21.8	25.5
9	19.9	2.66	16.5	17.2	18.3	19.5	21.1	23.4	25.8
10	20.4	2.46	17.3	18.0	18.8	20.1	21.7	23.9	25.5
11	21.6	2.85	18.2	18.8	19.7	21.2	23.3	25.6	27.3
12	23.0	3.00	18.9	19.5	20.9	22.6	24.8	27.2	28.7
13	24.3	3.27	20.1	20.7	21.8	23.7	26.2	28.6	30.4
14	25.8	3.41	20.8	21.8	23.6	25.4	27.8	30.4	31.7
15	27.2	3.08	22.8	23.6	25.1	26.8	29.0	31.5	33.0
16	27.9	3.03	23.4	24.4	25.8	27.6	29.5	31.7	33.5
17	28.7	3.15	24.2	25.0	26.5	28.3	30.7	32.7	34.4
Girls									
6	17.7	1.86	15.1	15.6	16.4	17.5	18.8	19.9	20.9
7	18.3	2.04	15.7	16.2	17.0	18.2	19.5	21.2	22.1
8	19.3	2.32	16.2	16.8	17.7	19.0	20.8	22.4	23.8
9	20.2	2.70	16.6	17.3	18.4	19.7	21.8	24.1	25.3
10	20.9	2.93	16.9	17.5	18.8	20.6	22.8	24.9	26.5
11	21.7	2.99	17.6	18.3	19.6	21.5	23.6	26.1	27.4
12	23.4	3.08	19.0	19.8	21.2	23.1	25.2	27.8	29.3
13	24.3	3.09	19.9	20.7	23.1	23.8	26.3	28.6	30.3
14	25.1	3.19	20.6	21.6	23.1	24.6	26.8	29.3	30.8
15	26.0	3.44	21.4	22.3	23.6	25.4	27.6	30.7	32.8
16	26.2	3.40	22.0	22.6	23.9	25.6	27.8	30.1	33.2
17	26.1	3.15	22.0	22.6	24.1	25.7	27.6	29.8	31.7

(data from Cycle II and Cycle III of the Health Examination Survey by the
National Center for Health Statistics, 1963-1965 and 1966-1970)

TABLE 1007

UPPER ARM CIRCUMFERENCE (cm) IN U.S. NEGRO CHILDREN

Age			Percentiles						
(years)	Mean	S.D.	5	10	25	50	75	90	95
				BOYS					
6	17.4	1.55	15.1	15.4	16.3	17.3	18.5	19.6	20.5
7	17.7	1.28	15.6	16.1	16.9	17.6	18.5	19.4	19.9
8	18.5	1.53	16.1	17.0	17.5	18.5	19.6	20.5	21.2
9	19.1	2.21	16.3	17.1	17.6	18.7	20.3	22.8	23.8
10	19.8	2.07	16.6	17.5	18.5	19.5	20.9	22.2	24.2
11	20.6	2.27	17.2	18.1	19.3	20.4	22.3	23.7	24.7
12	22.3	2.66	19.0	19.4	20.5	21.9	23.7	25.8	27.1
13	24.4	3.52	19.6	20.5	22.1	24.1	26.6	29.4	31.4
14	25.2	3.47	20.4	21.2	22.8	24.9	27.2	29.2	30.2
15	25.5	2.63	21.3	22.1	23.8	25.5	26.8	28.8	30.7
16	27.4	2.90	22.7	24.2	25.5	27.4	28.7	30.9	33.0
17	28.6	2.90	24.8	25.5	27.1	28.4	29.5	31.7	33.8
				GIRLS					
6	17.1	1.41	14.7	15.2	16.0	17.2	17.9	18.7	20.1
7	17.9	2.26	15.3	15.6	16.5	17.5	18.8	20.7	21.8
8	18.8	2.70	15.3	15.7	16.8	18.3	20.7	22.3	23.7
9	19.5	2.65	15.9	17.0	17.7	19.3	20.8	22.8	24.8
10	20.6	3.22	16.6	17.2	18.4	20.1	22.1	24.8	27.5
11	21.5	3.54	17.2	17.7	19.3	20.8	23.2	27.0	27.9
12	23.6	3.37	18.1	19.5	21.4	22.8	26.0	28.5	29.6
13	23.8	3.61	19.3	19.9	21.2	22.9	25.4	29.0	32.2
14	25.0	3.26	20.1	21.3	23.1	24.7	27.0	29.1	30.7
15	25.3	2.94	20.8	21.7	23.3	25.1	27.0	28.9	32.2
16	26.2	3.94	21.5	22.4	23.6	25.4	27.6	32.0	35.5
17	26.1	3.46	21.3	21.9	23.4	26.2	28.4	30.1	30.7

(data from Cycles II and III of the Health Examination Survey 1963-1965 and
1966-1970 by the National Center for Health Statistics).

TABLE 1008

UPPER ARM CIRCUMFERENCE (cm)
IN CHILDREN OF 8 STATES

Age	Boys		Girls	
(months)	Mean	S.D.	Mean	S.D.
0 - 3	10.9	1.5	11.0	1.3
4 - 6	13.1	0.9	12.7	1.5
7 - 9	13.8	1.1	13.5	1.1
10 - 12	14.3	1.1	13.7	0.9
13 - 18	14.4	1.0	14.1	1.3
19 - 24	15.1	1.1	14.4	1.2
25 - 30	15.0	1.1	15.0	1.2
31 - 36	15.5	1.1	15.1	1.3
37 - 42	15.6	1.1	15.4	1.2
43 - 48	15.7	1.1	15.6	1.0
49 - 54	16.0	1.1	15.8	1.1
55 - 60	16.2	1.1	16.2	1.3
61 - 66	16.5	1.1	16.2	1.6
67 - 72	16.5	1.3	16.4	1.6
73 - 78	16.7	1.1	16.6	1.5
79 - 84	17.4	1.7	16.9	1.4
85 - 96	17.8	2.1	17.6	1.8
97 - 108	18.4	1.8	18.2	2.0
109 - 120	19.1	2.0	19.1	2.2
121 - 132	20.4	2.2	20.0	2.3
133 - 144	21.1	1.8	21.3	2.9
145 - 156	21.4	2.5	22.2	2.4

(data from Snyder et al., 1975)

TABLE 1009

UPPER ARM CIRCUMFERENCE (cm)
IN COLORADO CHILDREN

Age (years)	Boys			Girls		
	Median	Mean	S.D.	Median	Mean	S.D.
0.08	10.64	10.65	1.10	10.62	10.59	0.80
0.13	11.03	11.22	1.06	10.98	10.89	0.77
0.25	12.27	12.46	0.99	12.27	12.19	0.82
0.38	13.21	13.39	1.11	13.24	13.21	0.85
0.50	13.97	14.06	1.22	13.99	13.86	1.05
0.63	14.79	14.86	1.15	14.62	14.56	1.06
0.75	14.92	14.96	1.19	14.85	14.86	1.08
0.88	15.11	15.32	1.26	15.19	15.15	1.07
1.00	15.37	15.38	1.25	15.03	15.10	1.10
1.50	15.71	15.67	1.30	15.54	15.48	1.16
2.00	15.82	15.90	1.13	15.52	15.50	1.14
2.50	16.08	16.10	1.05	15.70	15.77	1.14
3.00	16.21	16.25	0.99	15.90	15.88	1.18
3.50	16.31	16.29	1.01	16.03	16.10	1.13
4.00	16.35	16.36	1.07	16.25	16.25	1.09
4.50	16.41	16.48	1.01	16.61	16.61	1.07
5.00	16.58	16.70	1.05	16.79	16.74	1.20
5.50	16.71	16.84	1.10	16.92	16.93	1.35
6.00	17.07	17.12	1.26	16.98	17.20	1.32
6.50	17.22	17.32	1.37	17.34	17.35	1.39
7.00	17.47	17.69	1.40	17.53	17.71	1.46
7.50	17.63	17.84	1.34	17.99	18.13	1.65
8.00	18.01	18.26	1.56	18.41	18.47	1.69

TABLE 1009 (continued)

UPPER ARM CIRCUMFERENCE (cm)
IN COLORADO CHILDREN

Age (years)	Boys			Girls		
	Median	Mean	S.D.	Median	Mean	S.D.
8.50	18.25	18.60	1.61	18.65	18.92	1.95
9.00	18.83	19.14	1.64	19.13	19.28	1.96
9.50	19.19	19.61	1.73	19.43	19.69	1.95
10.00	19.65	20.04	1.93	19.81	19.88	2.00
10.50	19.97	20.32	1.95	20.26	20.29	2.12
11.00	20.41	20.71	2.12	20.51	20.78	2.10
11.50	20.58	21.05	2.06	20.56	21.06	2.19
12.00	21.10	21.54	2.15	21.27	21.57	2.30
12.50	21.82	21.81	2.23	21.66	21.89	2.34
13.00	22.31	22.58	2.39	21.89	22.00	2.27
13.50	22.65	22.80	2.34	22.34	22.35	2.31
14.00	23.39	23.44	2.20	22.72	22.86	2.41
14.50	23.48	23.51	2.24	22.78	22.90	2.09
15.00	24.12	24.40	2.39	23.08	23.27	2.19
15.50	24.48	24.71	2.24	22.80	22.96	1.80
16.00	25.49	25.66	2.43	23.82	23.89	2.23
16.50	25.63	25.91	2.20	24.92	24.03	2.05
17.00	25.61	26.15	2.25	23.58	24.02	2.29
17.50	26.41	26.92	1.97	--	--	--
18.00	26.30	26.75	2.07	23.87	24.22	2.38

(data from McCammon unpublished)

TABLE 1010

UPPER ARM CIRCUMFERENCE (mm) FOR PHILADELPHIA CHILDREN

Age (years)	WHITE BOYS		BLACK BOYS	
	Mean	S.D.	Mean	S.D.
6	181.16	18.42	171.37	23.05
7	187.12	22.37	176.74	20.30
8	194.78	22.00	184.19	18.03
9	202.75	27.18	197.03	21.35
10	216.00	32.57	202.37	30.48
11	220.77	29.12	212.31	35.79
12	228.97	27.40	214.60	26.22
13	--	--	225.96	30.83

Age (years)	WHITE GIRLS		BLACK GIRLS	
	Mean	S.D.	Mean	S.D.
6	179.78	12.92	173.94	19.95
7	190.55	19.73	184.19	22.88
8	195.90	23.75	192.47	24.28
9	206.15	24.04	198.81	28.83
10	216.54	26.07	208.87	29.65
11	219.18	22.11	215.48	31.18
12	--	--	224.87	34.73
13	--	--	222.86	27.61

Age is for completed years.

(data from Malina, unpublished)

TABLE 1011

UPPER ARM CIRCUMFERENCE (mm) AT EXAMINATIONS
ONE YEAR APART IN PHILADELPHIA CHILDREN

Age	Examination 1		Examination 2		Increase	
(years)	Mean	S.D.	Mean	S.D.	Mean	S.D.
NEGRO BOYS						
6- 7	164.7	16.2	173.1	23.4	8.3	8.6
7- 8	171.5	15.3	181.9	18.7	10.4	7.8
8- 9	181.5	17.4	192.4	22.5	10.9	8.2
9-10	187.4	17.3	200.7	21.9	13.3	9.7
10-11	197.8	24.2	211.3	32.0	13.5	11.4
11-12	206.3	26.5	220.1	33.5	13.8	9.8
12-13	203.7	13.9	215.3	15.0	11.6	6.7
WHITE BOYS						
6- 7	176.2	14.9	183.7	16.7	7.5	7.5
7- 8	185.2	20.2	193.0	27.5	7.8	10.1
8- 9	189.4	17.6	197.6	20.9	8.2	8.8
9-10	196.8	24.2	203.3	26.7	6.5	9.0
10-11	214.9	32.6	228.6	37.8	13.7	10.6
11-12	210.5	21.7	222.6	25.6	12.1	8.5
NEGRO GIRLS						
6- 7	165.1	9.2	174.2	13.9	9.0	7.4
7- 8	180.2	21.6	191.8	24.9	11.6	7.9
8- 9	184.5	19.6	198.5	23.8	14.0	9.7
9-10	194.9	28.4	208.6	34.3	13.7	12.2
10-11	207.4	33.7	220.2	35.8	12.8	8.0
11-12	202.3	23.7	215.9	29.8	13.6	9.5
12-13	215.9	33.1	232.0	37.2	16.1	9.7
WHITE GIRLS						
6- 7	176.0	11.2	185.1	13.6	9.1	5.1
7- 8	183.9	15.8	194.9	16.8	10.9	8.6
8- 9	187.9	22.2	197.4	26.7	9.5	8.7
9-10	198.6	21.4	208.6	25.0	9.9	7.8
10-11	206.3	19.2	217.2	23.3	10.9	8.9
11-12	216.0	24.3	224.1	22.7	8.1	7.3

(data from Malina, unpublished)

TABLE 1012

UPPER ARM "MUSCLE" CIRCUMFERENCE
(cm) IN U.S. CHILDREN

Age (years)	Mean	S.D.	5	10	25	Percentiles 50	75	90	95
						BOYS			
6	15.0	1.23	13.1	13.3	14.0	14.9	15.8	16.7	17.3
7	15.6	1.28	13.4	14.0	14.6	15.5	16.4	17.3	17.8
8	16.1	1.46	14.0	14.3	15.1	16.0	16.9	17.9	18.7
9	16.7	1.57	14.2	14.8	15.6	16.6	17.6	18.7	19.3
10	17.2	1.52	15.0	15.3	16.1	17.1	18.0	19.0	19.8
11	18.0	1.71	15.4	16.1	16.8	17.9	19.0	20.3	21.0
12	19.6	1.92	16.7	17.3	18.2	19.4	20.9	22.2	22.9
13	21.0	2.36	17.3	18.2	19.3	20.8	22.4	24.3	25.2
14	22.8	2.66	18.6	19.6	20.9	22.7	24.3	26.1	27.1
15	24.1	2.36	20.5	21.2	22.4	24.1	25.7	27.2	28.0
16	25.0	2.40	21.1	22.3	23.6	25.0	26.4	28.0	29.3
17	25.8	2.22	22.4	23.2	24.1	25.7	27.3	28.9	29.8
						GIRLS			
6	14.5	1.20	12.5	13.0	13.6	14.5	15.4	16.2	16.7
7	15.0	1.45	13.0	13.3	14.1	14.9	15.8	16.8	17.4
8	15.6	1.54	13.3	13.8	14.5	15.5	16.5	17.6	18.4
9	16.2	1.61	13.8	14.2	15.1	16.1	17.1	18.3	19.1
10	16.9	1.91	14.1	14.6	15.6	16.7	18.0	19.3	20.1
11	17.7	2.06	14.8	15.3	16.2	17.5	18.9	20.4	21.6
12	19.4	1.94	16.3	16.9	18.1	19.3	20.6	21.9	22.6
13	20.0	1.96	16.9	17.6	18.6	19.8	21.2	22.5	23.3
14	20.4	2.02	17.6	18.1	19.1	20.2	21.5	23.0	23.8
15	20.9	2.06	17.8	18.4	19.4	20.7	22.1	23.6	24.5
16	21.0	2.15	18.0	18.7	19.4	20.6	22.2	23.8	25.0
17	20.9	2.26	17.9	18.6	19.5	20.6	22.1	23.5	24.4

(data from Cycles II and III of the Health Examination Survey by the National
Center for Health Statistics, 1963-1965 and 1966-1970)

TABLE 1013

UPPER ARM "MUSCLE" CIRCUMFERENCE
(cm) IN U.S. WHITE CHILDREN

Age (years)	Mean	S.D.	5	10	25	Percentiles 50	75	90	95
						BOYS			
6	15.0	1.21	13.0	13.3	14.0	14.8	15.8	16.7	17.2
7	15.6	1.30	13.4	13.9	14.6	15.5	16.4	17.3	17.8
8	16.1	1.49	14.0	14.3	15.1	16.0	16.9	17.9	18.7
9	16.6	1.57	14.1	14.7	15.6	16.6	17.6	18.6	11.3
10	17.1	1.54	15.0	15.3	16.1	17.0	18.0	18.9	19.8
11	18.0	1.72	15.4	16.1	16.8	17.8	19.0	20.3	21.1
12	19.6	1.94	16.6	17.3	18.2	19.4	20.9	22.3	23.0
13	20.9	2.32	17.3	18.2	19.2	20.8	22.4	24.0	25.0
14	22.8	2.68	18.6	19.6	20.9	22.7	24.3	26.0	27.1
15	24.2	2.35	20.6	21.4	22.6	24.2	25.8	27.3	28.1
16	25.0	2.39	21.2	22.2	23.5	25.0	26.3	27.9	29.3
17	25.8	2.23	22.4	23.1	24.1	25.6	27.3	28.9	29.8
						GIRLS			
6	14.5	1.20	12.5	13.0	13.6	14.5	15.3	16.2	16.8
7	14.9	1.38	13.0	13.3	14.1	14.8	15.7	16.7	17.3
8	15.6	1.50	13.4	13.8	14.5	15.4	16.5	17.5	18.2
9	16.2	1.62	13.9	14.8	15.1	16.0	17.1	18.2	19.1
10	16.8	1.85	14.1	14.6	15.6	16.6	17.9	19.1	19.8
11	17.7	2.01	14.8	15.2	16.2	17.4	18.8	20.4	21.5
12	19.3	1.90	16.3	16.9	18.0	19.2	20.4	21.9	22.4
13	20.0	1.93	16.9	17.6	18.7	19.9	21.2	22.5	23.2
14	20.4	1.98	17.6	18.1	19.1	20.1	21.3	23.0	23.8
15	20.9	2.09	17.9	18.4	19.4	20.8	22.1	23.6	24.6
16	20.9	2.12	18.0	18.7	19.4	20.6	22.1	23.8	25.1
17	20.8	2.21	17.9	18.5	19.4	20.6	21.9	23.3	24.1

(data from Cycles II and III of the Health Examination Survey by the National
Center for Health Statistics, 1963-1965 and 1966-1970)

TABLE 1014

UPPER ARM "MUSCLE" CIRCUMFERENCE (cm) IN U.S. NEGRO CHILDREN

| Age (years) | Mean | S.D. | Percentiles | | | | | | |
			5	10	25	50	75	90	95
					BOYS				
6	15.2	1.31	13.1	13.4	14.1	15.2	16.7	17.1	17.7
7	15.7	1.18	13.5	14.1	14.9	15.6	16.5	17.1	17.8
8	16.2	1.19	14.3	14.9	15.4	16.8	16.9	17.9	18.5
9	16.8	1.58	14.6	15.1	15.6	16.6	18.1	18.8	19.7
10	17.4	1.43	15.0	15.6	16.5	17.4	18.3	19.3	19.8
11	18.1	1.60	15.3	15.9	17.2	18.1	19.1	20.4	20.9
12	19.6	1.76	17.3	17.0	18.3	19.5	20.7	21.6	22.3
13	21.8	2.55	18.0	19.0	19.7	21.5	23.6	25.2	26.0
14	22.8	2.56	18.6	19.4	21.2	22.7	24.5	26.2	26.8
15	23.4	2.30	19.3	20.4	22.0	23.4	24.8	26.5	27.1
16	25.2	2.39	21.4	22.4	23.8	25.0	26.5	28.5	29.4
17	26.2	2.16	22.7	23.5	24.5	26.4	27.6	28.8	29.8
					GIRLS				
6	14.6	1.25	12.5	12.8	13.6	14.7	15.6	16.3	16.7
7	15.3	1.73	13.0	13.3	14.2	15.2	16.2	17.2	17.7
8	15.8	1.74	13.2	13.5	14.4	15.6	16.8	18.2	18.8
9	16.3	1.61	14.0	14.3	15.2	16.3	17.2	18.6	19.3
10	17.4	2.16	14.2	15.0	15.8	17.0	18.6	20.3	21.5
11	18.1	2.34	15.0	15.6	16.6	17.7	19.2	20.8	23.1
12	19.8	2.12	16.0	16.8	18.6	19.7	21.1	22.6	24.1
13	20.0	2.13	17.1	17.7	18.4	19.8	21.2	22.7	23.6
14	20.7	2.22	17.3	18.3	19.4	20.6	22.1	23.5	24.6
15	20.8	1.79	18.2	18.5	19.7	20.7	21.9	23.3	24.2
16	21.5	2.25	18.5	19.2	20.2	21.1	22.7	23.8	25.0
17	21.6	2.43	18.2	18.9	20.0	21.4	22.8	24.4	25.5

(data from Cycles II and III of the Health Examination Survey by the National Center for Health Statistics, 1963-1965 and 1966-1970).

TABLE 1015

FOREARM CIRCUMFERENCE (cm) IN U.S. CHILDREN

Age (years)	Mean	S.D.	Percentiles						
			5	10	25	50	75	90	95
BOYS									
12	22.0	1.97	19.1	19.5	20.6	21.8	23.2	24.8	25.7
13	23.3	2.24	20.1	20.6	21.7	23.1	24.8	26.3	27.1
14	24.5	2.30	21.0	21.7	23.0	24.5	26.0	27.6	28.6
15	25.5	2.01	22.3	23.1	24.2	25.4	26.7	28.1	29.1
16	26.1	1.94	23.2	23.9	24.8	26.1	27.4	28.6	29.6
17	26.8	1.94	23.7	24.4	25.4	26.8	28.1	29.4	30.4
GIRLS									
12	21.7	1.98	18.6	19.3	20.4	21.6	22.9	24.5	25.4
13	22.3	1.83	19.4	20.1	21.1	22.3	23.5	24.8	25.5
14	22.8	1.85	20.1	20.7	21.6	22.7	23.9	25.3	26.0
15	23.3	1.94	20.4	21.0	22.0	23.1	24.4	25.8	26.8
16	23.4	2.02	20.6	21.2	22.1	23.2	24.5	25.8	27.3
17	23.3	1.74	20.9	21.3	22.2	23.3	24.4	25.6	26.6

(data from Cycle III of the Health Examination Survey 1966-1970 conducted by the National Center for Health Statistics).

TABLE 1016

FOREARM CIRCUMFERENCE (cm) IN U.S. WHITE CHILDREN

| | | | | | | Percentiles | | | |
Age (years)	Mean	S.D.	5	10	25	50	75	90	95
					BOYS				
6	17.4	1.32	15.4	16.0	16.6	17.4	18.2	18.9	19.8
7	18.0	1.40	16.0	16.3	17.2	18.0	18.8	19.8	20.6
8	18.7	1.59	16.3	16.9	17.6	18.6	19.6	20.8	21.6
9	19.4	1.74	16.9	17.3	18.3	19.3	20.4	21.7	22.8
10	19.9	1.61	17.4	18.0	18.7	19.7	20.9	22.1	22.9
11	20.8	1.83	18.2	18.7	19.6	20.6	21.8	23.5	24.3
12	22.0	1.99	19.1	19.5	20.6	21.8	23.3	24.9	25.7
13	23.2	2.18	20.1	20.7	21.7	23.1	24.7	26.1	26.8
14	24.5	2.29	21.1	21.7	23.1	24.5	26.0	27.6	28.6
15	25.6	1.95	22.4	23.2	24.3	25.5	26.8	28.0	29.2
16	26.1	1.91	23.2	24.0	24.9	26.1	27.4	28.6	29.5
17	26.8	1.95	23.6	24.3	25.4	26.8	28.1	29.3	30.3
					GIRLS				
6	17.1	1.30	15.1	15.5	16.3	17.1	17.8	18.8	19.6
7	17.5	1.42	15.5	16.1	16.6	17.5	18.5	19.5	20.1
8	18.3	1.47	16.1	16.5	17.4	18.3	19.3	20.4	21.2
9	19.0	1.78	16.4	17.1	17.9	18.8	20.2	21.5	22.4
10	19.7	1.91	17.1	17.4	18.4	19.7	20.9	22.1	22.9
11	20.5	1.90	17.5	18.2	19.1	20.4	21.8	23.1	23.9
12	21.6	1.96	18.6	19.3	20.4	21.5	22.8	24.3	25.3
13	22.2	1.76	19.5	20.2	21.1	22.3	23.5	24.7	25.4
14	22.8	1.82	20.2	20.6	21.6	22.6	23.8	25.2	25.8
15	23.3	1.96	20.4	21.1	22.0	23.1	24.5	25.8	26.8
16	23.4	1.99	20.6	21.1	22.1	23.2	24.5	25.7	26.8
17	23.2	1.67	20.8	21.3	22.2	23.2	24.2	25.2	26.2

(data from Cycles II and III of the Health Examination Survey by the National Center for Health Statistics, 1963-1965 and 1966-1970)

TABLE   1017

FOREARM CIRCUMFERENCE (cm) IN U.S. NEGRO CHILDREN

Age						Percentiles			
(years)	Mean	S.D.	5	10	25	50	75	90	95

BOYS

Age	Mean	S.D.	5	10	25	50	75	90	95
6	17.2	1.30	15.2	15.5	16.3	17.3	18.2	19.3	19.7
7	17.1	1.09	15.8	16.2	17.1	17.6	18.6	19.4	19.7
8	18.4	1.26	16.3	16.7	17.5	18.4	19.3	20.2	20.8
9	19.0	1.59	17.0	17.2	17.9	18.8	20.2	21.3	21.8
10	19.6	1.50	17.4	18.0	18.6	19.5	20.6	21.9	22.5
11	20.9	1.70	17.6	18.3	19.3	20.3	21.7	22.7	23.3
12	21.9	1.81	19.2	19.7	20.6	21.7	22.9	24.3	25.0
13	23.7	2.60	20.1	20.5	21.8	23.6	25.6	26.9	27.8
14	24.4	2.40	20.6	21.4	22.6	24.5	26.3	27.6	28.4
15	24.9	2.31	21.8	22.3	23.4	24.8	26.2	28.0	28.7
16	26.2	2.11	23.0	23.6	24.6	26.3	27.5	29.0	29.9
17	27.0	1.92	24.2	24.7	25.5	26.9	27.9	30.1	31.2

GIRLS

Age	Mean	S.D.	5	10	25	50	75	90	95
6	16.6	1.11	14.9	15.2	15.9	16.6	17.4	18.0	18.6
7	17.4	1.78	15.2	15.6	16.4	17.4	18.3	19.0	20.2
8	18.0	1.70	15.3	15.7	16.8	17.8	19.2	20.4	21.3
9	19.0	1.82	15.9	17.1	17.7	18.8	20.1	21.5	22.7
10	19.1	2.02	16.6	17.3	18.4	19.7	20.8	23.0	23.8
11	20.7	2.47	17.4	17.9	18.8	30.4	22.0	24.3	25.6
12	22.2	2.07	19.0	19.6	21.0	21.7	23.7	25.2	25.7
13	22.4	2.20	19.3	19.7	21.1	22.2	23.7	25.5	27.1
14	23.3	1.98	19.7	21.0	22.1	23.2	24.6	26.1	26.8
15	23.4	1.82	20.3	21.1	22.2	23.4	24.4	25.9	26.9
16	23.9	2.17	21.3	21.7	22.4	23.4	24.9	27.5	28.6
17	24.0	2.00	21.2	21.5	22.5	23.8	25.5	26.6	27.4

(data from Cycles II and III of the Health Examination Survey by the National Center for Health Statistics, 1963-1965 and 1966-1970)

TABLE 1018

FOREARM CIRCUMFERENCE (cm)
FOR CHILDREN IN 8 STATES

Age (months)	Boys Mean	S.D.	Girls Mean	S.D.
0 - 3	11.0	1.5	11.2	1.3
4 - 6	13.4	0.8	12.9	1.2
7 - 9	14.2	1.2	13.6	0.9
10 - 12	14.3	1.0	13.8	1.0
13 - 18	14.4	1.1	14.2	1.1
19 - 24	15.3	1.1	14.6	0.9
25 - 30	14.8	0.9	15.1	0.8
31 - 36	15.4	0.9	14.8	1.1
37 - 42	15.5	1.0	15.2	1.0
43 - 48	15.7	0.9	15.5	1.0
49 - 54	16.0	1.0	15.5	0.9
55 - 60	16.2	1.1	15.9	1.2
61 - 66	16.6	1.0	16.1	1.2
67 - 72	16.7	1.2	16.3	1.1
73 - 78	17.0	1.1	16.6	1.2
79 - 84	17.7	1.3	16.7	1.0
85 - 96	17.9	1.5	17.4	1.3
97 - 108	18.6	1.3	18.0	1.5
109 - 120	19.2	1.5	18.7	1.5
121 - 132	20.3	1.5	19.6	1.7
133 - 144	20.9	1.2	20.3	1.7
145 - 156	21.1	1.7	21.6	1.3

(data from Snyder et al., 1975)

TABLE 1019

FOREARM CIRCUMFERENCE (mm) FOR PHILADELPHIA CHILDREN

Age (years)	WHITE BOYS		BLACK BOYS	
	Mean	S.D.	Mean	S.D.
6	176.26	12.23	171.36	16.04
7	182.32	14.50	177.11	14.66
8	188.93	14.12	184.04	12.20
9	195.71	17.04	194.60	14.64
10	205.61	19.60	198.30	18.74
11	208.55	17.37	205.17	19.33
12	213.94	15.70	210.09	20.13
13	--	--	220.04	21.93

Age (years)	WHITE GIRLS		BLACK GIRLS	
	Mean	S.D.	Mean	S.D.
6	173.41	9.65	170.93	14.18
7	181.84	13.00	178.61	14.79
8	185.59	14.20	184.91	15.81
9	192.23	15.51	189.81	16.60
10	199.55	15.51	200.62	18.88
11	202.82	13.93	207.20	21.17
12	--	--	215.92	22.74
13	--	--	215.93	17.19

Age is for completed years.

(data from Malina, unpublished)

TABLE 1020

FOREARM CIRCUMFERENCE (mm) AT EXAMINATIONS
ONE YEAR APART IN PHILADELPHIA CHILDREN

Age	Examination 1		Examination 2		Increase	
(years)	Mean	S.D.	Mean	S.D.	Mean	S.D.

NEGRO BOYS

Age	Mean	S.D.	Mean	S.D.	Mean	S.D.
6- 7	166.5	15.2	174.0	16.8	7.5	4.2
7- 8	172.9	11.5	182.9	13.9	9.9	5.0
8- 9	180.6	13.1	190.6	14.8	10.0	4.6
9-10	185.7	12.0	197.2	14.4	11.5	5.6
10-11	194.6	17.8	206.1	20.8	11.5	6.4
11-12	200.4	16.0	212.2	22.0	11.8	9.9
12-13	200.5	10.5	211.1	12.3	10.6	5.0

WHITE BOYS

Age	Mean	S.D.	Mean	S.D.	Mean	S.D.
6- 7	171.4	10.7	180.0	11.0	8.6	4.7
7- 8	178.1	14.4	187.0	16.8	8.9	5.2
8- 9	183.4	12.0	192.1	12.9	8.7	4.8
9-10	189.4	16.0	198.4	17.1	8.9	4.9
10-11	201.7	19.5	214.3	22.2	12.6	5.3
11-12	199.3	12.7	210.3	14.9	11.1	5.3

NEGRO GIRLS

Age	Mean	S.D.	Mean	S.D.	Mean	S.D.
6- 7	163.3	8.4	171.5	10.0	8.3	4.6
7- 8	175.6	14.2	184.4	15.0	8.8	5.3
8- 9	179.3	13.1	188.9	13.6	9.6	5.6
9-10	186.8	18.2	196.9	20.1	10.1	7.3
10-11	198.6	22.6	208.9	23.3	10.3	4.1
11-12	198.9	17.5	210.2	20.1	11.3	6.1
12-13	208.8	23.4	218.8	23.0	10.1	5.9

WHITE GIRLS

Age	Mean	S.D.	Mean	S.D.	Mean	S.D.
6- 7	170.5	8.9	178.6	10.1	8.1	3.4
7- 8	176.5	11.8	186.2	12.5	9.7	5.1
8- 9	179.4	13.7	187.7	14.3	8.3	4.9
9-10	186.6	15.9	194.5	17.4	7.9	3.3
10-11	192.1	11.4	199.6	12.3	7.5	5.0
11-12	198.3	12.3	206.3	10.3	8.0	6.2

(data from Malina, unpublished)

TABLE 1021

UPPER LIMB CIRCUMFERENCES (cm) FOR BLACK BOYS IN TEXAS

Age	Upper arm				Forearm			
(years)	Lower Income		Middle Income		Lower Income		Middle Income	
	Mean	S.D.	Mean	S.D.	Mean	S.D.	Mean	S.D.
10.1 - 11	20.9	1.54	21.2	2.20	21.0	1.60	21.5	2.60
11.1 - 12	21.6	0.96	22.1	2.27	21.6	0.96	22.1	2.27
12.1 - 13	22.9	3.56	23.5	2.72	22.3	2.36	22.8	1.85
13.1 - 14	23.5	2.29	24.0	2.74	23.5	1.74	23.7	2.11
14.1 - 15	25.2	3.89	25.3	2.56	24.0	2.14	24.8	1.83
15.1 - 16	27.5	4.86	28.4	2.49	25.4	2.33	26.3	1.51
16.1 - 17	28.2	1.87	30.4	3.41	26.8	1.64	27.4	2.04
17.1 - 18	28.5	2.18	27.6	2.26	26.8	1.20	26.7	1.35

(data from Schutte, 1979)

TABLE 1022

UPPER ARM CIRCUMFERENCE (cm) AND WEIGHT (kg) IN NEW YORK
CHILDREN OF NEGRO OR PUERTO RICAN PARENTS (SEXES COMBINED)

Age	Arm Circumference		Mean
(months)	Mean	S.D.	Weight
1 - 3 mo.	12.3	1.3	4.9
3 - 6 mo.	13.9	1.2	6.8
6 - 9 mo.	14.9	1.4	8.5
9 - 12 mo.	15.3	1.2	9.7
1 - 2 yr.	15.5	1.2	10.9
2 - 3 yr.	16.0	1.2	12.4
3 - 4 yr.	16.6	1.2	15.5
4 - 5 yr.	16.7	1.3	17.7
5 - 6 yr.	17.5	1.5	20.8

(data from Choovivathanavanich and Kanthavichitra, 1970)

TABLE 1023

ARM CIRCUMFERENCES IN COLORADO GIRLS (cm)

Age (yr.-mo.)	Hand Percentiles			Bistyloid Percentiles			Maximum Forearm Percentiles			Biceps Percentiles		
	10th	50th	90th	10th	50th	90th	10th	50th	90th	10th	50th	90th
Birth	7.6	8.3	9.2	7.3	7.9	8.7	9.5	10.3	11.4	9.0	10.5	11.7
0 - 1	7.9	8.7	9.4	7.5	8.2	8.7	9.1	10.6	11.4	9.2	10.3	11.5
0 - 2	8.4	9.1	10.0	8.1	8.9	9.4	10.4	11.6	12.3	10.4	11.6	12.6
0 - 3	8.6	9.3	10.0	8.4	9.3	9.9	11.3	12.3	13.2	10.7	12.3	13.1
0 - 4	9.0	9.7	10.5	8.9	9.6	10.3	11.8	12.8	13.9	12.1	13.0	14.0
0 - 5	9.3	10.0	10.7	9.1	10.0	11.0	12.1	13.5	14.5	12.0	13.4	14.5
0 - 6	9.4	10.2	11.0	9.5	10.3	10.9	12.4	13.8	14.9	12.3	14.2	15.3
0 - 9	9.9	10.7	11.7	10.1	10.8	11.7	13.4	14.8	15.9	13.4	14.8	16.4
1 - 0	10.2	11.2	12.1	10.3	11.2	12.1	13.6	15.1	16.3	13.8	15.2	16.6
1 - 6	10.6	11.4	12.4	10.2	11.1	12.0	13.7	15.2	16.4	13.8	15.6	16.9
2 - 0	10.9	11.8	12.8	10.2	11.0	12.1	14.3	15.4	16.7	14.1	15.6	17.2
2 - 6	11.0	12.0	13.0	10.1	11.0	12.0	14.3	15.5	16.8	14.3	15.6	17.3
3 - 0	11.5	12.2	13.3	10.1	11.2	12.3	14.4	15.6	17.1	14.5	15.9	17.6
3 - 6	11.8	12.4	13.4	10.5	11.4	12.2	14.6	16.0	17.2	14.6	16.0	17.5
4 - 0	12.0	12.7	13.9	10.8	11.5	12.3	14.7	16.2	17.4	14.8	16.2	17.7
4 - 6	12.3	13.1	14.1	10.8	11.6	12.4	15.1	16.6	17.5	15.2	16.6	18.1
5 - 0	12.3	13.2	14.3	11.0	11.6	12.4	15.1	16.8	17.8	15.1	16.7	18.2
5 - 6	12.8	13.6	14.6	11.0	11.8	12.7	15.2	16.9	18.0	14.2	17.0	18.6
6 - 0	13.0	13.9	15.0	11.1	11.9	12.8	15.2	17.0	18.3	15.5	17.0	18.6
6 - 6	13.0	14.1	15.2	11.3	12.1	12.9	15.4	17.2	18.3	15.6	17.3	18.9
7 - 0	13.3	14.3	15.3	11.5	12.1	13.1	14.7	17.3	18.7	16.0	17.6	19.5
7 - 6	13.4	14.6	15.6	11.4	12.3	13.2	15.9	17.6	19.4	16.0	18.0	20.3
8 - 0	14.0	14.8	15.9	11.7	12.5	13.5	16.3	17.9	19.8	16.4	18.4	20.8
8 - 6	14.1	15.1	16.2	12.0	12.7	13.8	16.6	18.2	20.0	16.7	18.7	21.3
9 - 0	14.4	15.3	16.4	11.9	12.9	14.1	16.6	18.4	20.6	16.7	19.0	22.0
9 - 6	14.6	15.6	16.6	12.2	13.0	13.9	17.2	18.6	20.2	17.1	19.4	21.9
10 - 0	14.6	15.7	17.1	12.4	13.2	14.2	17.1	18.8	21.2	17.4	19.6	22.9
10 - 6	15.1	16.0	17.2	12.4	13.4	14.5	17.5	19.2	21.4	17.5	20.0	23.5
11 - 0	15.1	16.3	17.4	12.5	13.6	14.7	17.9	19.4	21.6	18.2	20.3	23.8
11 - 6	15.4	16.5	17.6	12.7	14.1	14.9	18.0	19.6	21.7	18.6	20.6	24.3
12 - 0	15.8	17.0	18.0	13.2	14.2	15.3	18.4	20.2	22.5	18.8	21.4	25.3
12 - 6	16.1	17.3	18.4	13.3	14.5	15.4	19.1	20.8	22.7	19.2	21.8	24.8
13 - 0	16.2	17.4	18.7	13.5	14.7	15.7	19.0	21.1	22.9	19.1	21.9	24.9
13 - 6	17.0	17.7	19.1	14.0	14.7	15.7	19.5	21.1	23.4	19.4	22.2	24.9
14 - 0	17.0	17.9	19.2	14.2	15.1	15.9	20.0	21.5	23.2	20.3	22.8	25.9
14 - 6	17.1	17.9	19.2	14.2	15.0	15.7	20.2	21.9	23.5	20.8	23.2	25.2

TABLE 1023   (continued)

ARM CIRCUMFERENCES IN COLORADO GIRLS (cm)

Age	Hand Percentiles			Bistyloid Percentiles			Maximum Forearm Percentiles			Biceps Percentiles		
(yr.-mo.)	10th	50th	90th	10th	50th	90th	10th	40th	90th	10th	40th	90th
15 - 0	17.2	18.1	19.2	14.5	15.1	16.1	20.5	22.2	23.9	20.8	23.2	25.6
15 - 6	17.2	17.9	19.2	14.3	14.9	16.5	20.8	22.0	23.1	20.6	23.2	25.4
16 - 0	17.3	18.2	19.2	14.6	15.2	16.1	20.7	22.4	24.3	21.0	23.8	26.9
17 - 0	17.1	18.2	19.3	14.4	15.2	15.8	21.2	22.4	24.2	21.6	24.0	26.9
18 - 0	17.3	18.4	19.1	14.4	15.2	15.8	21.0	23.0	24.5	21.0	24.3	28.1

ARM CIRCUMFERENCES IN COLORADO BOYS (cm)

Age	Hand Percentiles			Bistyloid Percentiles			Maximum Forearm Percentiles			Biceps Percentiles		
(yr.-mo.)	10	50	90	10	50	90	10	50	90	10	50	90
Birth	7.8	8.6	9.5	7.2	8.0	8.8	9.0	10.2	11.2	8.6	10.2	11.0
0 - 1	8.3	9.0	9.8	7.3	8.1	9.1	9.7	10.6	11.9	9.3	10.7	12.0
0 - 2	9.0	9.7	10.4	8.3	9.0	10.3	10.8	11.8	13.3	10.6	11.6	13.4
0 - 3	9.2	9.8	10.4	8.4	9.3	10.5	11.3	12.5	14.0	11.3	12.4	14.1
0 - 4	9.6	10.3	10.8	9.0	9.8	10.7	12.1	13.2	14.9	11.7	13.0	14.9
0 - 5	10.0	10.5	11.4	9.3	10.1	11.6	12.7	13.9	15.3	12.4	13.6	15.5
0 - 6	10.0	10.8	11.6	9.6	10.4	11.6	12.8	14.1	15.9	12.6	14.1	16.0
0 - 9	10.6	11.4	12.4	9.8	11.0	12.2	13.6	15.0	16.5	13.2	15.0	16.7
1 - 0	10.7	11.6	12.6	10.0	11.3	12.4	14.0	15.2	16.4	13.8	15.4	17.2
1 - 6	11.0	11.8	12.8	10.1	11.3	12.5	14.0	15.5	16.8	14.1	15.8	17.7
2 - 0	11.2	12.0	13.0	10.1	11.2	12.6	14.2	15.6	16.9	14.3	15.8	17.6
2 - 6	11.3	12.3	13.3	10.2	11.6	12.3	14.5	15.8	17.0	14.6	16.2	17.3
3 - 0	11.5	12.6	13.6	10.3	11.5	12.6	14.8	16.1	17.2	14.9	16.2	17.5
3 - 6	11.8	12.7	13.7	10.5	11.6	12.6	14.8	16.0	17.3	15.1	16.2	17.6
4 - 0	12.2	13.0	14.1	10.6	11.8	12.6	15.0	16.3	17.4	14.9	16.2	17.7
4 - 6	12.4	13.3	14.3	10.8	11.8	12.7	15.4	16.6	17.8	15.1	16.3	17.8
5 - 0	12.5	13.6	14.8	11.1	11.9	13.0	15.5	16.7	18.0	15.4	16.6	18.0
5 - 6	12.8	13.9	15.1	11.3	12.2	13.1	15.6	17.0	18.4	15.5	16.8	18.2
6 - 0	13.2	14.2	15.3	11.3	12.4	13.3	15.7	17.1	18.2	15.6	17.1	18.6
6 - 6	13.5	14.5	15.6	11.7	12.4	13.5	16.1	17.4	18.6	16.0	17.2	19.0
7 - 0	13.8	14.9	15.9	11.8	12.6	13.7	16.4	17.7	19.4	16.0	17.4	19.6
7 - 6	13.9	15.1	16.3	11.7	12.7	13.9	16.6	18.2	19.7	16.4	17.8	19.8
8 - 0	14.1	15.4	16.8	12.0	13.0	14.1	16.6	18.4	20.0	16.7	18.1	20.3
8 - 6	14.4	15.7	16.9	11.9	13.0	14.5	17.1	18.7	20.2	17.0	18.3	21.1

TABLE 1023   (continued)

ARM CIRCUMFERENCES IN COLORADO BOYS (cm)

Age (yr.-mo.)	Hand Percentiles			Bistyloid Percentiles			Maximum Forearm Percentiles			Biceps Percentiles		
	10	50	90	10	50	90	10	50	90	10	50	90
9 - 0	14.8	16.2	17.1	12.4	13.5	14.8	17.1	19.2	21.0	17.2	18.8	21.6
9 - 6	15.2	16.5	17.4	12.6	13.6	14.6	17.5	19.4	21.1	17.6	19.2	21.9
10 - 0	15.4	16.8	17.8	12.9	13.9	15.0	17.8	19.8	21.5	17.8	19.8	23.0
10 - 6	15.7	16.8	18.0	13.0	13.9	15.0	17.8	19.6	21.8	17.7	20.0	23.2
11 - 0	15.8	17.3	18.4	13.2	14.2	15.6	18.2	20.0	22.8	18.1	20.4	24.0
11 - 6	16.3	17.5	18.6	13.3	14.4	15.7	18.5	20.4	22.8	18.5	20.9	24.2
12 - 0	16.2	17.6	19.0	13.7	14.8	16.3	18.7	20.8	23.5	18.9	21.1	25.3
12 - 6	16.5	18.0	19.4	13.7	15.0	16.4	19.2	21.1	24.1	19.3	21.8	25.3
13 - 0	16.8	18.4	20.0	13.6	15.3	16.8	19.4	21.4	24.7	19.6	22.3	26.3
13 - 6	17.2	18.9	20.8	14.0	15.7	17.0	19.4	22.3	24.7	19.9	22.8	26.3
14 - 0	17.6	19.4	20.9	14.4	16.0	17.0	20.5	22.2	24.9	20.6	23.5	26.4
14 - 6	18.0	19.6	21.1	14.5	16.4	17.4	20.6	23.0	25.8	20.6	23.4	26.6
15 - 0	18.6	20.1	21.4	15.1	16.6	17.6	21.7	23.5	26.1	21.6	24.2	27.5
15 - 6	19.3	20.4	21.7	15.3	16.7	17.4	22.2	24.0	26.7	22.4	24.5	27.6
16 - 0	19.0	20.6	21.6	15.8	16.9	17.7	22.6	23.9	26.5	22.8	25.3	28.3
16 - 6	20.0	20.8	21.8	16.4	17.0	17.6	23.4	24.7	27.0	23.2	25.4	29.2
17 - 0	19.2	21.0	22.0	15.8	17.2	18.0	24.0	25.2	27.3	23.6	26.1	29.4
17 - 6	19.3	20.9	22.4	15.6	17.1	17.8	24.2	26.2	28.0	24.3	27.0	30.0
18 - 0	19.6	21.0	21.9	15.6	17.1	17.9	23.8	25.8	27.8	24.0	26.8	29.8

(data from McCammon, Human Growth and Development, 1970. Courtesy of Charles C Thomas, Publisher, Springfield, Illinois)

TABLE 1024

RATIOS BETWEEN ANTHROPOMETRIC VARIABLES FOR WHITE BOYS IN OREGON

				\multicolumn{6}{c}{Percentiles}
Ratio	Mean	S.E.	S.D.	5	10	30	70	90	95
\multicolumn{10}{c}{7 Years}
Arm Girth x 100 / Upper Limb Length	35.7	.15	2.46	32.0	32.6	34.2	36.8	38.6	40.0
Leg Girth x 100 / Lower Limb Length	45.1	.18	2.85	40.5	41.5	43.6	46.3	48.7	49.7
Chest Girth x 100 / Stem Length	86.4	.22	3.48	81.0	81.7	84.7	88.0	90.7	91.7
Biacromial Diameter x 100 / Bicristal Diameter	139.6	.37	5.95	128.9	131.5	136.2	142.9	147.3	149.0
Chest Girth x 100 / Abdomen Girth	103.8	.23	3.61	97.6	99.0	102.0	105.7	108.3	109.3
\multicolumn{10}{c}{10 Years}
Arm Girth x 100 / Upper Limb Length	34.5	.22	3.13	30.1	31.0	32.7	35.7	37.9	40.8
Leg Girth x 100 / Lower Limb Length	43.0	.21	3.03	38.9	39.6	41.2	44.3	46.5	47.9
Chest Girth x 100 / Stem Length	87.2	.32	4.54	81.1	82.1	84.9	88.6	93.1	96.1
Biacromial Diameter x 100 / Bicristal Diameter	140.2	.45	6.45	130.2	132.0	136.9	143.5	147.8	151.7
Chest Girth x 100 / Abdomen Girth	103.8	.26	3.75	97.0	98.9	102.2	105.8	109.1	110.3

(data from Meredith and Meredith, 1953)

TABLE 1025

ANTHROPOMETRIC RATIOS FOR WHITE BOYS IN OREGON
AGED 15 YEARS

Ratio	Mean	Median	S.D.
Arm Girth X 100 / Upper Limb Length	34.9	34.2	3.65
Chest Girth X 100 / Stem Length	91.1	90.1	5.73
Leg Girth X 100 / Lower Limb Length	42.8	42.5	3.59
Shoulder Width X 100 / Hip Width	138.9	139.1	7.69
Chest Girth X 100 / Abdomen Girth	105.6	106.0	4.90

(data from Trim and Meredith, 1952)

TABLE 1026

ELBOW BREADTH (cm) IN U.S. CHILDREN

Age						Percentiles			
(years)	Mean	S.D.	5	10	25	50	75	90	95
					BOYS				
12	6.3	0.45	5.6	5.7	6.0	6.3	6.7	6.9	7.2
13	6.6	0.47	5.8	6.0	6.2	6.6	6.9	7.3	7.5
14	6.9	0.47	6.1	6.2	6.6	6.8	7.2	7.6	7.8
15	7.0	0.41	6.2	6.5	6.7	7.1	7.3	7.7	7.8
16	7.1	0.39	6.5	6.6	6.8	7.1	7.3	7.7	7.8
17	7.1	0.42	6.5	6.6	6.8	7.1	7.4	7.7	7.8
					GIRLS				
12	6.0	0.36	5.5	5.6	5.7	6.1	6.3	6.6	6.8
13	6.1	0.36	5.5	5.6	5.8	6.2	6.3	6.7	6.8
14	6.2	0.35	5.6	5.7	6.0	6.2	6.4	6.8	6.8
15	6.2	0.41	5.5	5.6	6.0	6.2	6.4	6.8	6.9
16	6.2	0.38	5.6	5.6	6.0	6.2	6.4	6.8	6.9
17	6.2	0.40	5.6	5.6	5.9	6.2	6.4	6.8	6.9

(data from Cycle III of the Health Examination Survey by the National Center
of Health Statistics 1966-1970)

TABLE 1027

ELBOW BREADTH IN U.S. WHITE CHILDREN

			Percentiles						
Age (years)	Mean	S.D.	5	10	25	50	75	90	95

BOYS

Age	Mean	S.D.	5	10	25	50	75	90	95
12	6.3	0.45	5.6	5.7	6.0	6.3	6.7	6.9	7.2
13	6.6	0.46	5.7	6.0	6.2	6.6	6.9	7.3	7.4
14	6.9	0.47	6.1	6.2	6.6	6.8	7.2	7.6	7.8
15	7.0	0.41	6.3	6.5	6.7	7.1	7.3	7.7	7.3
16	7.1	0.38	6.5	6.6	6.8	7.2	7.3	7.7	7.8
17	7.1	0.42	6.5	6.6	6.8	7.1	7.4	7.7	7.8

GIRLS

Age	Mean	S.D.	5	10	25	50	75	90	95
12	6.0	0.35	5.4	5.6	5.7	6.1	6.3	6.6	6.7
13	6.1	0.36	5.5	5.6	5.8	6.2	6.3	6.7	6.8
14	6.2	0.35	5.6	5.6	6.0	6.2	6.4	6.7	6.8
15	6.2	0.41	5.5	5.6	6.0	6.2	6.4	6.8	6.9
16	6.2	0.38	5.6	5.6	6.0	6.2	6.4	6.8	6.9
17	6.2	0.40	5.6	5.6	5.8	6.2	6.4	6.7	6.8

ELBOW BREADTH (cm) IN U.S. NEGRO CHILDREN

			Percentiles						
Age (years)	Mean	S.D.	5	10	25	50	75	90	95

BOYS

Age	Mean	S.D.	5	10	25	50	75	90	95
12	6.3	0.41	5.6	5.8	6.1	6.3	6.6	6.8	6.9
13	6.7	0.51	5.8	6.1	6.4	6.7	7.1	7.5	7.6
14	6.8	0.43	6.1	6.2	6.6	6.9	7.2	7.4	7.6
15	7.0	0.43	6.2	6.4	6.6	6.9	7.3	7.6	7.8
16	7.1	0.45	6.4	6.5	6.7	7.1	7.3	7.7	7.8
17	7.1	0.44	6.3	6.5	6.7	7.1	7.4	7.7	7.8

GIRLS

Age	Mean	S.D.	5	10	25	50	75	90	95
12	6.1	0.39	5.5	5.6	5.8	6.2	6.4	6.8	6.8
13	6.2	0.38	5.5	5.6	6.0	6.2	6.4	6.7	6.8
14	6.3	0.33	5.6	5.8	6.1	6.3	6.6	6.8	6.9
15	6.2	0.40	5.5	5.6	5.9	6.2	6.6	6.8	6.9
16	6.3	0.39	5.6	5.7	6.0	6.2	6.6	6.8	6.9
17	6.3	0.38	5.6	5.8	6.1	6.3	6.7	6.9	7.0

(data from Cycle III of the Health Examination Survey by the National Center of Health Statistics 1966-1970)

TABLE 1028

BIEPICONDYLAR DIAMETER (mm) FOR PHILADELPHIA CHILDREN

Age (years)	WHITE BOYS		BLACK BOYS	
	Mean	S.D.	Mean	S.D.
6	50.04	3.04	49.90	3.03
7	51.94	2.71	51.49	2.93
8	53.76	2.99	54.05	3.11
9	55.50	3.64	56.32	3.30
10	57.64	3.67	57.61	3.71
11	58.73	3.48	59.17	3.72
12	60.06	3.52	61.33	4.31
13	--	--	63.81	4.62

Age (years)	WHITE GIRLS		BLACK GIRLS	
	Mean	S.D.	Mean	S.D.
6	48.00	2.35	48.17	3.15
7	49.82	2.50	50.05	3.02
8	50.90	3.07	51.73	3.39
9	53.03	3.04	52.83	3.09
10	54.39	2.51	55.92	3.51
11	56.54	3.19	58.24	4.29
12	--	--	60.08	4.27
13	--	--	61.07	3.50

Age is for completed years.

(data from Malina, unpublished)

TABLE 1029

WRIST BREADTH (cm) IN U.S. WHITE CHILDREN

Age (years)	Mean	S.D.	Percentiles						
			5	10	25	50	75	90	95
BOYS									
12	5.1	0.41	4.5	4.6	4.7	5.1	5.3	5.6	5.8
13	5.3	0.41	4.6	4.7	5.0	5.3	5.6	5.8	5.9
14	5.5	0.35	4.8	5.0	5.2	5.6	5.8	6.0	6.2
15	5.6	0.36	5.0	5.1	5.3	5.6	5.8	6.2	6.3
16	5.7	0.30	5.1	5.2	5.5	5.7	5.8	6.2	6.3
17	5.7	0.30	5.1	5.2	5.5	5.7	5.9	6.2	6.3
GIRLS									
12	4.9	0.28	4.4	4.5	4.6	4.8	5.2	5.3	5.4
13	4.9	0.28	4.5	4.6	4.7	4.9	5.2	5.3	5.4
14	5.0	0.27	4.5	4.6	4.7	5.0	5.2	5.3	5.4
15	5.0	0.29	4.5	4.6	4.7	5.0	5.2	5.4	5.5
16	5.0	0.28	4.5	4.6	4.7	5.0	5.2	5.4	5.4
17	5.0	0.28	4.5	4.6	4.7	5.0	5.2	5.3	5.4

WRIST BREADTH (cm) IN U.S. NEGRO CHILDREN

Age (years)	Mean	S.D.	Percentiles						
			5	10	25	50	75	90	95
BOYS									
12	5.0	0.32	4.5	4.6	4.7	5.0	5.2	5.4	5.7
13	5.3	0.46	4.5	4.7	5.1	5.3	5.7	5.9	6.2
14	5.5	0.30	5.0	5.1	5.2	5.6	5.8	5.9	6.0
15	5.6	0.36	5.0	5.1	5.3	5.6	5.8	6.1	6.3
16	5.7	0.31	5.1	5.2	5.5	5.7	5.9	6.2	6.3
17	5.7	0.38	5.1	5.2	5.5	5.7	6.0	6.3	6.4
GIRLS									
12	4.9	0.28	4.5	4.6	4.7	4.9	5.2	5.3	5.4
13	5.0	0.33	4.3	4.5	4.7	5.0	5.2	5.4	5.4
14	5.0	0.27	4.5	4.6	4.8	5.1	5.3	5.4	5.5
15	5.0	0.31	4.5	4.6	4.8	5.1	5.3	5.6	5.7
16	5.0	0.30	4.5	4.6	4.8	5.1	5.3	5.4	5.5
17	5.0	0.27	4.6	4.6	4.8	5.1	5.3	5.4	5.6

(data from Cycle III of the Health Examination Survey by the National Center of Health Statistics 1966-1970)

TABLE 1030

ELBOW–ELBOW BREADTH (cm) IN U.S. CHILDREN

Sex and age	5	10	25	50	75	90	95
Boys							
6 years..........21.7	22.5	23.7	25.3	26.8	28.0	28.8
7 years..........22.3	23.1	24.5	26.2	27.6	29.2	30.2
8 years..........23.1	23.8	25.3	26.8	28.6	30.1	31.6
9 years..........23.5	24.4	25.9	27.5	29.5	32.1	34.7
10 years.........24.3	25.3	27.0	28.5	30.5	32.6	34.4
11 years.........25.6	26.5	27.9	29.7	32.1	34.9	37.3
Girls							
6 years..........21.0	21.4	22.5	24.0	25.4	26.9	28.1
7 years..........21.3	22.0	23.1	24.6	26.4	28.3	29.5
8 years..........21.4	22.3	24.1	25.7	27.7	29.7	31.6
9 years..........23.0	23.5	24.8	26.5	28.8	31.7	34.2
10 years.........23.4	24.2	25.7	27.7	30.4	33.4	36.1
11 years.........24.5	25.3	26.8	29.2	32.1	35.2	37.4

(data from Malina et al., 1973)

TABLE 1031

WRIST BREADTH (cm) IN U.S. CHILDREN

Age (years)	Mean	S.D.	Percentiles						
			5	10	25	50	75	90	95
BOYS									
12	5.0	0.40	4.5	4.6	4.7	5.1	5.3	5.6	5.8
13	5.3	0.42	4.6	4.7	5.1	5.3	5.6	5.8	5.9
14	5.5	0.35	4.8	5.0	5.2	5.6	5.8	5.9	6.2
15	5.6	0.36	5.0	5.1	5.3	5.6	5.8	6.2	6.3
16	5.7	0.30	5.1	5.2	5.5	5.7	5.8	6.2	6.3
17	5.7	0.32	5.1	5.2	5.5	5.7	5.9	6.2	6.3
GIRLS									
12	4.9	0.28	4.4	4.5	4.6	4.8	5.2	5.3	5.4
13	4.9	0.29	4.5	4.6	4.7	4.9	5.2	5.4	5.4
14	5.0	0.27	4.5	4.6	4.7	5.0	5.2	5.4	5.4
15	5.0	0.30	4.5	4.6	4.7	5.0	5.2	5.4	5.6
16	5.0	0.28	4.5	4.6	4.7	5.0	5.2	5.4	5.4
17	5.0	0.28	4.5	4.6	4.7	5.0	5.2	5.4	5.4

(data from Cycle III of the Health Examination Survey by the National Center
of Health Statistics 1966-1970)

TABLE 1032

UPPER LIMB DIAMETERS (cm) FOR BLACK BOYS IN TEXAS

Age	Elbow				Wrist			
(years)	Lower Income		Middle Income		Lower Income		Middle Income	
	Mean	S.D.	Mean	S.D.	Mean	S.D.	Mean	S.D.
10.1 - 11	6.0	0.24	3.8	0.33	5.0	0.22	4.8	0.15
11.1 - 12	6.0	0.23	6.1	0.38	5.0	0.34	5.2	0.35
12.1 - 13	6.2	0.37	6.4	0.31	5.0	0.20	5.3	0.32
13.1 - 14	6.5	0.36	6.5	0.33	5.4	0.53	5.3	0.31
14.1 - 15	6.5	0.53	6.9	0.32	5.4	0.45	5.7	0.31
15.1 - 16	6.9	0.31	7.0	0.33	5.7	0.29	5.8	0.21
16.1 - 17	7.1	0.23	7.0	0.25	6.0	0.22	5.8	0.30
17.1 - 18	7.1	0.34	7.1	0.32	5.8	0.33	5.8	0.37

(data from Schutte, 1979)

TABLE 1033

HAND WIDTH (cm)
FOR CHILDREN IN 8 STATES

Age (months)	Boys		Girls	
	Mean	S.D.	Mean	S.D.
0 - 3	3.7	0.5	3.5	0.5
4 - 6	4.2	0.4	4.0	0.4
7 - 9	4.5	0.4	4.4	0.4
10 - 12	4.6	0.4	4.3	0.3
13 - 18	4.7	0.3	4.5	0.3
19 - 24	5.0	0.3	4.7	0.3
25 - 30	5.1	0.4	5.0	0.3
31 - 36	5.1	0.4	5.0	0.3
37 - 42	5.2	0.3	5.1	0.4
43 - 48	5.3	0.4	5.2	0.4
49 - 54	5.5	0.3	5.3	0.4
55 - 60	5.6	0.3	5.5	0.4
61 - 66	5.8	0.4	5.6	0.4
67 - 72	5.8	0.4	5.7	0.3
73 - 78	6.0	0.4	5.9	0.3
79 - 84	6.2	0.4	6.0	0.4
85 - 96	6.4	0.4	6.2	0.4
97 - 108	6.6	0.4	6.4	0.4
109 - 120	6.7	0.3	6.6	0.4
121 - 132	7.1	0.4	6.9	0.4
133 - 144	7.4	0.4	7.2	0.4
145 - 156	7.5	0.4	7.6	0.4

(data from Snyder et al., 1975)

TABLE 1034

HAND WIDTH (cm) IN U.S. CHILDREN

Age (years)	Percentiles						
	5	10	25	50	75	90	95
				BOYS			
6	5.1	5.3	5.7	6.2	6.6	6.8	6.9
7	5.2	5.4	6.1	6.4	6.7	6.8	6.9
8	5.4	5.8	6.2	6.5	6.8	7.2	7.4
9	6.0	6.1	6.3	6.7	7.2	7.6	7.8
10	6.1	6.1	6.4	6.8	7.4	7.7	7.8
11	6.1	6.3	6.7	7.3	7.6	7.8	7.9
				GIRLS			
6	5.1	5.1	5.4	5.8	6.4	6.7	6.8
7	5.1	5.3	5.7	6.2	6.6	6.8	6.8
8	5.2	5.4	6.1	6.4	6.7	6.9	7.1
9	5.6	6.0	6.2	6.5	6.8	7.2	7.6
10	6.0	6.1	6.3	6.6	7.2	7.6	7.8
11	6.1	6.2	6.5	7.1	7.5	7.8	7.9

(data from Malina et al., 1973)

TABLE 1035

THIRD FINGER DIAMETER (cm)
FOR CHILDREN IN 8 STATES

Age (months)	Boys		Girls	
	Mean	S.D.	Mean	S.D.
0 - 3	0.73	0.05	0.69	0.05
4 - 6	0.80	0.05	0.75	0.05
7 - 9	0.86	0.05	0.83	0.05
10 - 12	0.89	0.05	0.83	0.04
13 - 18	0.93	0.06	0.88	0.07
19 - 24	0.97	0.06	0.91	0.05
25 - 30	0.98	0.06	0.96	0.06
31 - 36	1.02	0.06	0.98	0.07
37 - 42	1.06	0.06	1.01	0.06
43 - 48	1.09	0.06	1.04	0.07
49 - 54	1.11	0.07	1.06	0.06
55 - 60	1.13	0.07	1.08	0.07
61 - 66	1.16	0.07	1.10	0.07
67 - 72	1.18	0.07	1.11	0.07
73 - 78	1.19	0.06	1.14	0.07
79 - 84	1.22	0.07	1.16	0.07
85 - 96	1.25	0.08	1.19	0.06
97 - 108	1.28	0.08	1.23	0.07
109 - 120	1.31	0.07	1.26	0.07
121 - 132	1.37	0.09	1.31	0.09
133 - 144	1.40	0.07	1.35	0.07
145 - 156	1.42	0.10	1.43	0.07

(data from Snyder et al., 1975)

TABLE 1036

FIFTH FINGER DIAMETER (cm)
FOR CHILDREN IN 8 STATES

Age (months)	Boys		Girls	
	Mean	S.D.	Mean	S.D.
0 - 3	0.64	0.05	0.61	0.05
4 - 6	0.71	0.05	0.66	0.04
7 - 9	0.76	0.05	0.73	0.04
10 - 12	0.77	0.04	0.74	0.04
13 - 18	0.81	0.05	0.77	0.05
19 - 24	0.84	0.05	0.79	0.05
25 - 30	0.84	0.04	0.83	0.05
31 - 36	0.89	0.05	0.84	0.05
37 - 42	0.91	0.05	0.86	0.05
43 - 48	0.92	0.06	0.88	0.06
49 - 54	0.94	0.06	0.90	0.05
55 - 60	0.97	0.06	0.91	0.06
61 - 66	0.99	0.07	0.94	0.06
67 - 72	1.00	0.06	0.94	0.06
73 - 78	1.02	0.05	0.96	0.06
79 - 84	1.04	0.06	0.98	0.06
85 - 96	1.07	0.07	1.00	0.06
97 - 108	1.09	0.07	1.03	0.06
109 - 120	1.12	0.06	1.06	0.06
121 - 132	1.16	0.07	1.10	0.07
133 - 144	1.19	0.06	1.13	0.06
145 - 156	1.22	0.09	1.19	0.05

(data from Snyder et al., 1975)

TABLE 1037

HAND MEASUREMENTS (cm) IN IOWA BOYS

Lat-erality	Grade	Right Hand						
		1 to 2	13 to 6	3 to 4	5 to 9	6 to 10	7 to 11	8 to 12
Right	K	6.518 0.380	7.636 0.512	4.181 0.596	5.013 0.361	5.635 0.366	5.208 0.363	4.168 0.353
	1	6.696 0.454	8.003 0.535	4.412 0.493	5.329 0.409	5.921 0.428	5.422 0.439	4.338 0.396
	2	6.975 0.483	8.285 0.539	4.583 0.623	5.517 0.389	6.168 0.450	5.678 0.504	4.519 0.408
	3	7.221 0.594	8.574 0.710	5.087 0.670	5.787 0.455	6.479 0.524	5.976 0.459	4.734 0.435
	4	7.444 0.521	8.916 0.618	5.161 0.622	5.972 0.421	6.635 0.451	6.154 0.419	4.971 0.436
	5	7.775 0.805	8.848 0.931	5.392 0.676	6.151 0.455	6.869 0.456	6.340 0.497	5.138 0.440
	6	6.247 0.464	7.238 0.512	4.089 0.402	4.775 0.378	5.376 0.404	4.923 0.444	3.913 0.390
Left	K	6.388 0.387	7.494 0.434	4.312 0.455	4.918 0.349	5.665 0.390	5.224 0.409	4.053 0.450
	1	6.900 0.476	8.285 0.517	4.585 0.566	5.324 0.345	5.975 0.512	5.535 0.475	4.420 0.471
	2	7.289 0.420	8.622 0.387	4.878 0.458	5.733 0.346	6.200 0.450	5.689 0.525	4.678 0.233
	3	7.279 0.607	8.533 0.868	5.050 0.595	5.763 0.481	6.450 0.523	5.942 0.519	4.758 0.544
	4	7.420 0.645	9.007 0.714	5.167 0.597	5.993 0.754	6.647 0.410	6.093 0.493	4.840 0.510
	5	7.473 0.550	9.255 0.885	5.764 0.700	6.145 0.534	6.955 0.532	6.527 0.613	5.245 0.532
	6	6.339 0.348	7.372 0.457	4.000 0.519	4.833 0.356	5.472 0.337	4.978 0.375	3.944 0.313

Lat-erality	Grade	Left Hand						
		1 to 2	13 to 6	3 to 4	5 to 9	6 to 10	7 to 11	8 to 12
Right	K	6.448 0.351	7.630 0.647	4.213 0.578	4.988 0.321	5.604 0.344	5.220 0.327	4.132 0.335

TABLE 1037 (continued)

HAND MEASURMENTS (cm) IN IOWA BOYS

Lat-erality	Grade	1 to 2	13 to 6	3 to 4	5 to 9	6 to 10	7 to 11	8 to 12
Right	1	6.660 0.468	7.973 0.557	4.496 0.505	5.304 0.421	5.911 0.417	5.446 0.416	4.387 0.386
	2	6.988 0.464	8.361 0.507	4.573 0.606	5.507 0.422	6.202 0.512	5.676 0.459	4.539 0.394
	3	7.154 0.596	8.613 0.679	5.054 0.551	5.736 0.449	6.446 0.521	5.957 0.452	4.781 0.414
	4	7.379 0.486	8.905 0.666	5.105 0.605	5.993 0.436	6.690 0.456	6.144 0.434	4.946 0.423
	5	7.808 0.841	8.903 0.993	5.426 0.731	6.200 0.459	6.887 0.480	6.396 0.465	5.108 0.429
	6	6.246 0.453	7.274 0.583	4.123 0.501	4.790 0.373	5.387 0.412	4.914 0.417	3.935 0.397
Left	K	6.335 0.424	7.465 0.398	4.306 0.455	4.971 0.372	5.682 0.411	5.212 0.364	4.171 0.351
	1	6.970 0.461	8.315 0.474	4.715 0.465	5.315 0.369	6.030 0.491	5.605 0.507	4.470 0.381
	2	7.422 0.447	8.600 0.332	4.900 0.490	5.656 0.343	6.378 0.533	5.856 0.456	4.622 0.254
	3	7.296 0.515	8.608 0.813	5.117 0.525	5.750 0.438	6.417 0.531	5.900 0.454	4.783 0.504
	4	7.373 0.615	8.967 0.720	5.320 0.621	5.907 0.512	6.607 0.440	6.093 0.413	4.727 0.440
	5	7.764 0.662	9.300 0.710	5.627 0.713	6.336 0.480	6.836 0.658	6.455 0.641	5.127 0.582
	6	6.289 0.334	7.428 0.430	4.056 0.482	5.839 0.303	5.411 0.386	4.961 0.348	3.972 0.407

(data from Burmeister et al., 1974)

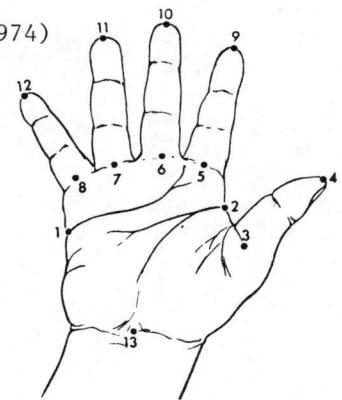

TABLE 1038

HAND MEASUREMENTS (cm) IN IOWA GIRLS

		Right-hand						
Laterality	Grade	1 to 2	13 to 6	3 to 4	5 to 9	6 to 10	7 to 11	8 to 12
Right	K	6.349	7.450	4.206	5.080	5.693	5.223	4.114
		0.394	0.562	0.651	0.405	0.398	0.429	0.336
	1	6.466	7.906	4.410	5.311	5.890	5.386	4.238
		0.492	0.495	0.528	0.392	0.435	0.428	0.361
	2	6.745	8.184	4.521	5.569	6.179	5.715	4.520
		0.554	0.564	0.579	0.470	0.541	0.464	0.459
	3	7.014	8.461	5.001	5.809	6.442	5.921	4.739
		0.560	0.664	0.559	0.513	0.541	0.533	0.449
	4	7.218	8.753	5.191	6.034	0.685	6.141	4.903
		0.615	0.747	0.660	0.451	0.514	0.479	0.442
	5	7.420	9.336	5.390	6.399	7.104	6.519	5.213
		0.516	0.923	0.704	0.510	0.528	0.473	0.420
	6	6.079	7.191	4.079	4.830	5.396	4.924	3.897
		0.387	0.412	0.361	0.332	0.347	0.350	0.330
Left	K	6.260	7.460	4.167	4.913	5.620	5.100	4.040
		0.385	0.447	0.546	0.304	0.375	0.409	0.352
	1	6.341	7.629	4.318	5.229	5.859	5.288	4.224
		0.348	0.412	0.414	0.297	0.336	0.312	0.266
	2	6.530	8.080	4.430	5.440	6.120	5.570	4.490
		0.501	0.471	0.633	0.438	0.459	0.450	0.456
	6	6.127	7.127	3.873	4.620	5.267	4.807	3.700
		0.394	0.369	0.322	0.334	0.327	0.417	0.248
		Left-hand						
Laterality	Grade	1 to 2	13 to 6	3 to 4	5 to 9	6 to 10	7 to 11	8 to 12
Right	K	6.301	7.470	4.136	5.053	5.698	5.217	4.123
		0.421	0.531	0.523	0.380	0.416	0.425	0.354
	1	6.420	7.904	4.476	5.283	5.881	5.401	4.292
		0.485	0.498	0.475	0.407	0.436	0.424	0.356
	2	6.700	8.257	4.600	5.517	6.180	5.708	4.561
		0.477	0.612	0.662	0.477	0.532	0.583	0.467

TABLE 1038 (continued)

HAND MEASUREMENTS (cm) IN IOWA GIRLS

Laterality	Grade	Left-hand						
		1 to 2	13 to 6	3 to 4	5 to 9	6 to 10	7 to 11	8 to 12
Right	3	6.942	8.518	5.017	5.787	6.436	5.936	4.753
		0.562	0.722	0.580	0.513	0.523	0.535	0.467
	4	7.091	8.797	5.141	5.995	6.682	6.129	4.895
		0.575	0.796	0.654	0.453	0.506	0.462	0.461
	5	7.337	9.346	5.392	6.360	7.087	6.520	5.205
		0.505	0.734	0.682	0.577	0.524	0.484	0.420
	6	6.067	7.208	4.080	4.826	5.382	4.943	3.877
		0.353	0.471	0.377	0.321	0.339	0.336	0.344
Left	K	6.280	7.440	4.160	4.973	5.607	5.040	4.020
		0.446	0.407	0.434	0.255	0.373	0.364	0.248
	1	6.306	7.641	4.471	5.206	5.776	5.400	4.300
		0.354	0.449	0.387	0.361	0.378	0.314	0.320
	2	6.690	8.060	4.440	5.540	6.110	5.570	4.490
		0.502	0.369	0.709	0.420	0.436	0.457	0.436
	6	6.140	6.933	3.967	4.700	5.273	4.793	3.773
		0.267	1.144	0.461	0.488	0.333	0.246	0.231

(data from Burmeister et al., 1974)

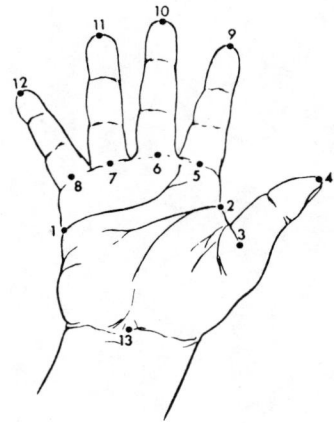

PHYSIQUE

TABLE 1039

DISTRIBUTION OF WETZEL PHYSIQUE TYPES BY AGE FOR BOYS IN CINCINNATI

Age (years)	Physique Type			
	Obese A_4 & > %	Stocky A_1-A_3 %	Medium M %	Thin B_1-B_4 %
White				
5	10.5	55.9	23.8	9.8
6	4.8	45.2	29.2	20.8
7	4.5	32.5	24.3	38.8
8	5.4	22.6	27.2	44.8
9	7.7	23.2	23.9	45.2
10	8.6	18.6	23.1	49.8
11	9.6	28.9	16.8	44.7
12	8.6	23.4	20.7	47.3
13	11.1	26.0	19.3	43.6
14	10.4	26.1	17.5	45.9
15	12.3	28.8	18.5	40.3
16	15.8	32.9	18.0	33.3
17	12.9	42.9	14.3	30.0
18	45.5	18.2	18.2	18.2
Nonwhite				
5	4.3	44.9	33.3	17.4
6	2.7	33.6	24.5	39.1
7	0.9	26.1	28.7	44.3
8	3.1	23.4	21.9	51.6
9	5.0	18.2	21.5	55.4
10	4.5	22.5	13.5	59.6
11	5.9	16.8	17.8	59.4

TABLE   1039   (continued)

DISTRIBUTION OF WETZEL PHYSIQUE TYPES BY AGE FOR BOYS IN CINCINNATI

| | Physique Type | | | |
| | Obese A_4 & > % | Stocky A_1–A_3 % | Medium M % | Thin B_1–B_4 % |
Age (years)				
Nonwhite				
12	6.9	26.7	13.9	52.5
13	11.6	10.5	18.6	59.3
14	5.4	22.7	18.2	53.6
15	7.0	36.6	12.7	43.7
16	5.4	33.9	17.9	42.9
17	15.8	34.2	18.4	31.6
18	7.7	38.5	15.4	38.5

(data from Rauh et al., 1967)

TABLE 1040

DISTRIBUTION OF WETZEL PHYSIQUE TYPES BY AGE FOR GIRLS IN CINCINNATI

Age (years)	Physique Type			
	Obese A_4 & > %	Stocky A_1-A_3 %	Medium M %	Thin B_1-B_4 %
White				
5	16.4	50.7	19.7	13.2
6	6.4	35.2	26.9	31.4
7	4.8	28.7	25.5	41.0
8	6.9	25.8	20.0	47.3
9	10.9	25.9	14.2	49.0
10	7.1	26.1	12.7	54.1
11	10.3	22.6	17.1	50.0
12	10.4	25.1	13.9	50.6
13	11.0	32.0	18.0	39.0
14	14.3	32.0	15.1	38.6
15	15.3	37.8	14.9	32.0
16	13.7	40.1	23.8	22.5
17	21.1	38.0	21.1	19.7
18	0.0	16.7	33.3	50.0
Nonwhite				
5	5.0	32.5	26.3	36.3
6	3.8	19.7	26.5	50.0
7	6.1	15.9	16.7	61.4
8	5.1	19.0	21.2	54.7
9	8.3	17.6	17.6	56.5
10	11.4	18.2	14.8	55.7
11	11.5	23.1	18.3	47.1

TABLE 1040 (continued)

DISTRIBUTION OF WETZEL PHYSIQUE TYPES BY AGE FOR GIRLS IN CINCINNATI

| | Physique Type | | | |
Age (years)	Obese A_4 & > %	Stocky A_1-A_3 %	Medium M %	Thin B_1-B_4 %
Nonwhite				
12	9.7	27.2	9.7	53.4
13	10.6	31.7	20.3	37.4
14	18.0	35.0	20.0	27.0
15	21.3	16.9	27.0	34.8
16	26.3	38.6	12.3	22.8
17	26.1	34.8	21.7	17.4
18	30.8	23.1	23.1	23.1

(data from Rauh et al., 1967)

TABLE 1041

PRESCHOOL PHYSIQUE AND LATE-ADOLESCENT SOMATOTYPE IN CONNECTICUT CHILDREN

| | Boys | | | Girls | | |
Method	Endo-morphy	Meso-morphy	Ecto-morphy	Endo-morphy	Meso-morphy	Ecto-morphy
Composite (Preschool)						
Mean	3.7	4.2	4.1	4.4	3.5	3.8
S.D.	0.9	0.9	0.9	0.8	0.9	1.0
Somatotypes (Young adult)						
Mean	4.1	4.3	3.6	5.1	3.3	3.4
S.D.	1.0	0.8	1.1	0.7	0.5	1.1

(data from Walker, 1978, Annals of Human Biology 5:113-129)

TABLE 1042

ANTHROPOSCOPIC RATINGS AND SHELDON SOMATOTYPE II FORECASTS AND RATINGS

Age (years)		Endomorphy		Mesomorphy		Ectomorphy	
		Anthropo-scopic	Somato-type II	Anthropo-scopic	Somato-type II	Anthropo-scopic	Somato-type II
5	Mean	3.5	4.1	3.8	4.5	3.8	3.1
	S.D.	0.8	0.5	0.7	0.6	1.1	0.8
8	Mean	3.5	3.9	3.9	4.3	4.0	3.2
	S.D.	0.8	0.6	0.9	0.7	1.0	0.9
11	Mean	3.5	3.8	3.8	4.0	4.0	3.3
	S.D.	0.9	0.8	0.8	0.6	1.1	0.9
14	Mean	3.5	4.0	3.8	4.0	4.2	3.3
	S.D.	0.9	0.8	0.9	0.6	1.2	0.9
18_p	Mean	3.5	3.7	4.0	4.3	3.9	3.2
	S.D.	0.7	0.6	0.8	0.8	1.1	1.1
18_s	Mean	--	3.6	--	4.2	--	3.2
	S.D.	--	0.7	--	0.7	--	1.1

18_p = somatotype II ratings based upon SPI predictions for that age alone; 18_s = ratings based upon SPI determined where possible from subjects' entire growth series.

(data from Walker and Tanner, 1980, Annals of Human Biology 7:213-224)

TABLE 1043

MEANS FOR ANTHROPOMETRIC INDICES FOR WHITE GIRLS IN ALABAMA

Measurement	Age	(years)		
	2	3	4	5
Skelic Index:				
Mean	66.0	71.5	74.2	77.0
Thorax/Abdomen Index:				
Mean	99.0	99.7	100.1	100.9
Hip/Stem Index:				
Mean	28.3	29.3	28.6	28.9
Thoracic Index:				
Mean	131.2	135.1	136.4	135.8

(Age is midpoint for group; skelic index = leg length/sitting height; thorax/
abdomen index = thoracic circumference at xiphoid/abdominal circumference at
umbilicus; hip/stem index = bicristal diameter/sitting height; thoracic index
= thoracic width/thoracic depth at xiphoid. All indices expressed as
percentages.)

(data from Wise and Meredith, 1942)

TABLE 1044

MEANS FOR FOUR BODY FORM RATIOS ON NORTH AMERICAN
WHITE BOYS OF NORTHWEST EUROPEAN ANCESTRY IN IOWA CITY (Iowa)

Age (years)	Mean Upper Limb Index	Mean Stem Index
4	39.4	88.7
5	37.6	88.2
6	36.3	87.4
7	35.5	86.9
8	34.4	86.4
	Lower Limb Index	Over-all Index
4	50.1	2.49
5	47.8	2.44
6	46.1	2.39
7	44.8	2.35
8	43.6	2.32

Upper limb index $= \dfrac{\text{upper arm circumference}}{\text{upper limb length}} \times 100;$

lower limb index $= \dfrac{\text{calf circumference}}{\text{lower limb length}} \times 100;$

stem index $= \dfrac{\text{chest circumference}}{\text{stem length}} \times 100;$

and over-all index $= \dfrac{\sqrt[3]{W}}{\text{stature}}$

(data from Meredith and Culp, 1951)

SKELETAL MEASUREMENTS

TABLE   1045

MEANS OF PERCENTAGES OF THE TOTAL WEIGHT
OF THE SKELETON CONTRIBUTED BY FOUR DIVISIONS
ACCORDING TO RACE, SEX AND SUCCESSIVE AGE PERIODS

	>0.5-3.0	Age (years) >3.0-13.0	>13.0
White Boys			
Skull	46	41	19
Postcranial	23	21	19
Superior limb	11	13	17
Inferior limb	20	25	45
White Girls			
Skull	46	33	19
Postcranial	26	22	21
Superior limb	10	14	17
Inferior limb	18	31	43
Negro Boys			
Skull	48	34	17
Postcranial	22	18	18
Superior limb	12	14	19
Inferior limb	18	34	46
Negro Girls			
Skull	48	30	20
Postcranial	21	19	19
Superior limb	11	14	17
Inferior limb	20	37	44

Note: Some groups have N < 10

(data from Trotter and Hixon, 1974)

TABLE 1046

MEAN WEIGHTS (gm) OF FOUR DIVISIONS OF THE SKELETON
AND THE TOTAL WEIGHT

	Age (years)			
	Birth-0.5	>0.5-3.0	>3.0-13.0	>13.0
White Boys				
Skull	48.2	133.4	365.6	697.3
Postcranial	19.3	71.8	192.0	756.2
Superior limb	11.5	36.3	127.8	703.6
Inferior limb	16.7	74.9	247.2	1847.3
Total	95.8	315.8	932.6	4004.4
White Girls				
Skull	30.1	129.3	312.0	524.0
Postcranial	16.9	62.2	250.4	559.0
Superior limb	10.1	27.5	160.9	468.3
Inferior limb	13.9	49.4	409.1	1173.0
Total	71.0	268.4	1132.4	2724.3
Negro Boys				
Skull	48.7	125.9	438.9	690.1
Postcranial	25.2	56.0	262.8	772.7
Superior limb	14.5	30.0	209.7	824.7
Inferior limb	21.7	48.4	545.2	1941.0
Total	110.0	260.2	1456.6	4228.5
Negro Girls				
Skull	45.4	182.4	379.5	629.8
Postcranial	24.6	75.7	241.0	632.0
Superior limb	12.9	44.2	182.2	549.3
Inferior limb	19.5	80.1	502.9	1445.1
Total	102.3	382.3	1305.6	3256.2

(data from Trotter and Hixon, 1974)

TABLE 1047

MEAN PERCENTAGE ASH WEIGHTS OF THE INDIVIDUAL BONES AND BONE SETS
AND THE TOTAL SKELETON OF THE WHITE MALE GROUP ACCORDING TO AGE

	Young[1] Age (years)			
	Birth-0.5	>0.5-3.0	>3.0-13.0	>13.0-22
Cranium	66.0	66.3	66.8	67.2
Mandible	64.9	70.2	72.0	71.1
Cervical v.	63.2	64.1	63.2	64.3
Thoracic v.	63.9	61.9	60.7	62.8
Lumbar v.	63.0	61.6	59.2	62.4
Sacrum	60.0	59.6	57.6	61.9
Sternum	62.4	58.8	56.2	59.7
Ribs	63.9	63.9	63.7	64.4
Scapula	63.1	62.3	62.9	64.1
Clavicle	63.3	64.4	63.6	65.2
Humerus	63.2	60.7	64.4	66.4
Radius	63.1	60.9	63.5	66.5
Ulna	62.3	63.0	63.0	66.0
Hand bones	58.3	51.7	58.3	60.8
Hip	62.8	60.8	60.2	62.8
Femur	64.0	59.3	63.4	66.0
Tibia	63.4	56.2	62.6	65.6
Fibula	62.9	58.0	64.0	66.3
Foot bones	59.4	49.9	55.0	60.3
Total	64.6	63.6	64.0	65.1

[1] Adapted from Trotter, M., 1973, Growth, 37:153-163; v = vertebrae.

(data from Trotter and Hixon, 1974)

TABLE   1048

MEANS OF PERCENTAGE ASH WEIGHT OF THE SKELETON

Age (years)	White Boys				White Girls			
	B-0.5	>0.5-3	>3-13	>13-22	B-0.5	>0.5-3	>3-13	>13-22
Mandible	64.92	70.15	71.96	71.11	67.94	70.58	71.70	66.89
Cranium	66.02	66.26	66.84	67.19	69.32	64.77	66.31	64.60
Humeri	63.16	60.68	64.38	66.30	64.62	62.00	63.98	59.09
Radii	63.09	60.88	63.52	66.28	62.60	62.76	63.26	61.74
Ulnae	62.34	62.99	63.03	66.04	63.58	62.47	63.47	61.46
Femora	63.97	59.32	63.41	65.96	64.48	60.63	64.17	59.81
Tibiae	63.40	56.15	62.61	65.63	63.04	61.27	63.72	57.34
Fibulae	62.88	58.01	63.96	66.60	65.44	62.54	64.19	61.30
Clavicles	63.30	64.38	63.58	65.16	66.02	63.17	63.10	62.17
Scapulae	63.09	62.32	62.94	64.14	65.90	60.45	62.82	60.40
Ribs	63.89	63.91	63.74	64.38	64.98	62.84	63.67	60.44
Hip bones	62.84	60.81	60.20	62.84	64.02	58.58	60.40	54.48
C. Vert.	63.17	64.06	63.23	64.32	67.16	62.86	62.26	61.80
T. Vert.	63.88	61.91	60.74	62.75	66.99	60.46	60.29	58.75
L. Vert.	62.98	61.59	59.18	62.39	66.66	58.54	59.70	54.55
Sacrum	60.01	59.58	57.59	61.93	67.82	59.56	59.20	53.24
Sternum	62.39	58.80	56.20	59.72	63.41	56.04	56.35	54.64
Hand bones	58.26	51.74	58.28	60.80	57.68	60.03	59.95	59.76
Foot bones	59.44	49.90	55.03	60.33	58.72	57.08	60.81	58.34
Total Skeleton	64.57	63.28	64.04	65.09	66.92	62.79	64.18	60.45

Percentage = $\dfrac{\text{ash weight x 100}}{\text{weight of dry, fat-free bone.}}$

(data from Trotter, 1973)

TABLE 1049

MEAN PERCENTAGE ASH WEIGHTS OF SEPARATE PARTS OF
WHITE MALE BONES AND OF THE COMBINED PARTS

Bones	Age Range (years)	Mean Percentage Ash Weight			
		Shaft	Epiphysis Proximal	Distal	Combined Parts
Humerus	6.6–17.6	64.7	59.4	55.4	64.1
Radius	10.0–18.0	64.3	55.4	54.6	63.7
Femur	0.6–18.7	63.5	60.0	55.6	62.9
Tibia	0.6–18.7	61.7	56.0	57.7	61.2

$$\text{Percentage} = \frac{\text{ash weight x 100}}{\text{weight of dry, fat-free bone}}$$

(data from Trotter, 1973)

TABLE 1050

GEOMETRIC MEAN DENSITIES OF BONES
IN EACH RACE-SEX GROUP

Group	Age range (years)	Bones			
		Humerus	Radius	Femur	Tibia
WM	Birth-12.7	0.702	0.818	0.609	0.627
WF	Birth-6.7	0.701	0.794	0.603	0.625
NM	Birth-11.0	0.766	0.861	0.686	0.691
NF	Birth-11.0	0.785	0.910	0.711	0.697

WM = white male; WF = white female; NM = Negro male and NF = Negro female.

(data from Trotter, 1971)

TABLE 1051

MEAN DENSITY OF SEPARATE OSSEOUS PARTS OF NEGRO MALE BONES
AND OF THE COMBINED PARTS OF EACH BONE DETERMINED BY
DIVIDING THE SUM OF THE INDIVIDUAL WEIGHTS BY THE SUM OF THE VOLUMES

Bones	Age range (years)	Shaft	Epiphysis Proximal	Distal	Sum of weights Sum of volumes
Humerus	3.1-11	.918	.483	.623	.836
Radius	6.0-14	1.003	.780	.617	.964
Femur	0.3-14	.708	.524	.389	.661
Tibia	0.3-17	.730	.395	.490	.669

(data from Trotter, 1971)

TABLE 1052

MEAN HEIGHTS (mm) OF VERTEBRAL BODIES IN OHIO BOYS:
THE STANDARD DEVIATIONS ARE IN PARENTHESES

Measurement	Age (years)					
	0.25	0.5	0.75	1.0	1.5	2.0
Dens to C4	27.61 (2.03)	30.33 (1.86)	32.05 (1.66)	33.85 (2.24)	35.69 (1.88)	38.15 (2.44)
Dens (Ant.)	14.71 (1.84)	16.57 (1.30)	17.61 (0.94)	18.85 (1.60)	20.21 (1.12)	21.36 (1.64)
Dens (Inter.)	14.70 (1.55)	16.29 (1.32)	17.36 (1.04)	18.18 (1.52)	19.56 (1.33)	20.65 (1.60)
Dens (Post.)	14.51 (1.46)	16.12 (1.24)	17.07 (1.15)	18.20 (1.55)	19.15 (1.25)	20.04 (1.47)
C3 (Ant.)	2.80 (0.63)	3.35 (0.41)	3.66 (0.74)	3.81 (0.87)	4.11 (0.73)	4.38 (0.67)
C3 (Inter.)	3.87 (0.58)	4.13 (0.60)	4.57 (0.78)	4.90 (0.74)	5.11 (0.73)	5.28 (0.99)
C3 (Post.)	3.54 (0.58)	4.04 (0.54)	4.35 (0.82)	4.72 (0.82)	5.04 (0.57)	5.20 (0.77)
C4 (Ant.)	3.04 (0.50)	3.40 (0.65)	4.09 (0.82)	4.20 (0.81)	4.31 (0.65)	4.35 (0.66)
C4 (Inter.)	3.56 (0.58)	4.39 (0.46)	4.62 (0.85)	4.93 (0.61)	5.04 (0.69)	5.30 (0.76)
C4 (Post.)	3.62 (0.74)	4.14 (0.35)	4.85 (0.87)	5.05 (0.69)	5.14 (0.81)	5.39 (0.87)
C5 (Ant.)	3.10 (0.43)	3.43 (0.45)	4.01 (0.59)	4.06 (0.57)	4.30 (0.60)	4.59 (0.85)
C5 (Inter.)	3.97 (0.56)	4.24 (0.67)	4.70 (0.99)	4.96 (0.56)	5.07 (0.57)	5.46 (0.94)
C5 (Post.)	3.52 (0.22)	3.71 (0.79)	4.52 (1.06)	4.69 (0.76)	4.82 (0.86)	5.20 (0.90)

Measurement	Age (years)					
	2.5	3.0	3.5	4.0	4.5	5.0
Dens to C4	40.03 (2.15)	40.34 (2.88)	41.67 (3.43)	42.58 (3.27)	43.94 (3.30)	44.69 (2.85)
Dens (Ant.)	22.69 (1.91)	23.28 (2.19)	24.25 (2.38)	24.67 (2.23)	25.71 (2.37)	26.55 (2.03)

TABLE 1052 (continued)

MEAN HEIGHTS (mm) OF VERTEBRAL BODIES IN OHIO BOYS:
THE STANDARD DEVIATIONS ARE IN PARENTHESES

| Measurement | Age (years) | | | | | |
	2.5	3.0	3.5	4.0	4.5	5.0
Dens (Inter.)	21.59 (1.76)	22.25 (2.18)	22.81 (2.27)	23.36 (2.31)	24.32 (2.56)	25.08 (2.43)
Dens (Post.)	21.11 (1.65)	21.67 (1.78)	22.01 (1.96)	22.75 (2.10)	23.75 (2.45)	24.39 (2.35)
C3 (Ant.)	4.61 (0.71)	4.57 (0.62)	4.66 (0.70)	4.86 (0.61)	4.99 (0.37)	5.14 (0.63)
C3 (Inter.)	5.59 (0.91)	5.48 (0.84)	5.64 (0.85)	5.80 (0.71)	5.96 (0.90)	6.07 (0.89)
C3 (Post.)	5.47 (0.87)	5.67 (0.74)	6.02 (0.75)	6.01 (0.89)	6.39 (0.80)	6.81 (1.02)
C4 (Ant.)	4.74 (0.70)	4.65 (0.50)	4.73 (0.55)	5.14 (0.59)	5.21 (0.55)	5.29 (0.64)
C4 (Inter.)	5.65 (0.83)	5.69 (0.69)	5.67 (0.76)	5.94 (0.78)	6.08 (0.72)	6.23 (0.61)
C4 (Post.)	5.82 (0.83)	5.79 (0.97)	6.19 (1.04)	6.24 (1.03)	6.44 (0.76)	6.90 (1.07)
C5 (Ant.)	5.22 (0.73)	4.91 (0.46)	5.01 (0.45)	5.36 (0.69)	5.53 (0.65)	5.52 (0.64)
C5 (Inter.)	5.65 (0.69)	5.72 (0.68)	5.70 (0.67)	5.83 (0.67)	6.27 (0.55)	6.22 (0.73)
C5 (Post.)	5.87 (0.83)	5.68 (0.77)	6.36 (0.76)	5.83 (0.86)	6.37 (0.67)	6.93 (0.92)

(data from Roche, unpublished b)

TABLE 1053

MEAN HEIGHTS (mm) OF VERTEBRAL BODIES IN OHIO BOYS:
THE STANDARD DEVIATIONS ARE IN PARENTHESES

Measurement		Age (years)				
	6.0	7.0	8.0	9.0	10.0	11.0
Dens to C4	47.34 (3.07)	48.99 (4.01)	51.75 (3.81)	54.17 (3.74)	56.19 (4.38)	57.77 (3.42)
Dens (Ant.)	27.66 (2.69)	28.65 (3.18)	30.39 (3.01)	32.11 (2.43)	33.16 (2.41)	34.33 (2.42)
Dens (Inter.)	26.05 (2.67)	26.91 (3.25)	28.47 (3.20)	29.69 (2.38)	30.92 (2.14)	32.02 (2.18)
Dens (Post.)	25.20 (2.26)	25.88 (2.76)	27.27 (2.77)	29.70 (2.46)	29.87 (2.35)	30.92 (2.25)
C3 (Ant.)	5.64 (0.78)	5.98 (0.97)	6.27 (1.00)	6.68 (1.16)	7.00 (0.95)	7.75 (1.05)
C3 (Inter.)	6.42 (0.96)	6.77 (1.06)	7.00 (1.08)	7.36 (0.88)	7.91 (1.02)	8.54 (1.31)
C3 (Post.)	7.27 (1.03)	7.51 (1.29)	8.17 (1.18)	8.37 (1.22)	9.28 (1.01)	9.61 (1.05)
C4 (Ant.)	5.79 (0.78)	5.99 (0.74)	6.39 (0.75)	6.80 (0.86)	7.09 (0.93)	7.71 (0.87)
C4 (Inter.)	6.61 (0.88)	6.91 (0.90)	7.14 (0.88)	7.57 (1.05)	7.86 (1.01)	8.45 (1.12)
C4 (Post.)	7.38 (1.04)	7.87 (1.10)	8.67 (1.28)	8.79 (1.15)	9.49 (1.24)	10.04 (1.16)
C5 (Ant.)	6.02 (0.90)	6.27 (0.62)	6.53 (0.79)	6.55 (0.78)	6.97 (0.84)	7.84 (1.01)
C5 (Inter.)	6.47 (0.74)	7.12 (0.93)	7.33 (0.72)	6.60 (0.14)	8.11 (0.78)	8.16 (0.83)
C 5 (Post.)	7.05 (0.81)	7.58 (0.97)	8.08 (1.19)	8.20 (0.95)	8.91 (1.14)	10.10 (0.88)
Measurement		Age (years)				
	12.0	13.0	14.0	15.0	16.0	17.0
Dens to C4	59.92 (3.64)	62.96 (4.31)	66.41 (5.26)	67.76 (5.40)	68.84 (5.21)	71.98 (5.21)
Dens (Ant.)	35.79 (2.48)	37.06 (3.30)	38.77 (3.47)	39.80 (3.59)	39.95 (3.79)	40.96 (3.63)

TABLE 1053 (continued)

MEAN HEIGHTS (mm) OF VERTEBRAL BODIES IN OHIO BOYS:
THE STANDARD DEVIATIONS ARE IN PARENTHESES

			Age (years)			
Measurement	12.0	13.0	14.0	15.0	16.0	17.0
Dens (Inter.)	32.84 (2.51)	33.61 (2.94)	34.84 (3.00)	35.64 (3.40)	35.85 (3.37)	36.29 (3.43)
Dens (Post.)	31.73 (2.18)	32.85 (2.78)	34.22 (2.64)	35.04 (3.01)	35.49 (3.34)	35.92 (3.16)
C3 (Ant.)	8.05 (1.40)	8.71 (1.84)	10.40 (2.30)	12.04 (2.53)	12.86 (2.02)	14.43 (1.49)
C3 (Inter.)	8.65 (1.46)	9.26 (1.76)	10.64 (1.98)	11.71 (1.76)	12.81 (1.45)	13.67 (1.36)
C3 (Post.)	10.07 (1.10)	11.03 (2.02)	12.16 (2.08)	13.09 (1.90)	14.32 (1.33)	15.34 (1.64)
C4 (Ant.)	7.88 (1.30)	8.88 (1.81)	9.57 (2.68)	11.12 (2.08)	11.69 (1.72)	13.76 (1.85)
C4 (Inter.)	8.64 (1.44)	9.28 (1.73)	10.06 (2.78)	11.27 (1.87)	11.94 (1.57)	12.95 (1.60)
C4 (Post.)	10.38 (1.26)	11.64 (2.07)	12.66 (1.75)	12.97 (1.93)	13.37 (1.29)	14.46 (1.63)
C5 (Ant.)	7.74 (1.01)	8.82 (2.09)	9.38 (1.28)	10.23 (2.34)	10.60 (0.14)	12.40 (2.12)
C5 (Inter.)	8.44 (1.18)	9.88 (1.57)	9.47 (1.43)	11.20 (1.86)	11.50 (0.28)	12.15 (0.35)
C5 (Post.)	10.01 (1.57)	11.24 (2.23)	9.68 (4.27)	12.87 (2.18)	13.55 (0.50)	15.30 (2.26)

(data from Roche, unpublished b)

TABLE 1054

MEAN HEIGHTS (mm) OF VERTEBRAL BODIES IN OHIO GIRLS:
THE STANDARD DEVIATIONS ARE IN PARENTHESES

Measurement	Age (years)					
	0.25	0.50	0.75	1.0	1.5	2.0
Dens to C4	26.17 (1.02)	28.11 (1.29)	30.14 (1.30)	31.71 (1.46)	34.21 (1.59)	36.48 (1.66)
Dens (Ant.)	13.37 (0.75)	14.96 (0.92)	16.27 (1.03)	17.45 (1.03)	18.86 (1.42)	20.75 (1.22)
Dens (Inter.)	13.12 (0.65)	14.75 (0.92)	15.85 (1.03)	17.00 (1.02)	18.41 (1.40)	20.16 (1.44)
Dens (Post.)	13.08 (0.63)	14.57 (0.97)	15.69 (1.02)	16.94 (1.06)	18.21 (1.28)	19.79 (1.53)
C3 (Ant.)	2.86 (0.50)	3.17 (0.48)	3.55 (0.44)	3.80 (0.61)	4.13 (0.64)	4.44 (0.80)
C3 (Inter.)	3.72 (0.40)	4.02 (0.37)	4.49 (0.49)	4.53 (0.50)	5.15 (0.46)	5.30 (0.60)
C3 (Post.)	3.26 (0.22)	3.67 (0.43)	4.11 (0.35)	4.21 (0.55)	4.72 (0.45)	4.97 (0.68)
C4 (Ant.)	3.21 (0.27)	3.34 (0.43)	3.76 (0.52)	3.81 (0.61)	4.14 (0.37)	4.39 (0.57)
C4 (Inter.)	3.73 (0.38)	4.22 (0.43)	4.65 (0.48)	4.64 (0.50)	5.10 (0.29)	5.15 (0.54)
C4 (Post.)	3.48 (0.49)	4.05 (0.45)	4.29 (0.59)	4.49 (0.63)	4.94 (0.51)	4.70 (0.55)
C5 (Ant.)	3.60 (0.44)	3.67 (0.54)	3.97 (0.42)	4.32 (0.50)	4.29 (0.31)	4.64 (0.62)
C5 (Inter.)	3.93 (0.38)	4.47 (0.48)	4.71 (0.43)	4.89 (0.32)	5.07 (0.30)	5.27 (0.45)
C5 (Post.)	3.60 (0.61)	4.07 (0.46)	4.22 (0.57)	4.54 (0.48)	4.80 (0.46)	4.94 (0.61)

Measurement	Age (years)					
	2.5	3.0	3.5	4.0	4.5	5.0
Dens to C4	38.25 (1.87)	39.37 (2.11)	40.94 (1.99)	41.60 (2.41)	43.18 (2.01)	44.74 (2.52)
Dens (Ant.)	22.24 (1.61)	22.89 (1.89)	23.90 (1.71)	24.25 (1.97)	25.41 (1.82)	26.55 (2.17)

TABLE 1054 (continued)

MEAN HEIGHTS (mm) OF VERTEBRAL BODIES IN OHIO GIRLS:
THE STANDARD DEVIATIONS ARE IN PARENTHESES

Measurement		Age (years)					
		2.5	3.0	3.5	4.0	4.5	5.0
Dens	(Inter.)	21.42	22.07	22.90	23.07	23.89	24.96
		(1.61)	(1.82)	(1.52)	(1.74)	(1.79)	(2.00)
Dens	(Post.)	20.95	21.46	22.11	22.31	23.33	24.24
		(1.49)	(1.67)	(1.66)	(2.14)	(2.01)	(2.02)
C3	(Ant.)	4.52	4.47	4.74	4.70	5.13	5.30
		(0.62)	(0.51)	(0.46)	(0.39)	(0.43)	(0.48)
C3	(Inter.)	5.55	5.49	5.75	5.79	6.03	6.14
		(0.58)	(0.53)	(0.64)	(0.64)	(0.59)	(0.56)
C3	(Post.)	5.07	5.12	5.56	5.63	5.89	6.07
		(0.63)	(0.59)	(0.81)	(0.67)	(0.75)	(0.68)
C4	(Ant.)	4.81	4.81	5.03	5.07	5.24	5.48
		(0.49)	(0.56)	(0.36)	(0.45)	(0.25)	(0.57)
C4	(Inter.)	5.48	5.42	5.63	5.80	5.96	6.26
		(0.48)	(0.56)	(0.48)	(0.54)	(0.54)	(0.65)
C4	(Post.)	5.15	5.19	5.51	5.87	5.82	6.11
		(0.61)	(0.57)	(0.50)	(0.57)	(0.61)	(0.63)
C5	(Ant.)	4.87	5.03	5.01	5.10	5.16	5.55
		(0.34)	(0.46)	(0.32)	(0.30)	(0.34)	(0.39)
C5	(Inter.)	5.51	5.77	5.72	5.61	6.12	6.46
		(0.35)	(0.38)	(0.42)	(0.67)	(0.44)	(0.51)
C5	(Post.)	5.13	5.55	5.45	5.46	5.95	5.98
		(0.26)	(0.68)	(0.58)	(0.61)	(0.65)	(0.55)

(data from Roche, unpublished b)

TABLE 1055

MEAN HEIGHTS (mm) OF VERTEBRAL BODIES IN OHIO GIRLS:
THE STANDARD DEVIATIONS ARE IN PARENTHESES

Measurement		Age (years)				
	6.0	7.0	8.0	9.0	10.0	11.0
Dens to C4	47.13 (2.39)	49.84 (2.50)	51.22 (2.65)	53.11 (2.53)	54.89 (2.47)	57.30 (3.16)
Dens (Ant.)	28.09 (1.97)	29.97 (2.22)	30.79 (2.24)	32.16 (1.96)	33.25 (1.81)	34.36 (2.19)
Dens (Inter.)	26.39 (1.86)	28.10 (2.18)	28.71 (2.24)	29.72 (1.81)	30.80 (1.62)	31.49 (1.59)
Dens (Post.)	25.31 (1.91)	27.06 (2.20)	27.46 (2.24)	28.59 (2.13)	29.68 (1.54)	30.58 (1.74)
C3 (Ant.)	5.59 (0.57)	6.16 (0.88)	6.32 (0.66)	7.03 (0.87)	7.07 (0.77)	7.74 (0.65)
C3 (Inter.)	6.68 (0.69)	7.01 (0.74)	7.39 (0.74)	7.95 (0.77)	8.16 (0.71)	8.68 (0.74)
C3 (Post.)	6.43 (0.82)	6.84 (0.64)	7.41 (0.75)	7.96 (1.11)	8.37 (1.05)	9.07 (1.03)
C4 (Ant.)	5.87 (0.99)	6.25 (0.59)	6.44 (0.75)	6.72 (0.70)	7.09 (0.93)	8.02 (1.01)
C4 (Inter.)	6.66 (0.93)	7.02 (0.79)	7.34 (0.76)	7.69 (0.79)	7.97 (0.86)	8.56 (0.83)
C4 (Post.)	6.64 (1.02)	7.17 (0.60)	7.29 (0.57)	7.84 (0.81)	8.42 (0.94)	8.96 (0.79)
C5 (Ant.)	5.82 (0.57)	6.57 (0.25)	6.57 (0.62)	6.57 (0.25)	6.84 (0.25)	7.64 (0.81)
C5 (Inter.)	6.53 (0.44)	7.25 (0.75)	7.60 (0.50)	7.55 (0.42)	7.88 (0.60)	8.40 (0.51)
C5 (Post.)	6.42 (0.62)	7.20 (0.22)	7.45 (0.52)	7.57 (0.69)	8.34 (0.75)	8.86 (0.50)

Measurement		Age (years)				
	12.0	13.0	14.0	15.0	16.0	17.0
Dens to C4	60.27 (2.91)	62.58 (3.39)	63.64 (2.94)	64.54 (2.83)	65.19 (2.67)	64.53 (3.29)
Dens (Ant.)	35.90 (1.69)	37.07 (2.30)	38.03 (2.15)	38.04 (1.88)	38.72 (1.25)	39.02 (2.43)

TABLE 1055   (continued)

MEAN HEIGHTS (mm) OF VERTEBRAL BODIES IN OHIO GIRLS:
THE STANDARD DEVIATIONS ARE IN PARENTHESES

			Age (years)			
Measurement	12.0	13.0	14.0	15.0	16.0	17.0
Dens (Inter.)	32.67 (1.53)	33.40 (1.59)	33.79 (1.75)	34.18 (1.51)	34.55 (1.57)	34.64 (2.76)
Dens (Post.)	31.89 (1.50)	32.85 (1.66)	33.41 (1.75)	33.40 (1.76)	34.32 (1.46)	34.39 (2.62)
C3 (Ant.)	9.11 (0.84)	10.09 (1.12)	11.36 (1.52)	12.40 (1.28)	12.48 (1.43)	12.15 (1.22)
C3 (Inter.)	9.83 (0.83)	10.41 (1.13)	11.25 (0.84)	11.75 (0.69)	11.61 (0.82)	11.69 (1.21)
C3 (Post.)	10.32 (1.33)	11.48 (1.44)	12.06 (1.27)	12.97 (1.24)	12.83 (1.34)	12.75 (1.25)
C4 (Ant.)	8.89 (1.28)	9.68 (1.11)	10.76 (1.33)	12.05 (1.24)	11.78 (1.35)	12.17 (0.90)
C4 (Inter.)	9.48 (0.73)	10.07 (1.05)	10.73 (0.99)	11.26 (0.93)	11.02 (1.03)	11.15 (0.78)
C4 (Post.)	10.19 (0.99)	11.21 (0.84)	11.98 (1.18)	12.78 (1.12)	12.84 (1.14)	12.62 (1.20)
C5 (Ant.)	8.45 (1.04)	10.28 (1.40)	10.87 (1.80)	11.47 (1.64)	11.90 (0.89)	-- --
C5 (Inter.)	9.36 (1.05)	10.43 (1.14)	10.18 (0.89)	11.12 (1.21)	11.80 (0.85)	-- --
C5 (Post.)	9.98 (1.58)	11.19 (1.49)	11.52 (1.51)	11.87 (1.12)	12.85 (0.65)	-- --

(data from Roche, unpublished b)

TABLE 1056

MEAN HEIGHTS (mm) OF INTERVERTEBRAL DISCS IN OHIO BOYS.
THE STANDARD DEVIATIONS ARE IN PARENTHESES

Measurement	Age (years)					
	0.25	0.50	0.75	1.0	1.5	2.0
C2-3 (Ant.)	3.80 (0.39)	3.98 (0.42)	3.80 (0.48)	3.98 (0.53)	4.25 (0.48)	4.51 (0.59)
C2-3 (Inter.)	2.61 (0.24)	2.86 (0.32)	2.78 (0.38)	3.04 (0.21)	3.28 (0.27)	3.39 (0.57)
C2-3 (Post.)	2.96 (0.35)	3.27 (0.44)	3.21 (0.58)	3.50 (0.45)	3.60 (0.47)	3.62 (0.52)
C3-4 (Ant.)	3.32 (0.65)	3.15 (0.82)	3.21 (0.85)	3.46 (0.99)	3.46 (0.96)	4.17 (1.04)
C3-4 (Inter.)	2.33 (0.75)	2.35 (0.71)	2.35 (0.50)	2.49 (0.79)	2.61 (0.62)	2.92 (0.89)
C3-4 (Post.)	2.68 (0.72)	2.60 (0.81)	2.22 (0.80)	2.25 (0.87)	2.44 (0.96)	2.80 (0.86)
C4-5 (Ant.)	3.65 (1.40)	3.70 (0.68)	3.67 (0.65)	3.65 (0.95)	3.79 (0.62)	3.81 (0.89)
C4-5 (Inter.)	2.05 (0.70)	2.44 (0.49)	2.41 (0.56)	2.42 (0.89)	2.77 (0.39)	2.64 (0.46)
C4-5 (Post.)	2.77 (0.62)	2.93 (0.81)	2.47 (0.95)	2.65 (1.05)	2.78 (0.79)	2.57 (0.83)

Measurement	Age (years)					
	2.5	3.0	3.5	4.0	4.5	5.0
C2-3 (Ant.)	4.69 (0.61)	4.64 (0.47)	4.74 (0.65)	5.05 (0.67)	5.02 (0.70)	5.22 (0.54)
C2-3 (Inter.)	3.56 (0.53)	3.53 (0.37)	3.83 (0.57)	4.06 (0.59)	4.00 (0.55)	4.12 (0.59)
C2-3 (Post.)	3.98 (0.71)	3.91 (0.54)	3.92 (0.56)	4.23 (0.68)	4.10 (0.65)	3.91 (0.83)
C3-4 (Ant.)	3.86 (0.92)	4.21 (1.01)	4.39 (0.88)	4.21 (0.99)	4.52 (0.78)	4.47 (0.89)
C3-4 (Inter.)	2.94 (0.56)	3.13 (0.65)	3.38 (0.68)	3.45 (0.62)	3.62 (0.55)	3.72 (0.69)
C3-4 (Post.)	2.90 (0.59)	2.96 (0.83)	2.91 (1.01)	3.14 (0.76)	3.10 (0.93)	3.02 (0.92)

TABLE 1056   (continued)

MEAN HEIGHTS (mm) OF INTERVERTEBRAL DISCS IN OHIO BOYS.
THE STANDARD DEVIATIONS ARE IN PARENTHESES

| Measurement | Age (years) | | | | | |
	2.5	3.0	3.5	4.0	4.5	5.0
C4-5 (Ant.)	3.82 (0.88)	3.77 (1.10)	3.89 (0.72)	4.18 (0.97)	4.36 (0.74)	4.32 (0.85)
C4-5 (Inter.)	2.78 (0.54)	2.67 (0.69)	2.79 (0.42)	3.13 (0.53)	3.14 (0.52)	3.30 (0.76)
C4-5 (Post.)	2.54 (0.74)	2.22 (1.11)	2.46 (1.05)	3.17 (0.80)	2.84 (0.81)	2.75 (0.90)

| Measurement | Age (years) | | | | | |
	6.0	7.0	8.0	9.0	10.0	11.0
C2-3 (Ant.)	5.46 (0.67)	5.68 (0.45)	5.79 (0.55)	5.93 (0.74)	5.77 (0.74)	6.01 (0.64)
C2-3 (Inter.)	4.37 (0.62)	4.39 (0.53)	4.61 (0.48)	4.41 (1.29)	4.60 (0.45)	4.79 (0.77)
C2-3 (Post.)	4.12 (0.73)	4.26 (1.07)	4.13 (0.88)	4.09 (0.93)	4.09 (1.04)	3.89 (0.92)
C3-4 (Ant.)	4.97 (0.69)	5.13 (0.79)	5.06 (0.73)	5.41 (0.77)	5.37 (0.76)	5.33 (1.03)
C3-4 (Inter.)	4.06 (0.71)	4.28 (0.84)	4.46 (0.79)	4.60 (0.98)	4.57 (1.06)	4.73 (1.24)
C3-4 (Post.)	3.11 (0.77)	2.96 (1.25)	2.82 (1.32)	2.98 (1.28)	2.62 (1.31)	2.73 (1.10)
C4-5 (Ant.)	4.75 (0.70)	4.75 (0.72)	4.72 (0.61)	4.75 (0.62)	5.35 (1.16)	4.92 (1.87)
C4-5 (Inter.)	3.52 (0.66)	3.62 (0.65)	3.68 (0.52)	3.84 (0.68)	3.95 (0.82)	4.27 (0.95)
C4-5 (Post.)	3.17 (1.05)	2.82 (1.53)	2.47 (1.09)	2.87 (0.95)	2.41 (1.41)	2.60 (0.97)

| Measurement | Age (years) | | | | | |
	12.0	13.0	14.0	15.0	16.0	17.0
C2-3 (Ant.)	5.94 (0.81)	5.91 (0.96)	5.69 (1.03)	5.48 (0.97)	5.17 (1.30)	4.49 (1.09)

TABLE 1056 (continued)

MEAN HEIGHTS (mm) OF INTERVERTEBRAL DISCS IN OHIO BOYS.
THE STANDARD DEVIATIONS ARE IN PARENTHESES

Measurement	Age (years)					
	12.0	13.0	14.0	15.0	16.0	17.0
C2-3 (Inter.)	4.98 (0.79)	5.27 (0.78)	5.22 (0.97)	5.28 (1.04)	4.91 (1.13)	4.91 (0.94)
C2-3 (Post.)	4.05 (0.77)	4.00 (0.69)	3.64 (0.64)	3.55 (0.74)	3.60 (1.53)	3.50 (1.44)
C3-4 (Ant.)	5.41 (0.86)	5.26 (0.96)	4.86 (1.34)	4.35 (1.62)	4.32 (1.06)	3.95 (0.92)
C3-4 (Inter.)	4.91 (0.87)	5.00 (0.94)	4.74 (0.89)	4.84 (0.74)	4.72 (1.15)	4.72 (0.89)
C3-4 (Post.)	2.98 (0.75)	2.76 (1.33)	2.92 (1.25)	3.07 (1.04)	3.01 (0.88)	2.94 (1.15)
C4-5 (Ant.)	4.99 (1.01)	5.09 (1.81)	4.93 (1.76)	5.25 (2.03)	6.30 (3.56)	5.18 (2.82)
C4-5 (Inter.)	4.05 (0.53)	4.51 (1.07)	4.42 (0.92)	4.15 (0.35)	4.58 (0.60)	4.22 (0.55)
C4-5 (Post.)	2.43 (0.71)	2.95 (1.19)	2.92 (1.12)	2.83 (1.27)	3.32 (0.57)	2.86 (0.95)

(data from Roche, unpublished b)

TABLE   1057

MEAN HEIGHTS (mm) OF INTERVERTEBRAL DISCS IN OHIO GIRLS
THE STANDARD DEVIATIONS ARE IN PARENTHESES

Measurement	Age (years)					
	0.25	0.50	0.75	1.0	1.5	2.0
C2-3 (Ant.)	3.60	3.81	3.85	3.98	4.16	4.33
	(0.48)	(0.55)	(0.52)	(0.43)	(0.61)	(0.80)
C2-3 (Inter.)	2.75	2.92	2.95	3.05	3.12	3.30
	(0.36)	(0.31)	(0.37)	(0.39)	(0.37)	(0.41)
C2-3 (Post.)	3.30	3.56	3.66	3.67	3.73	3.81
	(0.38)	(0.65)	(0.74)	(0.69)	(0.56)	(0.69)
C3-4 (Ant.)	3.31	3.44	3.35	3.70	3.75	4.04
	(0.47)	(0.36)	(0.75)	(0.58)	(0.58)	(0.59)
C3-4 (Inter.)	2.63	2.57	2.41	2.68	2.77	2.99
	(0.29)	(0.25)	(0.42)	(0.39)	(0.32)	(0.53)
C3-4 (Post.)	2.90	2.64	2.67	2.83	3.01	3.37
	(0.51)	(0.32)	(0.41)	(0.56)	(0.60)	(0.68)
C4-5 (Ant.)	2.97	3.44	3.29	3.72	3.89	3.58
	(0.25)	(0.62)	(0.46)	(0.76)	(0.48)	(0.81)
C4-5 (Inter.)	2.47	2.44	2.46	2.75	2.90	2.78
	(0.06)	(0.37)	(0.37)	(0.51)	(0.27)	(0.31)
C4-5 (Post.)	2.70	2.63	2.87	2.91	3.27	3.08
	(0.53)	(0.45)	(0.32)	(0.70)	(0.70)	(0.48)
Measurement	Age (years)					
	2.5	3.0	3.5	4.0	4.5	5.0
C2-3 (Ant.)	4.60	4.78	4.88	4.85	4.85	5.07
	(0.70)	(0.58)	(0.55)	(0.49)	(0.73)	(0.65)
C2-3 (Inter.)	3.25	3.50	3.70	3.71	3.93	3.83
	(0.50)	(0.43)	(0.32)	(0.52)	(0.65)	(0.57)
C2-3 (Post.)	3.81	4.07	4.21	4.26	4.31	4.31
	(0.50)	(0.72)	(0.63)	(0.62)	(0.62)	(0.71)
C3-4 (Ant.)	3.92	4.02	4.17	4.22	4.43	4.52
	(0.65)	(0.45)	(0.69)	(0.58)	(0.34)	(0.48)
C3-4 (Inter.)	2.97	3.25	3.31	3.42	3.58	3.67
	(0.42)	(0.33)	(0.42)	(0.41)	(0.35)	(0.47)
C3-4 (Post.)	3.29	3.41	3.41	3.43	3.60	3.71
	(0.58)	(0.55)	(0.55)	(0.63)	(0.48)	(0.78)

TABLE 1057   (continued)

MEAN HEIGHTS (mm) OF INTERVERTEBRAL DISCS IN OHIO GIRLS
THE STANDARD DEVIATIONS ARE IN PARENTHESES

Measurement	Age (years)					
	2.5	3.0	3.5	4.0	4.5	5.0
C4-5 (Ant.)	3.59 (0.52)	3.90 (0.59)	3.92 (0.87)	3.78 (0.61)	4.29 (0.48)	4.26 (0.45)
C4-5 (Inter.)	2.82 (0.29)	3.01 (0.20)	3.02 (0.51)	3.13 (0.62)	3.32 (0.49)	3.23 (0.45)
C4-5 (Post.)	2.91 (0.50)	3.15 (0.51)	3.13 (0.74)	3.09 (0.56)	3.45 (0.53)	3.31 (0.50)

Measurement	Age (years)					
	6.0	7.0	8.0	9.0	10.0	11.0
C2-3 (Ant.)	5.42 (0.80)	5.54 (0.91)	5.62 (0.87)	5.60 (1.19)	6.10 (0.93)	6.07 (0.59)
C2-3 (Inter.)	4.07 (0.54)	4.34 (0.66)	4.43 (0.61)	4.55 (0.76)	4.61 (0.68)	4.85 (0.67)
C2-3 (Post.)	4.54 (0.68)	4.51 (0.71)	4.57 (0.60)	4.31 (0.75)	4.29 (0.69)	4.21 (0.67)
C3-4 (Ant.)	4.77 (0.73)	5.01 (0.73)	5.12 (0.56)	5.31 (0.90)	5.23 (0.99)	5.23 (0.52)
C3-4 (Inter.)	3.80 (0.56)	4.02 (0.51)	4.00 (0.51)	4.11 (0.56)	4.23 (0.62)	4.48 (0.56)
C3-4 (Post.)	3.60 (0.57)	3.57 (0.77)	3.66 (0.48)	3.69 (0.68)	3.43 (0.64)	3.63 (0.73)
C4-5 (Ant.)	4.54 (0.69)	4.86 (0.40)	4.91 (0.47)	5.05 (0.44)	4.86 (0.51)	4.72 (0.59)
C4-5 (Inter.)	3.49 (0.66)	3.57 (0.55)	3.60 (0.48)	3.87 (0.51)	3.91 (0.51)	4.03 (0.53)
C4-5 (Post.)	3.32 (0.47)	3.49 (0.60)	3.59 (0.66)	3.74 (0.63)	3.53 (0.50)	3.58 (0.80)

Measurement	Age (years)					
	12.0	13.0	14.0	15.0	16.0	17.0
C2-3 (Ant.)	5.78 (0.82)	5.54 (0.84)	5.06 (1.32)	4.40 (1.31)	4.04 (0.94)	4.19 (0.94)

TABLE 1057   (continued)

MEAN HEIGHTS (mm) OF INTERVERTEBRAL DISCS IN OHIO GIRLS
THE STANDARD DEVIATIONS ARE IN PARENTHESES

| Measurement | Age (years) | | | | | |
	12.0	13.0	14.0	15.0	16.0	17.0
C2-3 (Inter.)	4.86 (0.63)	4.79 (0.70)	4.67 (0.79)	4.43 (0.92)	4.34 (0.83)	4.04 (0.83)
C2-3 (Post.)	3.83 (0.87)	3.22 (1.14)	3.26 (1.05)	2.85 (1.01)	2.94 (0.97)	2.59 (0.96)
C3-4 (Ant.)	5.19 (0.87)	4.63 (0.80)	4.17 (1.15)	3.68 (1.08)	3.82 (0.96)	3.72 (0.66)
C3-4 (Inter.)	4.36 (0.79)	4.57 (0.92)	4.31 (0.81)	4.26 (0.68)	4.61 (0.65)	4.21 (0.74)
C3-4 (Post.)	3.54 (0.49)	3.20 (0.92)	2.76 (0.77)	2.63 (0.77)	2.65 (0.71)	2.66 (0.87)
C4-5 (Ant.)	4.85 (0.70)	4.63 (0.74)	3.96 (0.87)	3.39 (0.70)	3.25 (1.01)	3.60 (0.75)
C4-5 (Inter.)	4.13 (0.47)	4.28 (0.60)	4.08 (0.63)	4.17 (0.67)	4.22 (0.39)	4.38 (0.57)
C4-5 (Post.)	3.40 (0.49)	3.40 (0.71)	3.10 (0.65)	3.25 (0.81)	3.02 (0.68)	2.74 (0.33)

(data from Roche, unpublished b)

TABLE 1058

RATIO HEIGHT/SAGITTAL DIAMETER
FOR SELECTED VERTEBRAL BODIES

Vertebral Body	Age Group	Mean	S.D.
D12	I	0.81	0.061
	II	0.91	0.077
	III	0.86	0.066
	IV f	0.86	0.062
	IV m	0.78	0.052
	V m	0.84	0.116
L1	I	0.87	0.060
	II f	1.02	0.066
	II m	0.96	0.043
	II m+f	0.98	0.055
	III	0.89	0.080
	IV f	0.87	0.068
	IV m	0.80	0.048
	V f	1.03	0.095
	V m	0.87	0.063
L2	I	0.92	0.060
	II	1.01	0.090
	III	0.91	0.060
	IV	0.82	0.076
	V f	1.03	0.096
	V m	0.88	0.086
L3	I	0.95	0.068
	II	0.98	0.084

TABLE 1058   (continued)

RATIO HEIGHT/SAGITTAL DIAMETER
FOR SELECTED VERTEBRAL BODIES

Vertebral Body	Age Group	Mean	S.D.
L3	III	0.88	0.081
	IV	0.79	0.072
	V f	1.00	0.101
	V m	0.87	0.094

f = girls; m = boys; D = dorsal thoracic; L = lumbar;
Age groups I = 0-1 month; II = 2-18 months; III = 19-36
months; IV = 4-12 years; V = 13 + years.

(data from Brandner, 1970, American Journal of Roentgenology
110:618-627. Copyright American Roentgen Ray Society, 1970)

TABLE 1059

INTERVERTEBRAL DISC THICKNESS/VERTEBRAL BODY HEIGHT
FOR SELECTED VERTEBRAE

	Age Group	Mean	S.D.
D 11/12	I	0.37	0.060
	II	0.30	0.065
	III	0.25	0.089
	IV	0.24	0.053
	V	0.18	0.042
D 12/L1	I	0.35	0.063
	II	0.28	0.068
	III	0.26	0.057
	IV	0.25	0.050
	V	0.19	0.043
L 1/2	I	0.35	0.046
	II	0.26	0.073
	III	0.27	0.055
	IV	0.28	0.047
	V	0.20	0.056
L 2/3	II	0.28	0.089
	III	0.30	0.083
	IV	0.30	0.049
	V	0.21	0.051

Age groups: I = 0-1 month; II = 2-18 months; III = 19-36 months;
IV = 4-12 years; V = 13 + years: D = dorsal thoracic; L = lumbar.

(data from Brandner, 1970, American Journal of Roentgenology
110:618-627. Copyright American Roentgen Ray Society, 1970)

TABLE 1060

MEAN INTERPEDICULATE DISTANCES (mm) OF EACH VERTEBRA FOR BOYS
IN OREGON

Age Group (years)	Age (years)						
	3,4,5	6,7,8	9,10	11,12	13,14	15,16	17,18
Mean age (years)	4.0	7.0	9.4	11.6	13.5	15.6	17.4
Vertebra C3	24.3	26.3	26.9	26.0	26.9	27.9	28.6
4	25.5	27.4	27.3	27.8	28.0	28.8	29.5
5	26.1	27.8	27.5	27.0	18.4	29.3	29.9
6	25.8	28.2	27.3	27.2	28.4	29.2	30.1
7	24.8	27.1	26.2	26.0	27.3	28.0	28.8
T1	22.5	22.7	23.8	23.8	24.2	25.2	24.5
2	19.4	19.7	20.4	20.6	20.6	21.7	21.2
3	17.9	18.1	19.0	18.8	19.1	19.2	19.4
4	16.9	17.6	18.2	17.6	18.3	18.0	18.7
5	16.4	17.3	17.4	17.4	18.3	17.7	18.4
6	16.5	17.3	18.8	17.4	18.3	17.3	18.2
7	16.7	17.5	16.5	17.4	18.5	17.6	17.9
8	17.0	18.1	16.7	17.6	18.9	18.3	18.4
9	17.0	18.5	17.0	17.8	19.3	18.6	18.7
10	16.9	18.8	17.3	18.1	19.5	18.1	18.9
11	18.2	20.1	18.8	19.2	20.9	19.8	20.2
12	20.4	22.5	21.3	21.8	23.5	23.2	23.4
L1	20.7	22.5	23.3	23.9	23.8	24.5	25.1
2	20.7	22.4	23.5	24.2	23.3	24.6	24.8
3	21.2	23.0	24.1	24.6	23.6	25.1	25.2
4	21.9	23.6	24.8	23.6	24.7	26.0	26.6
5	24.7	26.9	28.4	28.9	28.0	30.1	29.9

C = cervical; T = thoracic; L = lumbar.

(data from Hinck, Clark, and Hopkins, 1966, American Journal of Roentgenology 97: 141-153. Copyright American Roentgen Ray Society, 1966)

TABLE 1061

MEAN INTERPEDICULATE DISTANCES (mm) OF EACH VERTEBRA FOR GIRLS
IN OREGON

Age Group (years)	Age (years)							
	3,4,5	6,7,8	9,10	11,12	13,14	15,16	17,18	Adult
Mean age	4.2	7.2	9.6	11.6	13.7	15.5	17.6	>18
Vertebra								
C3	22.1	25.5	24.8	26.6	27.5	26.6	26.6	27.4
4	22.2	26.0	25.6	27.4	28.4	27.7	27.5	28.2
5	22.3	26.4	25.9	28.0	28.7	28.1	27.9	28.7
6	23.6	26.7	26.0	27.3	28.6	27.5	28.0	28.6
7	24.0	25.5	26.3	25.8	27.1	26.1	26.1	27.1
T1	19.8	22.1	23.1	23.0	22.6	22.3	22.4	23.1
2	16.6	18.2	19.7	18.8	19.6	18.9	19.1	19.8
3	15.2	17.0	17.9	17.5	18.3	17.6	17.9	18.2
4	14.6	16.3	17.1	16.6	17.8	16.9	17.4	17.4
5	14.6	16.3	16.6	16.0	17.2	16.8	17.1	17.1
6	14.5	16.1	16.3	15.9	17.4	16.5	16.9	16.9
7	14.7	16.3	16.5	16.1	17.8	16.6	17.4	17.0
8	15.0	16.9	16.6	16.4	18.0	16.8	17.8	17.4
9	15.1	17.1	16.9	16.8	18.1	17.3	18.2	17.9
10	15.4	17.3	17.1	17.1	18.2	17.7	18.7	18.4
11	16.5	18.0	18.2	18.4	19.8	19.2	19.9	20.0
12	19.1	20.2	21.1	20.9	22.8	21.7	22.5	22.9
L1	20.1	21.0	22.5	22.2	23.7	23.6	24.1	24.3
2	20.0	21.1	22.6	22.3	23.6	23.6	24.2	24.9
3	20.2	21.7	23.2	22.8	24.6	24.4	24.5	25.4
4	21.1	23.0	24.7	24.2	26.9	25.4	25.8	26.4
5	23.9	26.1	28.2	28.5	30.4	28.6	29.5	29.0

C = cervical; T = thoracic; L = lumbar.
(data from Hinck, Clark, and Hopkins, 1966, American Journal of Roentgenology 97:
141-153. Copyright American Roentgen Ray Society, 1966)

TABLE 1062

STANDARD DEVIATION OF INTERPEDICULATE DISTANCE (mm) OF EACH VERTEBRA BY AGE
FOR OREGON CHILDREN

	Age (years)						
	3,4,5	6,7,8	9,10	11,12	13,14	15,16	17,18
Vertebra							
C3	1.7	1.5	1.7	2.6	1.2	1.8	1.9
4	1.7	1.5	1.3	2.5	1.5	1.6	2.0
5	1.7	1.7	1.6	2.6	1.7	1.6	2.1
6	1.7	1.7	1.7	2.7	2.1	1.8	2.2
7	1.4	2.0	2.1	2.8	1.8	2.4	2.1
T1	2.5	1.9	1.4	2.2	1.5	2.3	1.6
2	2.2	1.6	1.2	2.0	1.8	2.4	1.4
3	1.6	1.5	0.9	1.6	1.6	1.5	1.3
4	1.5	1.4	1.0	1.7	1.8	1.3	1.3
5	1.4	1.2	0.8	1.6	1.8	1.4	1.3
6	1.6	1.2	1.0	1.5	1.8	1.3	1.6
7	1.6	1.2	1.3	1.4	1.9	1.6	1.5
8	1.6	1.4	1.0	1.5	1.9	1.6	1.5
9	1.5	1.3	1.3	1.6	1.8	1.8	1.6
10	1.3	1.4	1.6	1.7	2.0	1.6	1.5
11	1.5	1.9	1.8	1.8	2.3	1.7	1.7
12	1.5	2.1	1.6	1.8	2.3	1.7	1.7
L1	1.6	1.9	1.8	1.7	1.6	2.0	2.4
2	1.5	2.0	1.6	1.8	1.6	1.8	2.2
3	1.6	2.1	1.7	1.9	2.0	1.9	2.3
4	1.7	2.1	2.3	3.1	3.3	2.4	2.9
5	1.9	3.1	2.7	3.1	3.6	2.9	3.6

C = cervical; T = thoracic; L = lumbar

(data from Hinck, Clark, and Hopkins, 1966, American Journal of Roentgenology 97:
141-153. Copyright American Roentgen Ray Society, 1966)

TABLE   1063

SAGITTAL DIAMETER OF LUMBAR SPINAL CANAL (mm)
IN OREGON CHILDREN

Vertebrae	Boys		Girls		Boys		Girls	
	Mean	S.D.	Mean	S.D.	Mean	S.D.	Mean	S.D.
	Age 3, 4, 5 years				Age 6, 7, 8 years			
L1	20.3	1.8	--	--	20.3	1.9	19.3	2.6
2	19.6	1.2	--	--	19.9	1.7	19.5	1.6
3	18.4	1.4	--	--	19.1	1.8	18.4	1.6
4	18.8	1.1	--	--	19.0	1.7	19.1	1.8
5	19.0	1.6	--	--	19.0	2.4	19.1	2.3
	Age 13, 14 years				Age 15, 16 years			
L1	20.5	1.5	20.8	1.4	21.6	2.2	21.6	2.2
2	19.7	1.4	20.1	0.9	20.8	2.1	20.9	1.8
3	18.9	1.7	20.0	1.3	20.5	1.6	20.7	1.6
4	20.4	4.1	20.2	2.5	20.0	1.7	21.0	2.1
5	20.8	4.2	20.1	3.2	20.1	2.9	20.8	3.4
	Age 17, 18 years				Adults			
L1	23.9	1.9	21.7	1.7	--	--	--	--
2	22.4	2.3	21.3	1.9	--	--	--	--
3	22.6	2.3	22.0	3.1	--	--	--	--
4	22.9	2.8	21.9	2.6	--	--	--	--
5	22.6	3.4	21.4	2.2	--	--	--	--

(data from Hinck, Hopkins and Clark, 1965, RADIOLOGY 85:929-937)

TABLE   1064

TOLERANCE RANGES (mm) FOR THE SAGITTAL
DIAMETER OF THE CERVICAL SPINAL CANAL
IN OREGON CHILDREN

Diameter	Sex	Age (years)	Mean	S.D.	90% Tolerance Range Lower limit P_{05}	Upper limit P_{95}
C-1	Boys	3	19.5	1.7	15.2	23.8
		8	20.0	1.7	15.7	24.3
		13	20.7	1.7	16.4	25.0
		18	21.3	1.7	17.0	25.6
	Girls	3	16.8	1.5	12.9	20.7
		8	17.8	1.5	13.9	21.7
		13	18.8	1.5	14.9	22.7
		18	19.7	1.5	15.8	23.6
C-2	Boys and Girls	3	17.2	1.5	14.0	20.4
		8	17.8	1.5	14.6	21.0
		13	18.6	1.5	15.4	21.8
		18	19.4	1.5	16.2	22.6
C-3	Boys and Girls	3	15.0	1.4	12.0	18.0
		8	15.8	1.4	12.8	18.8
		13	16.6	1.4	13.6	19.6
		18	17.3	1.4	14.3	20.3
C-4	Boys and Girls	3	14.8	1.3	11.9	17.7
		8	15.6	1.3	12.7	18.5
		13	16.3	1.3	13.4	19.2
		18	17.1	1.3	14.2	20.0
C-5	Boys and Girls	3	15.0	1.2	12.3	17.7
		8	15.6	1.2	12.8	18.3
		13	16.1	1.2	13.3	18.8
		18	16.7	1.2	13.9	19.4

(data from Hinck, Hopkins and Savara, 1962, RADIOLOGY 79:97-108)

TABLE   1065

PERCENT PREVALENCE OF BRIDGING ON
THE FIRST CERVICAL VERTEBRA
IN OHIO CHILDREN

Group	Absent Bridge	Partial Bridge	Complete Bridge
Boys	72.9	14.6	12.5
Girls	71.7	22.6	5.7

(data from Selby et al., 1955)

TABLE 1066

PELVIC DIMENSIONS (mm) IN WHITE OHIO INFANTS

	BOYS			
	Birth		1 Month	
Item	Mean	S.D.	Mean	S.D.
Pelvis breadth	75.8	4.9	83.6	4.5
Inlet, sagittal	22.3	3.2	25.1	2.5
Inlet breadth	37.0	2.4	40.4	2.5
Inter-iliac breadth	27.6	2.9	30.4	2.4
Bi-ischial breadth	21.9	2.4	22.3	2.1
Pelvic index	74.3%	2.8	74.0%	2.4
Inlet index	60.2%	7.2	62.4%	6.4
Sacral index	36.4%	3.1	36.4%	2.8
Relative inlet breadth	48.9%	1.7	48.4%	1.9
Anterior segment index	125.4%	10.0	124.2%	6.6

	GIRLS			
Pelvis breadth	74.4	4.7	81.6	4.7
Sagittal inlet	23.0	2.9	25.2	3.0
Inlet breadth	36.8	2.5	40.2	2.2
Inter-iliac breadth	26.9	2.6	30.0	1.7
Bi-ischial breadth	23.1	2.8	23.5	2.9
Pelvic index	74.9%	3.6	74.9%	2.9
Inlet index	62.9%	7.8	62.8%	7.6
Sacral index	36.1%	2.8	36.9%	2.1
Relative inlet breadth	49.6%	2.0	49.2%	1.6
Anterior segment index	127.3%	8.2	130.0%	11.4

(data from Reynolds, 1945)

TABLE 1067

PELVIC DIMENSIONS (mm) IN WHITE OHIO INFANTS

Age	Boys		Girls	
	Mean	S.D.	Mean	S.D.
Pelvis Height				
Birth	56.1	3.3	55.7	3.2
1 month	61.8	3.4	61.1	3.0
3 months	72.9	3.2	70.8	3.9
6 months	82.8	3.5	80.5	3.8
9 months	89.9	3.5	87.6	4.1
12 months	94.9	3.9	92.7	4.0
Inter-pubic breadth				
Birth	7.2	1.6	7.7	1.8
1 month	7.6	1.3	8.2	1.1
3 months	8.5	1.3	8.4	1.4
6 months	8.6	1.6	8.1	1.7
9 months	8.3	1.2	8.1	1.4
12 months	7.8	1.7	7.6	1.6
Ilium length				
Birth	32.4	2.4	32.9	2.1
1 month	36.0	1.9	36.1	2.0
3 months	41.3	1.9	40.7	2.4
6 months	47.2	2.3	46.4	2.7
9 months	51.9	2.3	51.6	2.9
12 months	54.5	2.8	54.3	2.6

TABLE 1067 (continued)

PELVIC DIMENSIONS (mm) IN WHITE OHIO INFANTS

Age	Boys		Girls	
	Mean	S.D.	Mean	S.D.
Ilium breadth				
Birth	22.4	3.6	22.6	3.8
1 month	26.4	3.1	25.1	3.5
3 months	29.2	3.6	27.1	4.4
6 months	33.1	4.2	31.1	5.2
9 months	35.8	4.8	33.3	6.3
12 months	38.0	5.1	36.1	6.7
Ischium length				
Birth	19.6	1.5	19.7	1.6
1 month	22.0	1.9	22.3	1.5
3 months	26.8	1.8	26.7	2.1
6 months	30.8	2.0	31.0	2.0
9 months	34.2	2.2	34.4	2.0
12 months	36.7	1.9	36.6	1.7
Pubis length				
Birth	15.6	1.8	16.7	2.0
1 month	18.6	1.6	18.8	1.3
3 months	22.0	1.7	22.0	1.4
6 months	25.6	2.0	25.8	1.7
9 months	28.7	2.4	28.6	1.9
12 months	30.6	1.9	30.4	1.9

TABLE 1067   (continued)

PELVIC DIMENSIONS (mm) IN WHITE OHIO INFANTS

Age	Boys		Girls	
	Mean	S.D.	Mean	S.D.
Breadth of greater sciatic notch				
Birth	9.0	1.6	10.0	1.9
1 month	10.6	2.0	11.2	1.8
3 months	9.8	1.8	10.7	2.4
6 months	11.3	2.2	11.4	2.9
9 months	13.5	2.7	14.1	3.0
12 months	14.0	2.9	14.7	2.3
Iliac index				
Birth	69.3	10.7	68.8	11.3
1 month	72.7	8.0	69.5	8.2
3 months	70.7	8.3	66.8	10.0
6 months	70.3	9.2	66.7	10.5
9 months	69.1	9.2	65.2	11.0
12 months	70.0	9.3	67.4	12.0

(data from Reynolds, 1945)

TABLE 1068

PELVIC MEASUREMENTS (mm)
IN WHITE OHIO CHILDREN

Age (months)	Boys Mean	Boys S.D.	Girls Mean	Girls S.D.
Pelvis height (mm)				
22	110.0	4.4	106.5	5.0
34	121.9	4.7	119.2	5.7
45	130.9	6.7	129.5	6.2
57	140.4	8.1	138.4	7.5
69	148.9	8.0	147.9	7.7
78	156.9	8.9	153.0	9.2
90	164.5	7.6	159.5	7.9
Pelvis breadth (mm)				
22	143.4	5.8	138.2	7.0
34	159.0	6.2	156.2	7.5
45	170.7	7.7	168.6	9.4
57	181.1	8.0	177.4	9.2
69	190.1	8.0	187.5	9.3
78	195.4	9.7	191.7	10.3
90	201.2	8.7	196.5	8.7
Inlet breadth (mm)				
22	66.4	3.5	64.7	3.4
34	72.1	3.2	71.1	4.2
45	76.3	3.6	75.9	3.8
57	80.7	4.4	79.2	4.9
69	82.8	4.2	82.1	4.2
78	85.5	6.3	84.3	5.7
90	88.8	4.5	87.7	5.7
Inter-iliac breadth (mm)				
22	52.5	3.4	50.8	2.9
34	56.2	2.6	55.2	3.1
45	58.8	2.8	57.4	3.1
57	62.2	3.4	59.8	4.3
69	63.4	3.3	62.0	3.9
78	65.6	4.0	64.5	5.7
90	68.1	4.7	66.6	5.0
Inter-pubic breadth (mm)				
22	7.0	1.6	7.9	1.9
34	6.6	1.7	7.2	2.4
45	6.3	1.5	6.7	1.9
57	6.3	1.5	6.3	2.2
69	6.3	1.4	6.4	1.9
78	6.1	1.9	6.1	1.8
90	6.0	1.5	6.5	2.2

TABLE   1068   (continued)

PELVIC MEASUREMENTS (mm)
IN WHITE OHIO CHILDREN

Age	Boys		Girls	
(months)	Mean	S.D.	Mean	S.D.

Inter-tuberal breadth (mm)

Age	Mean	S.D.	Mean	S.D.
22	47.4	3.6	50.6	4.4
34	52.4	4.6	56.3	5.4
45	56.7	4.4	61.7	4.7
57	60.7	4.9	66.2	5.2
69	65.7	6.4	70.3	5.0
78	67.7	6.3	72.2	5.5
90	69.9	6.7	76.1	6.8

Ilium length (mm)

Age	Mean	S.D.	Mean	S.D.
22	60.5	3.0	58.8	4.0
34	66.7	2.9	65.4	4.4
45	71.6	3.7	70.5	4.5
57	77.1	5.7	74.8	5.4
69	80.9	5.5	79.4	6.0
78	83.5	5.7	82.5	6.7
90	89.3	5.4	85.8	5.1

Pubis length (mm)

Age	Mean	S.D.	Mean	S.D.
22	39.2	3.0	39.3	2.9
34	44.9	3.8	45.2	3.5
45	48.4	4.2	50.6	3.1
57	52.5	4.4	54.4	3.9
69	56.0	4.7	58.7	3.8
78	60.1	4.3	60.6	4.5
90	61.7	4.8	63.3	4.7

Inlet, sagittal diameter (mm)

Age	Mean	S.D.	Mean	S.D.
22	27.7	4.5	26.3	4.8
34	29.8	4.4	30.0	4.7
45	31.5	5.8	32.9	6.0
57	34.6	7.2	34.5	5.8
69	33.4	7.0	36.5	7.6
78	36.0	8.8	37.5	8.6
90	41.0	8.1	39.0	8.0

Ischium length (mm)

Age	Mean	S.D.	Mean	S.D.
22	49.1	2.7	47.7	3.5
34	55.2	3.7	53.7	3.3
45	59.4	4.3	59.1	3.4
57	63.5	4.3	63.7	4.3
69	67.5	4.6	68.5	4.6
78	72.7	4.4	70.9	5.3
90	75.9	5.2	73.9	5.4

TABLE 1068   (continued)

PELVIC MEASUREMENTS (mm)
IN WHITE OHIO CHILDREN

Age (month)	Boys		Girls	
	Mean	S.D.	Mean	S.D.

Breadth of iliac notch (mm)

Age (month)	Mean	S.D.	Mean	S.D.
22	12.7	3.9	12.1	3.4
34	13.7	2.8	14.4	3.7
45	13.9	3.3	15.5	3.6
57	15.5	4.0	16.6	4.2
69	15.0	3.6	16.7	4.7
78	15.0	4.6	17.3	5.3
90	17.1	4.2	18.6	4.9

Inter-obturator breadth (mm)

Age (month)	Mean	S.D.	Mean	S.D.
22	26.2	2.1	26.5	1.5
34	27.8	2.0	28.7	1.9
45	29.1	2.2	30.5	2.3
57	31.3	2.2	31.8	2.3
69	33.1	2.4	33.7	2.3
78	34.3	2.6	34.4	2.6
90	36.1	3.0	36.3	3.0

Bi-trochanteric breadth (mm)

Age (month)	Mean	S.D.	Mean	S.D.
22	148.6	6.6	143.0	7.0
34	161.1	7.2	157.8	8.5
45	171.2	8.9	169.5	8.1
57	181.4	9.1	178.8	9.1
69	191.7	10.3	187.8	9.4
78	197.0	9.1	192.1	9.0
90	202.8	10.0	199.4	8.8

Length of femoral neck (mm)

Age (month)	Mean	S.D.	Mean	S.D.
22	48.4	3.0	45.2	3.5
34	53.8	3.6	52.2	4.2
45	59.2	4.3	58.3	4.2
57	64.5	4.6	63.6	5.0
69	69.0	4.3	68.8	5.3
78	74.5	5.3	71.7	6.1
90	78.5	4.8	75.5	5.6

(data from Reynolds, 1947)

TABLE 1069

PELVIC INDICES
IN WHITE OHIO CHILDREN

Age	Boys		Girls	
(months)	Mean	S.D.	Mean	S.D.

Pelvic index (per cent)

22	76.6	2.8	77.0	2.9
34	76.4	2.3	76.2	2.8
45	76.4	2.8	77.0	2.9
57	77.3	3.1	78.1	3.0
69	78.3	3.3	78.9	3.6
78	80.1	3.6	79.3	3.6
90	81.6	3.3	80.7	3.0

Inlet index (per cent)

22	42.0	7.0	40.6	7.2
34	41.4	6.2	42.4	6.6
45	41.3	7.2	43.1	7.9
57	42.7	8.0	43.5	7.2
69	40.5	8.7	44.5	9.6
78	42.0	9.2	45.5	10.2
90	46.2	9.1	44.6	9.2

Relative sacrum breadth (per cent)

22	36.6	1.7	36.5	2.8
34	35.4	1.6	35.4	1.8
45	34.5	1.8	34.0	1.8
57	34.4	1.8	33.7	1.8
69	33.4	1.4	33.2	1.7
78	33.6	1.5	33.2	1.8
90	33.0	1.4	33.8	2.0

Relative inlet breadth (per cent)

22	46.2	1.7	46.9	2.2
34	45.4	1.5	45.6	1.8
45	44.8	1.7	45.0	1.9
57	44.3	1.9	44.6	1.9
69	43.4	1.4	43.7	1.9
78	43.8	2.5	43.6	2.4
90	43.8	1.6	44.4	2.1

(data from Reynolds, 1947)

TABLE 1070

PELVIC ANGLES (degrees)
IN WHITE OHIO CHILDREN

Age	Boys		Girls	
(months)	Mean	S.D.	Mean	S.D.

Pubic angle (degrees)

Age	Boys Mean	Boys S.D.	Girls Mean	Girls S.D.
22	61.5	7.3	66.1	10.6
34	56.5	7.1	62.4	9.4
45	51.8	8.4	58.5	9.3
57	51.7	8.8	55.4	10.7
69	45.4	7.4	52.0	11.2
78	47.2	8.7	52.9	12.2
90	46.1	7.9	50.8	12.7

Pelvic angle (degrees)

Age	Boys Mean	Boys S.D.	Girls Mean	Girls S.D.
22	32.9	3.8	30.0	3.5
34	33.4	3.6	31.8	3.2
45	33.5	3.1	32.3	3.5
57	33.6	3.3	31.3	4.8
69	33.4	3.9	31.6	4.5
78	33.7	4.1	31.4	4.7
90	33.2	4.3	30.7	4.7

Femoral angle (degrees)

Age	Boys Mean	Boys S.D.	Girls Mean	Girls S.D.
22	142.3	4.7	142.2	3.0
34	139.1	4.5	138.8	4.2
45	137.7	5.0	139.0	4.5
57	138.1	4.6	138.6	4.8
69	135.6	4.9	136.7	5.1
78	138.2	4.9	137.0	5.0
90	138.1	4.4	137.4	4.9

Femoral-pelvic angle (degrees)

Age	Boys Mean	Boys S.D.	Girls Mean	Girls S.D.
22	130.7	4.5	132.8	3.4
34	129.8	4.0	131.1	4.7
45	129.0	4.9	130.7	4.6
57	129.7	4.7	131.3	5.3
69	127.7	5.3	130.1	6.1
78	129.7	5.7	130.4	5.0
90	130.6	4.5	130.6	5.4

(data from Reynolds, 1947)

TABLE 1071

MEAN POINT GROWTH (mm)
OF THE PELVIS MEASURED ON RADIOGRAPHS OF OHIO CHILDREN

	Point	Boys	Girls
Sacrum	36.	25.07	28.68
	43.	37.78	38.56
	42.	36.21	37.75
	41.	13.85	15.37
	40.	13.93	11.87
Iliac crest	24.	22.43	29.50
	25.	41.29	42.19
	1.	64.36	53.56
	2.	60.86	54.19
	3.	51.92	41.19
Acetabular region	4.	40.36	36.68
	5.	37.92	36.81
	6.	35.71	36.00
	7.	32.14	33.62
	8.	25.43	28.69
	9.	21.75	26.79
	21.	22.40	27.43
	10.	27.41	30.63
	11.	27.07	30.56
Ischial tuberosity	12.	28.71	31.68
	13.	30.64	29.56
	14.	32.57	27.62
	27.	19.92	27.87
	28.	21.00	23.00
Ischial notch	23.	17.78	23.25
	22.	22.21	25.68
	20.	14.86	24.31
Ischio-pubic ramus	15.	18.00	22.18
	29.	16.50	20.00
	30.	16.64	18.87
Sup. pubic ramus	19.	14.29	19.50
	33.	17.21	20.94
Pubic region	32.	15.79	22.75
	31.	16.28	19.62
	16.	14.57	22.38
	17.	10.64	20.00
	18.	6.86	15.75

See original article for definition of points

(data from Coleman, 1969)

TABLE 1072

MEAN POINT DIRECTIONAL GROWTH (degrees)
OF THE PELVIS MEASURED ON RADIOGRAPHS OF OHIO CHILDREN

	Point	Boys	Girls
Sacrum	43.	20.93	20.56
	42.	33.57	41.56
	41.	132.00	66.31
Iliac crest	24.	13.93	25.00
	25.	18.29	22.69
	1.	44.29	43.50
	2.	56.14	63.62
	3.	68.36	82.81
Acetabular region	4.	82.57	87.50
	5.	90.50	92.69
	8.	80.43	89.38
	9.	95.57	92.44
	21.	75.14	88.19
	11.	108.86	98.88
Ischial tuberosity	12.	118.57	110.06
	13.	139.71	123.62
	14.	155.21	138.06
	27.	122.29	103.38
	28.	136.21	118.38
Ischial notch	23.	55.43	49.94
	22.	62.79	61.75
	20.	81.14	95.19
Ischio-pubic ramus	15.	172.21	139.88
	29.	148.21	131.75
Sup. pubic ramus	19.	110.00	107.88
	33.	119.50	116.31
Pubic region	31.	147.64	133.38

See original article for definition of points

(data from Coleman, 1969)

TABLE 1073

ACETABULAR ANGLES (degrees) AT BIRTH FOR INFANTS IN NEW YORK CITY

| Angles | White | | | | Negro | | | |
| | Boys | | Girls | | Boys | | Girls | |
	R	L	R	L	R	L	R	L
10	--	1	--	--	--	--	--	--
11	--	--	--	--	--	--	--	--
12	--	--	--	--	1	1	--	--
13	--	--	--	--	1	1	--	--
14	--	--	--	--	1	--	--	--
15	2	--	--	--	1	1	--	--
16	--	--	1	--	1	1	--	1
17	2	1	2	--	3	--	2	1
18	6	2	1	1	5	2	--	2
19	4	--	1	3	4	4	2	--
20	8	9	5	4	7	15	6	1
21	9	5	4	1	8	5	6	3
22	13	11	9	4	6	5	9	9
23	18	8	5	4	9	10	3	2
24	14	13	7	8	13	14	12	4
25	24	25	5	5	15	8	15	8
26	14	23	10	9	10	9	7	10
27	17	13	8	4	12	6	13	9
28	15	11	17	14	10	5	8	11
29	5	13	7	11	2	4	11	10
30	15	24	18	24	11	13	20	15
31	11	6	6	7	4	6	4	13
32	4	14	11	12	7	8	8	14
33	3	2	8	7	2	6	4	7
34	6	8	7	7	1	5	6	5
35	5	4	3	9	--	--	2	7
36	1	1	5	5	--	3	4	5
37	--	--	4	2	--	1	3	3
38	1	2	2	4	--	--	2	5
39	--	1	1	2	--	--	--	--
40	--	--	--	1	--	--	--	2
41	--	--	--	--	--	1	--	--
42	--	--	--	--	--	--	1	2
43	--	--	--	--	--	--	--	--
44	--	--	--	--	--	--	1	--
Mean	25.8	27.0	28.3	29.4	24.8	26.0	27.7	29.4
S.D.	4.5	4.3	4.9	4.7	4.5	5.2	4.9	5.9

(data from Caffey, Ames, Silverman, et al., 1956, Pediatrics 17:632-641. Copyright American Academy of Pediatrics 1956)

TABLE   1074

COMPARISON OF ACETABULAR ANGLES AT DIFFERENT AGES
FOR CHILDREN IN NEW YORK CITY

AGE	WMR*	WML	WFR	WFL	NMR	NML	NFR	NFL
MEAN VALUES								
Newborn	25.8	27.0	28.3	29.4	24.8	26.0	27.7	29.4
6 months	19.4	20.9	22.1	23.4	21.4	23.0	23.9	25.4
12 months	19.1	20.6	20.5	21.9	20.5	21.9	22.5	24.4
2-S.D. RANGE								
Newborn	34-17	37-17	38-18	39-20	34-15	36-16	38-18	39-19
6 months	26-12	28-13	30-14	32-15	31-12	32-14	32-16	33-18
12 months	26-12	28-13	28-13	29-14	29-12	30-14	30-15	32-16

*W--White   M--Male   R--Right
 N--Negro   F--Female   L--Left

(data from Caffey, Ames, Silverman, et al., 1956, Pediatrics 17:632-641. Copyright
American Academy of Pediatrics 1956)

TABLE   1075

ACETABULAR ANGLES OF INFANTS IN NEW YORK CITY AT SIX MONTHS
OF AGE: FREQUENCY DISTRIBUTION AND SUMMARY STATISTICS

Angles	White Boys		White Girls		Negro Boys		Negro Girls	
	R	L	R	L	R	L	R	L
10	--	--	--	--	--	--	--	--
12	2	1	1	2	1	--	--	--
14	17	7	2	--	3	1	1	1
16	28	23	10	5	18	11	3	--
18	27	19	14	9	18	14	10	5
20	45	43	23	18	24	16	17	11
22	23	26	19	17	19	19	24	10
24	23	28	23	26	17	16	29	32
26	4	16	16	19	13	34	27	31
28	3	5	2	13	7	8	10	26
30	--	3	5	5	3	3	9	14
32	--	1	1	2	3	3	4	3
34	--	--	1	1	--	--	--	2
36	--	--	1	1	--	1	--	--
38	--	--	--	--	--	--	--	--
40	--	--	--	--	--	--	1	--
42	--	--	--	--	--	--	--	--
44	--	--	--	--	--	--	--	--
Mean	19.4	20.9	22.1	23.4	21.4	23.0	23.9	25.4
S.D.	3.5	3.9	3.9	4.2	4.6	4.3	4.0	3.6

ACETABULAR ANGLES OF INFANTS IN NEW YORK CITY AT TWELVE MONTHS
OF AGE: FREQUENCY DISTRIBUTION AND SUMMARY STATISTICS

Angles	White Boys		White Girls		Negro Boys		Negro Girls	
	R	L	R	L	R	L	R	L
8	--	--	--	--	--	1	--	--
10	2	--	2	--	1	--	--	--
12	2	2	--	--	3	--	--	--
14	9	6	6	3	7	5	4	--
16	30	13	18	8	13	10	7	3
18	29	20	13	8	11	7	8	5
20	49	52	28	39	30	23	27	18
22	21	26	18	13	18	13	24	15
24	15	33	20	25	15	30	38	32
26	3	7	8	14	7	15	14	31
28	1	1	2	2	5	7	7	19
30	--	--	3	5	3	1	5	10
32	--	1	--	1	--	1	--	2
34	--	--	--	--	--	--	1	--
Mean	19.1	20.6	20.5	21.9	20.5	21.9	22.5	24.4
S.D.	3.5	3.7	3.8	3.8	4.2	4.0	3.9	4.0

(data from Caffey, Ames, Silverman, et al., 1956, Pediatrics 17:632-641. Copyright
American Academy of Pediatrics 1956)

TABLE 1076

MEAN PELVIC BREADTH (cm) IN RELATION TO TYPE OF ARTIFICIAL FEEDING
IN PHILADELPHIA CHILDREN

Group	Race	3 Months		1 Year		2 Years	
		Mean	S.D.	Mean	S.D.	Mean	S.D.
All	All White	10.1	±0.53	12.6	±0.62	14.4	±0.73
	All Negro	9.7	±0.55	11.9	±0.76	13.8	±0.81
Group I (irradiated evaporated milk)	White	10.1	±0.60	12.5	±0.70	14.4	±0.77
	Negro	9.8	±0.62	12.0	±0.73	13.8	±0.74
Group II (nonirradiated evaporated milk plus cod-liver oil)	White	10.0	±0.48	12.6	±0.60	14.4	±0.67
	Negro	9.6	±0.58	11.8	±0.80	13.7	±0.88
Group III (irradiated evaporated milk plus carotene)	White	10.2	±0.39	12.7	±0.49	14.4	±0.62
	Negro	9.8	±0.57	11.9	±0.71	13.9	±0.82
Group IV (irradiated evaporated milk plus carotene and yeast)	White	10.3	±0.55	12.8	±0.75	14.7	±0.84
	Negro	9.8	±0.40	12.0	±0.75	13.9	±0.87

Group	Race	3 Years		4 Years	
		Mean	S.D.	Mean	S.D.
All	All White	15.8	±0.79	16.9	±0.82
	All Negro	15.0	±0.88	16.1	±0.96
Group I (irradiated evaporated milk)	White	15.8	±0.82	16.8	±0.91
	Negro	15.0	±0.85	16.1	±0.85
Group II (nonirradiated evaporated milk plus cod-liver oil)	White	15.6	±0.81	16.8	±0.77
	Negro	15.0	±0.93	16.1	±1.03
Group III (irradiated evaporated milk plus carotene)	White	15.8	±0.60	16.9	±0.57
	Negro	14.9	±0.88	16.0	±0.95
Group IV (irradiated evaporated milk plus carotene and yeast)	White	16.1	±0.75	17.2	±1.04
	Negro	15.1	±0.91	16.2	±1.09

(data from Rhoads, Rapaport, Kennedy, et al., 1945, The Journal of Pediatrics 26:
415-454)

TABLE   1077

HUMERUS LENGTH (cm)
IN DENVER CHILDREN

Age	Boys		Girls	
Yr. Mo.	Mean	S.D.	Mean	S.D.
Diaphyseal Length				
0 - 2	7.24	.45	7.18	.36
0 - 4	8.06	.48	8.02	.38
0 - 6	8.84	.50	8.68	.46
1 - 0	10.55	.52	10.36	.48
1 - 6	11.88	.54	11.70	.51
2 - 0	13.00	.55	12.77	.58
2 - 6	13.90	.59	13.69	.61
3 - 0	14.75	.67	14.53	.67
3 - 6	15.50	.78	15.34	.71
4 - 0	16.27	.69	16.09	.77
4 - 6	16.98	.74	16.91	.83
5 - 0	17.74	.82	17.63	.87
5 - 6	18.46	.81	18.26	.90
6 - 0	19.09	.76	19.00	.96
6 - 6	19.73	.81	19.67	.97
7 - 0	20.36	.87	20.26	1.00
7 - 6	21.04	.89	20.93	1.05
8 - 0	21.73	.98	21.63	1.04
8 - 6	22.25	.92	22.13	1.12
9 - 0	22.87	.96	22.80	1.18
9 - 6	23.51	1.07	23.42	1.29
10 - 0	24.10	1.03	23.98	1.32
10 - 6	24.58	1.10	24.59	1.46
11 - 0	25.17	1.07	25.19	1.47
11 - 6	25.74	1.19	25.91	1.53
12 - 0	26.30	1.28	26.56	1.56
Length including Epiphyses				
10 - 0	25.83	1.12	25.61	1.46
10 - 6	26.37	1.16	26.29	1.61
11 - 0	27.00	1.15	26.96	1.64
11 - 6	27.63	1.27	27.84	1.73
12 - 0	28.20	1.38	28.75	1.82
12 - 6	28.92	1.31	29.40	1.77
13 - 0	29.66	1.53	30.10	1.75
13 - 6	30.50	1.66	30.57	1.74

TABLE 1077 (continued)

HUMERUS LENGTH (cm)
IN DENVER CHILDREN

Age	Boys		Girls	
Yr. Mo.	Mean	S.D.	Mean	S.D.
14 - 0	31.33	1.68	31.17	1.67
14 - 6	32.14	1.76	31.49	1.71
15 - 0	32.90	1.67	31.56	1.70
15 - 6	33.65	1.65	32.32	1.96
16 - 0	34.10	1.45	31.65	1.85
16 - 6	34.34	1.53	--	--
17 - 0	34.71	1.46	31.54	1.73
18 - 0	35.06	1.56	--	--

RADIUS LENGTH (cm)
IN DENVER CHILDREN

Age	Boys		Girls	
Yr. Mo.	Mean	S.D.	Mean	S.D.
Diaphyseal Length				
0 - 2	5.97	.33	5.78	.28
0 - 4	6.60	.33	6.34	.28
0 - 6	7.08	.35	6.76	.34
1 - 0	8.26	.40	7.89	.34
1 - 6	9.14	.44	8.75	.40
2 - 0	9.86	.47	9.50	.45
2 - 6	10.52	.48	10.14	.50
3 - 0	11.16	.53	10.77	.52
3 - 6	11.69	.62	11.38	.55
4 - 0	12.31	.56	11.92	.57
4 - 6	12.82	.56	12.52	.66
5 - 0	13.38	.61	13.02	.69
5 - 6	13.89	.64	13.46	.72
6 - 0	14.38	.59	14.00	.74
6 - 6	14.83	.64	14.47	.78
7 - 0	15.30	.67	14.93	.80
7 - 6	15.79	.69	15.43	.84
8 - 0	16.29	.71	15.89	.87
8 - 6	16.68	.66	16.28	.88

TABLE   1077   (continued)

RADIUS LENGTH (cm)
IN DENVER CHILDREN

Age	Boys		Girls	
Yr. Mo.	Mean	S.D.	Mean	S.D.
9 - 0	17.13	.74	16.76	.93
9 - 6	17.61	.77	17.22	1.02
10 - 0	18.05	.79	17.68	1.04
10 - 6	18.44	.84	18.18	1.18
11 - 0	18.87	.85	18.60	1.17
11 - 6	19.30	.92	19.20	1.21
12 - 0	19.74	.96	19.69	1.27
Length including Epiphyses				
10 - 0	19.30	.81	18.93	1.14
10 - 6	19.77	.89	19.50	1.30
11 - 0	20.26	.89	20.00	1.30
11 - 6	20.73	.97	20.67	1.35
12 - 0	21.23	1.03	21.35	1.42
12 - 6	21.80	1.02	21.88	1.42
13 - 0	22.37	1.18	22.36	1.31
13 - 6	23.02	1.29	22.78	1.27
14 - 0	23.69	1.35	23.14	1.18
14 - 6	24.28	1.41	23.35	1.17
15 - 0	24.87	1.34	23.45	1.17
15 - 6	25.50	1.28	23.74	1.52
16 - 0	25.77	1.17	23.50	1.18
16 - 6	25.98	1.13	--	--
17 - 0	26.18	1.12	23.38	1.18
18 - 0	26.32	1.28	--	--

(data from McCammon, Human Growth and Development,
1970. Courtesy of Charles C Thomas, Publisher,
Springfield, Illinois)

TABLE 1078

RADIUS LENGTH (mm)
IN OHIO CHILDREN

Age (months)	Boys		Girls	
	Mean	S.D.	Mean	S.D.
1	55.84	2.89	54.00	2.72
3	62.42	3.02	59.85	3.31
6	69.72	3.42	66.93	3.74
9	75.84	4.13	73.50	4.55
12	82.29	4.64	79.52	4.51
18	92.52	6.89	89.44	4.87
24	100.20	5.10	97.46	5.00
30	107.52	5.33	104.28	5.67
36	114.44	5.92	110.80	5.94
42	119.97	5.67	117.13	6.51
48	125.97	6.55	122.88	6.76
54	131.42	6.53	128.83	7.61
60	137.54	7.18	134.32	7.56
66	142.30	7.67	140.66	8.00
72	148.85	8.11	145.30	8.32
78	153.14	8.51	150.45	9.07
84	159.10	8.73	155.28	9.10
90	163.81	8.81	160.36	9.63
96	168.93	8.89	165.38	9.81
102	173.57	9.35	169.58	10.77
108	179.45	9.43	175.06	10.54

(data from Gindhart, 1971)

TABLE 1079

INCREMENTS (mm) IN RADIUS LENGTH
IN OHIO CHILDREN

Age Interval (months)	Boys		Girls	
	Mean	S.D.	Mean	S.D.
1-3	6.16	1.04	5.57	1.13
3-6	7.31	1.61	7.01	1.56
6-9	6.56	1.92	6.71	2.01
9-12	6.19	1.56	6.38	1.68
12-18	10.11	2.16	9.67	1.56
18-24	8.19	1.71	8.06	1.33
24-30	7.19	1.34	7.08	1.33
30-36	6.61	1.16	6.42	1.27
36-42	6.17	0.92	6.29	1.39
42-48	5.77	1.07	5.73	1.09
48-54	5.84	1.06	5.75	1.06
54-60	5.87	1.46	5.48	1.18
60-66	5.48	1.29	5.60	0.99
66-72	5.73	1.27	5.42	1.13
72-78	5.24	1.14	4.93	1.02
78-84	5.23	1.30	5.09	1.02
84-90	5.04	1.16	4.84	1.05
90-96	5.18	1.41	4.76	1.11
96-102	4.88	1.28	5.20	1.25
102-108	4.83	1.11	4.80	1.27

(data from Gindhart, 1971)

TABLE 1080

RADIOGRAPHIC MORPHOMETRY OF THE RADIUS SHAFT IN WHITE SCHOOL
CHILDREN IN WISCONSIN

Age (years)	Total		Thickness (mm) Medullary		Compact	
	F	M	F	M	F	M
Means						
6	9.4	10.0	4.4	4.6	5.0	5.4
7	9.7	10.3	4.4	4.8	5.2	5.5
8	9.6	10.6	4.2	4.9	5.4	5.7
9	9.9	10.8	4.4	5.2	5.5	5.6
10	10.3	11.2	4.6	5.0	5.7	6.2
11	11.2	11.4	4.8	5.2	6.3	6.2
12	11.5	12.0	5.2	5.7	6.3	6.3
13	11.6	12.2	4.9	5.5	6.7	6.6
Coefficients of Variation						
6	12	10	24	20	15	13
7	12	8	20	17	10	10
8	8	10	18	20	12	13
9	12	9	28	16	9	11
10	12	11	21	20	12	13
11	10	9	19	21	15	10
12	9	9	28	16	11	11
13	10	9	22	21	12	14

Age (years)	Area (mm^2) Total		Compact		Compact/Total (%) Thickness		Area	
	F	M	F	M	F	M	F	M
Means								
6	70	79	54	62	54	54	78	79
7	75	84	58	65	54	54	79	78
8	73	89	59	69	56	54	81	78
9	78	93	62	71	56	52	80	77
10	84	101	67	80	56	55	80	80
11	99	103	80	81	56	55	81	79
12	105	113	82	87	55	53	79	77
13	107	117	87	92	58	55	82	79
Coefficients of Variation								
6	25	20	23	20	15	12	10	7
7	23	15	20	14	9	10	6	7
8	16	20	16	19	11	12	7	8
9	23	18	17	17	14	10	9	6
10	24	22	22	21	11	11	7	7
11	20	19	22	16	12	12	7	7
12	18	17	11	17	17	10	11	6
13	19	18	18	18	12	14	7	9

(data from Mazess and Cameron, 1972)

TABLE 1081

DISTAL RADIAL METAPHYSEAL BAND WIDTH (mm)
IN RELATION TO STATURE VELOCITY (cm/yr)
IN THE PREPUBERTAL PERIOD
IN ENGLISH CHILDREN

Age	Boys				Girls			
	Band width		Stature		Band width		Stature	
(years)	Mean	S.D.	Mean	S.D.	Mean	S.D.	Mean	S.D.
3.0	1.84	.12	8.40	1.25	--	--	--	--
3.5	1.81	.17	7.61	1.16	--	--	--	--
4.0	1.78	.22	7.19	0.94	1.78	.23	7.60	0.94
4.5	1.70	.20	7.14	0.78	1.77	.21	7.10	0.84
5.0	1.79	.20	6.88	1.09	1.70	.26	6.78	0.72
5.5	1.64	.26	6.31	0.88	1.69	.21	6.30	0.89
6.0	1.64	.23	6.23	0.76	1.56	.23	6.08	1.08
6.5	1.68	.22	6.14	0.74	1.66	.22	6.04	0.90
7.0	1.71	.20	6.00	0.68	1.58	.24	5.95	0.82
7.5	1.58	.30	5.82	0.65	1.63	.24	5.69	0.78
8.0	1.55	.31	5.70	0.69	1.54	.21	5.67	0.66
8.5	1.57	.27	5.57	0.43	1.52	.28	5.54	0.64
9.0	1.63	.30	5.42	0.46	1.59	.27	5.51	0.66
9.5	1.57	.35	5.26	0.61	1.62	.22	5.13	0.52
10.0	1.61	.31	5.04	0.80	--	--	--	--
10.5	1.49	.41	4.82	0.68	--	--	--	--
11.0	1.55	.30	4.48	0.69	--	--	--	--
11.5	1.50	.25	5.24	0.77	--	--	--	--

(data from Edlin, Whitehouse and Tanner, 1976, American Journal of Diseases of Children 130:160-163, Copyright 1976, American Medical Association)

TABLE 1082

DISTAL RADIAL METAPHYSEAL BAND WIDTH (mm)
IN RELATION TO STATURE VELOCITY (cm/yr)
IN THE ADOLESCENT PERIOD
IN ENGLISH CHILDREN

| Age (years) | Boys | | | | Girls | | | |
| | Band width | | Stature | | Band width | | Stature | |
	Mean	S.D.	Mean	S.D.	Mean	S.D.	Mean	S.D.
9.00	--	--	--	--	1.62	.10	5.42	0.58
9.50	--	--	--	--	1.53	.31	5.38	0.78
10.00	--	--	--	--	1.62	.27	5.86	0.92
10.50	--	--	--	--	1.59	.30	5.52	1.14
10.75	--	--	--	--	1.70	.45	5.87	1.36
11.00	1.51	.20	4.87	0.78	1.73	.32	6.46	0.87
11.50	1.67	.31	4.95	0.66	1.75	.24	6.42	0.90
12.00	1.62	.19	5.24	1.01	2.04	.34	8.73	1.06
12.25	1.72	.29	5.50	1.09	2.00	.26	7.75	0.87
12.50	1.72	.33	5.42	1.24	1.82	.30	6.28	1.12
12.75	1.68	.29	6.23	1.36	1.71	.32	5.94	1.37
13.00	1.84	.36	7.48	1.38	1.61	.36	5.56	1.30
13.25	1.98	.32	7.55	1.36	1.48	.34	4.61	1.10
13.50	2.24	.38	8.86	1.45	1.31	.35	3.99	1.45
13.75	2.30	.32	10.05	1.53	1.05	.34	2.90	1.17
14.00	2.19	.28	9.41	1.34	0.90	.22	2.43	1.07
14.25	1.99	.34	8.04	1.21	0.76	.29	2.39	1.28
14.50	1.77	.33	6.94	1.29	0.76	.16	1.53	0.86
15.00	1.31	.42	5.24	1.05	0.61	.22	1.32	0.73
15.25	1.18	.20	4.43	0.96	--	--	--	--
15.50	1.17	.37	4.00	1.17	--	--	--	--
15.75	0.95	.26	3.03	0.82	--	--	--	--
16.00	0.94	.28	2.56	0.61	--	--	--	--

(data from Edlin, Whitehouse and Tanner, 1976, American Journal of Diseases of Children 130:160-163, Copyright 1976, American Medical Association)

TABLE 1083

ULNA LENGTH (cm)
IN DENVER CHILDREN

Age Yr. Mo.	Boys		Girls	
	Mean	S.D.	Mean	S.D.
0 - 2	6.70	.35	6.53	.31
0 - 4	7.38	.34	7.12	.31
0 - 6	7.91	.37	7.57	.38
1 - 0	9.26	.44	8.90	.40
1 - 6	10.23	.46	9.89	.44
2 - 0	10.97	.49	10.71	.48
2 - 6	11.66	.52	11.38	.52
3 - 0	12.34	.56	12.06	.54
3 - 6	12.91	.64	12.72	.57
4 - 0	13.56	.56	13.31	.58
4 - 6	14.10	.56	13.93	.66
5 - 0	14.70	.61	14.46	.71
5 - 6	15.26	.67	14.91	.72
6 - 0	15.75	.62	15.49	.74
6 - 6	16.22	.68	15.99	.79
7 - 0	16.73	.70	16.48	.83
7 - 6	17.22	.74	17.01	.85
8 - 0	17.73	.74	17.49	.87
8 - 6	18.16	.71	17.91	.88
9 - 0	18.64	.79	18.43	.95
9 - 6	19.17	.83	18.97	1.04
10 - 0	19.62	.85	19.44	1.06
10 - 6	20.04	.88	20.00	1.24
11 - 0	20.51	.92	20.47	1.20
11 - 6	20.98	.99	21.13	1.31
12 - 0	21.45	1.02	21.64	1.33

Length including Epiphyses

Age Yr. Mo.	Boys		Girls	
10 - 0	20.22	.90	20.38	1.23
10 - 6	20.80	.97	21.02	1.38
11 - 0	21.33	1.02	21.55	1.33
11 - 6	21.95	1.13	22.26	1.38
12 - 0	22.49	1.17	22.97	1.47
12 - 6	23.15	1.18	23.54	1.44
13 - 0	23.79	1.32	24.00	1.33
13 - 6	24.51	1.39	24.44	1.31

TABLE 1083 (continued)

ULNA LENGTH (cm)
IN DENVER CHILDREN

Age	Boys		Girls	
Yr. Mo.	Mean	S.D.	Mean	S.D.
14 - 0	25.23	1.46	24.81	1.21
14 - 6	25.90	1.47	25.02	1.18
15 - 0	26.51	1.40	25.10	1.22
15 - 6	27.19	1.31	25.50	1.51
16 - 0	27.48	1.22	25.23	.1.20
16 - 6	27.73	1.21	--	--
17 - 0	27.94	1.17	25.02	1.23
18 - 0	28.16	1.35	--	--

FEMUR LENGTH (cm)
IN DENVER CHILDREN

Age	Boys		Girls	
Yr. Mo.	Mean	S.D.	Mean	S.D.
Diaphyseal Length				
0 - 2	8.60	.54	8.72	.43
0 - 4	10.07	.48	10.08	.36
0 - 6	11.22	.50	11.11	.46
1 - 0	13.66	.58	13.46	.49
1 - 6	15.54	.68	15.39	.64
2 - 0	17.24	.73	17.08	.71
2 - 6	18.72	.78	18.52	.77
3 - 0	20.03	.85	19.84	.87
3 - 6	21.21	1.14	21.11	1.00
4 - 0	22.41	.99	22.32	1.01
4 - 6	23.57	1.05	23.55	1.14
5 - 0	24.75	1.11	24.70	1.15
5 - 6	25.82	1.17	25.70	1.22
6 - 0	26.97	1.20	26.89	1.35
6 - 6	28.03	1.26	27.90	1.38
7 - 0	29.11	1.33	28.88	1.36
7 - 6	30.12	1.35	29.98	1.52
8 - 0	31.21	1.46	30.98	1.56
8 - 6	32.10	1.46	31.89	1.58

TABLE   1083   (continued)

FEMUR LENGTH (cm)
IN DENVER CHILDREN

Age Yr. Mo.	Boys		Girls	
	Mean	S.D.	Mean	S.D.
9 - 0	33.04	1.46	32.87	1.68
9 - 6	34.00	1.58	33.88	1.86
10 - 0	34.93	1.57	34.79	1.91
10 - 6	35.74	1.62	35.65	2.14
11 - 0	36.70	1.65	36.70	2.24
11 - 6	37.58	1.81	37.80	2.34
12 - 0	38.61	1.90	38.76	2.29
Length including Epiphyses				
10 - 0	38.51	1.70	38.28	2.11
10 - 6	39.42	1.79	39.26	2.37
11 - 0	40.52	1.79	40.35	2.48
11 - 6	41.48	1.94	41.54	2.52
12 - 0	42.56	2.06	42.79	2.52
12 - 6	43.71	1.96	43.79	2.39
13 - 0	44.74	2.15	44.72	2.41
13 - 6	45.84	2.40	45.31	2.20
14 - 0	47.08	2.41	45.99	2.25
14 - 6	47.89	2.52	46.45	2.08
15 - 0	48.90	2.35	46.44	2.14
15 - 6	49.85	2.34	47.15	2.60
16 - 0	50.28	2.28	46.67	2.40
16 - 6	50.45	2.49	--	--
17 - 0	50.89	2.32	46.29	2.62
18 - 0	51.17	2.44	--	--

(data from McCammon, Human Growth and Development, 1970.
Courtesy of Charles C Thomas, Publisher, Springfield,
Illinois)

TABLE   1084

METACARPAL AND PHALANGEAL LENGTHS (mm)
IN OHIO CHILDREN

Bones		2 years		3 years		4 years		5 years		6 years	
		Mean	S.D.	Mean	S.D.	Mean	S.D.	Mean	S.D.	Mean	S.D.
BOYS											
Distal	5	8.8	--	8.4	0.6	9.0	0.7	9.9	0.6	10.7	0.6
	4	9.2	0.7	9.9	0.8	10.5	0.8	11.5	0.9	12.3	0.9
	3	8.7	0.9	9.5	0.8	10.2	0.8	11.1	0.8	11.8	0.9
	2	8.2	0.5	8.8	1.1	9.1	0.8	10.1	0.9	10.8	0.9
	1	11.1	0.6	12.3	0.8	13.2	1.0	14.4	0.9	15.4	0.9
Middle	5	8.8	0.9	9.8	0.8	10.6	1.0	11.2	1.0	12.0	1.0
	4	13.5	0.9	14.5	1.0	15.8	0.9	16.7	0.9	17.7	1.0
	3	14.1	0.8	15.1	1.1	16.5	1.0	17.6	1.0	18.7	1.1
	2	11.2	0.8	12.3	1.1	13.5	1.0	14.4	0.9	15.3	1.0
Proximal	5	16.1	0.7	17.8	0.9	19.2	1.0	20.6	1.0	21.8	1.0
	4	20.5	0.9	22.8	1.0	24.7	1.2	26.4	1.2	27.9	1.3
	3	21.8	1.0	24.2	1.1	26.3	1.4	28.1	1.4	29.8	1.4
	2	19.5	1.0	21.9	1.2	23.7	1.3	25.4	1.4	26.8	1.5
	1	15.2	--	15.9	1.1	17.2	1.1	18.3	1.2	19.6	1.2
Metacarpal	5	23.9	1.0	26.3	1.5	28.9	1.9	32.1	2.2	34.6	2.2
	4	25.5	1.1	28.9	1.5	31.7	2.1	35.0	2.5	37.9	2.7
	3	28.6	1.3	32.3	1.8	35.6	2.3	39.3	2.8	42.6	2.9
	2	30.6	1.5	34.5	1.7	37.9	2.3	41.6	2.7	44.9	2.9
	1	19.6	1.3	22.0	1.2	24.1	1.6	26.7	1.6	29.0	1.7
GIRLS											
Distal	5	7.8	0.6	8.4	0.6	9.1	0.7	9.9	0.7	10.6	0.8
	4	9.1	0.7	9.9	0.7	10.6	0.8	11.5	0.9	12.4	1.0
	3	8.8	0.7	9.9	0.8	10.2	0.7	11.1	0.9	12.2	1.3
	2	8.0	0.8	8.6	0.7	9.4	0.7	10.1	0.8	10.9	0.9
	1	11.3	0.8	12.5	0.9	13.2	0.8	14.4	1.0	15.4	1.1
Middle	5	9.0	1.2	9.8	1.1	10.5	1.1	11.2	1.1	12.2	1.2
	4	13.5	0.9	14.9	1.0	15.8	1.1	16.9	1.2	18.1	1.3
	3	14.2	0.9	15.6	1.1	16.6	1.2	17.9	1.2	19.2	1.3
	2	11.6	0.9	12.8	1.0	13.6	1.1	14.8	1.1	16.0	1.2
Proximal	5	16.3	1.0	17.9	1.1	19.1	1.1	20.6	1.3	22.0	1.4
	4	20.7	1.1	22.9	1.3	24.6	1.3	26.3	1.5	28.2	1.7
	3	22.2	1.2	24.5	1.3	26.4	1.4	28.3	1.8	30.4	1.8
	2	20.1	1.2	22.3	1.3	24.0	1.8	25.8	1.7	27.7	1.7
	1	14.9	1.0	16.3	1.1	17.2	1.3	18.8	1.3	20.2	1.3
Metacarpal	5	23.7	1.5	26.9	2.1	29.4	1.8	32.6	2.0	35.1	2.1
	4	26.0	1.9	29.6	2.7	32.2	2.0	35.6	2.5	38.4	2.7
	3	29.4	2.1	33.4	2.9	36.3	2.2	40.3	2.7	43.3	3.1
	2	31.3	1.9	35.2	2.7	38.2	2.3	42.2	2.7	45.6	3.2
	1	19.9	1.6	22.7	1.6	24.8	1.7	27.3	1.8	29.6	1.9

TABLE 1084   (continued)

METACARPAL AND PHALANGEAL LENGTHS (mm)
IN OHIO CHILDREN

Bones		7 years		8 years		9 years		10 years	
		Mean	S.D.	Mean	S.D.	Mean	S.D.	Mean	S.D.
					BOYS				
Distal	5	11.4	0.8	12.2	0.9	12.6	1.0	13.5	0.9
	4	13.1	1.0	13.9	1.0	14.4	1.0	15.3	1.2
	3	12.7	1.0	13.4	1.0	14.0	1.0	14.8	1.2
	2	11.6	1.0	12.4	1.0	13.0	1.0	13.7	1.1
	1	16.5	1.0	17.4	1.0	17.9	1.2	19.0	1.2
Middle	5	12.7	1.1	13.5	1.1	14.3	1.2	15.0	1.2
	4	18.7	1.1	19.8	1.1	20.9	1.3	21.6	1.4
	3	19.8	1.2	20.9	1.2	22.0	1.4	22.9	1.4
	2	16.1	1.1	17.1	1.1	18.1	1.2	18.8	1.2
Proximal	5	23.0	1.1	24.2	1.3	25.2	1.5	26.4	1.5
	4	29.5	1.4	31.0	1.6	32.3	1.9	33.9	1.8
	3	31.5	1.6	33.2	1.8	34.7	2.2	36.1	1.9
	2	28.3	1.6	29.7	1.8	31.4	1.9	32.5	1.9
	1	20.8	1.3	21.8	1.3	23.1	1.5	24.2	1.4
Metacarpal	5	36.7	2.1	38.8	2.5	40.6	2.5	42.7	2.9
	4	40.1	2.5	42.2	3.1	44.1	2.8	46.5	3.5
	3	45.3	2.8	47.6	3.5	49.8	3.0	52.3	3.7
	2	47.7	2.8	50.2	3.4	52.6	3.0	55.0	3.9
	1	30.9	1.8	32.7	2.1	34.4	2.1	36.3	2.3
					GIRLS				
Distal	5	11.4	0.9	12.1	1.0	12.7	1.1	13.5	1.2
	4	13.2	1.1	14.0	1.1	14.4	1.2	15.5	1.4
	3	12.7	1.1	13.5	1.1	14.1	1.1	15.0	1.4
	2	11.7	1.0	12.3	1.1	13.1	1.1	13.8	1.4
	1	16.3	1.2	17.3	1.3	17.8	1.3	19.0	1.6
Middle	5	12.9	1.3	13.6	1.4	14.2	1.4	15.2	1.6
	4	19.1	1.4	20.1	1.4	20.9	1.5	22.2	1.7
	3	20.3	1.4	21.4	1.4	22.1	1.6	23.6	1.8
	2	16.8	1.3	17.8	1.4	18.1	1.5	19.6	1.7
Proximal	5	23.1	1.6	24.4	1.6	25.2	1.6	27.1	2.0
	4	29.7	1.9	31.2	2.0	32.4	2.0	34.5	2.4
	3	32.1	2.0	33.7	2.2	35.0	2.2	37.3	2.6
	2	29.2	1.9	30.7	2.0	31.5	2.4	34.0	2.4
	1	21.4	1.5	22.7	1.6	23.5	2.0	25.5	2.1
Metacarpal	5	37.2	2.4	39.4	2.5	40.8	2.5	43.8	2.8
	4	40.5	2.8	34.1	3.0	44.3	2.8	47.5	3.5
	3	45.8	3.1	48.7	3.2	49.9	3.2	53.6	3.8
	2	48.1	3.3	51.2	3.3	52.6	3.4	56.6	4.1
	1	31.5	2.0	33.5	2.1	34.8	2.4	37.4	2.6

TABLE 1084   (continued)

METACARPAL AND PHALANGEAL LENGTHS (mm)
IN OHIO CHILDREN

Bones		11 years		12 years		13 years		14 years	
		Mean	S.D.	Mean	S.D.	Mean	S.D.	Mean	S.D.
				BOYS					
Distal	5	14.2	0.9	15.0	0.9	15.8	0.9	16.8	1.0
	4	16.1	1.2	17.0	1.3	17.8	1.4	18.8	1.3
	3	15.6	1.2	16.4	1.2	17.1	1.3	18.2	1.3
	2	14.3	1.1	15.0	1.0	15.7	1.4	16.7	1.2
	1	19.7	1.2	20.6	1.3	21.7	1.4	22.8	1.3
Middle	5	15.7	1.4	16.5	1.5	17.5	1.5	18.9	1.6
	4	22.6	1.5	23.6	1.5	24.8	1.7	26.5	1.6
	3	24.0	1.4	24.9	1.4	26.3	1.6	28.0	1.5
	2	19.8	1.8	20.4	1.3	21.6	1.6	23.2	1.5
Proximal	5	27.6	1.7	28.9	2.0	30.5	2.4	32.9	2.4
	4	35.3	2.0	37.0	2.4	38.8	2.8	41.6	2.8
	3	37.8	2.3	39.5	2.6	41.5	2.9	44.4	2.8
	2	33.9	2.1	35.5	2.4	37.2	2.6	39.8	2.6
	1	25.4	1.6	26.7	2.0	28.5	2.2	30.9	2.2
Metacarpal	5	44.6	2.8	47.1	3.2	49.1	4.0	52.2	3.9
	4	48.4	3.1	51.0	3.7	53.1	4.6	56.4	4.5
	3	54.6	3.4	57.3	4.0	59.5	5.1	63.1	4.9
	2	57.3	3.5	60.6	3.9	63.3	5.1	67.1	4.8
	1	38.2	2.4	40.2	2.7	42.5	3.0	45.1	2.8
				GIRLS					
Distal	5	14.2	1.3	15.0	1.3	15.4	1.3	15.6	1.3
	4	16.2	1.4	17.1	1.4	17.6	1.2	17.9	1.3
	3	15.8	1.3	16.6	1.4	17.1	1.4	17.3	1.3
	2	14.4	1.3	15.2	1.5	15.7	1.5	15.8	1.5
	1	20.0	1.7	20.9	1.7	21.4	1.6	21.7	1.6
Middle	5	16.2	1.7	17.2	1.7	17.9	1.8	18.1	1.6
	4	23.4	1.8	24.7	1.8	25.7	1.9	25.9	1.6
	3	24.9	1.9	26.2	1.9	27.2	2.0	27.5	1.7
	2	20.6	1.8	21.8	1.9	22.7	1.8	23.0	1.8
Proximal	5	28.7	2.1	30.5	2.2	31.9	2.2	32.3	2.1
	4	36.5	2.5	38.8	2.6	40.3	2.5	40.9	2.3
	3	39.5	2.7	41.7	2.8	43.5	2.8	44.1	2.4
	2	35.9	2.6	38.0	2.6	39.5	2.6	39.9	2.4
	1	27.2	2.3	29.2	2.4	30.6	2.2	31.1	1.9
Metacarpal	5	46.3	2.9	48.7	2.9	50.8	2.8	52.1	2.8
	4	50.2	3.8	52.8	3.7	55.1	3.6	56.2	3.6
	3	56.5	4.0	59.5	4.2	62.1	4.0	63.4	3.9
	2	59.9	4.3	63.2	4.4	66.2	4.2	67.4	3.9
	1	39.7	3.0	42.0	3.0	43.8	2.7	44.4	2.5

TABLE 1084   (continued)

METACARPAL AND PHALANGEAL LENGTHS (mm)
IN OHIO CHILDREN

Bones		15 years		16 years		17 years		18 years	
		Mean	S.D.	Mean	S.D.	Mean	S.D.	Mean	S.D.
BOYS									
Distal	5	17.6	1.1	17.9	1.0	18.1	1.0	18.1	1.2
	4	19.6	1.4	20.0	1.3	20.3	1.3	20.0	1.3
	3	19.0	1.4	19.3	1.4	19.5	1.3	19.4	1.3
	2	17.5	1.2	17.8	1.3	18.2	1.3	18.1	1.3
	1	24.1	1.4	24.5	1.4	24.9	1.4	24.8	1.5
Middle	5	19.9	1.4	20.5	1.4	20.6	1.4	21.0	1.4
	4	27.7	1.5	28.4	1.5	28.7	1.4	29.1	1.5
	3	29.2	1.5	30.0	1.6	30.2	1.6	30.6	1.8
	2	24.3	1.5	25.0	1.5	25.3	1.4	25.6	1.7
Proximal	5	34.7	2.0	35.6	1.8	36.1	1.8	35.9	2.0
	4	43.7	2.6	44.9	2.3	45.4	2.2	45.2	2.5
	3	46.6	2.5	47.8	2.4	48.3	2.3	48.2	2.7
	2	41.8	2.2	42.8	2.0	43.3	2.1	43.4	2.4
	1	32.9	1.8	33.8	1.5	34.6	2.6	34.7	1.8
Metacarpal	5	55.4	3.6	57.1	2.8	57.9	2.5	57.5	2.9
	4	59.5	4.1	61.5	3.7	62.6	3.1	61.7	3.4
	3	66.7	4.4	68.7	4.1	69.7	3.3	69.0	3.7
	2	70.6	4.3	73.2	3.8	74.2	2.9	73.9	3.5
	1	47.6	2.6	48.8	2.3	49.5	2.1	49.4	2.7
GIRLS									
Distal	5	15.9	1.4	15.9	1.4	16.2	1.3	16.0	1.2
	4	18.0	1.4	18.0	1.3	18.1	1.4	17.9	1.3
	3	17.6	1.5	17.5	1.4	17.6	1.4	17.4	1.3
	2	16.1	1.6	16.0	1.6	16.3	1.5	16.2	1.3
	1	22.0	1.7	22.0	1.7	22.1	1.8	22.0	1.6
Middle	5	18.4	1.7	18.5	1.7	18.5	1.9	18.6	1.7
	4	26.3	1.8	26.4	1.8	26.5	1.9	26.3	1.8
	3	28.1	1.8	28.0	1.9	28.0	1.8	27.8	1.8
	2	23.5	1.8	23.3	1.9	23.4	1.9	23.1	1.6
Proximal	5	32.9	2.2	32.8	2.3	32.8	2.3	32.5	2.0
	4	41.5	2.5	41.6	2.6	41.7	2.6	41.1	2.2
	3	44.8	2.6	44.8	2.7	44.8	2.5	44.2	2.4
	2	40.6	2.6	40.6	2.6	40.7	2.6	39.9	2.3
	1	31.8	2.0	31.7	2.1	31.9	2.2	31.3	1.9
Metacarpal	5	52.6	3.0	52.8	3.0	53.0	2.7	52.0	2.7
	4	56.9	3.6	57.2	3.9	57.2	3.5	56.1	2.9
	3	63.9	3.9	64.3	4.0	64.5	4.0	63.2	3.4
	2	68.1	4.2	68.6	4.3	68.9	4.1	67.5	3.4
	1	45.3	2.4	45.0	2.8	45.0	2.6	44.6	2.2

(data from Garn, Hertzog, Poznanski, et al., 1972, RADIOLOGY 105:375-381)

TABLE 1085

RATIOS BETWEEN THE LENGTHS OF THE BONES OF THE HAND
IN OHIO CHILDREN

	Boys						Girls					
	1 year		4 years		9 years		1 year		4 years		9 years	
Bone ratio	Mean	S.D.	Mean	S.D.	Mean	S.D.	Mean	S.D.	Mean	S.D.	Mean	S.D.
Proximal 1/metacarpal 5	0.62	.03	0.60	.04	0.57	.03	0.61	.04	0.59	.03	0.58	.03
/metacarpal 4	0.57	.03	0.54	.03	0.53	.03	0.56	.04	0.54	.03	0.53	.03
/metacarpal 3	0.51	.03	0.48	.03	0.47	.03	0.50	.03	0.48	.03	0.47	.03
/metacarpal 2	0.47	.02	0.45	.02	0.44	.02	0.47	.03	0.45	.02	0.45	.02
/metacarpal 1	0.77	.04	0.71	.03	0.67	.03	0.75	.04	0.70	.03	0.67	.03
Proximal 2 /proximal 1	1.35	.06	1.39	.05	1.36	.05	1.37	.07	1.39	.09	1.35	.08
/metacarpal 5	0.83	.05	0.83	.05	0.77	.03	0.84	.04	0.82	.05	0.77	.04
/metacarpal 4	0.76	.04	0.75	.04	0.71	.03	0.76	.04	0.74	.05	0.71	.04
/metacarpal 3	0.69	.04	0.67	.03	0.63	.03	0.68	.04	0.66	.04	0.63	.04
/metacarpal 2	0.64	.03	0.63	.03	0.60	.02	0.64	.03	0.63	.04	0.60	.03
/metacarpal 1	1.04	.06	0.99	.04	0.91	.04	1.02	.06	0.97	.06	0.91	.06
Proximal 3 /proximal 2	1.12	.04	1.11	.03	1.11	.04	1.13	.03	1.10	.04	1.11	.07
/proximal 1	1.52	.08	1.54	.07	1.50	.08	1.55	.09	1.54	.08	1.50	.09
/metacarpal 5	0.94	.05	0.91	.05	0.85	.04	0.95	.04	0.90	.04	0.86	.04
/metacarpal 4	0.86	.04	0.83	.04	0.79	.04	0.86	.04	0.82	.04	0.79	.04
/metacarpal 3	0.77	.04	0.74	.03	0.70	.03	0.77	.03	0.73	.03	0.70	.03
/metacarpal 2	0.71	.03	0.70	.03	0.66	.03	0.72	.03	0.69	.03	0.67	.03
/metacarpal 1	1.17	.06	1.09	.05	1.01	.05	1.15	.07	1.07	.05	1.01	.05
Metacarpal 2/metacarpal 1	1.64	.06	1.57	.06	1.53	.05	1.60	.09	1.54	.07	1.52	.06
Metacarpal 3/metacarpal 2	0.93	.03	0.94	.02	0.96	.02	0.94	.03	0.95	.02	0.95	.02
/metacarpal 1	1.52	.07	1.48	.06	1.45	.06	1.50	.08	1.47'	.07	1.44	.07
Metacarpal 4/metacarpal 3	0.90	.02	0.89	.02	0.89	.02	0.90	.02	0.89	.02	0.89	.02
/metacarpal 2	0.83	.02	0.83	.02	0.84	.02	0.84	.04	0.84	.03	0.84	.02
/metacarpal 1	1.36	.06	1.31	.05	1.28	.05	1.35	.07	1.30	.07	1.27	.06
Metacarpal 5/metacarpal 4	0.92	.03	0.91	.03	0.92	.02	0.91	.04	0.91	.02	0.92	.02
/metacarpal 3	0.82	.03	0.81	.03	0.82	.02	0.81	.03	0.81	.02	0.82	.02
/metacarpal 2	0.76	.03	0.76	.03	0.77	.02	0.76	.02	0.77	.03	0.77	.02
/metacarpal 1	1.25	.05	1.20	.05	1.18	.05	1.22	.07	1.19	.06	1.17	.05
Proximal 4 /proximal 3	0.95	.02	0.94	.02	0.93	.03	0.95	.02	0.93	.02	0.93	.03
/proximal 2	1.07	.04	1.04	.03	1.03	.03	1.07	.04	1.03	.04	1.03	.06
/proximal 1	1.45	.07	1.44	.07	1.40	.07	1.47	.08	1.43	.07	1.39	.08
/metacarpal 5	0.89	.04	0.86	.04	0.80	.03	0.90	.04	0.84	.04	0.80	.03
/metacarpal 4	0.82	.04	0.78	.04	0.73	.03	0.81	.04	0.76	.04	0.73	.03
/metacarpal 3	0.73	.03	0.69	.03	0.65	.02	0.73	.03	0.68	.03	0.65	.02
/metacarpal 2	0.68	.03	0.65	.03	0.62	.03	0.68	.03	0.64	.03	0.62	.02
/metacarpal 1	1.11	.06	1.02	.05	0.94	.04	1.09	.06	0.99	.05	0.93	.04
Proximal 5 /proximal 4	0.78	.02	0.78	.02	0.78	.02	0.77	.02	0.78	.02	0.78	.02
/proximal 3	0.74	.03	0.73	.02	0.73	.03	0.73	.03	0.72	.02	0.72	.03
/proximal 2	0.83	.03	0.81	.03	0.80	.03	0.83	.04	0.80	.03	0.80	.05
/proximal 1	1.12	.05	1.12	.05	1.09	.05	1.13	.06	1.11	.06	1.08	.06
/metacarpal 5	0.69	.03	0.67	.03	0.62	.02	0.69	.03	0.65	.03	0.62	.02
/metacarpal 4	0.63	.03	0.61	.03	0.57	.03	0.63	.04	0.59	.03	0.57	.02
/metacarpal 3	0.57	.03	0.54	.03	0.51	.02	0.56	.03	0.53	.02	0.51	.02
/metacarpal 2	0.53	.03	0.51	.02	0.48	.02	0.53	.03	0.50	.02	0.48	.02
/metacarpal 1	0.86	.05	0.80	.04	0.73	.03	0.84	.05	0.77	.04	0.73	.03

TABLE 1085   (continued)

RATIOS BETWEEN THE LENGTHS OF THE BONES OF THE HAND
IN OHIO CHILDREN

Bone ratio		Boys						Girls					
		1 year		4 years		9 years		1 year		4 years		9 years	
		Mean	S.D.	Mean	S.D.	Mean	S.D.	Mean	S.D.	Mean	S.D.	Mean	S.D.
Middle 2	/proximal 5	0.68	.05	0.70	.04	0.72	.04	0.69	.05	0.71	.04	0.72	.04
	/proximal 4	0.53	.04	0.55	.03	0.56	.03	0.53	.04	0.55	.03	0.56	.03
	/proximal 3	0.50	.03	0.51	.03	0.52	.03	0.50	.04	0.52	.03	0.52	.03
	/proximal 2	0.57	.03	0.57	.03	0.58	.03	0.57	.04	0.57	.03	0.58	.04
	/proximal 1	0.77	.05	0.79	.03	0.78	.03	0.78	.06	0.79	.04	0.77	.04
	/metacarpal 5	0.47	.04	0.47	.04	0.45	.03	0.48	.04	0.46	.03	0.44	.03
	/metacarpal 4	0.43	.04	0.43	.03	0.41	.03	0.43	.04	0.42	.03	0.41	.03
	/metacarpal 3	0.39	.03	0.38	.03	0.36	.02	0.39	.04	0.38	.02	0.36	.02
	/metacarpal 2	0.36	.03	0.36	.02	0.34	.02	0.36	.03	0.36	.02	0.34	.02
	/metacarpal 1	0.59	.04	0.56	.03	0.53	.03	0.58	.05	0.55	.03	0.52	.03
Middle 3	/middle 2	1.29	.10	1.22	.04	1.22	.04	1.27	.05	1.22	.04	1.22	.04
	/proximal 5	0.88	.07	0.86	.04	0.88	.04	0.88	.06	0.87	.04	0.88	.04
	/proximal 4	0.68	.05	0.67	.03	0.68	.03	0.67	.04	0.68	.03	0.68	.03
	/proximal 3	0.65	.05	0.63	.03	0.64	.03	0.64	.04	0.63	.03	0.63	.03
	/proximal 2	0.73	.06	0.69	.03	0.70	.03	0.72	.04	0.69	.03	0.70	.05
	/proximal 1	0.98	.09	0.96	.04	0.95	.04	0.99	.07	0.97	.05	0.94	.05
	/metacarpal 5	0.61	.05	0.57	.04	0.54	.03	0.60	.05	0.56	.03	0.54	.03
	/metacarpal 4	0.56	.05	0.52	.03	0.50	.03	0.55	.05	0.52	.03	0.50	.03
	/metacarpal 3	0.50	.04	0.46	.03	0.44	.02	0.49	.04	0.46	.03	0.44	.02
	/metacarpal 2	0.46	.04	0.44	.03	0.42	.02	0.46	.04	0.44	.02	0.42	.02
	/metacarpal 1	0.75	.07	0.69	.04	0.64	.03	0.74	.06	0.67	.04	0.63	.03
Middle 4	/middle 3	0.97	.05	0.96	.03	0.95	.03	0.97	.03	0.95	.03	0.95	.03
	/middle 2	1.24	.07	1.17	.05	1.15	.04	1.23	.07	1.16	.05	1.15	.05
	/proximal 5	0.84	.04	0.82	.03	0.83	.03	0.84	.04	0.83	.04	0.83	.04
	/proximal 4	0.66	.03	0.64	.03	0.65	.03	0.65	.03	0.64	.02	0.64	.02
	/proximal 3	0.62	.03	0.60	.02	0.60	.03	0.62	.03	0.60	.02	0.60	.03
	/proximal 2	0.70	.04	0.67	.03	0.67	.03	0.69	.04	0.66	.03	0.67	.04
	/proximal 1	0.95	.06	0.92	.04	0.90	.04	0.95	.05	0.92	.05	0.89	.04
	/metacarpal 5	0.59	.04	0.55	.03	0.51	.03	0.58	.04	0.54	.03	0.51	.03
	/metacarpal 4	0.54	.04	0.50	.03	0.47	.03	0.53	.04	0.49	.03	0.47	.03
	/metacarpal 3	0.48	.03	0.45	.03	0.42	.02	0.47	.03	0.44	.03	0.42	.02
	/metacarpal 2	0.45	.03	0.42	.02	0.40	.02	0.44	.03	0.42	.02	0.40	.02
	/metacarpal 1	0.73	.05	0.66	.04	0.61	.03	0.71	.05	0.64	.04	0.60	.03
Middle 5	/middle 4	0.66	.06	0.67	.05	0.69	.04	0.65	.06	0.66	.04	0.68	.04
	/middle 3	0.64	.07	0.64	.05	0.65	.04	0.63	.06	0.63	.04	0.65	.04
	/middle 2	0.81	.08	0.78	.06	0.79	.05	0.80	.08	0.77	.05	0.79	.05
	/proximal 5	0.55	.05	0.55	.04	0.57	.03	0.55	.05	0.55	.04	0.56	.04
	/proximal 4	0.43	.04	0.43	.04	0.44	.03	0.42	.04	0.43	.03	0.44	.03
	/proximal 3	0.41	.04	0.40	.04	0.41	.03	0.40	.04	0.40	.03	0.41	.03
	/proximal 2	0.46	.05	0.45	.04	0.46	.03	0.45	.04	0.44	.03	0.45	.04
	/proximal 1	0.62	.06	0.61	.05	0.62	.04	0.62	.05	0.61	.04	0.61	.04
	/metacarpal 5	0.38	.04	0.37	.03	0.35	.03	0.38	.04	0.36	.03	0.35	.03
	/metacarpal 4	0.35	.04	0.33	.03	0.33	.03	0.34	.04	0.33	.03	0.32	.03
	/metacarpal 3	0.32	.03	0.30	.03	0.29	.02	0.31	.03	0.29	.02	0.29	.02
	/metacarpal 2	0.29	.03	0.28	.03	0.27	.02	0.29	.03	0.27	.02	0.27	.02
	/metacarpal 1	0.48	.05	0.44	.04	0.42	.03	0.46	.05	0.42	.03	0.41	.03

TABLE 1085   (continued)

RATIOS BETWEEN THE LENGTHS OF THE BONES OF THE HAND
IN OHIO CHILDREN

	Boys						Girls					
	1 year		4 years		9 years		1 year		4 years		9 years	
Bone ratio	Mean	S.D.	Mean	S.D.	Mean	S.D.	Mean	S.D.	Mean	S.D.	Mean	S.D.
Distal 1 /middle 5	1.22	.13	1.27	.12	1.26	.10	1.22	.13	1.27	.10	1.26	.10
/middle 4	0.79	.05	0.84	.05	0.86	.04	0.79	.05	0.83	.04	0.86	.04
/middle 3	0.77	.06	0.80	.04	0.81	.04	0.76	.06	0.80	.04	0.81	.04
/middle 2	0.98	.08	0.98	.06	0.99	.05	0.97	.09	0.97	.05	0.99	.05
/proximal 5	0.67	.04	0.69	.04	0.71	.04	0.66	.04	0.69	.03	0.71	.04
/proximal 4	0.52	.03	0.54	.03	0.56	.03	0.51	.03	0.54	.02	0.55	.03
/proximal 3	0.49	.03	0.50	.03	0.52	.03	0.48	.03	0.50	.02	0.51	.03
/proximal 2	0.56	.03	0.56	.03	0.57	.03	0.55	.03	0.55	.03	0.57	.04
/proximal 1	0.75	.04	0.77	.04	0.78	.04	0.75	.05	0.77	.04	0.76	.05
/metacarpal 5	0.46	.03	0.46	.03	0.44	.02	0.46	.03	0.45	.02	0.44	.03
/metacarpal 4	0.42	.03	0.42	.03	0.41	.02	0.41	.03	0.41	.02	0.40	.02
/metacarpal 3	0.38	.02	0.37	.02	0.36	.02	0.37	.03	0.36	.02	0.36	.02
/metacarpal 2	0.35	.02	0.35	.02	0.34	.02	0.35	.02	0.35	.02	0.34	.02
/metacarpal 1	0.58	.03	0.55	.03	0.52	.03	0.56	.04	0.53	.03	0.51	.03
Distal 2 /distal 1	0.68	.05	0.69	.04	0.72	.03	0.68	.05	0.71	.04	0.73	.03
/middle 5	0.82	.11	0.87	.09	0.91	.08	0.83	.11	0.90	.09	0.92	.08
/middle 4	0.54	.04	0.58	.04	0.62	.04	0.53	.04	0.59	.04	0.63	.04
/middle 3	0.52	.04	0.55	.04	0.59	.03	0.51	.04	0.56	.04	0.59	.04
/middle 2	0.66	.05	0.67	.04	0.72	.04	0.65	.06	0.69	.04	0.72	.05
/proximal 5	0.45	.04	0.47	.04	0.52	.03	0.45	.04	0.49	.03	0.52	.04
/proximal 4	0.35	.03	0.37	.03	0.40	.03	0.35	.03	0.38	.02	0.40	.03
/proximal 3	0.33	.03	0.35	.03	0.37	.03	0.33	.03	0.35	.02	0.37	.03
/proximal 2	0.38	.03	0.38	.03	0.41	.02	0.37	.04	0.39	.03	0.42	.04
/proximal 1	0.51	.05	0.53	.04	0.56	.04	0.51	.05	0.54	.04	0.56	.04
/metacarpal 5	0.31	.03	0.32	.03	0.32	.02	0.31	.03	0.32	.02	0.32	.02
/metacarpal 4	0.29	.03	0.29	.02	0.29	.02	0.28	.03	0.29	.02	0.30	.02
/metacarpal 3	0.26	.02	0.26	.02	0.26	.02	0.25	.03	0.26	.02	0.26	.02
/metacarpal 2	0.24	.02	0.24	.02	0.25	.02	0.24	.02	0.25	.02	0.25	.02
/metacarpal 1	0.39	.03	0.38	.03	0.38	.02	0.38	.04	0.38	.03	0.38	.03
Distal 3 /distal 2	1.13	.07	1.13	.06	1.08	.04	1.16	.09	1.10	.04	1.08	.04
/distal 1	0.77	.05	0.77	.04	0.78	.03	0.78	.06	0.78	.03	0.79	.03
/middle 5	0.93	.11	0.98	.10	0.98	.08	0.94	.11	0.98	.09	0.99	.08
/middle 4	0.61	.04	0.64	.04	0.67	.04	0.61	.04	0.65	.04	0.67	.04
/middle 3	0.59	.05	0.62	.04	0.64	.03	0.59	.04	0.62	.03	0.64	.03
/middle 2	0.75	.06	0.76	.05	0.78	.05	0.78	.06	0.75	.04	0.78	.05
/proximal 5	0.51	.04	0.53	.03	0.56	.03	0.51	.04	0.54	.03	0.56	.03
/proximal 4	0.40	.03	0.41	.03	0.43	.03	0.40	.03	0.42	.02	0.43	.02
/proximal 3	0.38	.03	0.39	.02	0.40	.03	0.37	.03	0.39	.02	0.40	.03
/proximal 2	0.43	.03	0.43	.03	0.45	.02	0.42	.04	0.43	.03	0.45	.04
/proximal 1	0.58	.04	0.60	.04	0.61	.04	0.58	.05	0.60	.04	0.60	.04
/metacarpal 5	0.35	.03	0.35	.03	0.34	.02	0.35	.03	0.35	.02	0.35	.02
/metacarpal 4	0.33	.03	0.32	.02	0.32	.02	0.32	.03	0.32	.02	0.32	.02
/metacarpal 3	0.29	.02	0.29	.02	0.28	.02	0.29	.03	0.28	.02	0.28	.02
/metacarpal 2	0.27	.02	0.27	.02	0.27	.02	0.27	.02	0.27	.02	0.27	.02
/metacarpal 1	0.44	.04	0.42	.03	0.41	.02	0.43	.04	0.41	.03	0.41	.03

TABLE   1085   (continued)

RATIOS BETWEEN THE LENGTHS OF THE BONES OF THE HAND
IN OHIO CHILDREN

Bone ratio	Boys						Girls					
	1 year		4 years		9 years		1 year		4 years		9 years	
	Mean	S.D.	Mean	S.D.	Mean	S.D.	Mean	S.D.	Mean	S.D.	Mean	S.D.
Distal 4 /distal 3	1.04	.05	1.04	.03	1.03	.03	1.03	.05	1.03	.03	1.02	.03
/distal 2	1.18	.08	1.17	.06	1.11	.05	1.20	.11	1.13	.05	1.10	.05
/distal 1	0.80	.06	0.80	.04	0.80	.04	0.80	.07	0.80	.03	0.81	.03
/middle 5	0.97	.11	1.01	.10	1.01	.09	0.97	.12	1.02	.09	1.02	.09
/middle 4	0.63	.04	0.67	.04	0.69	.04	0.63	.05	0.67	.04	0.69	.04
/middle 3	0.61	.05	0.64	.04	0.65	.03	0.60	.04	0.64	.04	0.65	.04
/middle 2	0.78	.06	0.78	.05	0.80	.05	0.77	.07	0.78	.05	0.79	.05
/proximal 5	0.53	.04	0.55	.03	0.57	.03	0.53	.04	0.55	.03	0.57	.03
/proximal 4	0.41	.03	0.43	.03	0.45	.02	0.41	.03	0.43	.02	0.44	.02
/proximal 3	0.39	.03	0.40	.02	0.42	.03	0.39	.03	0.40	.02	0.41	.03
/proximal 2	0.44	.03	0.44	.03	0.46	.03	0.44	.04	0.44	.03	0.46	.04
/proximal 1	0.60	.05	0.62	.04	0.62	.04	0.60	.05	0.61	.04	0.61	.04
/metacarpal 5	0.37	.03	0.37	.03	0.35	.02	0.36	.03	0.36	.02	0.35	.02
/metacarpal 4	0.34	.03	0.33	.02	0.33	.02	0.33	.03	0.33	.02	0.33	.02
/metacarpal 3	0.30	.03	0.30	.02	0.29	.02	0.30	.03	0.29	.02	0.29	.02
/metacarpal 2	0.28	.02	0.28	.02	0.27	.02	0.28	.03	0.28	.02	0.27	.02
/metacarpal 1	0.46	.04	0.44	.03	0.42	.02	0.45	.04	0.43	.03	0.41	.02
Distal 5 /distal 4	0.83	.04	0.85	.04	0.88	.04	0.81	.06	0.86	.04	0.88	.04
/distal 3	0.87	.06	0.88	.05	0.90	.04	0.83	.06	0.89	.04	0.90	.04
/distal 2	0.98	.08	0.99	.06	0.97	.04	0.96	.11	0.98	.04	0.97	.05
/distal 1	0.66	.05	0.68	.04	0.70	.03	0.65	.06	0.69	.03	0.71	.04
/middle 5	0.81	.09	0.86	.08	0.88	.07	0.79	.09	0.88	.08	0.89	.08
/middle 4	0.52	.04	0.57	.03	0.60	.03	0.50	.05	0.58	.03	0.61	.04
/middle 3	0.51	.05	0.55	.03	0.57	.03	0.49	.05	0.55	.03	0.57	.04
/middle 2	0.65	.06	0.67	.04	0.70	.04	0.62	.06	0.67	.04	0.70	.05
/proximal 5	0.44	.04	0.47	.03	0.50	.03	0.43	.04	0.48	.03	0.50	.03
/proximal 4	0.34	.03	0.37	.02	0.39	.02	0.33	.03	0.37	.02	0.39	.03
/proximal 3	0.33	.03	0.34	.02	0.36	.03	0.31	.03	0.35	.02	0.36	.03
/proximal 2	0.37	.03	0.38	.02	0.40	.02	0.35	.03	0.38	.02	0.40	.04
/proximal 1	0.50	.04	0.53	.03	0.55	.03	0.48	.04	0.53	.04	0.54	.04
/metacarpal 5	0.31	.03	0.31	.02	0.31	.02	0.30	.03	0.31	.02	0.31	.02
/metacarpal 4	0.28	.02	0.29	.02	0.29	.02	0.27	.03	0.28	.02	0.29	.02
/metacarpal 3	0.25	.02	0.25	.02	0.25	.02	0.24	.02	0.25	.02	0.25	.02
/metacarpal 2	0.23	.02	0.24	.02	0.24	.02	0.23	.02	0.24	.02	0.24	.02
/metacarpal 1	0.38	.03	0.37	.02	0.37	.02	0.36	.04	0.37	.03	0.36	.03

(data from Poznanski, 1974)

TABLE 1086

PERCENTAGE PREVALENCE OF THE METACARPAL SIGN IN
NEW YORK PATIENTS LESS THAN 15 YEARS OLD

Metacarpal Sign	Boys	Girls
Negative	78.9	74.0
Borderline	15.6	16.3
Positive	5.5	9.7
less than 1 mm	1.6	2.0
between 1 and 2 mm	3.9	7.7
more than 2 mm	0.0	0.0

(data from Slater, 1970, Pediatrics 46:468-471.
Copyright American Academy of Pediatrics 1970)

TABLE 1087

METACARPAL INDEX OF ENGLISH CHILDREN

Age (months)	Sex	Mean	S.D.
6	M	5.23	0.46
	F	5.60	0.37
12	M	5.30	0.41
	F	5.75	0.41
18	M	5.28	0.40
	F	5.82	0.45
24	M	5.40	0.43
	F	5.84	0.43

This index is the sum of the lengths of metacarpals II-V
divided by the sum of the breadths of these bones at the
midpoints of the shafts.

(data from Joseph and Meadow, 1969, Archives of Diseases
in Childhood 44:515-516)

TABLE 1088

DISTRIBUTION STATISTICS FOR CORTICAL THICKNESS OF THE SECOND METACARPAL (mm) IN OHIO CHILDREN

Age (years)	Percentiles Boys 5	50	95	Girls 5	50	95
1	1.0	1.5	2.0	1.0	1.5	2.0
2	1.2	1.9	2.5	1.2	1.8	2.4
3	1.5	2.2	2.8	1.5	2.1	2.7
4	1.9	2.5	3.1	1.8	2.3	2.9
5	2.1	2.7	3.4	1.9	2.6	3.3
6	2.2	3.0	3.7	2.1	2.8	3.4
7	2.4	3.2	4.0	2.3	3.0	3.7
8	2.6	3.4	4.2	2.5	3.2	3.9
9	2.8	3.7	4.5	2.6	3.4	4.1
10	3.1	3.9	4.7	2.7	3.5	4.3
11	3.2	4.1	5.0	3.0	3.9	4.7
12	3.3	4.3	5.3	3.3	4.2	5.1
13	3.5	4.6	5.7	3.6	4.5	5.4
14	3.7	4.9	6.1	3.9	4.9	5.8
15	4.1	5.1	6.1	4.0	5.0	6.0
16	4.4	5.3	6.1	4.1	5.1	6.1
17	4.5	5.5	6.5	4.0	5.1	6.2
18	4.6	5.7	6.8	4.0	5.2	6.3

(data from MEDICAL RADIOGRAPHY AND PHOTOGRAPHY published by Health Sciences Markets Division, Eastman Kodak Company. Courtesy of S. M. Garn, C. G. Rohmann, and F. N. Silverman, 1967)

TABLE 1089

BONE DENSITY COEFFICIENTS OF THE FIFTH MIDDLE PHALANX (mid-shaft)
IN CHILDREN OF 8 WESTERN STATES OF THE U.S.

Age (years)	Males Percentiles					Females Percentiles				
	10	25	50	75	90	10	25	50	75	90
12	1.47	1.58	1.73	2.05	2.27	--	--	--	--	--
13	1.30	1.37	1.52	1.72	1.82	1.60	1.74	1.93	2.14	2.42
14	1.18	1.37	1.62	1.84	2.09	1.46	1.65	1.88	2.10	2.31
15	1.28	1.48	1.78	1.99	2.25	1.42	1.64	1.94	2.19	2.44
16	1.36	1.58	1.78	2.03	2.32	1.61	1.80	2.12	2.42	2.66
17	1.31	1.54	1.76	2.05	2.49	1.68	1.82	2.07	2.42	2.58

(data from Odland et al., 1958)

TABLE 1090

CARPAL ANGLE PERCENTILES (degrees)
FOR CHILDREN IN THE TEN-STATE NUTRITION SURVEY

Age Group	Percentiles			Percentiles		
	5	50	95	5	50	95
	White Males			White Females		
4 - 6 years	116.0	122.0	132.5	120.0	126.5	143.0
6 - 8 years	111.0	124.0	153.5	115.0	130.5	147.5
8 - 10 years	122.0	133.5	147.0	115.5	129.5	139.5
10 - 12 years	155.5	133.0	143.0	123.0	134.0	152.5
12 - 14 years	117.0	132.0	142.5	116.0	130.0	143.0
14 - 24 years	119.0	134.0	149.5	115.0	129.0	139.5
	Black Males			Black Females		
4 - 6 years	124.0	131.0	143.0	116.5	130.5	140.5
6 - 8 years	119.0	128.5	147.5	121.0	133.0	146.5
8 - 10 years	128.0	139.0	142.0	125.0	139.5	155.0
10 - 12 years	121.0	138.0	152.5	125.5	138.5	151.0
12 - 14 years	125.5	141.0	143.0	123.0	141.0	153.5
14 - 24 years	125.0	146.0	139.5	123.0	139.0	150.0

(data from Harper, Poznanski and Garn, 1974, Investigative Radiology 9:217-221,
by permission of Lippincott/Harper, Philadelphia, PA)

TABLE   1091

LENGTHS OF FEMUR AND TIBIA (cm) INCLUDING EPIPHYSES
IN BOSTON BOYS

| FEMUR | | | | TIBIA | | |
Age (years)	Mean	S.D.		Age (years)	Mean	S.D.
1	14.48	0.628		1	11.60	0.620
2	18.15	0.874		2	14.54	0.809
3	21.09	1.031		3	16.79	0.935
4	23.65	1.197		4	18.67	1.091
5	25.92	1.342		5	20.46	1.247
6	28.09	1.506		6	22.12	1.418
7	30.25	1.682		7	23.76	1.632
8	32.28	1.807		8	25.38	1.778
9	34.36	1.933		9	26.99	1.961
10	36.29	2.057		10	28.53	2.113
11	38.16	2.237		11	30.10	2.301
12	40.12	2.447		12	31.75	2.536
13	42.17	2.765		13	33.49	2.833
14	44.18	2.809		14	35.18	2.865
15	45.69	2.512		15	36.38	2.616
16	46.66	2.244		16	37.04	2.412
17	47.07	2.051		17	37.22	2.316
18	47.23	1.958		18	37.29	2.254

(data from Anderson et al., 1964)

TABLE 1092

LENGTHS OF FEMUR AND TIBIA (cm) INCLUDING EPIPHYSES
IN BOSTON GIRLS

FEMUR				TIBIA		
Age (years)	Mean	S.D.		Age (years)	Mean	S.D.
1	14.81	0.673		1	11.57	0.646
2	18.23	0.888		2	14.51	0.739
3	21.29	1.100		3	16.81	0.893
4	23.92	1.339		4	18.86	1.144
5	26.32	1.437		5	20.77	1.300
6	28.52	1.616		6	22.53	1.458
7	30.60	1.827		7	24.22	1.640
8	32.72	1.936		8	25.89	1.786
9	34.71	2.117		9	27.56	1.993
10	36.72	2.300		10	29.28	2.193
11	38.81	2.468		11	31.00	2.384
12	40.74	2.507		12	32.61	2.424
13	42.31	2.428		13	33.83	2.374
14	43.14	2.269		14	34.43	2.228
15	43.47	2.197		15	34.59	2.173
16	43.58	2.193		16	34.63	2.151
17	43.60	2.192		17	34.65	2.158
18	43.63	2.195		18	34.65	2.161

(data from Anderson et al., 1964)

TABLE 1093

LENGTH OF THE FEMUR AND TIBIA IN BOYS (cm)
IN BOSTON

Skeletal Age (years)	Percentiles			Mean	S.D.
	10	50	90		
FEMUR					
5	24.2	25.6	26.8	25.52	1.042
6	26.0	28.0	29.2	27.74	1.485
7	27.8	29.8	31.6	29.68	1.536
8	29.6	32.1	33.6	31.80	1.496
9	31.8	34.1	36.1	33.96	1.582
10	33.7	35.7	38.2	35.80	1.832
11	35.1	37.4	40.0	37.50	1.878
12	36.6	39.3	42.2	39.31	2.310
13	38.5	41.2	43.9	41.41	2.188
14	40.2	43.5	46.8	43.48	2.402
15	42.6	45.4	47.7	45.33	2.041
16	42.8	46.6	49.4	46.27	2.503
17	42.6	45.8	49.6	46.40	2.656
18	42.9	45.8	49.4	46.16	2.361
TIBIA					
5	18.8	20.3	21.5	20.25	1.046
6	20.3	21.9	23.7	21.92	1.204
7	21.8	23.5	25.4	23.43	1.245
8	23.4	25.1	27.0	25.00	1.321
9	24.8	26.5	28.6	26.54	1.470
10	26.0	28.0	30.1	27.94	1.529
11	26.9	29.2	31.2	29.20	1.617
12	28.3	30.6	33.2	30.72	1.877
13	29.8	32.8	35.0	32.46	1.961
14	30.9	34.6	36.5	34.16	2.148
15	33.2	36.0	38.1	35.87	1.984
16	33.4	37.0	39.2	36.53	2.300
17	32.6	37.0	39.1	36.49	2.344
18	32.5	37.0	39.0	36.42	2.627

(data from Anderson and Green, 1948, American Journal of Diseases of Children 75:279-290. Copyright, 1948, American Medical Association)

TABLE 1094

LENGTH OF THE FEMUR AND TIBIA IN GIRLS (cm)
IN BOSTON

Skeletal Age (years)	Percentiles				
	10	50	90	Mean	S.D.
FEMUR					
5	24.7	26.1	27.6	25.98	1.037
6	26.7	28.2	30.2	28.18	1.216
7	28.6	30.2	32.6	30.31	1.412
8	30.6	31.9	34.6	32.19	1.453
9	32.4	34.0	36.1	34.12	1.501
10	33.4	35.6	37.7	35.65	1.582
11	35.0	37.4	40.1	37.44	1.881
12	36.3	39.2	41.9	39.20	2.084
13	38.9	41.2	43.9	41.18	2.225
14	39.5	41.8	44.4	42.06	2.093
15	40.3	42.3	44.8	42.44	1.981
16	40.1	42.3	45.4	42.54	1.889
TIBIA					
5	19.2	20.4	21.8	20.52	0.768
6	21.0	22.4	23.8	22.36	1.044
7	22.6	24.3	25.7	24.07	1.170
8	24.0	25.2	27.3	25.49	1.196
9	25.1	26.9	28.9	26.94	1.355
10	26.0	28.2	30.5	28.24	1.520
11	27.5	29.4	32.2	29.75	1.684
12	28.8	31.0	33.7	31.24	1.947
13	30.3	32.8	35.6	32.88	2.076
14	31.0	33.0	36.2	33.52	2.040
15	31.4	33.0	36.6	33.83	1.915
16	31.2	33.6	36.9	33.95	1.924

(data from Anderson and Green, 1948, American Journal of Diseases of Children 75:279-290.
Copyright, 1948, American Medical Association)

TABLE 1095

ANNUAL INCREMENTS IN FEMORAL AND TIBIAL LENGTHS (cm)
IN BOSTON CHILDREN

Age Interval (Years)	Boys				Girls			
	Femur		Tibia		Femur		Tibia	
	Mean	S.D.	Mean	S.D.	Mean	S.D.	Mean	S.D.
8 – 9	2.0	0.27	1.6	0.22	2.0	0.28	1.7	0.29
9 – 10	1.8	0.32	1.5	0.27	2.0	0.32	1.8	0.36
10 – 11	1.8	0.34	1.5	0.28	2.1	0.35	1.8	0.38
11 – 12	1.9	0.42	1.7	0.42	1.9	0.52	1.6	0.56
12 – 13	2.1	0.50	1.8	0.49	1.4	0.67	1.0	0.63
13 – 14	2.0	0.52	1.7	0.58	0.6	0.50	0.4	0.41
14 – 15	1.5	0.79	1.1	0.68	0.2	0.30	0.1	0.24
15 – 16	0.8	0.73	0.5	0.77	0.1	0.20	0.0	0.14
16 – 17	0.3	0.38	0.2	0.25	0.0	0.06	0.0	0.04
17 - 18	0.1	0.17	0.0	0.08	0.0	0.00	0.0	0.00

(data from Anderson et al., 1963)

TABLE 1096

LENGTH OF TIBIAL SHAFT (cm)
IN BOSTON CHILDREN

Age (years)	Boys Percentiles			Girls Percentiles		
	10	50	90	10	50	90
6	19.68	21.25	22.91	20.12	21.87	23.70
7	20.95	23.05	24.67	21.52	22.98	25.45
8	22.44	24.61	26.36	22.80	24.50	27.15
9	23.58	26.17	28.15	24.21	25.82	28.58
10	24.70	27.41	29.17	25.10	27.43	29.80

(data from Stuart and Dwinell, 1942)

TABLE 1097

TIBIAL LENGTH (mm) IN CLEVELAND CHILDREN

Age (years)	BOYS		GIRLS	
	Mean	S.D.	Mean	S.D.
2 years	168.52	8.02	166.17	10.77
2½ years	179.98	10.42	181.25	11.54
3 years	192.22	9.91	193.07	11.02
3½ years	204.38	10.62	203.55	10.78
4 years	213.10	10.94	213.11	12.28
4½ years	221.15	12.51	224.72	13.01
5 years	231.21	12.32	232.93	14.10
6 years	248.85	13.29	251.15	15.12
7 years	266.03	15.25	268.91	18.60
8 years	283.26	16.90	285.89	17.24
9 years	299.76	17.72	301.97	18.26
10 years	316.17	18.50	319.18	18.56
11 years	332.32	20.60	336.52	19.41
12 years	347.03	22.43	351.60	19.26
13 years	361.48	24.20	361.88	19.39
14 years	380.34	23.72	367.50	19.02
15 years	393.19	20.88	367.78	18.07
16 years	398.44	19.26	367.12	19.03
17 years	401.31	22.83	367.28	21.12

(data from Simmons, 1944)

TABLE   1098

TIBIA LENGTH (cm)
IN DENVER CHILDREN

| Age | Boys | | Girls | |
Yr. Mo.	Mean	S.D.	Mean	S.D.
Diaphyseal Length				
0 - 2	7.08	.54	7.03	.46
0 - 4	8.19	.53	8.08	.46
0 - 6	9.10	.52	8.89	.53
1 - 0	11.03	.52	10.85	.48
1 - 6	12.61	.60	12.40	.56
2 - 0	14.01	.65	13.82	.65
2 - 6	15.25	.68	15.01	.70
3 - 0	16.35	.77	16.11	.82
3 - 6	17.28	.98	17.12	.87
4 - 0	18.28	.90	18.08	.95
4 - 6	19.18	.92	19.09	1.05
5 - 0	20.14	.99	19.99	1.14
5 - 6	21.03	1.07	20.79	1.25
6 - 0	21.89	1.00	21.74	1.26
6 - 6	22.78	1.16	22.63	1.36
7 - 0	23.62	1.18	23.41	1.41
7 - 6	24.42	1.24	24.32	1.50
8 - 0	25.33	1.29	25.17	1.56
8 - 6	26.06	1.23	25.91	1.56
9 - 0	26.87	1.34	26.75	1.71
9 - 6	27.69	1.44	27.66	1.87
10 - 0	28.49	1.42	28.43	1.93
10 - 6	29.20	1.51	29.24	2.14
11 - 0	29.98	1.50	30.08	2.12
11 - 6	30.68	1.65	31.05	2.14
12 - 0	31.59	1.70	31.82	2.17
Length including Epiphyses				
10 - 0	32.00	1.57	32.11	2.17
10 - 6	32.80	1.70	33.00	2.37
11 - 0	33.86	1.71	34.01	2.31
11 - 6	34.74	1.85	35.04	2.32
12 - 0	35.73	1.91	36.09	2.38
12 - 6	36.75	1.86	36.73	2.30
13 - 0	37.67	2.06	37.45	2.22
13 - 6	38.82	2.20	37.90	2.18

TABLE 1098   (continued)

TIBIA LENGTH (cm)
IN DENVER CHILDREN

| Age | Boys | | Girls | |
Yr. Mo.	Mean	S.D.	Mean	S.D.
14 - 0	39.74	2.19	38.43	2.14
14 - 6	40.60	2.31	38.69	2.05
15 - 0	41.22	2.14	38.57	2.08
15 - 6	42.05	2.23	39.04	2.85
16 - 0	42.26	2.18	38.68	2.26
16 - 6	42.51	2.42	--	--
17 - 0	42.65	2.32	38.07	2.36
18 - 0	42.95	2.56	--	--

(data from McCammon, Human Growth and Development,
1970. Courtesy of Charles C Thomas, Publisher, Springfield,
Illinois)

TABLE 1099

TIBIA LENGTH (mm)
IN OHIO CHILDREN

| Age | Boys | | Girls | |
(months)	Mean	S.D.	Mean	S.D.
1	72.14	4.90	71.34	4.53
3	84.83	4.20	84.95	18.14
6	99.26	5.34	97.06	5.01
9	110.06	5.02	109.49	17.32
12	119.57	5.81	117.08	5.82
18	135.53	6.87	134.24	6.98
24	150.14	7.43	149.08	7.49
30	162.74	7.53	163.04	19.04
36	174.15	9.31	173.07	9.93
42	183.98	9.11	183.73	10.47
48	194.00	10.67	193.68	11.25
54	203.59	10.29	203.62	11.96
60	212.37	11.66	213.18	12.54
66	221.89	11.76	223.62	13.85
72	232.95	13.09	231.15	15.16
78	240.88	14.19	241.19	16.61
84	250.42	14.26	250.29	16.91
90	258.93	15.37	261.01	19.31
96	268.36	15.86	270.47	20.15
102	276.88	16.16	278.56	20.91
108	287.97	17.43	290.74	21.51

(data from Gindhart, 1971)

TABLE 1100

INCREMENTS (mm) IN TIBIA LENGTH
IN OHIO CHILDREN

Age Interval (months)	Boys		Girls	
	Mean	S.D.	Mean	S.D.
1-3	12.14	2.55	11.86	2.28
3-6	14.94	2.46	13.98	2.48
6-9	10.84	2.08	11.09	1.76
9-12	9.24	2.02	9.23	1.61
12-18	16.18	2.67	16.90	1.86
18-24	14.18	2.02	14.56	2.19
24-30	12.50	1.91	12.78	2.32
30-36	11.43	1.61	11.46	2.21
36-42	10.33	1.50	10.86	1.43
42-48	9.92	1.46	10.10	1.80
48-54	9.81	1.74	9.56	1.71
54-60	9.57	1.90	9.89	2.90
60-66	9.68	1.62	9.19	2.85
66-72	9.40	1.88	9.10	2.83
72-78	9.10	1.82	10.49	3.17
78-84	9.26	2.09	8.25	3.47
84-90	9.03	1.85	10.08	2.48
90-96	9.11	1.91	8.89	2.85
96-102	9.01	2.26	10.41	2.99
102-108	9.74	2.73	9.22	3.00

(data from Gindhart, 1971)

TABLE   1101

TOTAL ELONGATION (cm) OF FEMUR AND TIBIA AFTER PARTICULAR CHRONOLOGICAL AGES
IN BOSTON CHILDREN

| | Boys | | | | Girls | | | |
| | Femur | | Tibia | | Femur | | Tibia | |
Age (years)	Mean	S.D.	Mean	S.D.	Mean	S.D.	Mean	S.D.
8	14.2	1.44	11.4	1.20	10.2	1.54	8.4	1.32
9	12.2	1.50	9.9	1.17	8.3	1.55	6.6	1.29
10	10.5	1.46	8.5	1.17	6.3	1.61	4.8	1.38
11	8.7	1.48	7.0	1.18	4.2	1.71	3.0	1.45
12	6.8	1.66	5.3	1.40	2.3	1.42	1.4	1.15
13	4.7	1.90	3.5	1.63	0.9	0.91	0.5	0.71
14	2.7	1.78	1.8	1.42	0.2	0.53	0.1	0.40
15	1.2	1.21	0.7	0.88	0.1	0.24	0.0	0.17
16	0.4	0.54	0.2	0.33	--	--	--	--
17	0.1	0.19	0.0	0.10	--	--	--	--

(data from Anderson et al., 1963)

TABLE 1102

ELONGATION (cm) AT THE DISTAL END OF THE FEMUR AND THE
PROXIMAL END OF THE TIBIA AFTER PARTICULAR GREULICH-PYLE (1959) SKELETAL AGES
IN BOSTON CHILDREN

Skeletal	Femur				Tibia			
Age (Years)	Boys		Girls		Boys		Girls	
	Mean	S.D.	Mean	S.D.	Mean	S.D.	Mean	S.D.
8.25	--	--	6.54	1.14	--	--	4.25	0.83
9.25	--	--	5.30	0.92	--	--	3.39	0.75
10.25	7.21	1.28	4.15	0.78	4.65	0.83	2.58	0.62
11.25	6.01	1.14	2.82	0.53	3.83	0.75	1.65	0.53
12.25	4.65	0.91	1.66	0.40	2.92	0.62	0.86	0.35
13.25	3.09	0.78	0.75	0.30	1.80	0.53	0.32	0.12
14.25	1.48	0.50	0.27	0.18	0.74	0.35	0.09	0.06
15.25	0.45	0.23	0.05	0.08	0.16	0.12	0.02	0.02
16.25	0.15	0.12	--	--	0.04	0.06	--	--
17.25	0.04	0.06	--	--	0.01	0.02	--	--

(data from Anderson et al., 1963)

TABLE   1103

GROWTH AT DISTAL END OF NORMAL FEMUR
AND PROXIMAL END OF NORMAL TIBIA OBSERVED
IN LONGITUDINAL SERIES FOLLOWING GIVEN SKELETAL AGES
IN BOSTON CHILDREN (growth in cm; skeletal age in years
and months by method of Greulich and Pyle, 1959)

Skeletal age	BOYS							
	10^3	11^3	12^3	13^3	14^3	15^3	16^3	17^3
Distal End of the Femur (Total Growth Femur X 71%)								
Mean	7.21	6.01	4.65	3.09	1.48	0.45	0.15	0.04
S.D.	1.28	1.14	0.91	0.78	0.50	0.23	0.12	0.06
Percentiles								
90th	8.9	7.8	5.7	4.2	2.2	0.8	0.3	0.1
75th	8.3	6.7	5.2	3.5	1.8	0.6	0.2	0.1
50th	7.2	6.1	4.8	2.9	1.4	0.4	0.1	0.0
25th	6.3	5.2	4.1	2.6	1.2	0.3	0.1	0.0
10th	5.3	4.4	3.4	2.3	1.0	0.2	0.0	0.0
Proximal End of the Tibia (Total Growth Tibia X 57%)								
Mean	4.65	3.83	2.92	1.80	0.74	0.16	0.04	0.01
S.D.	0.83	0.75	0.62	0.53	0.35	0.12	0.06	0.02
Percentiles								
90th	5.8	4.8	3.6	2.5	1.1	0.3	0.1	0.0
75th	5.3	4.3	3.3	2.0	0.8	0.2	0.0	0.0
50th	4.6	3.8	3.0	1.8	0.7	0.2	0.0	0.0
25th	4.0	3.2	2.6	1.4	0.5	0.0	0.0	0.0
10th	3.4	2.7	2.0	1.1	0.3	0.0	0.0	0.0

(data from Anderson et al., 1963)

TABLE 1104

GROWTH AT DISTAL END OF NORMAL FEMUR
AND PROXIMAL END OF NORMAL TIBIA OBSERVED
IN LONGITUDINAL SERIES FOLLOWING GIVEN SKELETAL AGES
IN BOSTON CHILDREN (growth in cm; skeletal age in years
and months by method of Greulich and Pyle, 1959)

Skeletal age	GIRLS							
	8^3	9^3	10^3	11^3	12^3	13^3	14^3	15^3
Distal End of the Femur (Total Growth Femur X 71%)								
Mean	6.54	5.30	4.15	2.82	1.66	0.75	0.27	0.05
S.D.	1.14	0.92	0.78	0.53	0.40	0.30	0.18	0.08
Percentiles								
90th	8.4	6.7	5.0	3.4	2.1	1.1	0.6	0.1
75th	7.2	5.8	4.6	3.2	1.9	1.0	0.4	0.1
50th	6.5	5.2	4.1	2.8	1.7	0.7	0.3	0.1
25th	5.8	4.8	3.7	2.4	1.4	0.6	0.1	0.0
10th	5.0	4.3	3.3	2.2	1.1	0.4	0.0	0.0
Proximal End of the Tibia (Total Growth Tibia X 57%)								
Mean	4.25	3.39	2.58	1.65	0.86	0.32	0.09	0.02
S.D.	0.74	0.58	0.50	0.32	0.26	0.17	0.06	0.03
Percentiles								
90th	5.5	4.2	3.2	1.9	1.2	0.6	0.2	0.1
75th	4.6	3.7	2.7	1.8	1.0	0.4	0.1	0.1
50th	4.1	3.3	2.6	1.6	0.8	0.3	0.0	0.0
25th	3.8	3.0	2.3	1.5	0.7	0.2	0.0	0.0
10th	3.3	2.8	2.0	1.2	0.6	0.1	0.0	0.0

(data from Anderson et al., 1963)

TABLE 1105

ANNUAL INCREMENTS (mm) IN TIBIA AND RADIUS LENGTHS
IN OHIO CHILDREN

Age Interval (years)	Boys		Girls	
	Mean	S.D.	Mean	S.D.
Tibia				
1-2	30.40	3.38	31.83	2.89
2-3	23.99	2.78	24.09	3.75
3-4	20.44	2.24	20.92	2.76
4-5	19.38	2.44	19.72	3.18
5-6	18.80	2.44	18.23	4.48
6-7	18.43	2.87	18.87	3.12
7-8	18.39	3.11	18.91	3.41
8-9	18.42	3.49	19.65	3.39
Radius				
1-2	18.12	2.43	17.79	1.87
2-3	13.97	1.99	13.48	1.83
3-4	11.95	1.90	12.16	1.60
4-5	11.71	1.97	11.28	1.55
5-6	11.12	2.30	10.81	1.53
6-7	10.46	1.45	10.12	1.39
7-8	10.27	1.45	9.83	1.50
8-9	9.70	1.47	9.91	1.66

(data from Gindhart, 1971)

TABLE 1106

TIBIAL SHAFT WIDTH (cm)
AT GREATEST WIDTH OF CALF
IN BOSTON CHILDREN

Age (years)	Boys Percentiles			Girls Percentiles		
	10	50	90	10	50	90
6	1.62	1.72	1.99	1.60	1.75	1.94
7	1.77	1.84	2.04	1.66	1.84	1.99
8	1.76	1.94	2.13	1.71	1.90	2.07
9	1.85	2.01	2.29	1.77	1.97	2.15
10	1.91	2.15	2.35	1.83	2.04	2.23

(data from Stuart and Dwinell, 1942)

TABLE 1107

FIBULA LENGTH (cm) IN
WHITE CHILDREN IN DENVER (CO)

Age Yr. Mo.	Boys		Girls	
	Mean	S.D.	Mean	S.D.
Diaphyseal Length				
0 - 2	6.81	.53	6.68	.44
0 - 4	7.86	.49	7.71	.41
0 - 6	8.72	.48	8.49	.52
1 - 0	10.71	.55	10.50	.51
1 - 6	12.39	.62	12.13	.59
2 - 0	13.81	.67	13.60	.68
2 - 6	15.07	.71	14.79	.71
3 - 0	16.21	.77	15.94	.79
3 - 6	17.16	.96	16.96	.83
4 - 0	18.18	.87	17.95	.91
4 - 6	19.08	.88	18.94	1.02
5 - 0	20.04	.96	19.86	1.11
5 - 6	20.90	1.02	20.65	1.17
6 - 0	21.75	.96	21.60	1.22
6 - 6	22.60	1.05	22.43	1.34

TABLE 1107 (continued)

FIBULA LENGTH (cm) IN
WHITE CHILDREN IN DENVER (CO)

Age	Boys		Girls	
Yr. Mo.	Mean	S.D.	Mean	S.D.
7 - 0	23.42	1.13	23.21	1.34
7 - 6	24.21	1.18	24.08	1.45
8 - 0	25.10	1.24	24.88	1.48
8 - 6	25.77	1.18	25.61	1.52
9 - 0	26.56	1.30	26.37	1.63
9 - 6	27.38	1.38	27.22	1.76
10 - 0	28.13	1.39	27.94	1.83
10 - 6	28.78	1.46	28.72	2.04
11 - 0	29.49	1.46	29.44	1.98
11 - 6	30.17	1.60	30.38	2.07
12 - 0	31.01	1.64	31.11	2.08

Length including Epiphyses

Age	Boys		Girls	
10 - 0	31.04	1.52	30.79	1.95
10 - 6	21.80	1.62	31.67	2.18
11 - 0	32.62	1.59	32.47	2.15
11 - 6	33.40	1.76	33.46	2.21
12 - 0	34.28	1.80	34.46	2.27
12 - 6	35.19	1.68	35.10	2.22
13 - 0	36.02	1.98	35.85	2.19
13 - 6	37.11	2.14	36.34	2.14
14 - 0	38.03	2.13	36.79	2.06
14 - 6	38.85	2.25	36.89	2.15
15 - 0	39.53	2.15	37.02	2.00
15 - 6	40.44	2.21	37.57	2.58
16 - 0	40.63	2.17	37.24	2.15
16 - 6	40.86	2.28	-	-
17 - 0	41.04	2.26	36.68	2.42
18 - 0	41.28	2.42	-	-

(data from McCammon, Human Growth and Development, 1970.
Courtesy of Charles C Thomas, Publisher, Springfield,
Illinois)

TABLE 1108

DENSITY COEFFICIENTS OF THE OS CALCIS AND FIFTH MIDDLE PHALANX (mid shaft)
IN U.S. CHILDREN

| Age Group (years) | Os Calcis | | | | Phalanx Center | | | |
| | Males | | Females | | Males | | Females | |
	Mean	S.D.	Mean	S.D.	Mean	S.D.	Mean	S.D.
7 – 9	0.60	0.09	0.60	0.09	1.02	0.23	1.10	0.21
10 – 12	0.67	0.10	0.63	0.08	1.03	0.28	1.25	0.30
13 – 15	0.74	0.10	0.68	0.10	1.32	0.30	1.64	0.22
16 – 20	0.92	0.05	0.68	0.09	1.43	0.24	1.74	0.31

(data from Schraer, 1958, Journal of Pediatrics 52:416-423)

TABLE 1109

BONE DENSITY COEFFICIENTS OF THE OS CALCIS
IN CHILDREN OF 8 WESTERN STATES OF THE U.S.

| Age (years) | Males Percentiles | | | | | Females Percentiles | | | | |
	10	25	50	75	90	10	25	50	75	90
12	0.65	0.71	0.75	0.85	0.91	--	--	--	--	--
13	0.72	0.75	0.78	0.85	1.07	0.65	0.70	0.78	0.88	0.90
14	0.67	0.71	0.77	0.81	0.89	0.58	0.66	0.70	0.77	0.82
15	0.66	0.73	0.78	0.86	0.92	0.60	0.65	0.70	0.78	0.84
16	0.68	0.73	0.80	0.85	0.93	0.60	0.64	0.70	0.77	0.84
17	0.67	0.72	0.78	0.88	0.92	0.60	0.66	0.69	0.80	0.96

(data from Odland et al., 1958)

MATURATION

TABLE 1110

MEDIAN AGE OF ONSET OF OSSIFICATION IN SELECTED HAND-WRIST BONES
IN U.S. CHILDREN

Race, region, and annual family income	Boys				Girls	
	Ulna	Trapezium	Trapezoid	Pisiform	Pisiform	Adductor
Race	Chronological age (years)					
White..............	7.2	6.4	6.4	11.2	8.7	10.6
Negro.............	7.0	4.9	6.1	11.3	8.9	10.7
Region						
Northeast..........	7.0	6.3	6.2	11.0	8.5	10.7
Midwest...........	7.0	6.5	6.5	11.5	8.8	10.7
South.............	7.3	6.2	6.4	11.0	8.8	10.6
West..............	7.3	6.0	6.0	11.4	8.9	10.7
Annual family income						
Less than $5,000...	7.2	6.2	6.5	11.2	8.9	10.6
$5,000-$9,999......	7.1	6.3	6.4	11.3	8.8	10.9
$10,000 or more....	7.0	6.3	6.0	11.3	8.7	10.5

(data from Roche, Roberts, and Hamill, 1975)

TABLE 1111

OSSIFICATION OF ULNAR SESAMOID AND
ADOLESCENT SPURT IN STATURE
IN AUSTRALIAN CHILDREN

Ulnar sesamoid related to Timing	Number of children			Mean Interval (years)
	Before	At	After	
Beginning of Adolescent Spurt				
Boys	2	4	30	0.99
Girls	3	3	30	1.08
Peak Stature Velocity				
Boys	31	4	3	-1.01
Girls	32	9	5	-0.67

(data from Bowden, 1971)

TABLE   1112

PERCENTAGE OF OSSIFICATION CENTERS PRESENT IN
NEWLY-BORN INFANTS IN NEW YORK CITY

			Boys			Girls	
	Total	White	Negro	Total	White	Negro	Total
Anterior arch of atlas	(27)	(24)	(29)	(28)	(25)	(22)	(25)
Hyoid							
Body	(60)	(61)	(59)	(59)	(60)	(52)	(57)
Horns	(59)	(71)	(48)	(63)	(53)	(46)	(51)
Body and horns	(53)	(61)	(42)	(54)	(51)	(44)	(48)
Anterior arch of atlas and both horns and body of hyoid	(18)	(16)	(19)	(18)	(18)	(18)	(17)
Interparietal	(12)	(10)	(11)	(10)	(15)	(8)	(12)
Occipital ossicle							
Single	(0.7)	(0.7)	0	(0.4)	(0.7)	(1)	(0.9)
Multiple	(0.1)	(0.3)	0	(0.2)	0	0	0
Coracoid							
Right	(20)	(17)	(26)	(20)	(13)	(28)	(18)
Left	(20)	(17)	(25)	(20)	(13)	(28)	(19)
Bilateral	(20)	(17)	(25)	(20)	(13)	(28)	(19)
Humeral head							
Right	(18)	(21)	(13)	(18)	(16)	(19)	(17)
Left	(18)	(21)	(13)	(18)	(16)	(19)	(17)
Bilateral	(18)	(21)	(13)	(18)	(16)	(19)	(17)
Coracoid and humeral head: bilateral	(5)	(6)	(6)	(6)	(6)	(5)	(5)

(data from Dedick & Caffey, 1953, RADIOLOGY 61:13-20, by permission of the Radiological
Society of North America, Inc.)

TABLE 1113

PERCENTAGE OF OSSIFICATION CENTERS IN NEWLY-BORN
MALE WHITE INFANTS IN NEW YORK CITY
ACCORDING TO MOTHER'S AGE AND INFANT'S WEIGHT

	Total	Mother's Age (years)			Infant's Weight (gm)		
		25	26-35	36-	2,500	3,000	4,000
Total male white	--	(38)	(49)	(13)	(18)	(74)	(8)
Anterior arch of atlas	(24)	(26)	(23)	(22)	(24)	(25)	(20)
Hyoid							
Body	(61)	(60)	(61)	(71)	(62)	(61)	(72)
Horns	(71)	(60)	(82)	(61)	(49)	(75)	(76)
Body and horns	(61)	(58)	(65)	(61)	(48)	(66)	(60)
Anterior arch of atlas and							
horns and body of hyoid	(16)	(18)	(14)	(13)	(16)	(16)	(12)
Interparietal	(10)	(12)	(9)	(7)	(3)	(11)	(8)
Occipital ossicle							
Single	(0.7)	(0.2)	(0)	(0)	(0)	(1)	(0)
Multiple	(0.3)	(0)	(0)	(0.2)	(0)	(0.5)	(0)
Coracoid							
Right	(17)	(22)	(15)	(10)	(7)	(20)	(16)
Left	(17)	(22)	(15)	(10)	(7)	(20)	(16)
Bilateral	(17)	(22)	(15)	(10)	(7)	(20)	(16)
Humeral head							
Right	(21)	(26)	(20)	(15)	(18)	(21)	(28)
Left	(21)	(26)	(20)	(15)	(18)	(21)	(28)
Bilateral	(21)	(26)	(20)	(15)	(18)	(21)	(28)
Both coracoids and humeral							
heads	(6)	(8)	(4)	(2)	(1.8)	(0.6)	(0.8)

(data from Dedick & Caffey, 1953, RADIOLOGY 61:13-20, by permission of The Radiological
Society of North America, Inc.)

TABLE 1114

PERCENTAGE OF OSSIFICATION CENTERS IN NEWLY-BORN
FEMALE WHITE INFANTS IN NEW YORK CITY
ACCORDING TO MOTHER'S AGE AND INFANT'S WEIGHT

	Total	Mother's Age (years)			Infant's Weight (gm)		
		25	26-35	36-	2,500	3,000	4,000
Total female whites	--	(36)	(50)	(10)	(26)	(62)	(7)
Anterior arch of atlas	(25)	(29)	(21)	(50)	(18)	(27)	(35)
Hyoid							
Body	(60)	(70)	(49)	(81)	(64)	(58)	(70)
Horns	(53)	(61)	(45)	(70)	(53)	(54)	(61)
Body and Horns	(51)	(60)	(42)	(67)	(50)	(51)	(61)
Anterior arch of atlas and							
horns and body of hyoid	(18)	(18)	(16)	(28)	(13)	(20)	(22)
Interparietal	(15)	(11)	(18)	(17)	(11)	(18)	(9)
Occipital ossicle							
Single	(17)	0	(1)	0	0	(0.8)	0
Multiple	0	0	0	0	0	0	0
Coracoid							
Right	(13)	(18)	(11)	(11)	(15)	(13)	(13)
Left	(13)	(18)	(11)	(11)	(15)	(13)	(13)
Bilateral	(13)	(18)	(11)	(11)	(15)	(13)	(13)
Humeral head							
Right	(16)	(12)	(18)	(19)	(11)	(17)	(26)
Left	(16)	(12)	(18)	(19)	(11)	(17)	(26)
Bilateral	(16)	(12)	(18)	(19)	(11)	(17)	(26)
Both coracoids and humeral							
heads	(6)	(4)	(8)	(3)	(4)	(6)	(4)

(data from Dedick & Caffey, 1953, RADIOLOGY 61:13-20, by permission of The Radiological
Society of North America, Inc.)

TABLE 1115

PERCENTAGE OF OSSIFICATION CENTERS IN NEWLY-BORN
MALE NEGRO INFANTS IN NEW YORK CITY
ACCORDING TO MOTHER'S AGE AND INFANT'S WEIGHT

	Total	Mother's Age (years)			Infant's Weight (gm)		
		25	26-35	36-	2,500	3,000	4,000
Total male Negroes	--	(39)	(46)	(15)	(29)	(65)	(6)
Anterior arch of atlas	(29)	(26)	(25)	(18)	(29)	(29)	(45)
Hyoid							
Body	(52)	(45)	(61)	(48)	(67)	(46)	(45)
Horns	(50)	(72)	(34)	(48)	(82)	(21)	(45)
Body and horns	(42)	(38)	(34)	(43)	(58)	(19)	(45)
Anterior arch of atlas and							
horns and body of hyoid	(19)	(11)	(27)	(14)	(18)	(18)	(36)
Interparietal	(11)	(13)	(8)	(18)	(13)	(12)	0
Occipital ossicle							
Single	0	0	0	0	0	0	0
Multiple	0	0	0	0	0	0	0
Coracoid							
Right	(26)	(4)	(50)	(7)	(11)	(28)	(27)
Left	(26)	(4)	(49)	(7)	(11)	(27)	(27)
Bilateral	(26)	(4)	(50)	(7)	(11)	(28)	(27)
Humeral head							
Right	(13)	(4)	(24)	(7)	(11)	(14)	(27)
Left	(13)	(4)	(24)	(7)	(11)	(14)	(27)
Bilateral	(13)	(4)	(24)	(7)	(11)	(14)	(27)
Both coracoids and humeral							
heads	(7)	(3)	(9)	(7)	(3)	(7)	(9)

(data from Dedick & Caffey, 1953, RADIOLOGY 61:13-20, by permission of The Radiological
Society of North America, Inc.)

TABLE 1116

PERCENTAGE OF OSSIFICATION CENTERS IN NEWLY-BORN
FEMALE NEGRO INFANTS IN NEW YORK CITY
ACCORDING TO MOTHER'S AGE AND INFANT'S WEIGHT

	Total	Mother's Age (years)			Infant's Weight (gm)		
		25	26-35	36-	2,500	3,000	4,000
Total female Negroes	--	(43)	(40)	(13)	(37)	(59)	(4)
Anterior arch of atlas	(22)	(34)	(31)	(29)	(28)	(26)	(38)
Hyoid							
Body	(52)	(61)	(46)	(50)	(71)	(41)	(75)
Horns	(46)	(50)	(44)	(50)	(47)	(45)	(62)
Body and horns	(44)	(50)	(37)	(50)	(45)	(41)	(62)
Anterior arch of atlas and							
horns and body of hyoid	(18)	(22)	(13)	(25)	(16)	(19)	(38)
Interparietal	(8)	(3)	(9)	(17)	(4)	(9)	(25)
Occipital ossicle							
Single	(1)	(2)	(1)	0	0	0	0
Multiple	0	0	0	0	0	0	0
Coracoid							
Right	(28)	(29)	(30)	(20)	(26)	(28)	(50)
Left	(28)	(29)	(30)	(20)	(26)	(28)	(50)
Bilateral	(28)	(29)	(30)	(20)	(26)	(28)	(50)
Humeral head							
Right	(19)	(18)	(21)	(17)	(8)	(24)	(38)
Left	(19)	(18)	(21)	(17)	(8)	(24)	(38)
Bilateral	(19)	(18)	(30)	(17)	(8)	(24)	(38)
Both coracoids and							
humeral heads	(5)	(5)	(3)	(13)	(2)	(5)	(25)

(data from Dedick & Caffey, 1953, RADIOLOGY 61:13-20, by permission of The Radiological
Society of North America, Inc.)

TABLE 1117

AGES (months) OF ONSET AND COMPLETION AND SPAN OF THE OSSEOUS STAGE OF DEVELOPMENT OF
SELECTED GROWTH CENTERS IN BOSTON BOYS

Bone Growth Center In	Onset		Completion		Span	
	Mean	S.D.	Mean	S.D.	Mean	S.D.
Tibia, Distal epiphysis	4.5	2.4	195	11	191	11
Lateral Cuneiform	4.7	4.8	176	11	172	11
Humerus, Capitulum	6.8	4.6	175	12	168	13
Fibula, Distal epiphysis	12.7	4.8	197	12	184	12
Humerus, Greater tubercle	12.9	7.1	56	13	43	11
Finger 3, Prox. phal. epiph.	16.0	3.9	190	11	174	12
Toe I, Distal phal. epiph.	17.0	5.4	177	14	160	15
Medial Cuneiform	24.6	10.4	176	11	152	11
Triquetral	31.0	16.4	182	12	150	18
Toe V, Prox. phal. epiph.	32.9	7.9	188	12	155	14
Navicular (Foot)	34.9	13.5	176	11	141	15
Finger I, Prox. phal. epiph.	35.4	8.0	189	12	154	14
Finger 2, Distal phal. epiph.	42.5	9.1	182	13	140	14
Lunate	44.7	15.6	181	10	137	15
Fibula, Proximal epiphysis	46.3	13.2	204	14	159	16
Metatarsal V, Epiphysis	52.6	11.3	189	11	136	16
Radius, Proximal epiphysis	66.7	17.8	186	13	119	21
Scaphoid	71.4	14.9	181	11	110	16
Calcaneus, Epiphysis	88.6	14.7	186	12	98	15
Ulna, Olecranon	116.2	18.4	184	11	68	18
Adductor Sesamoid (Thumb)	144.7	16.3	190	16	45	17

(data from Pyle et al., 1961)

TABLE   1118

AGES (months) OF ONSET AND COMPLETION AND SPAN OF THE OSSEOUS STAGE OF DEVELOPMENT
SELECTED GROWTH CENTERS IN BOSTON GIRLS

	Onset		Completion		Span	
Bone Growth Center In	Mean	S.D.	Mean	S.D.	Mean	S.D.
Lateral Cuneiform	4.0	3.9	147	15	143	15
Tibia, Distal epiphysis	4.3	1.8	177	12	172	13
Humerus, Capitulum	5.7	4.1	146	11	140	11
Humerus, Greater tubercle	7.8	3.5	38	9	30	8
Fibula, Distal epiphysis	9.4	3.0	181	13	171	12
Toe I, Distal phal. epiph.	10.2	3.3	148	13	138	12
Finger 3, Prox. phal. epiph.	10.9	3.4	167	13	156	12
Medial Cuneiform	16.2	7.6	147	13	130	15
Toe V, Prox. phal. epiph.	21.3	5.7	168	13	145	13
Finger I, Prox. phal. epiph.	22.2	6.7	166	12	144	10
Navicular	25.0	10.9	146	13	121	15
Finger 2, Distal phal. epiph.	25.3	6.7	159	11	133	10
Triquetral	25.4	12.8	160	9	135	15
Fibula, Proximal epiphysis	35.7	11.8	188	12	152	13
Lunate	36.1	15.0	160	9	125	15
Metatarsal V, epiphysis	40.3	7.8	170	14	129	14
Scaphoid	52.8	13.2	160	9	107	15
Radius, proximal epiphysis	57.3	17.2	157	15	100	20
Calcaneus, epiphysis	65.7	11.3	168	14	102	14
Ulna, Olecranon	97.6	12.9	159	12	62	13
Adductor Sesamoid (Thumb)	126.2	11.7	169	14	43	14

(data from Pyle et al., 1961)

TABLE   1119

AGES OF APPEARANCE OF OSSEOUS CENTERS IN THE HAND AND FOOT
IN BOSTON CHILDREN (ages given in years and months)

| | | BOYS | | | | GIRLS | | | |
| | | Percentiles | | | | Percentiles | | | |
		10	25	75	90	10	25	75	90
HAND BONES									
Metacarpal	1	1- 8	2- 1	3- 2	3-11	1- 1	1- 3	1-10	2- 1
	2	1- 1	1- 3	1-11	2- 5	0- 9	0-11	1- 4	1- 4
	3	1- 1	1- 5	2- 0	2- 9	0-10	0-11	1- 4	1- 9
	4	1- 3	1- 7	2- 6	2-11	0-11	1- 1	1- 5	1-10
	5	1- 5	1- 8	2- 9	3- 4	1- 1	1- 2	1- 8	1-11
Prox. Phalanx	1	1-11	2- 3	3- 5	3-11	1- 2	1- 4	1-11	2- 1
	2	0-11	1- 1	1- 8	1-11	0- 7	0- 9	1- 0	1- 4
	3	0-10	1- 1	1- 6	2- 0	0- 7	0- 8	1- 0	1- 3
	4	1- 0	1- 2	1- 9	2- 0	0- 7	0- 9	1- 0	1- 1
	5	1- 3	1- 6	2- 4	2-10	0-10	1- 0	1- 5	1- 1
Mid. Phalanx	2	1- 5	1- 8	2- 7	3- 0	0-10	1- 0	1- 7	1-11
	3	1- 3	1- 7	2- 6	2-11	0- 9	0-11	1- 5	1-11
	4	1- 3	1- 7	2- 6	2-11	0-10	0-11	1- 6	1- 4
	5	2- 2	2- 7	4- 3	5- 0	1- 2	1- 5	2- 3	2- 6
Dist. Phalanx	1	1- 0	1- 2	1-11	2- 7	0- 7	0- 9	1- 3	1- 4
	2	2- 3	2- 8	3-10	4- 6	1- 3	1- 7	2- 5	2-10
	3	1- 8	2- 0	2-10	3- 4	0-11	1- 2	1-10	2- 1
	4	1- 8	2- 1	2-11	3- 5	1- 0	1- 2	1-10	2- 3
	5	2- 3	2- 8	3-10	4- 7	1- 3	1- 7	2- 5	2-10
FOOT BONES									
Metatarsal	1	1- 7	1-11	2-10	3- 5	1- 2	1- 4	1-10	2- 0
	2	2- 1	2- 5	3- 6	4- 0	1- 5	1- 8	2- 6	2-11
	3	2- 6	3- 1	4- 1	4-10	1- 9	2- 1	2-10	3- 2
	4	3- 2	3- 6	4-10	5- 3	2- 1	2- 4	3- 3	3-10
	5	3- 5	4- 0	5- 2	5-11	2- 3	2- 7	3- 9	4- 5
Prox. Phalanx	1	1- 9	2- 1	2-10	3- 1	1- 1	1- 3	1-10	2- 0
	2	1- 2	1- 4	2- 3	2- 9	0-10	0-11	1- 5	1- 6
	3	1- 1	1- 3	1-11	2- 7	0- 7	0-10	1- 2	1- 4
	4	1- 2	1- 4	2- 0	2- 8	0- 8	1-10	1- 3	1- 6
	5	1- 9	2- 0	3- 0	3- 8	1- 1	1- 3	2- 2	2- 7
Mid. Phalanx	2	1- 1	1- 6	3- 0	3-10	0- 7	0- 9	1- 5	2- 0
	3	0- 8	1- 0	2- 0	3- 7	0- 5	0- 7	1- 3	1-11
	4	0-10	1- 1	2- 8	4- 9	0- 7	0- 9	1-11	2- 9
	5	--	--	--	--	--	--	--	--

TABLE 1119 (continued)

AGES OF APPEARANCE OF OSSEOUS CENTERS IN THE HAND AND FOOT
IN BOSTON CHILDREN (ages given in years and months)

| | | BOYS | | | | GIRLS | | | |
| | | Percentiles | | | | Percentiles | | | |
		10	25	75	90	10	25	75	90
Dist. Phalanx	1	0-10	1- 1	1- 6	1-10	0- 6	0- 7	0-11	1- 7
	2	3- 6	4- 2	5- 8	6- 3	2- 1	2- 5	3- 7	1- 2
	3	3- 3	3-10	5- 1	5-11	2- 0	2- 3	3- 4	3-10
	4	3- 3	3-10	5- 1	5-11	1- 9	2- 1	3- 0	3- 4
	5	3- 4	4- 2	5-10	6- 9	1- 4	1- 9	2- 9	3- 4

(data from Harding, 1952)

TABLE 1120

AGE AT ONSET OF OSSIFICATION IN THE DISTAL EPIPHYSIS OF THE ULNA,
THE ADDUCTOR SESAMOID OF THE THUMB AND THE CREST OF THE ILIUM,
AND THE AGE AT ONSET OF FUSION IN THE SECOND DISTAL PHALANX,
AND AGE AT MENARCHE IN CLEVELAND (Ohio) CHILDREN

	Ulnar Age (yr.)	Sesamoid Age (yr.)	Iliac Age (yr.)	Fusion Dist. II (yr.)	Menarche (yr.)
BOYS					
Mean	7.6	12.6	14.3	14.7	--
S.D.	1.2	1.1	1.6	1.1	--
GIRLS					
Mean	5.6	10.1	12.7	12.5	12.7
S.D.	1.2	1.1	1.3	1.0	1.4

(data from Buehl and Pyle, 1942, The Journal of Pediatrics 21:335-342)

TABLE   1121

AGES (months) AT ONSET AND COMPLETION OF
OSSIFICATION, AND THE SPAN FROM ONSET TO
FUSION IN THE HAND AND WRIST
IN BOSTON CHILDREN

Bone Growth Center	Mean Onset Order	Onset		Completion		Span	
		Mean	S.D.	Mean	S.D.	Mean	S.D.
BOYS							
Capitate	1	2.9	1.7	183	12	180	12
Hamate	2	4.2	2.7	183	12	179	11
Radius, distal epiphysis	3	12.3	5.3	208	8	196	11
Finger 3, prox. phal. epiph.	4	16.4	4.6	191	12	175	12
Finger 2, prox. phal. epiph.	5	17.2	4.8	191	13	174	13
Finger 4, prox. phal. epiph.	6	18.1	5.0	191	13	173	13
Metacarpal 2, epiphysis	7	19.3	5.5	194	13	175	13
Finger 1, distal phal. epiph.	8	20.6	6.8	184	13	163	14
Metacarpal 3, epiphysis	9	21.8	6.9	195	13	173	14
Finger 5, prox. phal. epiph.	10	23.8	6.7	191	13	167	14
Metacarpal 4, epiphysis	11	24.8	7.3	194	13	170	14
Finger 3, middle phal. epiph.	12	25.1	6.2	192	13	167	14
Finger 4, middle phal. epiph.	13	26.0	6.7	192	13	166	14
Finger 2, middle phal. epiph.	14	27.1	7.1	191	12	164	14
Metacarpal 5, epiphysis	15	27.1	8.6	196	14	168	15
Triquetral	16	29.5	16.2	183	12	153	20
Finger 3, distal phal. epiph.	17	30.7	7.1	186	13	156	13
Finger 4, distal phal. epiph.	18	31.2	7.4	186	13	155	14
Metacarpal 1, epiphysis	19	34.8	11.1	187	14	152	16
Finger 1, prox. phal. epiph.	20	36.3	9.1	191	13	155	15
Finger 2, distal phal. epiph.	21	41.2	9.0	185	13	144	14
Finger 5, distal phal. epiph.	22	41.9	10.1	186	13	144	15
Lunate	23	43.5	14.7	183	11	140	16
Finger 5, middle phal. epiph.	24	44.4	11.9	190	13	146	15
Scaphoid	25	69.6	15.4	183	11	113	17
Trapezoid	26	72.0	16.1	183	11	111	15
Trapezium	27	72.7	18.4	183	11	110	19
Ulna, distal epiphysis	28	80.3	13.4	205	10	124	16
Adductor sesamoid (thumb)	29	150.8	13.7	192	14	41	14
GIRLS							
Capitate	1	2.5	1.8	159	10	157	10
Hamate	2	3.1	2.2	159	10	156	10
Radius, distal epiphysis	3	10.0	4.1	200	12	190	12
Finger 3, prox. phal. epiph.	4	10.6	3.2	166	12	155	11
Finger 2, prox. phal. epiph.	5	10.9	2.9	166	12	155	11
Finger 4, prox. phal. epiph.	6	11.0	3.2	166	12	155	11
Metacarpal 2, epiphysis	7	13.1	3.3	170	13	156	13
Finger 1, distal phal. epiph.	8	13.2	5.5	157	11	144	10
Metacarpal 3, epiphysis	9	14.4	3.9	170	13	156	13
Finger 5, prox. phal. epiph.	10	14.7	3.8	164	12	150	11
Metacarpal 4, epiphysis	11	15.5	3.8	169	14	153	14
Finger 3, middle phal. epiph.	12	15.7	5.4	167	12	152	11
Finger 4, middle phal. epiph.	13	16.0	5.2	168	12	152	11
Finger 2, middle phal. epiph.	14	17.0	5.4	166	12	149	11
Metacarpal 5, epiphysis	15	17.0	5.0	170	13	153	12

TABLE 1121   (continued)

AGES (months) AT ONSET AND COMPLETION OF
OSSIFICATION, AND THE SPAN FROM ONSET TO
FUSION IN THE HAND AND WRIST
IN BOSTON CHILDREN

Bone Growth Center	Mean Onset Order	Onset		Completion		Span	
		Mean	S.D.	Mean	S.D.	Mean	S.D.
			GIRLS				
Triquetral	23	26.6	14.0	160	9	133	15
Finger 3, distal phal. epiph.	16	19.1	6.0	159	11	140	11
Finger 4, distal phal. epiph.	17	19.6	5.9	159	12	139	11
Metacarpal 1, epiphysis	18	19.9	5.4	164	13	144	12
Finger 1, prox. phal. epiph.	19	21.6	6.4	165	11	144	11
Finger 2, distal phal. epiph.	20	25.0	6.8	158	11	133	11
Finger 5, distal phal. epiph.	21	25.0	6.7	160	15	135	15
Lunate	24	36.1	17.3	160	9	124	16
Finger 5, middle phal. epiph.	22	25.9	8.3	165	12	139	12
Scaphoid	27	53.7	13.8	160	9	106	15
Trapezoid	26	51.8	12.3	160	9	108	13
Trapezium	25	51.6	16.4	160	9	108	18
Ulna, distal epiphysis	28	72.4	12.1	191	12	119	12
Adductor sesamoid (thumb)	29	127.8	10.3	167	14	39	13

(data from Stuart, Pyle, Cornoni, et al., 1962, Pediatrics 29:237-249. Copyright American
Academy of Pediatrics 1962)

TABLE   1122

PERCENTILES OF AGES (weeks) AT EPIPHYSEAL
OSSIFICATION IN MICHIGAN NEONATES

| Center | Percentiles | |
	5	95
Humeral head	37th week	16 postnatal weeks*
Distal femoral	31st week	39th week (female)
		40th week (male)
Proximal tibial	34th week	2 postnatal weeks (female)*
		5 postnatal weeks (male)*
Calcaneus	22nd week	25th week
Talus	25th week	31st week
Cuboid	37th week	8 postnatal weeks (female)*
		16 postnatal weeks (male)*

* Based on data of Garn et al. (1967) for postnatal ossification

(data from Kuhns and Finnström, 1976, RADIOLOGY 119:655-650)

TABLE   1123

AGES (years) OF APPEARANCE OF SESAMOID BONES OF THE HAND

| | | BOYS | | | | GIRLS | | | |
| | | Chronological Age | | Skeletal Age | | Chronological Age | | Skeletal Age | |
Ray	Sesamoid	Mean	S.D.	Mean	S.D.	Mean	S.D.	Mean	S.D.
I	Ulnar (adductor)	13.29	0.85	13.34	0.51	11.29	0.92	11.12	0.65
	Radial (Flexor)	14.43	0.79	14.37	0.70	12.72	1.06	12.82	1.12
	Inter Phalangeal	14.90	0.71	14.73	0.63	13.55	1.37	13.82	1.14
II	Metacarpo Phalangeal	15.06	1.05	14.89	0.57	13.46	0.77	13.57	1.17
III	Metacarpo Phalangeal	16.21	--	15.53	--	--	--	--	--
IV	Metacarpo Phalangeal	--	--	--	--	--	--	--	--
V	Metacarpo Phalangeal	15.11	0.84	14.93	0.59	13.48	0.94	13.69	1.09

(data from Bowden, 1971)

TABLE 1124

PERCENTAGES FOR OSSIFICATION OF
THE PISIFORM IN WHITE GIRLS
IN NEW YORK CITY

Age	Present
5 yrs. to 5 yrs. 9 mos.	4.55
5 yrs. 10 mos. to 6 yrs. 9 mos.	5.41
6 yrs. 10 mos. to 7 yrs. 9 mos.	21.62
7 yrs. 10 mos. to 8 yrs. 9 mos.	45.95
8 yrs. 10 mos. to 9 yrs. 9 mos.	70.27
9 yrs. 10 mos. to 10 yrs. 9 mos.	94.28
10 yrs. 10 mos. to 11 yrs. 9 mos.	96.55
11 yrs. 10 mos. to 12 yrs. 9 mos.	100.00

(data from "Studies in the physical development.
V. The ossification time of the os pisiforme,"
by N. Michelson, Human Biology 17:143-146, 1945, by
permission of the Wayne State University Press.
Copyright 1945, Wayne State University Press,
Detroit, Michigan 48202)

TABLE 1125

AGES AT ONSET OF OSSIFICATION (years) IN OHIO CHILDREN

Ossification Center	Percentiles					
	Boys			Girls		
	5	50	95	5	50	95
Shoulder						
Head of humerus	--	.03	.32	--	.03	.30
Coracoid process of scapula	--	.04	.36	--	.03	.42
Acromion of clavicle	12.15	13.74	15.48	10.32	11.92	13.79
Accessory epiphysis, coracoid process of scapula	12.74	14.35	16.31	10.37	12.21	14.37
Elbow						
Capitulum of humerus	.06	.33	1.07	.05	.26	.77
Greater tubercle of humerus	.25	.83	2.33	.20	.51	1.14
Head of radius	3.00	5.21	7.97	2.26	3.87	6.28
Medial epicondyle of humerus	4.27	6.25	8.41	2.05	3.40	5.07
Olecranon of ulna	7.78	9.67	11.90	5.62	8.01	9.93
Lateral epicondyle of humerus	9.23	11.24	13.70	7.14	9.24	11.28
Carpus						
Capitate	--	.25	.60	--	.15	.56
Hamate	.03	.31	.82	--	.18	.59
Lunate	1.53	4.07	6.77	1.08	2.62	5.65
Navicular	3.59	5.63	7.81	2.35	4.12	5.99
Trapezium	3.53	5.87	8.97	1.94	4.08	6.36
Trapezoid	3.12	6.22	8.50	2.38	4.17	6.01
Triquetral	.49	2.43	5.47	.29	1.70	3.73
Hand						
Distal epiphysis of ulna	5.25	7.10	9.07	3.29	5.37	7.63
Distal epiphysis of radius	.53	1.10	2.30	.38	.82	1.70
Epiphysis of 1st metacarpal	1.45	2.59	4.32	.92	1.60	2.67
Epiphysis of 2nd metacarpal	.93	1.61	2.82	.64	1.09	1.69
Epiphysis of 3rd metacarpal	.95	1.79	3.01	.65	1.13	1.94
Epiphysis of 4th metacarpal	1.09	2.03	3.60	.75	1.29	2.17
Epiphysis of 5th metacarpal	1.27	2.17	3.82	.86	1.37	2.35
Epiphysis, proximal segment of 1st finger	1.84	3.00	4.57	.93	1.71	2.84
Epiphysis, proximal segment of 2nd finger	.78	1.41	2.17	.40	.87	1.64
Epiphysis, proximal segment of 3rd finger	.77	1.37	2.15	.41	.85	1.61
Epiphysis, proximal segment of 4th finger	.80	1.49	2.40	.41	.90	1.66
Epiphysis, proximal segment of 5th finger	1.00	1.85	2.82	.65	1.19	2.07
Epiphysis, middle segment of 2nd finger	1.30	2.19	3.31	.67	1.36	2.54
Epiphysis, middle segment of 3rd finger	1.01	1.97	3.31	.63	1.28	2.36
Epiphysis, middle segment of 4th finger	1.00	2.05	3.24	.63	1.24	2.43
Epiphysis, middle segment of 5th finger	1.94	3.40	5.84	.88	1.97	3.54
Epiphysis, distal segment of 1st finger	.75	1.51	2.70	.42	.99	1.73
Epiphysis, distal segment of 2nd finger	1.80	3.17	4.97	1.06	2.50	3.29
Epiphysis, distal segment of 3rd finger	1.31	2.41	3.72	.72	1.46	2.69
Epiphysis, distal segment of 4th finger	1.37	2.44	3.73	.73	1.52	2.82
Epiphysis, distal segment of 5th finger	2.06	3.29	4.98	1.01	1.96	3.45
Adductor sesamoid of 1st finger	11.03	12.76	14.62	8.67	10.72	12.68

TABLE 1125 (continued)

AGES AT ONSET OF OSSIFICATION (years) IN OHIO CHILDREN

| | Percentiles | | | | | |
| | Boys | | | Girls | | |
Ossification Center	5	50	95	5	50	95
Hip						
Head of femur	.06	.35	.64	.04	.33	.62
Os acetabulum	11.90	13.54	15.32	9.60	11.47	13.39
Epiphysis, iliac crest	12.03	14.03	15.91	10.81	12.79	15.31
Greater trochanter of femur	1.92	2.96	4.35	.96	1.85	3.03
Ischial tuberosity	13.57	15.26	17.08	11.71	13.89	16.00
Knee						
Proximal epiphysis of tibia	--	.04	.10	--	.01	.04
Tubercle of tibia	9.92	11.81	13.38	7.89	10.25	11.82
Proximal epiphysis of fibula	1.86	3.47	5.24	1.33	2.61	3.92
Patella	2.55	4.00	5.96	1.47	2.48	4.01
Tarsus						
Epiphysis of calcaneus	5.17	7.59	9.55	3.54	5.37	7.30
Navicular	1.12	3.02	5.40	.77	1.94	3.58
Lateral cuneiform	.05	.46	1.58	--	.23	1.23
Middle cuneiform	1.19	2.65	4.21	.81	1.80	3.00
Medial cuneiform	.89	2.17	3.77	.50	1.43	2.82
Cuboid	--	.07	.30	--	.05	.16
Foot						
Epiphysis of 1st metatarsal	1.39	2.18	3.12	.96	1.58	2.23
Epiphysis of 2nd metatarsal	1.93	2.86	4.33	1.22	2.14	3.43
Epiphysis of 3rd metatarsal	2.33	3.48	5.00	1.42	2.48	3.68
Epiphysis of 4th metatarsal	2.92	4.02	5.74	1.77	2.84	4.05
Epiphysis of 5th metatarsal	3.12	4.37	6.34	2.08	3.24	4.93
Epiphysis, proximal segment of 1st toe	1.45	2.35	3.31	.89	1.55	2.47
Epiphysis, proximal segment of 2nd toe	.97	1.74	2.65	.63	1.19	2.05
Epiphysis, proximal segment of 3rd toe	.90	1.58	2.52	.51	1.05	1.88
Epiphysis, proximal segment of 4th toe	.95	1.64	2.65	.61	1.24	2.06
Epiphysis, proximal segment of 5th toe	1.53	2.45	3.65	.97	1.73	2.67
Epiphysis, middle segment of 2nd toe	.89	2.04	4.05	.49	1.18	2.24
Epiphysis, middle segment of 3rd toe	.41	1.40	4.27	.21	1.02	2.47
Epiphysis, middle segment of 4th toe	.40	1.21	2.88	.40	.92	3.00
Primary center, middle segment of 5th toe	--	1.04	3.81	--	.74	2.08
Epiphysis, distal segment of 1st toe	.71	1.21	2.10	.39	.78	1.68
Epiphysis, distal segment of 2nd toe	3.24	4.64	6.75	1.50	2.93	4.50
Epiphysis, distal segment of 3rd toe	2.99	4.36	6.19	1.37	2.73	4.11
Epiphysis, distal segment of 4th toe	2.95	4.38	6.40	1.36	2.58	4.09
Epiphysis, distal segment of 5th toe	2.34	3.94	6.30	1.17	2.31	4.07

(data from MEDICAL RADIOGRAPHY AND PHOTOGRAPHY published by Health Sciences Markets Division, Eastman Kodak Company. Courtesy of S. M. Garn, C. G. Rohmann, and F. N. Silverman, 1967)

TABLE   1126

AGES AT ONSET OF OSSIFICATION IN CLEVELAND (OH) BOYS (months)

Bone Growth Center	Mean	S.D.
Shoulder		
Humerus, greater tuberosity	12.4	6.7
Elbow		
Humerus, capitulum	6.9	3.9
Humerus, medial epicondyle	75.6	15.3
Radius, proximal epiphysis	66.0	16.1
Carpus		
Lunate	44.4	18.9
Triquetral	29.3	16.4
Trapezium	68.4	18.6
Trapezoid	69.1	13.8
Scaphoid	67.8	15.3
Hand		
Radius, distal epiphysis	13.2	5.4
Ulna, distal epiphysis	82.4	14.2
Metacarpal 1, epiphysis	31.8	8.4
Metacarpal 2, epiphysis	18.0	5.1
Metacarpal 3, epiphysis	20.5	5.4
Metacarpal 4, epiphysis	23.3	6.2
Metacarpal 5, epiphysis	26.0	7.0
Finger 1, proximal phalanx epiphysis	33.2	7.7
Finger 2, proximal phalanx epiphysis	16.1	4.3
Finger 3, proximal phalanx epiphysis	15.0	4.2
Finger 4, proximal phalanx epiphysis	16.6	4.9
Finger 5, proximal phalanx epiphysis	21.5	5.3
Finger 2, middle phalanx epiphysis	25.8	6.6
Finger 3, middle phalanx epiphysis	23.4	6.3
Finger 4, middle phalanx epiphysis	23.8	6.1
Finger 5, middle phalanx epiphysis	39.2	10.3
Finger 1, distal phalanx epiphysis	18.3	6.3
Finger 2, distal phalanx epiphysis	37.6	7.9
Finger 3, distal phalanx epiphysis	27.4	6.4
Finger 4, distal phalanx epiphysis	27.7	6.5
Finger 5, distal phalanx epiphysis	37.3	8.5
Hip		
Femur, head	4.7	1.8
Femur, greater trochanter	51.9	8.6
Knee		
Patella	47.7	11.8
Fibula, proximal epiphysis	47.0	12.5

TABLE 1126 (continued)

AGES AT ONSET OF OSSIFICATION IN CLEVELAND (OH) BOYS (months)

Bone Growth Center	Mean	S.D.
Tarsus		
Calcaneus, epiphysis	90.3	12.6
Navicular	33.8	13.0
Medial cuneiform	24.1	10.4
Intermediate cuneiform	29.3	9.0
Foot		
Tibia, distal epiphysis	4.4	1.6
Fibula, distal epiphysis	12.6	4.2
Metatarsal 1, epiphysis	28.5	4.9
Metatarsal 2, epiphysis	35.3	6.8
Metatarsal 3, epiphysis	42.1	7.7
Metatarsal 4, epiphysis	47.8	7.7
Metatarsal 5, epiphysis	53.6	9.2
Toe 1, proximal phalanx epiphysis	27.7	5.3
Toe 2, proximal phalanx epiphysis	20.7	5.1
Toe 3, proximal phalanx epiphysis	18.1	4.8
Toe 4, proximal phalanx epiphysis	20.0	5.0
Toe 5, proximal phalanx epiphysis	30.5	6.4
Toe 1, distal phalanx epiphysis	15.5	5.1
Toe 2, distal phalanx epiphysis	58.5	11.9
Toe 3, distal phalanx epiphysis	53.7	12.4
Toe 4, distal phalanx epiphysis	52.2	11.6

(data from Hoerr, Pyle and Francis, Radiographic Atlas of Skeletal Development of the Foot and Ankle: A Standard of Reference, 1962. Courtesy of Charles C Thomas, Publisher, Springfield, Illinois)

TABLE   1127

AGES AT ONSET OF OSSIFICATION IN CLEVELAND (OH) GIRLS (months)

Bone Growth Center	Mean	S.D.
Shoulder		
Humerus, greater tuberosity	6.7	3.1
Elbow		
Humerus, capitulum	4.8	2.8
Humerus, medial epicondyle	43.0	11.9
Radius, proximal epiphysis	50.3	14.6
Carpus		
Lunate	36.0	13.4
Triquetral	22.7	13.4
Trapezium	47.4	14.6
Trapezoid	49.4	12.7
Scaphoid	50.4	12.6
Hand		
Radius, distal epiphysis	9.8	4.1
Ulna, distal epiphysis	68.3	13.1
Metacarpal 1, epiphysis	19.1	5.1
Metacarpal 2, epiphysis	12.3	2.9
Metacarpal 3, epiphysis	13.8	3.7
Metacarpal 4, epiphysis	15.6	4.1
Metacarpal 5, epiphysis	17.1	4.8
Finger 1, proximal phalanx epiphysis	20.9	5.6
Finger 2, proximal phalanx epiphysis	10.8	3.2
Finger 3, proximal phalanx epiphysis	10.0	3.1
Finger 4, proximal phalanx epiphysis	10.8	3.4
Finger 5, proximal phalanx epiphysis	14.3	4.0
Finger 2, middle phalanx epiphysis	17.2	5.6
Finger 3, middle phalanx epiphysis	15.4	5.5
Finger 4, middle phalanx epiphysis	15.6	5.6
Finger 5, middle phalanx epiphysis	23.6	8.4
Finger 1, distal phalanx epiphysis	12.1	4.6
Finger 2, distal phalanx epiphysis	24.1	6.5
Finger 3, distal phalanx epiphysis	18.4	5.2
Finger 4, distal phalanx epiphysis	18.7	5.4
Finger 5, distal phalanx epiphysis	23.7	6.3
Hip		
Femur, head	3.9	1.9
Femur, greater trochanter	28.4	5.4
Knee		
Patella	30.3	7.6
Fibula, proximal epiphysis	34.3	10.6

TABLE   1127   (continued)

AGES AT ONSET OF OSSIFICATION IN CLEVELAND (OH) GIRLS (months)

Bone Growth Center	Mean	S.D.
Tarsus		
Calcaneus, epiphysis	61.6	12.7
Navicular	23.3	10.9
Medial cuneiform	15.7	6.9
Intermediate cuneiform	20.0	6.7
Foot		
Tibia, distal epiphysis	4.0	1.5
Fibula, distal epiphysis	9.0	2.8
Metatarsal 1, epiphysis	19.9	3.9
Metatarsal 2, epiphysis	24.3	5.0
Metatarsal 3, epiphysis	28.6	5.6
Metatarsal 4, epiphysis	33.4	7.1
Metatarsal 5, epiphysis	38.9	8.6
Toe 1, proximal phalanx epiphysis	18.8	4.8
Toe 2, proximal phalanx epiphysis	13.6	4.0
Toe 3, proximal phalanx epiphysis	11.5	3.8
Toe 4, proximal phalanx epiphysis	12.7	4.1
Toe 5, proximal phalanx epiphysis	20.9	5.8
Toe 1, distal phalanx epiphysis	9.4	3.0
Toe 2, distal phalanx epiphysis	36.2	9.1
Toe 3, distal phalanx epiphysis	34.0	8.8
Toe 4, distal phalanx epiphysis	30.3	7.6

(data from Hoerr, Pyle, and Francis, Radiographic Atlas of Skeletal Development of the Foot and Ankle: A Standard of Reference, 1962. Courtesy of Charles C Thomas, Publisher, Springfield, Illinois)

TABLE 1128

BLACK-WHITE DIFFERENCES IN TIMING OF ONSET OF OSSIFICATION IN THE HAND
FOR CHILDREN IN THE TEN-STATE NUTRITION SURVEY

Ossification Center	BOYS				GIRLS			
	Black Mean age	White age	% Dif-ference	Z-score difference	Black Mean age	White age	% Dif-ference	Z-score difference
Proximal 3	1.15	1.33	+ 9	+0.52	0.72	0.94	+13	+0.69
Proximal 2	1.28	1.42	+ 6	+0.35	0.75	0.91	+10	+0.49
Proximal 4	1.22	1.51	+13	+0.71	0.79	0.96	+10	+0.55
Distal 1	1.26	1.55	+13	+0.62	0.72	1.05	+18	+1.00
Metacarpal 2	1.29	1.57	+12	+0.55	0.88	1.05	+ 9	+0.61
Metacarpal 3	1.37	1.67	+12	+0.53	0.86	1.05	+11	+0.68
Proximal 5	1.72	1.73	0	+0.02	0.92	1.17	+13	+0.84
Metacarpal 4	1.53	1.82	+11	+0.52	1.05	1.16	+ 6	+0.44
Middle 3	1.65	1.85	+11	+0.52	0.94	1.19	+13	+0.59
Middle 4	1.65	1.88	+ 9	+0.46	0.95	1.18	+12	+0.61
Middle 2	1.93	1.96	+ 1	+0.06	0.99	1.22	+12	+0.50
Metacarpal 5	1.69	2.08	+14	+0.55	1.05	1.19	+ 7	+0.54
Distal 3	1.99	2.11	+ 4	+0.21	0.99	1.40	+19	+0.84
Distal 4	2.03	2.12	+ 3	+0.15	1.14	1.43	+13	+0.58
Metacarpal 1	2.67	2.56	- 3	-0.12	1.22	1.33	+ 5	+0.27
Triquetral	2.33	2.70	+11	+0.22	1.36	1.71	+14	+0.31
Proximal 1	2.50	2.74	+ 7	+0.26	1.18	1.69	+21	+0.85
Distal 2	2.46	2.97	+14	+0.59	1.34	1.86	+20	+0.83
Distal 5	2.49	3.01	+14	+0.60	1.35	1.85	+19	+0.85
Middle 5	2.90	3.01	+ 3	+0.10	1.22	1.74	+21	+0.89
Lunate	3.66	4.01	+ 7	+0.21	2.46	2.99	+14	+0.42
Scaphoid	5.48	6.07	+ 9	+0.42	3.92	4.38	+ 9	+0.41
Trapezoid	5.68	6.16	+ 7	+0.41	4.14	4.40	+ 5	+0.23
Trapezium	5.83	6.22	+ 6	+0.25	4.16	4.24	+ 2	+0.06
Distal ulna	6.72	7.21	+ 6	+0.39	5.59	5.79	+ 3	+0.17

(data from Garn et al., 1972, The Journal of Pediatrics 80:965-969)

TABLE 1129

NUMBER OF HAND-WRIST CENTERS OSSIFIED
AT PARTICULAR AGES IN OHIO BOYS

Chronological age (months)	Mean	S.D.
6	1.92	0.75
12	3.28	2.01
18	6.76	3.99
24	11.78	4.43
30	16.43	4.25
36	19.61	3.25
42	21.61	2.19
48	22.77	1.86
60	24.12	1.60
72	25.44	1.68
84	26.74	1.42

(data from Yarbrough, Habicht, Klein, et al., 1973, Investigative Radiology 8:233-243. Courtesy of Lippincott/Harper Company, Philadelphia, PA)

TABLE 1130

NUMBER OF OSSIFICATION CENTERS PRESENT AT
VARIOUS AGE-LEVELS IN THE FOOT
IN OHIO CHILDREN

Development of Foot	Age in Months									
	Birth	1	3	6	9	12	18	24	30	36
BOYS										
Very fast	3	4	4	6	6-8	6-10	9-12	13-16	16-17	17-18
Fast	3	3	4	4-5	4-5	5	8	11-12	14-15	16
Average	2-3	3	3-4	4	4	4	6-7	9-10	11-13	14-15
Slow	2	2-3	3	3-4	4	4	5	7-8	9-10	12-13
Very slow	2	2	3	3	3	3	3-4	3-6	7-8	8-11
GIRLS										
Very fast	4	4	4	5-7	8-11	10-12	15-16	18-21	21	21
Fast	3-4	3-4	4	4	6-7	9	13-14	16-17	19-20	21
Average	3-4	3	3-4	4	4-5	6-8	10-12	13-15	16-18	18-20
Slow	2	3	3	4	4	5	8-9	11-12	14-15	16-17
Very slow	2	2	3	3	3	4	4-7	6-10	10-13	11-15

NUMBER OF OSSIFICATION CENTERS PRESENT AT
VARIOUS AGE-LEVELS IN THE HAND
IN OHIO CHILDREN

Development of Hand	Age in Months									
	Birth	1	3	6	9	12	18	24	30	36
BOYS										
Very fast	1-2	2	3	3-5	5-9	8-14	16-19	21-23	23	24
Fast	0	1-2	2	2	3-4	5-7	10-15	16-20	21-22	23
Average	0	0	1-2	2	2	3-4	6-9	12-15	16-20	19-22
Slow	0	0	0-1	2	2	2	4-5	8-11	13-15	16-18
Very slow	0	0	0	0-1	0-1	1	1-3	1-7	5-12	6-15
GIRLS										
Very fast	2-3	3	3-4	4-8	11-13	16-18	22-24	24-25	25-26	26-27
Fast	0-1	2	2	3	6-10	12-15	20-21	23	24	25
Average	0	0-1	2	2	3-5	6-11	14-19	20-22	22-23	23-24
Slow	0	0	1-2	2	2	3-5	12-13	17-19	20-21	22
Very slow	0	0	0	0-1	2	2	2-11	6-16	14-19	17-21

(data from Reynolds and Asakawa, 1951, American Journal of Roentgenology 65:403-410.
Copyright American Roentgen Ray Society, 1951)

TABLE 1131

PERCENTILES FOR HAND-WRIST OSSIFICATION (number of centers) INCREMENTS
IN OHIO CHILDREN

Age interval (years)	Boys Percentiles				Girls Percentiles			
	5	50	95	Skew	5	50	95	Skew
1.0-1.5	0	3	10	2.3*	2	8	13	0.8*
1.5-2.0	0	5	10	1.0	0	4	8	1.0
2.0-2.5	1	4	8	1.3	0	2	7	2.5
2.5-3.0	0	3	7	1.3	0	1	4	3.0
3.0-3.5	0	2	5	1.5	0	1	3	2.0
3.5-4.0	0	1	4	3.0	0	1	3	2.0
4.0-4.5	0	1	4	3.0	0	1	2	1.0
4.5-5.0	0	1	3	2.0	0	1	3	2.0
5.0-5.5	0	0	2	--	0	1	2	1.0
5.5-6.0	0	1	3	2.0	0	0	2	--
6.0-6.5	0	1	3	2.0	0	0	2	--
6.5-7.0	0	1	3	2.0	0	0	1	--
7.0-7.5	0	0	2	--	0	0	1	--

$$*\frac{P_{95} - P_{50}}{P_{50} - P_5}$$

(data from Garn et al., 1961)

TABLE 1132

PERCENT OF BOYS AND GIRLS SHOWING THE NUMBER OF RADIO-OPAQUE HAND-WRIST BONES--
OSSIFYING OR ADULT BY CHRONOLOGICAL AGE IN YEARS AT LAST BIRTHDAY
IN U.S. CHILDREN

Number of Bones	BOYS Age (years)						GIRLS Age (years)					
	6	7	8	9	10	11	6	7	8	9	10	11
Children with 1 or more bones not yet adult---	100.0	100.0	100.0	100.0	100.0	100.0	100.0	100.0	99.8	99.0	96.7	88.0
10 or fewer---	–	–	–	–	–	–	–	–	0.2	0.2	–	0.2
15 or fewer---	0.1	0.2	0.2	0.1	–	–	–	–	0.2	0.2	–	0.5
20 or fewer---	0.3	0.4	0.2	0.5	0.2	0.2	0.2	0.1	0.2	0.2	0.3	1.7
21 or fewer---	0.9	0.5	0.2	0.5	0.2	0.2	–	0.1	0.2	0.2	0.3	1.7
22 or fewer---	3.0	0.7	0.4	0.5	0.2	0.2	0.2	0.1	0.2	0.2	0.3	2.3
23 or fewer---	9.4	3.1	1.0	0.8	0.2	0.2	0.8	0.1	0.2	0.2	0.9	2.7
24 or fewer---	27.9	9.6	2.4	1.2	0.2	0.2	3.0	0.3	0.7	0.4	1.2	4.6
25 or fewer---	42.7	17.6	5.3	1.4	0.2	0.2	6.0	1.0	1.5	0.4	1.7	5.7
26 or fewer---	53.6	26.7	9.1	2.6	0.6	0.4	10.8	2.2	2.3	1.1	2.2	7.2
27 or fewer---	79.7	48.5	24.9	6.5	1.5	1.1	30.4	12.3	3.9	2.4	3.4	8.4
28 or fewer---	99.6	99.0	95.4	83.4	61.0	36.9	94.0	81.9	55.1	26.6	12.8	11.6
29 or fewer---	100.0	100.0	100.0	99.4	99.4	89.4	99.8	99.8	96.5	84.6	56.3	30.8
30 or fewer---	100.0	100.0	100.0	99.7	99.8	97.4	100.0	99.9	98.9	95.3	83.3	65.1
31 or fewer---	100.0	100.0	100.0	100.0	100.0	100.0	100.0	100.0	99.8	99.0	96.7	88.0

TABLE 1132 (continued)

PERCENT OF BOYS AND GIRLS SHOWING THE NUMBER OF RADIO-OPAQUE HAND-WRIST BONES--
OSSIFYING OR ADULT BY CHRONOLOGICAL AGE IN YEARS AT LAST BIRTHDAY
IN U.S. CHILDREN

Number of Bones	BOYS Age (years)						GIRLS Age (years)					
	6	7	8	9	10	11	6	7	8	9	10	11
Children with 1 or more adult bones	-	-	-	-	-	-	-	-	0.2	1.0	3.3	12.0
Only 1	-	-	-	-	-	-	-	-	-	-	1.5	1.5
2 or fewer	-	-	-	-	-	-	-	-	-	0.5	2.1	5.4
3 or fewer	-	-		-	-	-	-	-	-	0.5	2.3	6.2
4 or fewer	-	-	-	-	-	-	-	-	-	0.8	2.5	6.8
5 or fewer	-	-	-	-	-	-	-	-	-	1.0	2.7	7.8
6 or fewer	-	-	-	-	-	-	-	-	-	1.0	2.9	8.9
7 or fewer	-	-	-	-	-	-	-	-	-	1.0	3.3	10.1
8 or fewer	-	-	-	-	-	-	-	-	-	1.0	3.3	10.6
12 or fewer	-	-	-	-	-	-	-	-	-	1.0	3.3	11.2
15 or fewer	-	-	-	-	-	-	-	-	-	1.0	3.3	11.4
16 or fewer	-	-	-	-	-	-	-	-	-	1.0	3.3	11.5
17 or fewer	-	-	-	-	-	-	-	-	-	1.0	3.3	11.8
23 or fewer	-	-	-	-	-	-	-	-	-	1.0	3.3	12.0
28 or fewer	-	-	-	-	-	-	-	-	0.2	1.0	3.3	12.0

(data from Roche et al., 1974)

TABLE 1133

PRESENCE OF SELECTED CENTERS OF OSSIFICATION
IN NEWBORN INFANTS BY WEIGHT GROUPS*

Center of Ossification	Weight groups (gm)					
	Less than 2,000	2,000 to 2,499	2,500 to 2,999	3,000 to 3,499	3,500 to 3,999	4,000 or more
Calcaneum	Percent					
White boys	100.0	100.0	100.0	100.0	100.0	100.0
White girls	100.0	100.0	100.0	100.0	100.0	100.0
Negro boys	100.0	100.0	100.0	100.0	100.0	100.0
Negro girls	100.0	100.0	100.0	100.0	100.0	100.0
Talus						
White boys	72.7	100.0	100.0	99.1	100.0	100.0
White girls	83.3	100.0	100.0	100.0	100.0	100.0
Negro boys	90.9	100.0	100.0	100.0	100.0	100.0
Negro girls	100.0	100.0	100.0	100.0	100.0	100.0
Distal epiphysis of femur						
White boys	9.1	75.0	85.3	100.0	100.0	100.0
White girls	50.0	91.7	98.0	100.0	100.0	100.0
Negro boys	18.2	88.5	90.7	94.0	100.0	100.0
Negro girls	50.0	93.8	99.0	100.0	100.0	100.0
Proximal epiphysis of tibia						
White boys	0.0	18.8	52.9	78.8	84.1	97.1
White girls	0.0	54.2	75.5	85.7	90.7	90.5
Negro boys	0.0	38.5	62.7	76.0	80.0	92.9
Negro girls	14.3	40.6	76.7	88.1	86.4	100.0
Cuboid bone						
White boys	0.0	6.2	14.7	39.8	44.3	60.0
White girls	0.0	37.5	57.1	65.2	70.4	76.2
Negro boys	0.0	23.1	43.8	58.0	68.2	100.0
Negro girls	21.4	37.5	68.0	78.2	81.8	75.0
Head of humerus						
White boys	0.0	7.7	13.8	41.9	49.0	59.1
White girls	0.0	5.6	25.8	41.9	69.4	86.7
Negro boys	0.0	0.0	15.2	27.6	48.4	63.6
Negro girls	0.0	10.7	22.7	52.6	38.9	100.0
Capitate bone						
White boys	0.0	0.0	0.0	8.0	15.9	17.6
White girls	0.0	0.0	14.9	15.1	20.8	38.1
Negro boys	0.0	7.7	16.4	20.8	26.2	30.8
Negro girls	0.0	12.5	19.8	41.6	40.9	100.0
Hamate bone						
White boys	0.0	6.7	6.2	6.2	10.3	11.4
White girls	0.0	0.0	10.6	13.0	20.8	33.3
Negro boys	0.0	16.4	16.4	17.7	44.2	28.6
Negro girls	0.0	9.4	22.5	41.0	54.5	66.7
Third cuneiform bone						
White boys	0.0	0.0	0.0	2.7	2.3	3.0
White girls	0.0	0.0	0.0	0.0	5.6	9.5
Negro boys	0.0	3.8	8.1	15.0	14.0	14.3
Negro girls	0.0	6.2	13.6	16.8	18.2	25.0

TABLE 1133 (continued)

PRESENCE OF SELECTED CENTERS OF OSSIFICATION
IN NEWBORN INFANTS BY WEIGHT GROUPS*

Center of Ossification	Weight groups (gm)					
	Less than 2,000	2,000 to 2,499	2,500 to 2,999	3,000 to 3,499	3,500 to 3,999	4,000 or more
	Percent					
Head of femur	0.0	0.0	0.0	0.0	0.0	0.0
White boys	0.0	0.0	0.0	1.0	0.0	0.0
White girls	0.0	0.0	0.0	1.0	0.0	0.0
Negro boys	0.0	0.0	0.0	0.0	0.0	0.0
Negro girls	0.0	0.0	1.0	1.0	0.0	0.0

*These infants were born alive to mothers delivered in the ward service of the Johns Hopkins Hospital during parts of 1936, 1937 and 1938 and were examined roentgenographically within seventy-two hours of birth.

(data from Christie, 1949, American Journal of Diseases of Children 77:355-361. Copyright 1949, American Medical Association)

TABLE 1134

EARLIEST OBSERVED AGE (weeks) OF ONSET
OF EPIPHYSEAL OSSIFICATION IN SELECTED BONES
IN MICHIGAN CHILDREN

Bone	Female	Male
Humeral head	36.0	37.0
Distal femoral	31.0	31.0
Proximal tibial	34.0	34.5

Ages refer to the prenatal period

(data from Kuhns and Finnström, 1976, Radiology 119:655-660)

TABLE 1135

ABSOLUTE AND RELATIVE VARIABILITY IN POSTNATAL ONSET OF OSSIFICATION (years)
IN OHIO CHILDREN

Center of Ossification	Standard Deviation		Coefficient of variation	
	Boys	Girls	Boys	Girls
Head, humerus	0.14	0.13	0.19	0.17
Proximal epiphysis, tibia	0.03	0.02	0.04	0.02
Coracoid process, scapula	0.16	0.19	0.20	0.25
Cuboid, tarsus	0.11	0.05	0.14	0.07
Capitate, carpus	0.17	0.20	0.17	0.23
Hamate, carpus	0.20	0.20	0.19	0.22
Capitulum, humerus	0.25	0.18	0.23	0.18
Head, femur	0.14	0.14	0.13	0.13
Lateral (3rd) cuneiform, tarsus	0.38	0.50	0.32	0.51
Greater tubercle, humerus	0.52	0.23	0.33	0.19
Primary center, middle phalanx 5th toe	1.38	0.67	0.77	0.45
Distal epiphysis, radius	0.44	0.33	0.24	0.21
Epiphysis, distal phalanx 1st toe	0.35	0.32	0.18	0.21
Epiphysis, middle phalanx 4th toe	0.62	0.65	0.32	0.39
Epiphysis, proximal phalanx 3rd finger	0.35	0.30	0.16	0.19
Epiphysis, middle phalanx 3rd toe	0.96	0.56	0.45	0.32
Epiphysis, proximal phalanx 2nd finger	0.35	0.31	0.16	0.19
Epiphysis, proximal phalanx 4th finger	0.40	0.31	0.18	0.19
Epiphysis, distal phalanx 1st finger	0.49	0.33	0.22	0.22
Epiphysis, proximal phalanx 3rd toe	0.40	0.34	0.17	0.19
Epiphysis, 2nd metacarpal	0.47	0.26	0.20	0.14
Epiphysis, proximal phalanx 4th toe	0.42	0.36	0.18	0.18
Epiphysis, proximal phalanx 2nd toe	0.42	0.36	0.17	0.18
Epiphysis, 3rd metacarpal	0.51	0.32	0.20	0.17
Epiphysis, proximal phalanx 5th finger	0.45	0.36	0.17	0.18
Epiphysis, middle phalanx 3rd finger	0.58	0.43	0.21	0.21
Epiphysis, 4th metacarpal	0.63	0.35	0.23	0.17
Epiphysis, middle phalanx 2nd toe	0.79	0.44	0.28	0.23
Epiphysis, midddle phalanx 4th finger	0.56	0.45	0.20	0.23
Epiphysis, 5th metacarpal	0.64	0.37	0.22	0.18
Medial (1st) cuneiform, tarsus	0.72	0.58	0.25	0.27
Epiphysis, 1st metatarsal	0.43	0.32	0.15	0.14
Epiphysis, middle phalanx 2nd finger	0.50	0.47	0.17	0.22
Epiphysis, proximal phalanx 1st toe	0.46	0.39	0.15	0.17
Epiphysis, distal phalanx 3rd finger	0.60	0.49	0.19	0.22
Triquetral, carpus	1.24	0.86	0.39	0.35
Epiphysis, distal phalanx 4th finger	0.59	0.52	0.18	0.23
Epiphysis, proximal phalanx 5th toe	0.53	0.42	0.17	0.17
Epiphysis, 1st metacarpal	0.72	0.44	0.21	0.19
Intermediate (2nd) cuneiform, tarsus	0.75	0.55	0.22	0.21
Epiphysis, 2nd metatarsal	0.60	0.55	0.17	0.19
Greater trochanter, femur	0.61	0.52	0.16	0.20
Epiphysis, proximal phalanx 1st finger	0.68	0.48	0.18	0.19
Navicular, tarsus	1.07	0.70	0.28	0.26
Epiphysis, distal phalanx 2nd finger	0.79	0.56	0.20	0.17
Epiphysis, distal phalanx 5th finger	0.73	0.61	0.18	0.23
Epiphysis, middle phalanx 5th finger	0.97	0.67	0.23	0.24
Proximal epiphysis, fibula	0.85	0.65	0.20	0.19

TABLE 1135 (continued)

ABSOLUTE AND RELATIVE VARIABILITY IN POSTNATAL ONSET OF OSSIFICATION (years)
IN OHIO CHILDREN

Center of Ossification	Standard Deviation		Coefficient of variation	
	Boys	Girls	Boys	Girls
Epiphysis, 3rd metatarsal	0.67	0.57	0.16	0.17
Epiphysis, distal phalanx 5th toe	0.99	0.73	0.21	0.24
Patella, knee	0.85	0.63	0.18	0.20
Epiphysis, 4th metatarsal	0.71	0.57	0.15	0.16
Lunate, carpus	1.31	1.14	0.27	0.34
Epiphysis, distal phalanx 3rd toe	0.80	0.68	0.16	0.20
Epiphysis, 5th metatarsal	0.80	0.71	0.16	0.18
Epiphysis, distal phalanx 4th toe	0.86	0.68	0.17	0.20
Epiphysis, distal phalanx 2nd toe	0.88	0.75	0.16	0.20
Head (capitulum), radius	1.24	1.01	0.21	0.22
Scaphoid, carpus	1.05	0.91	0.17	0.19
Trapezium, carpus	1.36	1.10	0.21	0.23
Trapezoid, carpus	1.34	0.91	0.19	0.18
Medial epicondyle, humerus	1.03	0.76	0.15	0.18
Distal epiphysis, ulna	0.95	1.08	0.12	0.18
Epiphysis, calcaneus	1.09	0.94	0.13	0.15
Olecranon, ulna	1.03	1.08	0.10	0.12
Lateral epicondyle, humerus	1.12	1.03	0.09	0.10
Tubercle, tibia	0.87	0.98	0.07	0.09
Adductor sesamoid, 1st finger	0.90	1.00	0.07	0.09
Acetabulum, hip	0.85	0.95	0.06	0.08
Acromial extremity, clavicle	0.83	0.87	0.06	0.07
Epiphysis, iliac crest, hip	0.97	1.12	0.07	0.08
Accessory epiphysis, coracoid process, scapula	0.89	1.00	0.06	0.08
Ischial tuberosity, hip	0.88	1.07	0.05	0.07

(data from Garn and McCreery, 1970)

TABLE 1136

PERCENTAGE FREQUENCY OF SEQUENCE POLYMORPHISM
IN HAND-WRIST OSSIFICATION
IN OHIO CHILDREN

Predominant sequence	Frequency
Lunate - Navicular	95
Dist. III - Prox. I	84
Mid. II - Dist. IV	74
Prox. V - Met. IV	73
Prox. I - Dist. II	72
Mid. V - Lunate	68
Dist. III - Met. I	68
Dist. Radius - Prox. III	67
Prox. V - Mid. III	67
Prox. IV - Dist. I	60
Mid. II - Met. V	60
Dist. II - Dist. V	58
Dist. V - Mid. V	58
Mid. IV - Mid. III	57
Dist. IV - Triquet.	55
Trapezoid - Trapezium	52
Trapezoid - Navicular	51
Triquet. - Met. I	51

(data from Garn, Rohmann, and Blumenthal, 1966)

TABLE 1137

PERCENTAGE FREQUENCY OF DIFFERENT SEQUENCES FOR SIX CARPAL CENTERS
EXCLUDING THE SCAPHOID
IN OHIO CHILDREN

Order of Appearance						Percentage Frequency	
Capitate	Hamate	Triquetral	Lunate	Trapezium	Trapezoid	Boys %	Girls %
1	2	3	4	5	6	49	61
1	2	3	4	6	5	27	23
1	2	3	5	4	6	4	1
1	2	3	6	4	5	3	4
1	2	3	5	6	4	1	1
1	2	3	6	5	4	1	0
1	2	4	3	5	6	4	3
1	2	4	3	6	5	4	1
1	2	4	5	3	6	0	1
1	2	4	6	3	5	1	0
1	2	6	3	4	5	1	0
2	1	3	4	5	6	1	3
2	1	3	4	6	5	1	0
2	1	4	3	5	6	1	0
2	1	4	3	6	5	0	3

(data from Garn, Rohmann, and Blumenthal, 1966)

TABLE 1138

PERCENTAGE FREQUENCY OF DIFFERENT SEQUENCES FOR FIVE TARSAL CENTERS
IN OHIO CHILDREN

| Order of Appearance | | | | | Percentages | |
Cuboid	Lateral cuneiform	Medial cuneiform	Middle cuneiform	Navicular	Boys %	Girls %
1	2	3	4	5	66	75
1	2	3	5	4	15	12
1	2	4	3	5	9	3
1	2	4	5	3	6	5
1	2	5	3	4	3	5
1	2	5	4	3	1	0

SEXUAL DIMORPHISM AND SEQUENCE POLYMORPHISM
IN ELBOW OSSIFICATION ORDER
IN OHIO CHILDREN

| Elbow sequence | Percent | |
	Boys %	Girls %
Capitulum radius- medial epicondyle	75	20
Indeterminate	11	11
Medial epicondyle- capitulum radius	14	69

PERCENTAGE FREQUENCY OF SEQUENCE POLYMORPHISM
IN SHOULDER OSSIFICATION ORDER
IN OHIO CHILDREN

Shoulder sequence	Boys %	Girls %
Acromial process - accessory centers of coracoid process	65	45
Indeterminate	35	45
Accessory centers of coracoid process - acromial process	0	10

PERCENTAGE FREQUENCY OF SEQUENCE POLYMORPHISM
AND SEXUAL DIMORPHISM IN KNEE OSSIFICATION ORDER
IN OHIO CHILDREN

Knee sequence	Boys %	Girls %
Prox. fibula-patella	56	27
Indeterminate	14	23
Patella-prox. fibula	30	50

(data from Garn, Rohmann, and Blumenthal, 1966)

TABLE   1139

MAJOR SEX DIFFERENCES IN DICHOTOMOUS OSSIFICATION SEQUENCES
FOR CHILDREN IN THE TEN-STATE NUTRITION SURVEY

Sequence and centers		Percent with Sequence		Sex difference in percent*
		Boys	Girls	
Dist IV	Met V	49.2	14.3	34.9
Dist IV	Met I	89.0	60.7	28.2
Dist V	Met I	32.5	8.0	24.5
Dist IV	Met IV	27.5	4.3	23.2
Dist II	Met I	33.1	10.7	22.4
Prox I	Met I	47.2	25.6	21.6
Triq	Mid V	49.7	29.6	20.1
Triq	Met I	31.4	12.8	18.6
Mid V	Met I	26.7	9.4	17.3
Dist III	Met I	90.3	74.1	16.2
Dist IV	Prox I	94.6	79.1	15.6
Triq	Dist II	42.7	29.9	12.8
Triq	Dist V	42.3	29.7	12.6
Triq	Prox I	34.1	22.1	12.0
Lunate	Mid V	16.6	5.2	11.4
Scaphoid	Trpzm	55.6	44.6	11.0
Dist Ulna	Trpzm	13.1	4.3	8.8
Lunate	Dist V	11.8	3.8	8.0
Dist Ulna	Trpzd	9.5	1.7	7.7
Lunate	Dist II	10.9	3.4	7.5
Dist Ulna	Scaphoid	8.5	2.3	6.1
Lunate	Prox I	9.2	3.4	5.8
Prox V	Prox I	100.0	94.4	5.6
Lunate	Met I	7.5	2.0	5.5
Dist I	Dist II	99.6	95.4	4.2
Lunate	Trpzd	98.3	95.4	2.9
Lunate	Trpzm	98.7	96.3	2.4

*Number of individuals showing pairs of sequences involving the centers in
question, i.e., distal IV/metacarpal V; metacarpal V/distal IV.

(data from Garn et al., 1975)

TABLE 1140

SEASONAL DIFFERENCES IN APPEARANCE OF OSSIFICATION CENTERS IN WHITE OHIO CHILDREN

	BOYS		GIRLS	
Period	Gain	Per Cent of Yearly Gain	Gain	Per Cent of Yearly Gain
July to Jan.	4.8	49	4.0	48
Jan to July	4.9	51	4.3	52
Aug. to Feb.	4.2	40	3.8	43
Feb. to Aug.	6.3	60	5.0	57
Sept. to March	5.7	53	4.0	45
March to Sept.	5.0	47	4.8	55
Oct. to April	5.1	50	4.4	47
April to Oct.	5.0	50	5.0	53
Nov. to May	4.7	52	5.2	53
May to Nov.	4.3	48	4.6	47
Dec. to June	4.9	51	5.0	57
June to Dec.	4.8	49	3.8	43

(data from Reynolds and Sontag, 1944, The Journal of Pediatrics 24:524-535)

TABLE 1141

INCOME AND OSSIFICATION TIMING OF WRIST AREA CENTERS IN U.S.
CHILDREN INCLUDED IN THE TEN-STATE NUTRITION SURVEY

| | BOYS Income-needs group | | | | GIRLS Income-needs group | | | |
| | Lower[1] | | Higher[2] | | Lower[1] | | Higher[2] | |
Center	N	M	N	M	N	M	N	M
Triquetral	467	2.81	159	2.24	270	1.71	83	1.33
Lunate	458	4.16	248	3.78	352	3.20	80	2.85
Scaphoid	448	6.12	242	5.98	337	4.46	163	4.30
Trapezium	573	6.36	242	6.14	411	4.39	135	4.20
Trapezoid	403	6.19	242	6.09	399	4.55	127	4.40
Distal ulna	500	7.33	231	7.07	371	6.02	143	5.61

[1] = Income-needs group through 1.49, mean per-capita $724.00

[2] = Income-needs group 2.25 and above, mean per-capita income $2920.00.

N = number; M = mean age of onset (years)

(data from Garn et al., 1973)

TABLE 1142

THE NUMBER OF CENTERS OF OSSIFICATION THAT APPEARED IN THE LEFT HAND
IN SIXTY AUSTRALIAN BOYS BETWEEN THE AGES OF TWO AND FOUR YEARS

Bone	Number of centers in each boy						Total with multiple centers	Multiple centers showing some fusion	Total with no center at first: later	
	0	1	2	3	4	5			Single	Multiple
M. 1	3	44	11	1	1	--	13	9	17	9
M. 2	--	54	5	1	--	--	6	6	2	1
M. 3	--	58	2	--	--	--	2	2	9	1
M. 4	--	54	4	2	--	--	6	6	18	3
M. 5	--	55	5	--	--	--	5	5	23	3
P.P.1	5	9	20	17	5	4	46	29	3	28
P.P.2	--	58	2	--	--	--	2	1	1	1
P.P.3	--	57	3	--	--	--	3	3	3	2
P.P.4	--	59	1	--	--	--	1	1	1	--
P.P.5	1	48	9	2	--	--	11	10	10	7
M.P.2	2	56	2	--	--	--	2	2	2	--
M.P.3	1	54	3	2	--	--	5	5	5	2
M.P.4	2	55	3	--	--	--	3	3	3	1
M.P.5	21	35	4	--	--	--	4	3	3	4
D.P.1	--	60	--	--	--	--	--	--	--	--
D.P.2	10	48	2	--	--	--	2	1	1	2
D.P.3	3	56	1	--	--	--	1	1	1	1
D.P.4	4	54	2	--	--	--	2	2	2	2
D.P.5	11	47	2	--	--	--	2	1	1	2

M = metacarpal; P.P. = proximal phalanx; M.P. = middle phalanx; D.P. = distal
phalanx. The thumb, index, middle, ring and little fingers are designated by
the numbers 1, 2, 3, 4 and 5 respectively.

(data from Roche and Sunderland, 1959)

TABLE 1143

THE NUMBER OF CENTERS OF OSSIFICATION THAT APPEARED IN THE LEFT HAND
IN SIXTY AUSTRALIAN GIRLS BETWEEN THE AGES OF TWO AND FOUR YEARS

Bone	Number of centers in each girl				Total with multiple centers	Mutltiple centers showing some fusion	Total with no center at first; later	
	0	1	2	3			Single	Multiple
M. 1	--	59	--	1	1	1	4	1
M. 2	--	60	--	--	--	--	1	--
M. 3	--	60	--	--	--	--	1	--
M. 4	--	59	1	--	1	1	3	--
M. 5	--	60	--	--	--	--	6	--
P.P.1	2	43	11	4	15	12	6	5
P.P.2	--	60	--	--	--	--	1	--
P.P.3	--	60	--	--	--	--	--	--
P.P.4	--	60	--	--	--	--	--	--
P.P.5	--	58	2	--	2	2	1	--
M.P.2	1	58	1	--	1	1	3	1
M.P.3	1	58	1	--	1	1	2	1
M.P.4	1	58	1	--	1	1	4	1
M.P.5	4	56	--	--	--	--	14	--
D.P.1	--	60	--	--	--	--	4	--
D.P.2	1	58	1	--	1	1	21	--
D.P.3	1	58	1	--	1	1	3	1
D.P.4	1	59	--	--	--	--	7	--
D.P.5	1	59	--	--	--	--	27	--

M = metacarpal; P.P. = proximal phalanx; M.P. = middle phalanx; D.P. = distal
phalanx. The thumb, index, middle, ring and little fingers are designated by
the numbers 1, 2, 3, 4 and 5 respectively.

(data from Roche and Sunderland, 1959)

TABLE 1144

THE NUMBER OF CENTERS OF OSSIFICATION THAT APPEARED IN THE LEFT FOOT IN SIXTY AUSTRALIAN BOYS BETWEEN THE AGES OF TWO AND FOUR YEARS

Bone	Number of centers in each boy							Total with multiple centers	Multiple centers showing some fusion	Total with no center at first; later	
	0	1	2	3	4	5	6-8			Single	Multiple
M. 1	--	3	13	18	10	14	2	57	32	2	23
M. 2	7	30	19	4	--	--	--	23	11	26	21
M. 3	16	29	12	3	--	--	--	15	4	27	14
M. 4	35	15	9	1	--	--	--	10	3	15	10
M. 5	44	11	4	1	--	--	--	5	--	10	5
P.P.1	--	6	3	10	13	15	13†	54	47	1	20
P.P.2	--	43	12	5	--	--	--	17	17	1	4
P.P.3	--	46	13	1	--	--	--	14	14	2	4
P.P.4	--	49	11	--	--	--	--	11	9	6	3
P.P.5	4	25	24	6	1	--	--	31	14	10	20
M.P.2	9	49	2	--	--	--	--	2	2	15	1
M.P.3	15	45	--	--	--	--	--	--	--	11	--
M.P.4	23	37	--	--	--	--	--	--	--	10	--
M.P.5	57	3	--	--	--	--	--	--	--	--	--
D.P.1	--	51	7	2	--	--	--	9	8	1	2
D.P.2	52	8	--	--	--	--	--	--	--	7	--
D.P.3	47	13	--	--	--	--	--	--	--	10	--
D.P.4	48	12	--	--	--	--	--	--	--	10	--
D.P.5	42	18	--	--	--	--	--	--	--	17	--

† Two boys had seven centres and one had eight.

M = metatarsal; P.P. = proximal phalanx; M.P. = middle phalanx; D.P. = distal phalanx. The thumb, index, middle, ring and little fingers are designated by the numbers 1, 2, 3, 4 and 5 respectively.

(data from Roche and Sunderland, 1959)

TABLE 1145

THE NUMBER OF CENTERS OF OSSIFICATION THAT APPEARED IN THE LEFT FOOT IN
SIXTY AUSTRALIAN GIRLS BETWEEN THE AGES OF TWO AND FOUR YEARS

Bone	Number of centers in each girl					Total with multiple centers	Multiple centers showing some fusion	Total with no center at first; later	
	0	1	2	3	4			Single	Multiple
M. 1	--	17	33	8	2	43	21	--	20
M. 2	1	46	13	--	--	13	8	11	21
M. 3	1	30	29	--	--	29	9	13	14
M. 4	9	29	16	5	1	22	9	20	10
M. 5	14	32	12	2	--	14	5	29	5
P.P.1	--	33	9	9	9	27	24	--	20
P.P.2	--	60	--	--	--	--	--	--	4
P.P.3	--	60	--	--	--	--	--	--	4
P.P.4	--	60	--	--	--	--	--	--	3
P.P.5	--	55	5	--	--	5	5	8	1
M.P.2	2	58	--	--	--	--	--	7	1
M.P.3	10	50	--	--	--	--	--	6	--
M.P.4	28	32	--	--	--	--	--	7	--
M.P.5	59	1	--	--	--	--	--	--	--
D.P.1	--	60	--	--	--	--	--	--	--
D.P.2	21	39	--	--	--	--	--	30	--
D.P.3	12	48	--	--	--	--	--	39	--
D.P.4	8	52	--	--	--	--	--	38	--
D.P.5	28	32	--	--	--	--	--	26	--

M = metatarsal; P.P. = proximal phalanx; M.P. = middle phalanx; D.P. = distal
phalanx. The thumb, index, middle, ring and little fingers are designated by
the numbers 1, 2, 3, 4 and 5 respectively.

(data from Roche and Sunderland, 1959)

TABLE 1146

PERCENTAGE FREQUENCY OF MISSING FOOT CENTERS IN OHIO CHILDREN

Ossification Center	Boys %	Girls %
Middle 2	0.9	1.0
Middle 3	16.0	24.0
Middle 4	54.0	70.0
Middle 5	98.0	99.0
Distal 3	0.0	1.0
Distal 4	1.0	1.0
Distal 5	35.0	31.0

(data from Garn, Rohmann and Silverman, "Missing secondary ossification centers of the foot. Inheritance and developmental meaning," Annales de Radiologie 8:629-644, 1965)

TABLE 1147

SEPARATE CENTERS OF OSSIFICATION OF THE TIP OF THE MEDIAL MALLEOLUS IN OHIO CHILDREN

Age of appearance of centers

Age (years)	6.0	6.5	7.0	7.5	8.0	8.5	9.0	9.5	10.0	10.5	11.0	Combined Total
Boys	1	--	1	1	1	2	5	1	2	--	1	15
Girls	--	6	1	11	3	6	3	--	--	--	--	30

Age at which the center joined tibia

Age (years)	6.0	7.5	8.0	8.5	9.0	9.5	10.0	10.5	11.0	11.5	12.0	Combined Total
Boys		1	--	1	1	--	4	--	4	--	1	12
Girls		3	3	10	6	2	4	--	--	1	--	29

(data from Selby, 1961, American Journal of Roentgenology 86:496-501. Copyright American Roentgen Ray Society, 1961)

TABLE 1148

SKELETAL AGES (months) IN CLEVELAND CHILDREN*

	BOYS		GIRLS	
Age	Mean	S.D.	Mean	S.D.
3 mos.	3.01	0.69	3.02	0.72
6 "	6.09	1.13	6.04	1.16
9 "	9.56	1.43	9.05	1.36
12 "	12.74	1.97	12.04	1.77
18 "	19.36	3.52	18.22	3.49
2 yrs.	25.97	3.92	24.16	4.64
2½ "	32.40	4.52	30.96	5.37
3 "	38.21	5.08	36.63	5.97
3½ "	43.89	5.40	43.50	7.48
4 "	49.04	6.66	50.14	8.98
4½ "	56.00	8.36	60.06	10.73
5 "	62.43	8.79	66.21	11.65
6 "	75.46	9.17	78.50	10.23
7 "	88.20	8.91	89.30	9.64
8 "	101.38	9.10	100.66	10.23
9 "	113.90	9.00	113.86	10.74
10 "	125.68	9.79	125.66	11.73
11 "	137.32	10.09	137.87	11.94
12 "	148.82	10.38	149.62	10.24
13 "	158.39	10.44	162.28	10.67
14 "	170.02	10.72	174.25	11.30
15 "	182.72	11.32	183.62	9.23
16 "	195.32	12.86	189.44	7.31
17 "	206.21	13.05	--	--

* assessed using Todd (1937) standards.
(data from Simmons, 1944)

TABLE 1149

SELECTED PERCENTILES FOR SKELETAL AGE (HAND-WRIST; GREULICH-PYLE)
BY CHRONOLOGICAL AGE IN YEARS AT LAST BIRTHDAY
IN U.S. CHILDREN

Percentiles	Chronological age (years)					
	6	7	8	9	10	11
Boys						
P_{95}	94.7	108.2	113.4	122.1	131.0	150.8
P_{75}	83.8	95.0	107.1	113.9	120.1	131.6
P_{50}	75.6	87.0	98.3	107.6	113.6	121.6
P_{25}	68.5	79.1	89.5	98.8	108.3	115.5
P_5	57.2	69.6	75.5	87.4	96.1	108.9
	12	13	14	15	16	17
P_{95}	168.3	183.3	199.6	207.9	216.2	223.1
P_{75}	154.4	170.6	182.1	198.0	206.2	212.8
P_{50}	139.4	158.9	174.0	185.7	198.9	206.8
P_{25}	127.2	145.2	165.0	176.6	188.7	199.3
P_5	115.5	123.5	147.5	163.5	172.0	186.3
	6	7	8	9	10	11
Girls						
P_{95}	96.9	102.6	112.9	126.4	133.6	145.5
P_{75}	86.1	96.0	103.0	112.3	126.2	134.9
P_{50}	76.7	90.1	97.4	105.1	115.9	129.9
P_{25}	67.5	79.7	92.3	99.3	108.6	123.6
P_5	57.1	67.0	78.3	88.3	97.9	110.6
	12	13	14	15	16	17
P_{95}	167.8	179.4	188.8	201.0	201.7	201.8
P_{75}	152.5	166.8	177.1	187.0	197.3	197.7
P_{50}	143.2	156.3	170.4	178.1	186.2	186.9
P_{25}	134.7	145.3	162.4	170.4	178.4	176.2
P_5	124.0	134.1	144.7	160.5	164.8	163.4

(data from Roche et al., 1974, 1976)

TABLE 1150

SKELETAL AGE (HAND-WRIST; GREULICH-PYLE)
IN U.S. CHILDREN

Chronological age at last birthday	Boys		Girls	
	Mean	S.D.	Mean	S.D.
Age in years	Skeletal age hand-wrist (months)			
12 years...............	140.2	17.02	142.9	12.12
13 years...............	157.4	18.05	155.2	10.79
14 years...............	173.6	15.12	168.0	9.18
15 years...............	186.5	14.29	177.6	7.91
16 years...............	196.4	13.81	185.6	8.74
17 years...............	205.4	11.08	186.0	8.75
Age in 6-month intervals				
12 years:				
0-5 months..........	134.4	16.08	139.6	11.40
0-11 months.........	145.0	16.26	145.9	12.24
13 years:				
0-5 months..........	153.1	18.15	151.2	10.45
6-11 months.........	162.3	16.60	159.4	10.26
14 years:				
0-5 months..........	171.7	14.91	165.6	9.41
6-11 months.........	175.5	15.09	171.6	8.42
15 years:				
0-5 months..........	183.7	14.75	176.2	7.77
6-11 months.........	189.5	13.12	178.5	7.95
16 years:				
0-5 months..........	193.7	13.78	184.3	8.29
6-11 months.........	199.5	13.17	187.8	9.16
17 years:				
0-5 months..........	204.7	11.09	184.9	7.95
6-11 months.........	206.2	11.00	187.4	9.57

(data from Roche et al., 1976)

TABLE 1151

GREULICH-PYLE SKELETAL AGES (months) FOR CHILDREN IN
THE HARVARD SCHOOL OF PUBLIC HEALTH GROWTH STUDY

Chronological Age (months)	Boys		Girls	
	Mean	S.D.	Mean	S.D.
3	3.4	1.7	2.9	1.1
6	6.5	2.0	5.9	1.7
9	9.7	2.4	9.5	2.1
12	12.7	2.1	12.7	2.7
18	17.5	2.7	18.4	3.4
24	22.6	4.0	23.7	4.0
30	28.1	5.4	29.0	4.8
36	33.8	6.0	34.5	5.6
42	39.5	6.6	40.6	6.5
48	44.8	7.0	46.4	7.2
54	50.3	7.8	52.3	8.0
60	56.2	8.4	58.1	8.6
66	62.4	9.1	63.9	8.9
72	68.4	9.3	70.4	9.0
84	80.6	10.1	82.0	8.3
96	92.5	10.8	94.0	8.8
108	104.9	11.0	105.9	9.3
120	118.0	11.4	119.0	10.8
132	132.1	10.5	132.9	12.3
144	144.5	10.4	147.2	14.0
156	156.4	11.1	160.3	14.6
168	168.5	12.0	172.4	12.6
180	180.7	14.2	184.3	11.2
192	193.0	15.1	196.7	10.6
204	206.0	15.4	205.1	8.2
216	212.8	12.2	211.5	5.6

(data from Pyle, Reed and Stuart, 1959, Pediatrics 24:886-903.
Copyright American Academy of Pediatrics 1962)

TABLE 1152

SKELETAL AGE (years) IN OHIO CHILDREN

Age	Boys			Girls		
(years)	Median	Mean	S.D.	Median	Mean	S.D.
1.50	--	--	--	1.50	1.57	0.23
2.00	--	--	--	1.99	1.93	0.32
2.50	2.50	2.53	0.25	2.49	2.34	0.34
3.00	3.00	2.83	0.41	2.98	2.81	0.39
3.50	3.26	3.23	0.45	3.38	3.28	0.51
4.00	3.75	3.72	0.50	3.99	3.77	0.57
4.50	4.25	4.19	0.57	4.50	4.28	0.58
5.00	4.76	4.76	0.57	4.99	4.80	0.58
5.50	5.49	5.23	0.68	5.50	5.35	0.62
6.00	5.99	5.77	0.64	6.00	5.90	0.61
6.50	6.49	6.29	0.64	6.49	6.37	0.66
7.00	6.99	6.78	0.62	7.00	6.94	0.58
7.50	7.50	7.34	0.55	7.50	7.46	0.65
8.00	7.99	7.82	0.55	8.00	8.01	0.61
8.50	8.50	8.38	0.51	8.50	8.48	0.64
9.00	9.00	8.83	0.59	9.00	9.02	0.63
9.50	9.50	9.31	0.58	9.50	9.54	0.67
10.00	10.00	9.88	0.59	10.00	10.08	0.67
10.50	10.49	10.32	0.34	--	--	--
11.00	11.00	10.90	0.56	11.00	11.02	0.67
12.00	12.01	11.92	0.47	12.00	12.06	0.59
13.00	12.99	12.92	0.48	13.01	13.08	0.52
14.00	14.00	13.99	0.50	13.99	13.77	0.47
15.00	15.00	14.97	0.49	14.50	14.44	0.63
16.00	15.99	15.87	0.51	--	--	--

Skeletal age by bone-specific method of Greulich-Pyle (1959) for hand-wrist.

(data from Roche, unpublished a)

TABLE 1153

SEX-ASSOCIATED DIFFERENCES IN SKELETAL AGE
(months) IN AUSTRALIAN CHILDREN WHEN MALE
STANDARDS FOR AREAS ARE ASSESSED AS FEMALE

Male skeletal age range (months)	Hand-wrist		Knee	Foot-ankle
	Greulich-Pyle [1959]	Tanner et al. [1962]	Pyle and Hoerr [1955]	Hoerr et al. [1962]
0- 3	-	-	0.5	0.4
4- 6	2	-	1	1
7- 9	3	-	-	1
10- 12	-	-	2	2
13- 18	4	4	3	4
19- 21	5	5	-	-
22- 24	6	7	4	6
25- 27	-	8	-	-
28- 30	6	8	-	7
31- 33	-	7	-	-
34- 36	-	10	12	8
37- 39	-	10	-	-
40- 42	10	11	15	-
43- 48	12	13	16	9
49- 54	-	14	16	12
55- 60	15	11	18	15
61- 66	15	12	-	15
67- 72	-	12	16	17
73- 78	16	13	-	18
79- 84	-	13	20	18
85- 96	17	14	20	22
97-108	-	14	24	24
109-120	17	13	28	28
121-132	18	16	34	30
133-144	19	19	36	34
145-156	23	23	36	36
157-168	24	24	36	36
169-180	24	23	32	36
181-192	23	20	-	34
193-204	24	23	30	-

(data from Roche, 1968)

TABLE 1154

SKELETAL AGE (years) OF OREGON BOYS
CLASSIFIED INTO PUBESCENT DEVELOPMENT GROUPS

Age (years)	Pubescent Development Group	Low	High	S.D.
10	1	6.90	12.75	8.79
10	2	8.33	11.66	1.259
13	2	11.08	13.25	.518
13	3	12.25	14.44	.610
13	4+5	13.75	14.86	.412
16	4	14.08	16.90	.633
16	5	14.17	19.00	1.156

Skeletal age was assessed by the Clarke and Hayman (1962) modification of the Greulich-Pyle method (1959)

(data from Clarke and Degutis, 1962, Research Quarterly 33:356-368, by permission of the American Alliance for Health, Physical Education, Recreation and Dance, 1900 Association Drive, Reston VA 22091)

TABLE 1155

DIFFERENCES BETWEEN MEAN SKELETAL AGES (years) OF
OREGON BOYS CLASSIFIED INTO PUBESCENT DEVELOPMENT GROUPS

Ages (years)	Pubescent Development Groups					
	1	2	3	4	4+5	5
Contrasts within chronological ages						
10	9.926	10.138	--	--	--	--
13	--	12.423	13.280	--	--	--
13	--	12.423	--	--	14.345	--
13	--	--	13.280	--	14.345	--
16	--	--	--	14.974	--	16.583
Contrasts between chronological ages						
10	--	10.138	--	--	--	--
13	--	12.423	--	--	--	--
13	--	--	--	--	14.345	--
16	--	--	--	14.974	--	--

Skeletal age was assessed by the Clarke and Hayman (1962) modification
of the Greulich-Pyle method (1959)

(data from Clarke and Degutis, 1962, Research Quarterly 33:356-368,
by permission of the American Alliance for Health, Physical Education,
Recreation and Dance, 1900 Association Drive, Reston VA 22091)

TABLE 1156

RELATIVE SKELETAL AGE
(skeletal age less chronological age in years)
(Greulich-Pyle) IN COLORADO CHILDREN

Age (years)	BOYS		GIRLS	
	Mean	S.D.	Mean	S.D.
0.50	--	--	0.22	0.29
1.00	-0.07	0.37	0.00	0.36
1.50	-0.09	0.38	0.06	0.36
2.00	-0.20	0.46	0.02	0.40
2.50	-0.31	0.57	-0.08	0.40
3.00	-0.31	0.55	-0.17	0.41
3.50	-0.29	0.72	-0.27	0.55
4.00	-0.30	0.60	-0.42	0.65
4.50	-0.29	0.63	-0.42	0.69
5.00	-0.41	0.65	-0.47	0.82
5.50	-0.46	0.77	-0.58	0.81
6.00	-0.50	0.80	-0.53	0.86
6.50	-0.49	0.94	-0.47	0.87
7.00	-0.51	0.86	-0.56	0.82
7.50	-0.63	0.89	-0.68	0.83
8.00	-0.75	0.88	-0.75	0.80
8.50	-0.64	0.93	-0.77	0.90
9.00	-0.57	0.92	-0.80	0.99
9.50	-0.53	1.07	-0.84	1.11
10.00	-0.40	1.01	-0.81	1.22
10.50	-0.40	1.08	-0.70	1.31
11.00	-0.17	1.04	-0.66	1.30
11.50	-0.19	1.04	-0.51	1.27
12.00	-0.27	1.04	-0.28	1.18
12.50	-0.24	0.88	-0.17	1.19
13.00	-0.29	0.97	-0.04	1.11
13.50	-0.37	0.98	-0.14	1.01
14.00	-0.35	0.90	-0.03	1.12
14.50	-0.45	0.93	-0.51	0.75
15.00	-0.40	1.09	0.13	1.09
15.50	-0.50	1.11	-0.01	0.51
16.00	-0.36	1.22	-0.10	1.15
16.50	-0.95	1.17	-0.98	0.09
17.00	-0.53	1.02	-0.29	0.76
17.50	-1.13	1.01	-0.03	1.36
18.00	-0.79	0.9	-0.73	0.61

(data from McCammon, unpublished)

TABLE   1157

DEVIATIONS OF SKELETAL (Greulich-Pyle) FROM
CHRONOLOGICAL AGE IN PHILADELPHIA WHITE CHILDREN

Sex	Age (years)	SA-CA	S.D.
BOYS	7	-0.06	0.72
	8	+0.15	0.82
	9	+0.30	1.06
	10	+0.58	0.94
	11	+0.65	0.79
	12	+0.59	0.74
	13	+0.45	0.91
	14	+0.32	0.98
	15	+0.41	1.05
	16	+0.37	0.89
	17	+0.42	1.11
GIRLS	7	-0.15	0.73
	8	-0.21	0.84
	9	-0.29	1.00
	10	-0.13	1.12
	11	+0.07	1.13
	12	+0.42	1.07
	13	+0.53	1.19
	14	+0.59	0.92
	15	+0.62	1.21
	16	+0.56	0.95
	17	+0.58	0.55

(data from Johnston, 1962)

TABLE   1158

MEAN SKELETAL AGE (hand-wrist Greulich-Pyle; months) IN U.S. CHILDREN

Chronological age at last birthday	White		Negro		Other	
	Mean	S.D.	Mean	S.D.	Mean	S.D.
Boys						
6 years	74.9	11.49	79.0	10.78	91.2	46.68
7 years	86.7	11.57	88.1	12.94	85.7	43.70
8 years	96.3	11.94	100.8	10.11	95.1	48.65
9 years	105.6	11.39	106.3	10.58	111.6	4.73
10 years	113.6	10.23	112.2	10.27	109.9	55.17
11 years	123.8	13.06	125.8	13.66	133.0	67.08
12 years	140.4	17.17	138.7	16.04	149.3	105.58
13 years	156.8	17.77	160.6	19.69	164.5	52.37
14 years	173.7	14.84	172.7	17.04	170.7	54.06
15 years	187.1	13.91	181.7	15.91	197.3	5.02
16 years	197.0	13.12	192.7	16.93	188.9	74.35
17 years	205.8	10.97	202.4	11.37	209.1	1.05
Girls						
6 years	76.7	10.69	78.9	11.54	76.3	32.18
7 years	87.9	10.56	88.9	10.24	86.2	27.45
8 years	96.3	10.43	96.3	12.35	97.6	49.95
9 years	107.1	12.57	109.2	15.99	120.9	41.88
10 years	116.5	14.69	119.2	14.87	123.9	63.07
11 years	128.1	12.04	128.8	14.12	131.1	50.78
12 years	142.3	11.72	146.4	13.72	152.5	7.79
13 years	154.8	10.89	156.5	10.10	156.2	110.50
14 years	167.9	8.87	169.1	10.81	168.3	119.00
15 years	177.3	8.01	179.9	7.24	175.9	55.63
16 years	185.8	8.78	184.2	8.20	178.6	126.26
17 years	186.0	8.63	184.5	9.06	197.0	139.30

(data from Roche, Roberts, and Hamill, 1975, 1978)

TABLE 1159

SELECTED PERCENTILES IN THE DISTRIBUTION OF THE INDIVIDUAL
RANGE IN BONE-SPECIFIC SKELETAL AGES (Greulich-Pyle)
RADIOPAQUE (not adult) BONES IN THE HAND-WRIST FOR WHITE
AND NEGRO U.S. YOUTHS

Chronological age (years)	White			Negro		
	75	50	25	75	50	25

Bone-specific skeletal age range (months)
male standard

Boys

12 years..........	20.4	15.9	12.3	20.5	15.0	12.1
13 years..........	18.8	14.6	10.9	18.1	13.3	10.6
14 years..........	16.8	12.3	8.9	18.2	13.8	8.9
15 years..........	16.0	11.0	6.7	13.4	10.0	6.5
16 years..........	14.9	8.7	3.7	15.3	9.2	4.2
17 years..........	11.5	4.9	0.9	16.2	8.4	2.1

Girls

12 years..........	19.0	14.4	11.0	17.9	11.6	7.8
13 years..........	18.9	12.8	8.7	18.7	12.5	9.1
14 years..........	18.5	10.8	5.0	18.5	10.0	4.0
15 years..........	17.0	6.8	1.6	12.9	4.6	0.8
16 years..........	10.1	2.1	0.5	13.9	3.2	0.7
17 years..........	14.5	1.7	0.5	13.8	2.7	0.6

(data from Roche et al., 1978)

TABLE   1160

DIFFERENCES (months) BETWEEN MEAN SKELETAL AGES FOR PARTICULAR BONES IN GROUPS
OF BLACK CHILDREN AND OF ORIENTAL CHILDREN CONTRASTED WITH MEANS OF
CORRESPONDING SKELETAL AGES IN WHITE CHILDREN OF MEDIUM STATURE (Greulich-Pyle)
IN CALIFORNIA

Bone		BOYS				GIRLS			
		Tall White	Short White	Medium Black	Oriental	Tall White	Short White	Medium Black	Oriental
Radius		-1.4	-0.8	2.0	-1.6	0.5	-0.3	5.3	0.2
Capitate		0.9	-1.4	1.3	0.7	2.0	0.3	5.5	1.3
Hamate		0.2	-0.7	2.8	0.7	3.4	-1.6	6.6	1.1
Metacarpal	I	1.6	4.1	3.3	4.1	0.8	-3.3	4.1	-0.4
Proximal	I	1.2	3.5	0.1	3.2	0.2	0.6	-1.0	2.0
Distal	I	-2.4	2.2	-5.4	2.6	2.0	0.1	-2.1	2.0
Metacarpal	II	4.2	-2.2	2.6	-1.4	2.0	-3.8	1.2	0.4
Metacarpal	III	2.0	0.5	4.9	-0.3	1.4	-4.1	4.7	-1.1
Metacarpal	IV	1.4	-0.9	8.1	-1.5	1.5	-4.3	4.9	-2.2
Metacarpal	V	5.3	-3.1	7.7	-0.8	2.0	-4.3	5.7	-0.3
Proximal	II	-1.0	1.2	-6.0	-2.8	-1.6	-2.1	-5.6	-1.5
Proximal	III	0.1	-0.8	-6.3	-4.3	-1.2	-1.6	-5.1	-1.9
Proximal	IV	0.5	-0.9	-5.3	-3.2	0.7	-0.6	-4.0	-1.7
Proximal	V	0.3	-0.5	-3.6	-3.4	-0.3	-0.4	-3.8	-0.9
Middle	II	-0.8	1.8	-3.7	0.5	-2.9	1.1	-3.7	0.0
Middle	III	-1.2	1.4	-2.2	0.7	-1.5	3.1	-2.4	1.1
Middle	IV	-0.6	0.0	-1.7	0.4	-0.8	3.8	-2.2	1.2
Middle	V	-0.2	-0.8	-1.5	1.4	-0.3	3.4	-1.8	1.9
Distal	II	-3.7	-0.4	0.1	2.7	-2.9	4.3	-1.3	0.6
Distal	III	-2.5	0.3	0.2	1.2	-2.6	3.3	-1.6	-0.6
Distal	IV	-1.6	-0.3	1.4	0.9	-1.1	3.4	-2.0	-1.0
Distal	V	-2.2	-2.3	1.1	-0.1	-1.4	3.0	-1.3	-0.1

(data from Wingerd, Peritz, and Sproul, 1974, Annals of Human Biology 1:201-209)

TABLE 1161

DIFFERENCES (months) BETWEEN MEAN SKELETAL AGES FOR PARTICULAR BONES IN
GROUPS OF BLACK CHILDREN CONTRASTED WITH MEANS OF CORRESPONDING SKELETAL AGES IN
WHITE CHILDREN OF MEDIUM STATURE
(Greulich-Pyle) IN CALIFORNIA

Bone		BOYS		GIRLS	
		Tall Black	Short Black	Tall Black	Short Black
Radius		−3.2	−2.1	−1.8	−2.8
Capitate		0.7	1.1	−1.8	−2.7
Hamate		0.3	−1.0	−1.3	−4.7
Metacarpal	I	0.6	6.0	1.8	1.1
Proximal	I	−2.6	1.4	1.6	3.7
Distal	I	1.3	2.3	3.3	−0.8
Metacarpal	II	1.9	−2.5	5.2	−2.8
Metacarpal	III	−1.1	−0.2	−1.1	−5.6
Metacarpal	IV	−0.2	−2.6	−0.8	−6.3
Metacarpal	V	−0.1	−4.0	−0.1	−5.7
Proximal	II	−0.9	2.1	3.7	1.8
Proximal	III	3.6	2.9	3.1	1.0
Proximal	IV	2.0	1.5	2.5	−0.5
Proximal	V	0.8	2.0	2.5	−0.1
Middle	II	−1.9	−0.7	−1.2	1.8
Middle	III	−1.7	−0.9	−2.0	3.0
Middle	IV	−3.0	1.0	−2.5	2.6
Middle	V	−2.3	0.7	−1.5	1.5
Distal	II	1.7	0.5	−3.5	3.9
Distal	III	1.4	−1.1	−2.2	3.6
Distal	IV	1.9	−2.8	−1.3	4.1
Distal	V	0.4	−3.6	−3.8	3.7

(data from Wingerd, Peritz, and Sproul, 1974, Annals of Human Biology 1:201-209)

TABLE 1162

SKELETAL AGE (GREULICH-PYLE; YEARS)
OF AMERICAN-JAPANESE CHILDREN
IN CALIFORNIA

Age Group (years)	Skeletal Age		Skeletal Age less Chronological Age	
	Mean	S.E.	Mean	S.E.
BOYS				
5	4.64	.21	−0.59	.22
6	5.58	.13	−0.45	.13
7	6.26	.14	−0.72	.14
8	7.65	.24	−0.35	.23
9	8.79	.29	−0.18	.28
10	9.56	.22	−0.41	.21
11	11.34	.21	+0.32	.21
12	12.39	.23	+0.41	.22
13	13.63	.19	+0.65	.18
14	14.61	.19	+0.61	.19
15	15.54	.23	+0.58	.22
16	17.24	.19	+1.17	.20
17	17.55	.15	+0.54	.15
18	17.83	.08	+0.04	.08
GIRLS				
5	5.39	.25	+0.26	.23
6	5.85	.16	−0.13	.17
7	6.98	.15	−0.08	.16
8	8.11	.18	+0.12	.18
9	8.99	.27	+0.11	.26
10	10.61	.16	+0.64	.16
11	11.59	.21	+0.61	.20
12	12.62	.23	+0.62	.22
13	14.32	.25	+1.38	.23
14	14.98	.20	+1.00	.17
15	16.08	.16	+1.00	.15
16	17.13	.15	+1.16	.16
17	17.65	.10	+0.65	.10
18	17.78	.07	0	.08

(data from Greulich, 1957)

TABLE 1163

MEAN SKELETAL AGE (HAND-WRIST: GREULICH-PYLE) BY ANNUAL FAMILY INCOME
IN U.S. CHILDREN

Chronological age at last birthday	Less than $5,000		$5,000- $9,999		$10,000 or more		Less than $3,000		$15,000 or more	
	Mean	S.D.	Mean	S.D.	Mean	S.D.	Mean	S.D.	Mean	S.D.
Boys				Skeletal age in (months)						
6 years	75.1	11.50	76.6	11.15	73.6	12.00	73.4	1.21	74.3	1.46
7 years	85.9	12.14	87.6	11.82	87.9	10.58	85.1	2.42	87.2	1.99
8 years	96.7	12.06	96.9	11.25	96.9	12.60	96.5	0.92	97.6	4.72
9 years	104.3	11.04	106.2	10.85	107.9	12.00	102.5	1.41	107.8	3.54
10 years	112.9	10.52	114.4	10.25	112.2	9.50	112.5	1.14	113.9	1.93
11 years	124.9	14.00	123.6	13.20	124.8	12.04	124.5	1.49	125.8	2.54
12 years	139.5	16.34	139.7	17.59	141.6	16.31	138.8	2.51	141.3	2.28
13 years	156.2	19.27	159.2	17.66	156.5	17.70	154.4	3.66	155.8	2.42
14 years	174.0	15.90	174.3	15.57	172.1	14.15	175.7	2.26	173.9	1.49
15 years	185.1	15.42	186.3	14.15	187.9	13.48	183.6	2.17	189.3	1.81
16 years	195.6	13.91	195.6	14.32	198.7	12.20	193.3	1.25	199.9	2.64
17 years	203.9	11.18	205.7	9.75	206.8	11.78	204.7	1.80	205.7	2.51
Girls										
6 years	77.0	11.34	77.4	10.47	76.7	10.33	77.4	0.80	78.6	3.24
7 years	87.9	11.10	88.1	10.24	88.3	10.43	88.2	1.13	93.2	2.54
8 years	95.6	9.79	97.5	11.22	95.6	9.66	93.9	0.93	95.8	1.97
9 years	105.4	13.30	108.6	13.65	109.4	12.32	105.4	0.96	108.9	1.62
10 years	116.8	15.23	116.6	14.40	117.2	15.64	117.5	1.60	119.2	3.23
11 years	128.2	12.66	128.4	12.33	128.0	11.91	128.1	1.23	129.4	2.84
12 years	144.2	13.34	141.5	12.76	143.7	10.05	146.1	1.94	141.2	1.13
13 years	153.3	10.23	156.0	11.25	156.3	10.61	153.5	1.09	153.0	1.82
14 years	167.5	9.71	166.6	9.95	171.0	7.40	168.0	1.02	170.3	0.93
15 years	178.3	7.53	177.1	7.35	177.0	8.47	178.5	1.01	175.1	1.29
16 years	184.9	10.42	185.0	7.83	186.8	7.95	185.8	1.12	185.6	1.05
17 years	186.2	9.33	186.6	8.42	185.8	8.53	188.8	1.84	181.7	1.45

(data from Roche, Roberts, and Hamill, 1975, 1978)

TABLE   1164

MEAN SKELETAL AGES (HAND-WRIST; GREULICH-PYLE) OF WHITE
AND NEGRO BOYS AND GIRLS, BY ANNUAL FAMILY INCOME

Chronological age at last birthday	Less than $5,000				$5,000 - $9,999			
	White		Negro		White		Negro	
	Mean	S.D.	Mean	S.D.	Mean	S.D.	Mean	S.D.
Boys			Skeletal age in (months)					
6 years......	73.5	1.08	79.8	1.59	76.4	0.83	77.6	1.68
7 years......	85.4	1.62	87.4	1.82	87.3	0.74	92.0	3.81
8 years......	94.4	0.85	102.1	0.89	96.9	0.69	95.7	4.37
9 years......	103.5	1.57	106.5	1.33	106.2	0.62	105.7	1.81
10 years.....	113.0	1.08	112.7	1.11	114.7	0.54	110.0	1.50
11 years.....	124.0	0.96	127.3	1.34	123.7	0.92	122.0	2.09
Girls								
6 years......	76.3	0.93	78.6	1.40	77.2	0.73	83.0	2.01
7 years......	87.1	1.26	88.9	1.46	88.2	0.54	88.2	1.69
8 years......	95.8	0.71	95.4	1.39	97.1	0.84	100.9	3.20
9 years......	103.8	0.86	108.4	2.05	108.5	1.35	110.5	2.91
10 years.....	116.3	1.76	118.9	2.46	116.4	0.79	119.3	3.18
11 years.....	127.0	1.20	130.0	1.47	128.5	0.70	127.1	2.29

Chronological age at last birthday	$10,000 or more				Under $3,000				$15,000 or more	
	White		Negro		White		Negro		White	
	Mean	S.D.	Mean	S.D.	Mean	S.D.	Mean	S.D.	Mean	S.D.
Boys			Skeletal age in (months)							
6 years......	73.3	1.58	87.6	43.87	74.9	0.64	79.0	1.29	74.3	1.46
7 years......	87.9	1.14	--	--	86.7	0.60	88.1	1.41	87.2	1.99
8 years......	97.0	1.35	105.0	74.24	96.3	0.55	100.8	1.35	98.2	4.43
9 years......	107.9	1.11	--	--	105.6	0.47	106.3	0.67	107.8	3.54
10 years.....	112.6	0.85	101.0	50.49	113.6	0.52	112.2	1.04	114.4	2.03
11 years.....	124.8	1.05	--	--	123.8	0.59	125.8	0.76	125.8	2.54
Girls										
6 years......	76.7	1.15	--	--	76.7	0.56	78.9	1.32	78.6	3.23
7 years......	88.2	1.48	92.3	46.27	87.9	0.46	88.9	0.82	93.2	2.54
8 years......	95.6	1.31	103.7	52.04	96.3	0.51	96.3	0.99	95.8	1.97
9 years......	108.5	1.10	--	--	107.1	0.75	109.2	1.42	108.9	1.62
10 years.....	117.2	1.54	--	--	116.5	0.65	119.2	1.70	119.2	3.23
11 years.....	127.9	1.35	--	--	128.1	0.49	128.8	1.05	129.2	3.02

(data from Roche, Roberts, and Hamill, 1975)

TABLE   1165

THE MEAN SKELETAL AGE (hand-wrist; Greulich-Pyle; months)
OF WHITE AND NEGRO U.S. YOUTHS BY ANNUAL FAMILY INCOME

Chronological age at last birthday	Less than $3,000				Less than $5,000			
	White		Negro		White		Negro	
	Mean	S.D.	Mean	S.D.	Mean	S.D.	Mean	S.D.
Boys:								
12 years	139.4	3.45	137.8	3.39	139.9	2.32	138.5	4.53
13 years	154.4	5.69	154.5	5.07	154.7	2.57	158.8	4.94
14 years	173.6	2.14	178.4	3.69	173.5	2.51	175.1	3.42
15 years	184.1	3.07	182.8	3.24	186.2	1.89	182.3	2.05
16 years	194.2	1.61	191.8	1.75	196.4	1.20	193.5	1.91
17 years	206.8	1.70	200.9	3.10	205.0	1.19	201.3	1.89
Girls:								
12 years	145.8	2.67	147.0	2.08	143.1	2.32	147.4	2.30
13 years	155.0	1.50	152.0	2.21	153.1	0.98	153.8	1.63
14 years	168.0	1.43	168.2	2.28	166.8	1.05	170.0	1.41
15 years	178.5	1.42	178.5	1.07	178.1	0.78	179.0	0.71
16 years	189.3	1.06	181.2	1.69	185.6	1.81	184.2	0.69
17 years	190.5	1.32	184.5	4.00	187.6	1.28	184.4	2.72

Chronological age at last birthday	$5,000-$9,999				$10,000 or more			
	White		Negro		White		Negro	
	Mean	S.D.	Mean	S.D.	Mean	S.D.	Mean	S.D.
Boys:								
12 years	139.8	0.96	138.4	3.10	141.5	1.36	144.2	10.04
13 years	158.4	0.93	164.1	3.46	156.4	1.56	162.5	51.95
14 years	175.0	0.88	168.6	2.65	172.4	1.19	159.6	8.72
15 years	186.8	1.02	180.7	3.66	188.1	1.06	177.2	8.27
16 years	196.4	1.19	190.1	7.03	198.6	2.01	205.9	103.16
17 years	205.7	0.70	206.0	2.32	206.6	1.19	211.8	5.51
Girls:								
12 years	141.2	0.87	144.0	3.67	143.1	0.76	151.9	10.23
13 years	155.4	0.92	158.1	2.09	156.1	1.07	170.0	4.37
14 years	166.4	0.76	170.1	2.94	170.9	0.45	171.8	4.63
15 years	176.8	0.54	179.2	1.79	176.6	1.02	189.9	3.15
16 years	184.9	0.82	187.4	3.09	186.8	0.85	183.6	58.19
17 years	186.4	0.99	187.6	2.16	186.0	0.89	181.3	46.72

(data from Roche, 1978)

TABLE   1166

MEAN DIFFERENCES BETWEEN SKELETAL (GREULICH-PYLE) AND
CHRONOLOGICAL AGES IN YEARS AMONG INCOME GROUPS
IN U.S. CHILDREN, 6-11 YEARS

	Income					
Sex and bone	Less than $3,000	$3,000-$4,999	$5,000-$6,999	$7,000-$9,999	$10,000-$14,999	$15,000 or more
Boys	Mean difference					
Triquetral..................	-0.6	0.9	0.7	-0.8	0.1	0.5
Metacarpal III.............	-1.0	0.6	0.7	-0.1	-0.2	0.1
Metacarpal V...............	-1.2	0.7	0.6	0.1	-0.4	0.3
Proximal phalanx III.......	-2.4	-0.2	0.7	1.1	0.6	1.2
Proximal phalanx IV........	-2.5	-0.2	0.7	1.0	0.8	1.4
Middle V..................	-2.2	0.2	0.9	0.9	-0.2	0.1
Girls						
Triquetral..................	-0.1	-0.5	-0.5	0.8	-0.6	2.1
Metacarpal III.............	-0.5	-0.3	-0.4	1.0	-0.6	3.0
Metacarpal V...............	-0.4	-0.3	-0.5	1.2	-0.6	2.8
Proximal phalanx III.......	-1.2	-1.1	-0.6	2.2	0.1	2.6
Proximal phalanx IV........	-1.3	-1.0	-0.5	2.2	0.1	2.6
Middle V..................	-1.4	-1.3	-0.7	2.6	-0.2	4.0

MEAN SKELETAL AGE (hand-wrist; Greulich-Pyle)
OF U.S. BOYS AND GIRLS BY EDUCATION OF FIRST PARENT

Chronological age at last birthday	Education of first parent							
	Less than 5 years	5-7 years	8 years	9-11 years	12 years	13-15 years	16 years	17 or more years
Boys	Mean skeletal age (months)							
6 years......	72.5	75.8	77.1	74.6	76.4	75.3	77.1	74.1
7 years......	83.7	82.5	87.7	87.1	87.3	89.3	89.4	84.4
8 years......	96.2	97.2	97.5	97.6	95.5	99.1	96.3	99.3
9 years......	103.8	105.1	105.9	105.5	105.9	107.1	107.5	105.2
10 years.....	112.6	112.9	113.8	113.5	113.7	114.5	113.0	112.4
11 years.....	122.0	126.7	123.2	122.7	123.7	127.8	124.7	126.8
Girls								
6 years......	75.6	74.3	78.9	76.0	77.8	77.4	79.5	73.3
7 years......	85.4	88.0	88.1	87.8	88.2	90.6	85.5	89.8
8 years......	93.8	95.8	95.2	97.8	96.6	98.5	97.2	95.2
9 years......	105.0	105.5	107.2	108.2	107.8	108.9	108.6	110.0
10 years.....	118.0	116.0	115.8	116.9	117.7	114.4	120.3	116.9
11 years.....	126.6	128.2	127.8	128.9	127.2	129.1	129.0	129.8

(data from Roche, Roberts, and Hamill, 1975)

TABLE 1167

MEAN SKELETAL AGE (hand-wrist; Greulich-Pyle)
OF U.S. WHITE AND NEGRO YOUTHS BY EDUCATION OF FIRST PARENT

Chronological age at last birthday	Education of parent							
	Less than 5 years				5-8 years			
	White		Negro		White		Negro	
	Mean	S.D.	Mean	S.D.	Mean	S.D.	Mean	S.D.
Boys								
12 years......	140.4	0.79	138.6	3.12	138.0	0.62	136.3	3.07
13 years......	156.9	0.97	161.0	3.19	154.5	1.20	159.0	3.55
14 years......	174.4	0.79	173.4	2.09	174.8	0.92	173.2	2.49
15 years......	187.5	0.77	183.0	1.21	188.6	0.81	183.8	1.26
16 years......	197.2	0.82	192.7	2.93	197.8	0.84	192.9	2.95
17 years......	206.1	0.54	203.4	1.42	206.4	0.56	203.4	1.42
Girls								
12 years......	142.4	0.64	147.6	1.94	143.1	0.76	147.5	2.34
13 years......	155.0	0.64	157.4	1.16	157.2	0.74	158.7	1.00
14 years......	168.4	0.49	170.2	1.50	170.0	0.53	172.3	1.22
15 years......	177.5	0.49	179.9	0.70	177.8	0.44	179.9	0.70
16 years......	186.0	0.45	184.7	0.81	186.0	0.44	184.7	0.81
17 years......	187.0	0.60	184.9	1.88	187.0	0.60	184.9	1.88

Chronological age at last birthday	9-12 years				13 years or more			
	White		Negro		White		Negro	
	Mean	S.D.	Mean	S.D.	Mean	S.D.	Mean	S.D.
Boys								
12 years......	138.8	0.61	137.3	2.97	138.6	0.77	137.2	3.26
13 years......	155.1	1.09	159.4	3.56	155.2	1.15	159.7	3.80
14 years......	174.6	0.84	172.8	2.42	175.4	0.87	173.5	2.33
15 years......	188.2	0.79	182.3	1.24	188.7	0.81	183.9	1.15
16 years......	197.5	0.80	192.9	2.95	197.7	0.82	192.9	2.95
17 years......	206.0	0.54	202.4	1.65	206.3	0.55	203.2	1.51
Girls								
12 years......	143.0	0.72	146.8	2.26	143.2	0.78	147.8	2.11
13 years......	156.9	0.68	157.7	0.89	157.3	0.74	158.4	0.89
14 years......	168.9	0.56	171.6	1.23	170.1	0.52	172.2	1.20
15 years......	177.6	0.45	179.9	0.70	177.7	0.44	179.9	0.70
16 years......	185.8	0.46	184.2	0.84	185.9	0.44	184.4	0.79
17 years......	186.0	0.56	184.5	1.88	186.2	0.59	184.9	1.88

(data from Roche et al., 1978)

TABLE 1168

MEANS AND STANDARD DEVIATIONS FOR WITHIN-CHILD VARIANCE OF THE
DIFFERENCES BETWEEN BONE-SPECIFIC SKELETAL AGES (Greulich-Pyle)
AND THE MEAN SKELETAL AGE OF THE HAND-WRIST IN CALIFORNIA CHILDREN

Chronological age at last birthday	Less than $5,000			$5,000-$9,999			$10,000 or more		
	75	50	25	75	50	25	75	50	25
Boys				Bone-specific skeletal age range in (months)					
6 years........	36.4	29.5	23.4	38.9	30.2	22.8	36.9	28.7	22.8
7 years........	35.4	26.6	20.1	36.8	29.9	21.9	41.0	30.3	22.5
8 years........	34.3	26.4	19.2	36.6	27.9	20.5	34.4	26.4	20.1
9 years........	30.5	23.1	14.8	30.9	23.0	16.4	30.1	23.0	14.9
10 years.......	27.4	20.6	14.1	30.2	22.5	16.0	25.7	20.1	14.5
11 years.......	26.0	19.8	14.4	29.0	21.5	15.1	26.6	18.5	13.6
Girls									
6 years........	34.8	27.1	20.4	34.5	27.6	20.7	33.0	27.4	23.2
7 years........	32.8	24.0	16.7	31.7	24.2	16.7	31.1	18.2	12.2
8 years........	25.9	19.8	12.1	28.9	20.8	14.4	28.1	20.7	14.5
9 years........	28.0	21.8	15.1	32.3	23.1	15.2	29.9	21.3	16.2
10 years.......	30.2	23.7	17.9	32.1	24.5	16.7	34.5	24.9	18.3
11 years.......	32.7	24.9	18.0	34.1	23.2	17.5	32.2	24.5	16.5

(data from Roche, Roberts, and Hamill, 1975)

TABLE 1169

MEANS AND STANDARD DEVIATIONS FOR WITHIN-CHILD
VARIANCE OF THE DIFFERENCES BETWEEN BONE-SPECIFIC
SKELETAL AGES (Greulich-Pyle) AND THE MEAN
SKELETAL AGE OF THE HAND-WRIST
IN CALIFORNIA CHILDREN

Group	Boys		Girls	
	Mean	S.D.	Mean	S.D.
White				
Tall	63.0	37.0	43.5	25.1
Medium	72.3	31.4	44.2	25.5
Short	81.2	40.4	43.7	24.0
Black				
Tall	60.5	39.1	66.7	29.3
Medium	58.2	29.4	69.7	33.9
Short	55.2	26.2	70.4	33.1
Chinese	71.7	42.5	38.5	18.6
Japanese	71.8	32.0	48.5	27.7

(data from Wingerd, Peritz, and Sproul, 1974, Annals of Human Biology 1:201-209)

TABLE 1170

MEAN SKELETAL AGE (HAND-WRIST; GREULICH-PYLE) OF BOYS AND GIRLS, BY
POPULATION SIZE IN URBAN AREAS AND LAND USE IN RURAL AREAS OF RESIDENCE
IN THE U.S.

Chronological age at last birthday	Urban total	Urbanized Areas				Urban outside urbanized areas		
		3 million or more	1.0-2.9 million	250,000-999,999	Less than 250,000	25,000 or more	10,000-24,999	2,500-9,999
Boys				Mean skeletal age	(months)			
6 years	75.6	77.3	77.1	74.3	73.8	74.2	76.6	74.4
7 years	86.7	87.5	86.3	86.5	88.8	85.6	84.0	85.7
8 years	97.4	98.6	97.6	93.9	97.7	98.4	99.3	97.0
9 years	106.1	106.6	105.6	106.8	106.6	104.1	104.2	105.8
10 years	113.1	113.4	113.6	111.0	111.6	113.8	116.7	114.1
11 years	124.2	126.8	125.4	124.0	123.1	122.5	120.6	120.1
12 years	140.2	140.2	139.7	141.8	138.6	137.6	138.8	141.3
13 years	157.9	159.0	155.7	156.7	160.3	163.0	160.4	155.3
14 years	173.4	173.4	172.8	174.0	175.5	172.1	169.6	173.9
15 years	187.1	188.4	185.1	190.8	186.3	186.2	182.6	185.9
16 years	196.4	196.1	194.8	197.5	197.1	199.1	197.4	196.1
17 years	205.6	206.9	206.5	205.3	205.6	208.3	202.5	201.3
Girls								
6 years	76.6	76.9	78.8	74.7	76.9	77.8	73.0	73.7
7 years	88.3	89.5	88.0	88.6	86.7	85.2	87.9	87.8
8 years	96.4	98.8	96.6	95.1	95.2	93.6	102.8	95.2
9 years	107.9	110.4	105.4	105.4	106.3	107.2	111.9	102.8
10 years	116.8	117.8	116.5	115.2	118.6	117.5	119.9	116.2
11 years	128.3	128.6	125.8	127.8	129.4	128.7	129.7	131.6
12 years	143.0	146.5	141.2	141.9	141.9	138.5	139.4	140.0
13 years	155.0	156.5	152.9	154.3	157.0	159.4	154.9	154.3
14 years	168.4	170.6	164.2	169.0	169.2	166.7	170.4	170.1
15 years	177.8	179.8	174.4	178.1	178.2	175.2	176.2	178.4
16 years	184.7	187.4	179.9	187.0	185.9	184.3	181.8	184.7
17 years	187.0	182.9	185.3	191.6	189.6	188.2	182.0	189.3

Chronological age at last birthday	Rural total	Rural-farm		Rural-nonfarm	
		10 acres or more	Less than 10 acres	10 acres or more	Less than 10 acres
Boys			Mean skeletal age	(months)	
6 years	75.1	76.2	69.4	78.1	75.0
7 years	87.4	89.3	--	85.0	86.9
8 years	95.9	95.6	--	93.5	96.4
9 years	105.2	104.8	103.0	104.9	105.3
10 years	114.0	116.0	118.0	111.1	114.0
11 years	124.1	122.4	153.0	122.1	124.4
12 years	140.4	137.0	--	134.1	141.5
13 years	156.4	153.2	161.8	152.6	157.4
14 years	173.9	174.9	190.3	173.3	173.6
15 years	185.3	183.5	163.0	184.2	186.2
16 years	196.2	200.4	191.2	197.7	195.2
17 years	204.9	203.6	--	204.7	205.3

TABLE 1170   (continued)

MEAN SKELETAL AGE (HAND-WRIST: GREULICH-PYLE) OF BOYS AND GIRLS, BY
POPULATION SIZE IN URBAN AREAS AND LAND USE IN RURAL AREAS OF RESIDENCE
IN THE U.S.

Chronological age at last birthday	Rural total	Rural-farm		Rural-nonfarm	
		10 acres or more	Less than 10 acres	10 acres or more	Less than 10 acres
Girls					
6 years	78.0	79.9	66.0	77.5	77.6
7 years	87.5	87.2	71.0	82.2	88.1
8 years	96.3	97.4	89.0	97.2	95.9
9 years	107.6	105.9	96.8	109.7	107.6
10 years	116.9	117.9	--	115.2	117.1
11 years	128.0	128.9	129.6	127.7	127.7
12 years	143.1	141.0	158.2	143.6	143.3
13 years	154.8	148.6	167.5	156.1	157.3
14 years	167.7	167.2	176.0	166.0	167.9
15 years	177.2	174.8	162.6	184.7	177.2
16 years	187.6	186.0	--	185.3	187.8
17 years	184.6	181.1	187.0	181.4	185.8

(data from Roche, Roberts, and Hamill, 1975)

TABLE 1171

MEAN SKELETAL AGES (HAND-WRIST; GREULICH-PYLE)
BY GEOGRAPHIC REGION OF THE U.S.

Chronological age at last birthday	Northeast		Midwest		South		West	
	Mean	S.D.	Mean	S.D.	Mean	S.D.	Mean	S.D.
Boys			Skeletal age in (months)					
6 years......	76.7	11.09	75.3	12.23	74.8	10.97	75.4	11.65
7 years......	88.1	12.05	88.9	12.57	85.1	10.81	85.5	11.14
8 years......	98.2	12.42	96.9	10.87	97.5	11.77	95.4	12.12
9 years......	106.4	12.45	107.0	11.27	105.7	10.38	103.7	10.72
10 years.....	114.3	10.74	113.9	10.26	113.7	11.06	111.8	8.75
11 years.....	127.2	13.24	123.2	12.59	123.0	12.72	123.7	13.80
Girls								
6 years......	78.5	10.96	80.3	11.15	75.3	10.25	·74.4	10.51
7 years......	89.2	10.59	88.4	9.78	86.6	10.84	87.0	10.65
8 years......	97.0	11.30	97.7	10.32	95.7	11.20	95.5	10.06
9 years......	110.9	14.39	107.4	11.26	106.8	13.69	103.8	13.24
10 years.....	116.4	15.89	118.0	14.67	116.8	14.31	116.5	14.64
11 years.....	128.8	12.30	127.1	11.61	128.0	13.51	129.0	11.49

(data from Roche, Roberts, and Hamill, 1975)

TABLE 1172

MEAN SKELETAL AGES (HAND-WRIST; GREULICH-PYLE) OF WHITE AND
NEGRO BOYS AND GIRLS, BY GEOGRAPHIC REGION OF THE U.S.

Chronological age at last birthday	Northeast				Midwest			
	White		Negro		White		Negro	
	Mean	S.D.	Mean	S.D.	Mean	S.D.	Mean	S.D.
Boys			Skeletal age in (months)					
6 years......	76.4	1.03	79.0	3.19	74.8	1.31	80.3	3.12
7 years......	88.2	0.69	87.4	2.66	88.9	1.54	88.9	4.12
8 years......	98.0	1.37	99.7	3.06	96.5	1.19	100.6	6.73
9 years......	106.3	0.72	107.6	3.30	106.9	0.62	108.3	2.57
10 years.....	114.2	0.79	114.7	2.21	114.2	0.58	111.3	2.12
11 years.....	127.3	0.91	126.4	2.70	122.9	0.91	127.2	6.38
Girls								
6 years......	78.3	0.82	79.9	4.32	80.2	0.72	85.0	27.59
7 years......	89.0	0.61	90.8	1.39	88.3	0.90	88.7	1.49
8 years......	96.6	0.48	100.0	0.89	97.1	1.25	105.0	6.51
9 years......	110.4	1.34	111.8	3.95	107.4	0.80	107.5	2.76
10 years.....	115.8	1.61	120.7	1.99	117.5	1.55	122.1	12.90
11 years.....	128.8	1.35	128.2	4.54	126.9	0.75	128.8	2.58

Chronological age at last birthday	South				West			
	White		Negro		White		Negro	
	Mean	S.D.	Mean	S.D.	Mean	S.D.	Mean	S.D.
Boys			Skeletal age in (months)					
6 years......	73.4	1.41	79.2	1.96	75.0	1.12	77.4	24.70
7 years......	83.5	1.30	89.0	2.33	85.7	1.54	83.8	26.63
8 years......	95.8	0.62	101.0	0.59	95.0	1.27	101.5	5.34
9 years......	105.9	1.23	105.2	0.45	103.3	1.31	110.3	42.79
10 years.....	114.5	0.93	110.8	2.26	111.7	1.46	113.9	5.79
11 years.....	122.3	0.62	124.9	1.97	123.3	1.96	126.1	40.23
Girls								
6 years......	74.3	1.11	77.6	1.52	74.0	0.65	76.1	24.62
7 years......	85.4	1.45	89.0	1.79	87.0	0.73	87.1	8.04
8 years......	96.1	1.45	94.4	1.70	95.7	1.14	91.4	2.21
9 years......	105.4	1.64	109.6	2.29	103.6	1.84	107.2	0.96
10 years.....	116.3	1.82	118.4	2.67	116.4	1.02	118.2	37.54
11 years.....	127.4	0.70	128.8	0.75	128.9	1.70	130.2	4.29

(data from Roche, Roberts, and Hamill, 1975)

TABLE   1173

SELECTED PERCENTILES IN THE DISTRIBUTION OF THE INDIVIDUAL CHILD'S
RANGE IN BONE-SPECIFIC SKELETAL AGES (Greulich-Pyle)
FOR BONES IN THE HAND-WRIST BY GEOGRAPHIC REGION OF THE U.S.

Chronological age at last birthday	Northeast			Midwest		
	75	50	25	75	50	25
Boys	Bone-specific skeletal age range (months)					
6 years..........	36.7	29.2	22.1	37.4	30.2	22.1
7 years..........	35.6	28.3	22.0	36.3	28.3	20.3
8 years..........	36.6	24.1	17.6	33.5	26.6	20.3
9 years..........	27.7	20.5	14.8	30.3	23.8	16.1
10 years.........	25.5	20.3	14.3	26.1	20.3	14.3
11 years.........	27.1	18.6	13.6	25.5	20.1	14.5
Girls						
6 years..........	34.4	27.8	20.1	32.5	26.9	22.4
7 years..........	29.7	21.8	14.7	31.4	23.6	15.4
8 years..........	30.1	20.9	14.3	31.3	22.6	16.2
9 years..........	30.2	22.2	16.2	33.2	23.9	15.4
10 years.........	29.4	23.1	16.6	33.5	25.8	18.6
11 years.........	28.9	22.6	17.2	35.8	24.9	18.6

Chronological age at last birthday	South			West		
	75	50	25	75	50	25
Boys	Bone-specific skeletal age range (months)					
6 years..........	36.5	27.4	22.5	37.0	31.0	24.3
7 years..........	36.7	28.3	20.4	37.4	28.3	20.8
8 years..........	34.9	26.2	19.6	34.3	27.6	20.2
9 years..........	29.6	20.3	14.2	30.6	24.1	14.8
10 years.........	29.3	22.4	14.9	24.8	17.9	13.3
11 years.........	24.5	18.7	14.4	28.3	19.4	13.3
Girls						
6 years..........	35.7	29.3	21.8	34.4	26.4	19.9
7 years..........	34.7	26.6	18.5	30.2	22.3	14.2
8 years..........	27.7	20.8	13.6	28.6	22.2	15.4
9 years..........	30.1	22.7	16.2	27.0	18.5	13.9
10 years.........	30.4	24.9	18.8	32.6	22.6	16.0
11 years.........	33.1	24.7	15.7	32.8	24.2	17.4

(data from Roche, Roberts, and Hamill, 1975)

TABLE 1174

MEAN SKELETAL AGES (hand-wrist; Greulich-Pyle) FOR REGIONS OF THE UNITED STATES

Chronological age at last birthday	Northeast				Midwest			
	White		Negro		White		Negro	
	Mean	S.D.	Mean	S.D.	Mean	S.D.	Mean	S.D.

Skeletal age in (months)

BOYS:

12 years	141.1	1.24	140.7	1.56	139.6	1.22	131.1	2.66
13 years	157.4	2.63	158.8	7.86	157.2	1.60	164.5	13.84
14 years	173.4	0.82	174.9	3.05	172.9	1.43	162.9	4.52
15 years	189.6	1.63	184.5	8.39	184.2	1.18	176.1	3.35
16 years	196.6	1.44	196.0	3.40	196.9	2.08	190.4	2.82
17 years	207.3	1.24	200.0	8.88	206.2	0.61	206.8	4.51

GIRLS:

12 years	145.0	1.09	147.4	4.01	141.1	1.22	142.9	8.04
13 years	156.8	1.00	157.5	2.14	155.2	1.65	156.5	2.04
14 years	166.8	1.08	171.5	3.15	168.4	1.24	174.1	2.56
15 years	176.6	0.92	176.8	2.04	176.2	1.49	185.4	1.91
16 years	185.2	0.87	180.8	0.60	186.4	0.88	189.4	1.70
17 years	182.6	0.99	181.6	3.95	185.2	1.53	185.2	2.57

Chronological age at last birthday	South				West			
	White		Negro		White		Negro	
	Mean	S.D.	Mean	S.D.	Mean	S.D.	Mean	S.D.

Skeletal age in (months)

BOYS:

12 years	142.0	1.19	139.1	5.85	139.9	1.76	142.4	7.66
13 years	154.5	1.19	159.4	4.34	157.4	1.55	166.5	12.96
14 years	174.0	2.13	174.9	3.50	174.7	1.17	174.8	4.44
15 years	188.5	1.08	181.4	1.10	187.3	1.58	186.7	41.90

TABLE 1174 (continued)

MEAN SKELETAL AGES (hand-wrist; Greulich-Pyle) FOR REGIONS OF THE UNITED STATES

Chronological age at last birthday	South				West			
	White		Negro		White		Negro	
	Mean	S.D.	Mean	S.D.	Mean	S.D.	Mean	S.D.

Skeletal age in (months)

BOYS:

16 years	196.8	1.14	196.1	1.76	197.4	1.27	176.0	57.46
17 years	206.1	0.92	199.1	2.14	203.8	1.44	208.3	46.75

GIRLS:

12 years	142.9	1.09	145.8	3.07	140.8	1.81	150.7	47.74
13 years	155.6	0.94	155.2	1.54	153.2	0.77	157.7	2.29
14 years	168.0	0.42	166.4	1.96	168.8	1.33	170.9	1.03
15 years	178.6	0.60	180.8	1.22	177.4	0.94	178.1	2.38
16 years	187.8	1.34	185.2	1.17	184.9	1.07	175.0	4.26
17 years	189.3	1.15	187.6	3.59	189.1	0.86	182.4	41.55

(data from Roche et al., 1978)

TABLE 1175

SELECTED PERCENTILES IN THE DISTRIBUTION OF THE INDIVIDUAL CHILD'S
RANGE IN BONE-SPECIFIC SKELETAL AGES (Greulich-Pyle) IN
THE HAND-WRIST OF U.S. CHILDREN

Chronological age at last birthday	White			Negro		
	75	50	25	75	50	25
Boys	Bone-specific skeletal age range in (months)					
6 years..........	36.7	29.2	22.7	40.5	28.8	22.4
7 years..........	36.1	28.4	20.8	35.9	26.9	19.2
8 years..........	34.2	26.4	19.0	35.4	25.8	16.8
9 years..........	29.9	22.6	15.1	36.7	25.1	14.2
10 years.........	26.4	20.1	14.2	28.5	21.9	14.3
11 years.........	26.4	18.7	14.2	27.2	20.8	14.7
Girls						
6 years..........	34.3	27.1	20.6	36.4	28.1	19.9
7 years..........	31.4	23.4	15.1	32.5	24.9	17.7
8 years..........	28.9	21.5	14.5	28.7	22.0	14.9
9 years..........	30.5	22.3	15.4	27.8	22.1	14.5
10 years.........	32.2	24.3	17.3	30.9	24.7	20.2
11 years.........	33.9	24.3	17.0	32.3	24.4	17.9

SELECTED PERCENTILES, MEAN AND STANDARD DEVIATIONS IN THE DISTRIBUTION
OF THE INDIVIDUAL YOUTH'S RANGE IN BONE-SPECIFIC SKELETAL
AGES (GREULICH-PYLE) IN THE HAND-WRIST FOR YOUTHS
FOR U.S. YOUTHS

Chronological age at last birthday	95	75	50	25	5	Mean	S.D.
Boys	Bone-specific skeletal age range (months)						
12 years..........	30.2	20.4	15.8	12.3	7.1	16.3	6.83
13 years..........	27.5	18.7	14.5	10.9	6.3	15.0	7.00
14 years..........	26.6	16.9	12.4	8.9	4.7	13.1	6.77
15 years..........	24.8	15.6	10.8	6.7	1.8	11.5	7.34
16 years..........	28.0	15.0	8.8	3.8	0.4	10.3	8.94
17 years..........	25.4	12.4	5.4	0.9	0.2	7.7	8.76
Girls							
12 years..........	26.9	18.8	14.1	10.5	5.0	14.8	7.42
13 years..........	30.1	18.9	12.8	8.7	2.6	14.1	8.90
14 years..........	30.8	18.5	10.7	4.9	0.6	12.4	10.22
15 years..........	34.2	16.5	6.7	1.4	0.2	10.3	11.39
16 years..........	34.8	10.7	2.3	0.5	0.1	7.6	12.25
17 years..........	34.5	14.3	2.0	0.5	0.1	8.3	11.86

(data from Roche et al., 1976)

TABLE   1176

DIFFERENCES IN SKELETAL MATURITY (years) BETWEEN
THE KNEE AND THE HAND WITHIN INDIVIDUALS
IN AUSTRALIAN CHILDREN

Areas Compared	Chronologic Age (years)	BOYS Percentiles			GIRLS Percentiles		
		25	50	75	25	50	75
Knee less hand	12	-.04	.06	.15	-.19	-.12	-.05
	13	-.05	.06	.17	-.22	-.07	.00
	14	-.04	.02	.16	-.07	-.01	.08
	15	-.04	.04	.20	-.09	.00	.13
Tibia less hand	12	-.06	.04	.12	-.11	-.04	.08
	13	-.11	.00	.14	-.24	-.04	.09
	14	-.09	-.02	.17	-.10	-.01	.07
	15	-.11	.07	.21	-.09	-.04	.13
Femur less hand	12	-.04	.09	.22	-.39	-.21	-.08
	13	-.11	.09	.35	-.24	-.12	-.01
	14	-.05	.03	.17	-.05	.03	.14
	15	-.12	.02	.20	-.13	.01	.13

(data from Roche and French, 1970, American Journal of Roentgenology 109:307-312.
Copyright American Roentgen Ray Society, 1970)

TABLE   1177

ROCHE–WAINER–THISSEN SKELETAL AGES (years)
IN OHIO CHILDREN

Age (years)	Boys		Girls	
	Mean	S.D.	Mean	S.D.
.1	-0.05	0.24	0.12	0.11
.3	0.16	0.25	0.32	0.16
.5	0.54	0.28	0.54	0.19
.8	0.80	0.25	0.84	0.25
1.0	1.06	0.31	1.04	0.26
1.5	1.50	0.42	1.52	0.36
2.0	2.08	0.41	2.05	0.47
2.5	2.52	0.46	2.55	0.47
3.0	3.00	0.49	3.12	0.72
3.5	3.55	0.59	3.67	0.79
4.0	3.95	0.63	4.17	0.92
4.5	4.73	0.79	4.63	0.89
5.0	5.24	0.96	5.08	0.98
5.5	5.72	0.98	5.76	0.91
6.0	6.14	0.98	6.08	0.95
6.5	6.68	0.96	6.51	1.01
7.0	7.15	0.95	7.13	0.91
7.5	7.45	0.93	7.72	0.98
8.0	8.24	0.99	8.15	0.94
8.5	8.45	1.08	8.38	1.18
9.0	9.15	1.08	8.88	1.09
9.5	9.09	1.47	9.35	1.03
10.0	9.98	.99	9.91	1.14
11.0	10.81	1.05	11.02	1.16
12.0	11.70	1.05	12.11	1.02
13.0	13.02	1.03	12.79	0.83
14.0	13.95	1.12	13.88	1.10
15.0	14.80	1.13	15.10	1.30
16.0	15.87	1.03	15.84	1.13
17.0	16.86	1.13	16.42	0.81
18.0	17.49	.88	16.88	0.60

(data from Roche et al., 1975a)

TABLE   1178

PERCENTILES OF STANDARD ERRORS (years) OF THE ESTIMATES FOR
ROCHE–WAINER–THISSEN SKELETAL AGES
IN OHIO CHILDREN

Chronological Age (years)	Boys			Girls		
	10	50	90	10	50	90
.1	.21	.31	.42	.13	.15	.18
.3	.19	.21	.42	.13	.15	.16
.5	.19	.21	.25	.13	.15	.18
.8	.19	.21	.23	.16	.18	.23
1.0	.20	.23	.29	.16	.20	.30
1.5	.21	.26	.32	.19	.21	.42
2.0	.26	.27	.33	.20	.32	.43
2.5	.27	.28	.32	.21	.35	.46
3.0	.27	.30	.38	.25	.42	.49
3.5	.29	.35	.48	.32	.43	.53
4.0	.31	.43	.50	.38	.47	.57
4.5	.40	.49	.56	.42	.51	.59
5.0	.45	.52	.64	.44	.55	.69
5.5	.49	.55	.67	.51	.58	.67
6.0	.50	.60	.77	.54	.63	.92
6.5	.58	.72	.89	.59	.74	.88
7.0	.57	.66	.84	.60	.68	.82
7.5	.66	.79	.92	.73	.82	1.00
8.0	.64	.70	.82	.64	.69	.94
8.5	.72	.84	1.05	.69	.81	1.00
9.0	.65	.71	.89	.65	.69	.84
9.5	.75	1.02	1.38	.84	.96	1.33
10.0	.63	.70	.87	.66	.72	.96
11.0	.64	.69	.78	.67	.71	.88
12.0	.65	.71	.85	.64	.72	.93
13.0	.58	.69	.81	.59	.71	.91
14.0	.52	.65	.85	.54	.65	.87
15.0	.51	.61	.76	.52	.64	.89
16.0	.51	.58	.80	.54	.71	1.17
17.0	.52	.65	.88	.57	.81	1.18
18.0	.55	.75	1.02	.63	.80	1.24

(data from Roche et al., 1975a)

TABLE   1179

TANNER–WHITEHOUSE SKELETAL AGES (years) AND ANNUAL INCREMENTS
IN THESE AGES IN NEGRO AND WHITE PHILADELPHIA BOYS

| Age Group (years) | Visit 1 | | | | Visit 2 | | | | Rate (years/year) | | |
| | Chronological Age (CA) | | Skeletal Age (SA) | | Chronological Age | | Skeletal Age | | CA | SA | |
	Mean	S.D.	Mean	S.D.	Mean	S.D.	Mean	S.D.	Mean	Mean	S.D.
					Negro Boys						
6-7	6.31	0.13	6.54	1.37	7.29	0.13	7.67	1.51	0.98	1.13	0.54
7-8	7.00	0.31	7.13	1.46	7.96	0.31	8.25	1.56	0.96	1.12	0.43
8-9	8.01	0.28	8.26	1.37	8.99	0.27	9.52	1.37	0.98	1.27	0.60
9-10	8.90	0.28	9.29	1.17	9.89	0.28	10.33	1.12	0.99	1.04	0.54
10-11	10.03	0.32	10.49	1.30	11.03	0.32	11.51	1.12	1.00	1.03	0.47
11-12	11.02	0.27	11.44	1.36	12.00	0.27	12.27	1.20	0.98	0.84	0.56
12-13	11.80	0.29	12.15	1.18	12.80	0.28	12.90	0.90	1.00	0.75	0.45
					White Boys						
6-7	6.22	0.15	6.08	1.24	7.22	0.15	6.83	1.34	1.00	0.75	0.32
7-8	7.00	0.28	6.71	1.00	7.98	0.28	7.91	1.04	0.98	1.20	0.48
8-9	7.99	0.28	8.19	1.01	8.96	0.28	9.25	1.01	0.97	1.07	0.47
9-10	8.99	0.26	9.13	1.15	9.95	0.25	10.00	1.11	0.96	0.87	0.33
10-11	10.03	0.31	10.62	1.14	11.00	0.30	11.50	1.08	0.97	0.88	0.37
11-12	10.98	0.31	11.40	1.21	11.90	0.24	12.30	1.08	0.92	0.91	0.52

(data from "Skeletal maturation studied longitudinally over one year in American whites and Negroes six through thirteen years of age," Human Biology 42:377-390, 1970, by R. M. Malina, by permission of the Wayne State University Press. Copyright 1970, Wayne State University Press, Detroit, Michigan 48202)

TABLE 1180

TANNER-WHITEHOUSE SKELETAL AGES (years) AND ANNUAL INCREMENTS
IN THESE AGES IN NEGRO AND WHITE PHILADELPHIA GIRLS

| Age Group (years) | Visit 1 | | | | Visit 2 | | | | Rate (years/year) | | |
| | Chronological Age (CA) | | Skeletal Age (SA) | | Chronological Age | | Skeletal Age | | CA | SA | |
	Mean	S.D.	Mean	S.D.	Mean	S.D.	Mean	S.D.	Mean	Mean	S.D.
					Negro Girls						
6-7	6.31	0.13	7.10	1.23	7.30	0.13	8.23	1.18	0.99	1.14	0.50
7-8	6.99	0.28	7.94	1.25	7.96	0.27	8.93	1.11	0.97	0.99	0.51
8-9	7.98	0.27	8.29	1.30	8.96	0.28	9.30	1.01	0.98	1.01	0.49
9-10	8.97	0.30	9.60	1.01	9.96	0.29	10.45	1.00	0.99	0.85	0.40
10-11	9.97	0.32	10.65	1.24	10.94	0.32	11.42	1.14	0.97	0.77	0.33
11-12	11.02	0.27	11.59	0.88	12.01	0.26	12.26	0.82	0.99	0.66	0.33
12-13	11.90	0.25	12.39	1.15	12.88	0.25	13.26	1.13	0.98	0.86	0.52
					White Girls						
6-7	6.32	0.12	7.19	1.03	7.29	0.11	7.89	0.95	0.97	0.70	0.35
7-8	7.01	0.27	7.57	1.25	8.00	0.28	8.59	1.30	0.99	1.01	0.45
8-9	8.06	0.28	8.54	0.93	9.02	0.27	9.30	0.66	0.96	0.76	0.41
9-10	9.00	0.28	9.16	1.04	9.97	0.28	10.01	1.04	0.97	0.84	0.38
10-11	10.07	0.27	10.18	0.67	11.05	0.28	11.09	0.75	0.98	0.91	0.38
11-12	10.95	0.29	10.85	0.89	11.91	0.28	11.80	0.91	0.96	0.95	0.39

(data from "Skeletal maturation studied longitudinally over one year in American whites
and Negroes six through thirteen years of age," Human Biology 42:377-390, 1970, by R.
M. Malina, by permission of the Wayne State University Press. Copyright 1970, Wayne
State University Press, Detroit, Michigan 48202)

TABLE 1181

MEAN RATES OF SKELETAL MATURATION
(TANNER-WHITEHOUSE POINTS/YEAR)
IN PHILADELPHIA CHILDREN

Age (years)	Boys				Girls			
	Negro		White		Negro		White	
	Mean	S.D.	Mean	S.D.	Mean	S.D.	Mean	S.D.
6-7	45.9	21.7	31.5	14.3	49.1	20.9	28.5	14.6
7-8	46.7	17.2	50.2	21.0	50.5	20.9	46.5	22.3
8-9	52.1	24.1	44.0	19.7	56.1	30.2	41.3	16.2
9-10	47.2	23.3	37.6	14.6	81.4	48.5	68.3	44.0
10-11	59.6	27.5	53.7	29.5	93.0	42.1	107.3	53.0
11-12	67.6	50.6	68.8	48.0	91.2	45.9	128.2	59.7
12-13	62.4	34.8	--	--	95.2	59.3	--	--

(data from Malina, 1969, by permission from Nature, 223:1075. Copyright (c) 1969 Macmillan Journals Limited)

TABLE 1182

MEAN CHRONOLOGICAL AND SKELETAL AGES
OF JAPANESE-AMERICAN CHILDREN IN LOS ANGELES

Chronological age group (years)	Boys				Girls			
	Chronological age		Skeletal age		Chronological age		Skeletal age	
	Mean	S.D.	Mean	S.D.	Mean	S.D.	Mean	S.D.
6	5.93	0.14	6.71	0.79	5.92	0.26	6.61	0.94
7	6.93	0.25	6.70	1.40	7.04	0.33	7.46	1.18
8	7.99	0.35	7.24	0.82	8.00	0.33	8.80	1.02
9	8.92	0.26	8.38	0.90	8.82	0.25	9.46	0.83
10	9.93	0.26	9.56	0.90	10.11	0.32	11.59	0.81
11	11.01	0.28	10.28	0.98	11.06	0.25	12.41	0.67
12	11.92	0.26	11.72	1.34	11.98	0.22	12.23	0.42
13	13.93	0.57	13.00	0.61	12.96	0.28	13.11	0.43
14	13.90	0.30	13.52	0.93	13.88	0.30	14.26	1.01

Skeletal ages are for Tanner-Whitehouse method, 1962

(data from Kondo and Eto, 1975, pp. 13-45 in Comparative Studies on Human Adaptability of Japanese, Caucasians and Japanese Americans, S. Horvath et al., eds. University of Tokyo Press)

TABLE 1183

SKELETAL MATURITY (units) OF MONTREAL CHILDREN,
ACCORDING TO THE 20 BONE TW2 SYSTEM
INTERPOLATED AT EXACT MEAN AGE

Age Group ±0.5 Years	Girls Median	10-90 percentile range	Boys Median	10-90 percentile range
6	454	136	365	168
7	489	162	390	157
8	567	174	468	174
9	624	223	524	206
10	711	239	573	160
11	782	235	630	202
12	892	157	719	226
13	958	102	838	274
14	982	41	900	174
15	992	26	955	72
16	1000	--	977	50
17	1000	--	993	47

(data from "Skeletal maturity standards for French-Canadian children of
school-age with a discussion of the reliability and validity of such measures,"
Human Biology 51:353-370, 1979, by Baughan, Demirjian and Levesque, by
permission of the Wayne State University Press. Copyright 1979, Wayne State
University Press, Detroit, Michigan 48202)

TABLE 1184

MATURITY SCORES (points) FOR THE HIP AND PELVIS OF
CLEVELAND (Ohio) BOYS, ASSESSED BY THE OXFORD METHOD

Age (years)	Mean	S.D.
0.25	4.45	1.01
0.50	6.31	1.06
0.75	7.71	0.87
1	8.37	0.82
1.50	9.34	0.85
2	10.11	0.89
2.50	11.14	1.10
3	12.05	1.26
3.50	12.94	1.30
4	13.98	1.33
4.50	14.98	1.47
5	15.97	1.42
6	17.25	1.43
7	18.39	1.40
8	19.57	1.35
9	20.71	1.32
10	21.90	1.52
11	23.35	1.73
12	24.98	2.06
13	26.85	2.38
14	29.72	2.61
15	32.56	3.28
16	36.13	3.96
17	39.54	3.17
18	40.95	3.15

The maximum score, attained by the mature individual, is
45 points.

(data from "The relationship between physique and rate
of skeletal maturation in boys," Human Biology 29:167-
193, 1957, by Acheson and Dupertuis, by permission
of the Wayne State University Press. Copyright 1957,
Wayne State University Press, Detroit, Michigan 48202)

TABLE   1185

AGES OF FUSION OF SELECTED OSSEOUS CENTERS
(Ages given in years and months)
IN BOSTON CHILDREN

Centers	BOYS Percentiles				GIRLS Percentiles			
	10	25	75	90	10	25	75	90
Talus	10-0	10-5	12-2	13-2	8-1	8-5	9-10	10-10
V. Metatarsal	12-6	13-3	14-8	15-3	10-5	11-2	12-7	13-6
Trochlea	13-2	13-8	14-10	15-6	11-1	11-5	12-9	13-5
Lat. Epicondyle Humerus	13-7	14-3	15-5	15-10	11-4	11-10	13-4	13-11
Med. Epicondyle Humerus	14-4	--	--	--	12-3	12-11	14-4	14-10
Prox. Ulna (2 centers)	14-4	15-1	16-1	--	12-2	12-8	14-2	14-8
Prox. Radius	14-6	15-2	16-6	--	12-3	12-9	14-3	14-10
Dist. Tibia	14-7	15-2	--	--	12-5	13-2	14-7	14-11
Calcaneus (2 centers)	14-2	14-8	--	--	12-6	13-3	14-11	15-11
Dist. Ulna	--	--	--	--	14-9	15-5	--	--

(data from Harding, 1952)

TABLE   1186

ORDER OF FUSION OF THE EPIPHYSES OF THE HAND IN NEW YORK CHILDREN

Order	Digit	Phalanx
1	1	Distal
2	5	Distal
3	2	Distal
4	3	Distal
5	4	Distal
6	5	Proximal
7	5	Middle
8	2	Proximal
9	1	Metacarpus
10	1	Middle
11	3	Proximal
12	2	Middle
13	4	Proximal
14	4	Middle
15	3	Middle

(data from Lavine, Moss and Noback, 1962, The Journal of Pediatrics 61:571-575)

TABLE 1187

MODAL AGES FOR EPIPHYSEAL FUSION IN THE HAND-WRIST OF U.S. WHITE YOUTHS

Hand-wrist bone	BOYS		GIRLS	
	Modal age			
Radius	18.0[1]	--	15.8[1]	--
Ulna	17.8[1]	--	15.9[1]	16.2[4]
Metacarpal I	16.3[1,2]	15.8[4]	14.1[1,2]	13.8[4]
Metacarpal II	16.4[2],16.5[1]	16.5[4]	14.5[1],14.6[2]	14.8[4]
Metacarpal III	16.4[2],16.5[1]	16.6[4]	14.5[1],14.6[2]	14.8[4]
Metacarpal IV	16.4[1,2]	16.6[4]	14.4[1],14.6[2]	14.9[4]
Metacarpal V	16.5[1,2]	16.6[4]	14.4[1],15.0[2]	15.0[4]
Proximal phalanx I	16.2[5,1],16.3[2]	16.3[4]	14.2[1],14.3[5],14.4[2]	14.0[4]
Proximal phalanx II	16.3[2],16.4[1]	15.9[4]	14.2[1,2]	14.0[4]
Proximal phalanx III	16.3[1,2]	16.1[4]	14.2[1],14.5[2]	14.1[4]
Proximal phalanx IV	16.2[2],16.5[1]	16.1[4]	14.2[1,2]	14.1[4]
Proximal phalanx V	16.2[1,2]	15.8[4]	14.2[1,2]	14.0[4]
Middle phalanx II	16.4[1,2]	16.1[4]	14.2[1,2]	13.9[4]
Middle phalanx III	16.4[2],16.5[1]	16.3[4]	14.4[1],14.5[2]	14.1[4]
Middle phalanx IV	16.4[1,2]	16.3[4]	14.3[1],14.5[2]	14.1[4]
Middle phalanx V	16.3[1],16.4[2]	16.3[4]	14.2[1],14.3[2]	14.0[4]
Distal phalanx I	15.7[2],15.9[1]	15.7[4]	13.5[2],13.6[1]	13.5[4]
Distal phalanx II	15.8[1],16.0[2]	15.8[4]	12.5[3],13.6[1,2]	13.5[4]
Distal phalanx III	16.0[1,2]	15.8[4]	13.6[1,2]	13.5[4]
Distal phalanx IV	15.8[1],16.0[2]	15.6[4]	13.6[1,2]	13.4[4]
Distal phalanx V	15.9[1],16.0[2]	15.7[4]	13.6[1,2]	13.4[4]

1 = Hansman, 1962
2 = Garn et al., 1961
3 = Buehl and Pyle, 1942
4 = Roche et al., 1978
5 = Pyle et al., 1961

(data from Roche et al., 1978)

TABLE 1188

MEDIAN AGE IN MONTHS AT EPIPHYSEAL FUSION FOR SELECTED
HAND-WRIST BONES OF WHITE AND NEGRO BOYS AND GIRLS

Hand-wrist bone	White		Negro	
	Boys	Girls	Boys	Girls
	Median age in months			
Radius........................	--	--	--	--
Ulna.........................	--	194	--	192
Metacarpal I..................	190	165	194	167
Metacarpal II.................	198	178	199	176
Metacarpal III................	199	178	198	175
Metacarpal IV.................	199	179	198	174
Metacarpal V..................	199	180	199	175
Proximal phalanx I............	195	168	198	170
Proximal phalanx II...........	191	168	194	171
Proximal phalanx III..........	193	169	196	171
Proximal phalanx IV...........	193	169	196	171
Proximal phalanx V............	190	168	196	171
Middle phalanx II.............	193	167	197	171
Middle phalanx III............	196	169	199	171
Middle phalanx IV.............	196	169	198	172
Middle phalanx V..............	195	168	198	171
Distal phalanx I..............	188	162	188	160
Distal phalanx II.............	189	162	188	160
Distal phalanx III............	189	162	188	161
Distal phalanx IV.............	187	161	187	159
Distal phalanx V..............	188	161	188	161

(data from Roche et al., 1978)

TABLE 1189

PERCENTILES FOR AGES AT COMPLETE UNION OF METACARPAL AND DIGITAL EPIPHYSES

Center		Boys			Girls		
		5	50	95	5	50	95
Metacarpal	5	15.1	16.5	17.9	12.4	15.0	17.2
	4	15.1	16.4	17.9	12.3	14.6	16.8
	3	15.0	16.4	18.0	12.2	14.7	16.9
	2	15.0	16.4	18.0	12.2	14.6	16.9
	1	14.6	16.3	17.9	11.7	14.1	16.6
Proximal	5	14.3	16.2	17.9	11.8	14.2	16.2
	4	14.6	16.2	17.9	11.3	14.2	16.4
	3	14.4	16.3	17.8	11.8	14.3	16.4
	2	14.5	16.3	17.9	11.8	14.2	16.3
	1	14.7	16.3	17.9	12.1	14.4	16.5
Middle	5	15.1	16.4	17.9	11.8	14.3	16.6
	4	15.1	16.4	17.0	11.9	14.5	16.7
	3	15.1	16.4	17.9	12.1	14.5	16.6
	2	15.0	16.4	17.9	12.1	14.4	16.5
Distal	5	14.2	16.0	17.8	11.8	13.6	15.7
	4	14.2	16.0	17.7	11.9	13.6	15.8
	3	14.2	16.0	17.6	11.6	13.6	15.7
	2	14.2	16.0	17.8	11.7	13.6	15.7
	1	14.1	15.7	17.6	11.4	13.5	15.7

(data from Garn, Rohmann, and Apfelbaum, 1961)

TABLE   1190

FREQUENCY OF BILATERAL ASYMMETRIES IN THE MATURATION OF
THE BONES OF THE HAND AND WRIST: BOYS
IN MICHIGAN

Bone Center	S.D. (months)	Pairs of Hands Symmetrical	Pairs of Hands Asymmetrical	Percentage of Asymmetry
Capitate	1.8	2	1	--
Hamate	2.2	2	0	--
Distal radius	4.7	0	2	--
Proximal 3rd finger	5.3	3	0	--
Proximal 2nd finger	5.0	3	0	--
Proximal 4th finger	5.4	3	2	--
Metacarpal II	5.1	1	1	--
Distal 1st finger	6.2	5	0	--
Metacarpal III	6.4	3	3	--
Proximal 5th finger	5.6	12	2	14.3
Metacarpal IV	7.1	18	5	21.7
Middle 3rd finger	7.6	40	7	14.9
Middle 4th finger	7.8	41	8	16.3
Metacarpal V	8.0	24	11	31.4
Middle 2nd finger	7.5	48	10	17.2
Triquetral	15.9	76	24	24.0
Distal 3rd finger	6.4	33	8	19.5
Distal 4th finger	7.0	39	8	17.0
Metacarpal I	7.3	42	24	36.4
Proximal 1st finger	7.9	76	16	17.4
Distal 2nd finger	7.9	59	26	30.6
Distal 5th finger	7.4	73	18	19.8
Middle 5th finger	11.7	69	18	20.7
Lunate	19.3	63	33	34.4
Navicular	14.1	15	18	54.5
Greater multangular	19.7	12	20	62.5
Lesser multangular	15.2	12	14	53.8
Distal ulna	10.6	3	7	70.0

(data from Baer and Durkatz, 1957)

TABLE 1191

FREQUENCY OF BILATERAL ASYMMETRIES IN THE MATURATION OF
THE BONES OF THE HAND AND WRIST: GIRLS
IN MICHIGAN

Bone Center	S.D. (months)	Pairs of Hands Symmetrical	Pairs of Hands Asymmetrical	Percentage of Asymmetry
Capitate	2.1	1	0	--
Hamate	2.3	1	0	--
Proximal 3rd finger	3.1	4	0	--
Distal radius	4.4	1	3	--
Proximal 2nd finger	3.0	4	0	--
Proximal 4th finger	3.2	4	0	--
Distal 1st finger	5.0	4	1	--
Metacarpal II	3.7	4	1	--
Metacarpal III	4.0	6	0	--
Proximal 5th finger	4.2	4	1	--
Middle 4th finger	4.8	12	0	0.0
Middle 3rd finger	4.9	12	0	0.0
Metacarpal IV	4.1	4	1	--
Metacarpal V	4.7	2	2	--
Middle 2nd finger	5.2	13	2	13.3
Distal 4th finger	5.9	6	4	40.0
Distal 3rd finger	3.9	8	0	--
Metacarpal I	5.3	15	6	28.6
Proximal 1st finger	5.1	17	6	26.1
Triquetral	13.7	46	20	30.3
Middle 5th finger	7.9	35	10	22.2
Distal 5th finger	7.0	26	4	13.3
Distal 2nd finger	6.9	22	7	24.1
Lunate	14.2	81	23	22.1
Greater multangular	14.8	24	42	63.6
Navicular	12.3	30	27	47.4
Lesser multangular	14.8	37	29	43.9
Distal ulna	15.3	5	18	78.3

(data from Baer and Durkatz, 1957)

TABLE 1192

RELATIONSHIPS AMONG SKELETAL AGE, SECONDARY SEX CHARACTERISTICS AND SERUM URIC
ACID CONCENTRATION IN U.S. BOYS 12 TO 17 YEARS OF AGE

Skeletal Age (years)	Percentage at each skeletal age with corresponding pubic hair and genital stage*		
	I	II	III
9	71.4 (71.4)	28.6 (28.6)	0.0 (0.0)
10	63.0 (54.3)	35.9 (42.4)	1.1 (3.3)
11	45.2 (34.6)	48.8 (48.4)	5.5 (16.1)
12	30.6 (22.7)	53.7 (51.4)	15.3 (24.1)
13	19.2 (10.4)	47.2 (43.7)	29.2 (35.8)
14	5.7 (4.0)	23.5 (19.4)	40.0 (32.0)
15	0.2 (0.8)	3.9 (3.0)	17.1 (12.4)
16	0.0 (0.5)	0.7 (0.2)	2.0 (1.2)
17	0.0 (0.3)	0.0 (0.0)	0.5 (1.0)
18	0.0 (0.0)	0.0 (0.0)	0.0 (1.0)
>18	0.0 (0.0)	0.0 (0.0)	0.0 (0.0)

Skeletal Age (years)	Percentage at each skeletal age with corresponding pubic hair and genital stage *		Uric acid (mg/dl) Confidence limits Mean
	IV	V	
9	0.0 (0.0)	0.0 (0.0)	3.7 (3.0-4.4)
10	0.0 (0.0)	0.0 (0.0)	4.0 (3.8-4.2)
11	0.0 (0.5)	0.5 (0.5)	4.0 (3.9-4.2)
12	0.5 (1.9)	0.0 (0.0)	4.2 (4.1-4.4)
13	4.4 (8.8)	0.0 (1.3)	4.6 (4.5-4.7)
14	28.8 (35.5)	2.0 (9.1)	5.1 (5.0-5.2)
15	54.3 (42.1)	24.5 (41.7)	5.6 (5.5-5.7)
16	42.7 (27.6)	54.6 (70.5)	5.8 (5.6-5.9)
17	15.3 (12.1)	84.2 (86.5)	6.0 (5.9-6.1)

TABLE 1192   (continued)

RELATIONSHIPS AMONG SKELETAL AGE, SECONDARY SEX CHARACTERISTICS AND SERUM URIC
ACID CONCENTRATION IN U.S. BOYS 12 TO 17 YEARS OF AGE

Skeletal Age (years)	Percentage at each skeletal age with corresponding pubic hair and genital stage*		Uric acid (mg/dl) Confidence limits Mean
	IV	V	
18	7.1 (6.7)	92.9 (92.4)	6.0 (5.9-6.1)
>18	0.9 (3.4)	99.1 (96.6)	6.1 (5.8-6.3)

* Genital stage in parentheses.

(data from Harlan, Grillo, Cornoni-Huntley, et al., 1979, The Journal
of Pediatrics 95:293-297)

TABLE   1193

SKELETAL AGE, STATURE, WEIGHT AND SELECTED HEAD
MEASUREMENTS IN ST. LOUIS (MO)   FOR BOYS AGED 8.0 to 9.5 YEARS

Variable	Mean	S.D.
CA (mo)	106.80	5.25
Skeletal age (mo)	97.44	14.23
Stature (in)	52.15	2.62
Weight (lb)	65.52	12.60
Head:		
circumference (cm)	52.78	1.53
A-P diameter (cm)	18.99	0.77
Width (cm)	14.66	0.52

(Skeletal age by method of Garn et al., 1967)

(data from Weinberg, Dietz, Penick, et al., 1974. The Journal
of Pediatrics 85:482-489)

TABLE   1194

AGE (years) FOR STAGES OF FORMATION OF THE DECIDUOUS MANDIBULAR LEFT CANINE IN BOSTON CHILDREN

Stage	BOYS Percentiles					GIRLS Percentiles				
	10	25	50	75	90	10	25	50	75	90
$Cr\frac{1}{2}$	--	0.25	0.34	0.43	0.51	--	0.27	0.37	0.49	0.60
Cr_C	0.49	0.58	0.68	0.81	0.92	0.36	0.50	0.62	0.76	0.89
$R\frac{1}{4}$	0.65	0.75	0.88	1.01	1.14	0.49	0.60	0.73	0.88	1.02
$R\frac{1}{2}$	0.89	1.02	1.16	1.32	1.47	0.75	0.88	1.04	1.22	1.39
$R\frac{2}{3}$	1.24	1.39	1.56	1.76	1.94	0.91	1.06	1.24	1.44	1.63
$R\frac{3}{4}$	--	--	--	--	--	1.05	1.21	1.41	1.62	1.83
R_C	1.79	1.98	2.21	2.45	2.68	1.42	1.61	1.85	2.10	2.36
$A\frac{1}{2}$	2.13	2.34	2.59	2.87	3.14	1.74	1.96	2.23	2.53	2.82
$A\frac{3}{4}$	2.62	2.87	3.16	3.49	3.80	--	--	--	--	--
A_C	2.69	2.94	3.24	3.58	3.89	--	--	--	--	--

Cr = crown; R = root; A = apex. Cr_C and A_C = crown and apex complete respectively.

(data from Fanning, 1961)

TABLE   1195

AGE (years) FOR STAGES OF FORMATION OF THE
DECIDUOUS MANDIBULAR LEFT FIRST MOLAR
IN BOSTON CHILDREN

	BOYS Percentiles					GIRLS Percentiles				
Stage	10	25	50	75	90	10	25	50	75	90
Cr_C	0.36	0.49	0.62	0.76	0.90	0.39	0.47	0.57	0.67	0.77
$R\frac{1}{2}$	0.63	0.76	0.92	1.10	1.27	0.65	0.75	0.86	0.99	1.11
$R\frac{2}{3}$	0.76	0.90	1.06	1.26	1.45	--	--	--	--	--
$R\frac{3}{4}$	0.96	1.12	1.32	1.53	1.75	0.95	1.07	1.21	1.36	1.51
R_C	1.27	1.46	1.69	1.95	2.20	--	--	--	--	--
$A\frac{1}{2}$	1.42	1.63	1.88	2.15	2.42	1.47	1.63	1.81	2.01	2.21

Cr = crown, R = root and A = apex. Cr_C and R_C = crown and root- complete respectively.

AGES (years) FOR STAGES OF ROOT RESORPTION
OF THE DECIDUOUS MANDIBULAR LEFT FIRST INCISOR
IN BOSTON CHILDREN

	BOYS Percentiles					GIRLS Percentiles				
Stage	10	25	50	75	90	10	25	50	75	90
$Res\frac{1}{3}$	--	--	--	--	--	4.05	4.38	4.77	5.20	5.61
$Res\frac{1}{2}$	3.78	4.67	5.23	5.85	6.47	4.56	4.93	5.38	5.84	6.30
$Res\frac{2}{3}$	4.89	5.41	6.05	6.76	7.45	5.10	5.50	5.99	6.51	7.01
$Res\frac{3}{4}$	5.08	5.62	6.28	7.01	7.80	5.23	5.65	6.14	6.67	7.19
Exf	5.23	5.79	6.46	7.21	7.95	5.35	5.77	6.27	6.81	7.34

Exf = exfoliated.

(data from Fanning, 1961)

TABLE 1196

AGE (years) FOR STAGES OF FORMATION OF THE
DECIDUOUS MANDIBULAR LEFT SECOND MOLAR
IN BOSTON CHILDREN

	Boys Percentiles					Girls Percentiles				
Stage	10	25	50	75	90	10	25	50	75	90
$Cr\frac{1}{4}$	--	0.31	0.38	0.46	0.54	--	0.27	0.34	0.41	0.47
$Cr\frac{2}{3}$	--	--	--	--	--	0.49	0.56	0.64	0.73	0.81
Cr_C	0.68	0.77	0.88	1.00	1.11	0.66	0.74	0.83	0.93	1.02
$R\frac{1}{4}$	0.93	1.04	1.16	1.31	1.44	--	--	--	--	--
$R\frac{1}{2}$	--	--	--	--	--	--	--	--	--	--
$R\frac{2}{3}$	1.41	1.55	1.71	1.89	2.06	--	--	--	--	--
$R\frac{3}{4}$	1.70	1.86	2.04	2.25	2.44	--	--	--	--	--
R_C	1.92	2.10	2.30	2.52	2.73	1.82	1.96	2.13	2.31	2.48
$A\frac{1}{2}$	2.35	2.55	2.79	3.04	3.29	2.28	2.44	2.64	2.86	3.06

Cr = crown; R = root; A = apex. Cr_C and R_C = crown and root complete respectively.

AGES (years) FOR STAGES OF ROOT RESORPTION
OF THE DECIDUOUS MANDIBULAR LEFT SECOND INCISOR
IN BOSTON CHILDREN

	BOYS Percentiles					GIRLS Percentiles				
Stage	10	25	50	75	90	10	25	50	75	90
Res_i	--	--	--	--	--	3.98	4.35	4.80	5.28	5.76
$Res\frac{1}{4}$	4.57	5.05	5.64	6.29	6.92	4.50	4.91	5.41	5.95	6.48
$Res\frac{1}{3}$	4.69	5.18	5.78	6.44	7.09	4.76	5.19	5.71	6.28	6.84
$Res\frac{1}{2}$	5.07	5.60	6.24	6.95	7.64	5.34	5.82	6.40	7.05	7.64
$Res\frac{2}{3}$	5.76	6.35	7.06	7.86	8.64	5.65	6.16	6.76	7.42	8.07
$Res\frac{3}{4}$	6.10	6.72	7.47	8.30	9.12	5.89	6.41	7.04	7.93	8.40
Exf	6.41	7.06	7.85	8.72	9.58	6.23	6.78	7.45	8.16	8.87

Res_i = root shows blunting or rounding at apex. $Res\frac{1}{4}$, $\frac{1}{3}$, etc. refer to the amount of the root that is resorbed. Exf = exfoliated.

(data from Fanning, 1961)

TABLE 1197

AGE (years) FOR THREE STAGES OF PRIMARY INCISOR ROOT RESORPTION
AND FOR GINGIVAL EMERGENCE OF PERMANENT SUCCESSORS
IN IOWA CHILDREN

				Percentiles		
	10	50	90	10	50	90
Tooth Stage		Males			Females	
Maxillary Central Incisor						
Initial	3.9	5.8	6.7	3.8	5.4	6.6
One-half	5.5	6.3	7.2	5.3	6.2	7.1
Three-fourths	5.8	6.5	7.4	5.6	6.4	7.4
Successor Emergence	6.3	7.0	7.9	6.1	6.8	7.8
Maxillary Lateral Incisor						
Initial	5.4	6.2	7.4	4.8	6.0	7.2
One-half	6.3	7.1	8.0	5.8	6.7	8.1
Three-fourths	6.4	7.3	8.3	6.0	7.1	8.2
Successor Emergence	7.1	8.0	8.9	6.9	7.8	8.8
Mandibular Central Incisor						
Initial	4.5	5.5	6.5	4.5	5.5	6.4
One-half	4.8	5.8	6.7	4.9	5.8	6.7
Three-fourths	5.0	6.0	6.8	5.1	5.9	6.8
Successor Emergence	5.2	6.1	7.0	5.3	6.1	6.8
Mandibular Lateral Incisor						
Intital	5.1	6.2	7.4	5.1	6.0	7.1
One-half	5.7	6.7	7.7	5.6	6.7	7.7
Three-fourths	5.9	6.9	7.9	5.8	6.8	7.8
Successor Emergence	6.2	7.2	8.3	6.2	7.2	8.1

(data from Knott and O'Meara, 1967)

TABLE 1198

TIME (months) FROM THREE STAGES OF PRIMARY INCISOR ROOT RESORPTION
TO GINGIVAL EMERGENCE OF PERMANENT SUCCESSORS
IN IOWA CHILDREN

Interval	Incisor Tooth	Percentiles				
		10	25	50	75	90
Initial Root Resorption to Successor Emergence	Mandibular					
	Central	3	5	7	9	12
	Lateral	7	9	13	16	21
	Maxillary					
	Central	10	12	15	27	35
	Lateral	11	15	20	26	35
One-half Root Resorption to Successor Emergence	Mandibular					
	Central	1	2	3	5	7
	Lateral	3	4	6	7	11
	Maxillary					
	Central	5	6	8	11	16
	Lateral	5	7	11	15	20
Three-fourths Root Resorption to Successor Emergence	Mandibular					
	Central	0	1	2	3	4
	Lateral	1	2	3	5	9
	Maxillary					
	Central	2	4	5	8	11
	Lateral	3	5	7	12	15

(data from Knott and O'Meara, 1967)

TABLE   1199

ROOT DEVELOPMENT OF PERMANENT MANDIBULAR CANINES
AND PREMOLARS ATTAINED AT CLINICAL EMERGENCE
IN BOSTON CHILDREN

		Percentage of individuals having attained			
Tooth	Sex	1/4 Root	1/2 Root	3/4 Root	Full root Length open apex
Canine	Male	0	10	80	10
	Female	0	2	77	22
First premolar	Male	0	27	71	2
	Female	2	25	65	8
Second premolar	Male	6	34	54	6
	Female	6	23	69	2

(data from Moorrees et al., 1962)

TABLE   1200

DEGREE OF ROOT FORMATION AT THE TIME OF ALVEOLAR EMERGENCE
AND FIRST RECORDED APPEARANCE OF PERMANENT
MANDIBULAR FIRST MOLAR TEETH IN ORAL CAVITY

	Alveolar emergence		First presence noted in oral cavity	
Root Stage	No. boys	No. girls	No. boys	No. girls
Total series	21	21	21	17
Beginning of root to 1/4 complete	10	8	0	1
1/4 complete	7	10	5	1
1/4 to 1/3 complete	0	0	1	1
1/3 complete	3	3	2	4
1/3 to 1/2 complete	0	0	0	1
1/2 complete	1	0	9	6
1/2 to 2/3 complete	0	0	1	0
2/3 complete	0	0	2	0
3/4 complete	0	0	0	3
3/4 to almost complete	0	0	1	0

(data from Gleiser and Hunt, 1955)

TABLE   1201

THE MATURATION OF PERMANENT TEETH
IN MICHIGAN CHILDREN

Mandibular Teeth (Growth Stage)

Age (years)	1T1	2T2	3T3	4T4	5T5	6T6	7T7	8T8
Boys								
3	5.2	4.5	3.2	2.6	1.1	5.0	.7	--
4	6.5	5.7	4.2	3.5	2.2	6.2	2.0	--
5	7.5	6.8	5.1	4.4	3.3	7.0	3.0	--
6	8.2	7.7	5.9	5.2	4.3	7.7	4.0	--
7	8.8	8.5	6.7	6.0	5.3	8.4	5.0	.8
8	9.3	9.1	7.4	6.8	6.2	9.0	5.9	1.4
9	9.7	9.5	8.0	7.5	7.0	9.5	6.7	1.8
10	10.0	9.8	8.6	8.2	7.7	9.8	7.4	2.0
11	--	--	9.1	8.8	8.3	9.9	7.9	2.7
12	--	--	9.6	9.4	8.9	--	8.4	3.5
13	--	--	9.8	9.7	9.4	--	8.9	4.5
14	--	--	--	10.0	9.7	--	9.3	5.3
15	--	--	--	--	10.0	--	9.7	6.2
16	--	--	--	--	--	--	10.0	7.3
17	--	--	--	--	--	--	--	7.6
Girls								
3	5.3	4.7	3.4	2.9	1.7	5.0	1.6	--
4	6.6	6.0	4.4	3.9	2.8	6.2	2.8	--
5	7.6	7.2	5.4	4.9	3.8	7.3	3.9	--
6	8.5	8.1	6.3	5.8	4.8	8.1	5.0	--
7	9.3	8.9	7.2	6.7	5.7	8.7	5.9	1.8
8	9.8	9.5	8.0	7.5	6.6	9.3	6.7	2.1
9	10.0	9.9	8.7	8.3	7.4	9.7	7.4	2.3
10	--	10.0	9.2	8.9	8.1	10.0	8.1	3.2
11	--	--	9.7	9.4	8.6	--	8.6	3.7
12	--	--	10.0	9.7	9.1	--	9.1	4.7
13	--	--	--	10.0	9.4	--	9.5	5.8
14	--	--	--	--	9.7	--	9.7	6.5
15	--	--	--	--	10.0	--	9.8	6.9
16	--	--	--	--	--	--	10.0	7.5
17	--	--	--	--	--	--	--	8.0

Maxillary Teeth (Growth Stage)

Age (years)	1|1	2|2	3|3	4|4	5|5	6|6	7|7	8|8
Boys								
3	4.3	3.4	3.0	2.0	1.0	4.2	1.0	--
4	5.4	4.5	3.9	3.0	2.0	5.3	2.0	--
5	6.4	5.5	4.8	4.0	3.0	6.4	3.0	--
6	7.3	6.4	5.6	4.9	4.0	7.4	4.0	--
7	8.2	7.2	6.3	5.7	4.9	8.2	5.0	--
8	8.8	8.0	7.0	6.5	5.8	8.9	5.8	1.0
9	9.4	8.7	7.7	7.2	6.6	9.4	6.5	1.8

TABLE   1201   (continued)

THE MATURATION OF PERMANENT TEETH
IN MICHIGAN CHILDREN

Maxillary Teeth (Growth Stage)

Age (years)	1 1	2 2	3 3	4 4	5 5	6 6	7 7	8 8
				Boys				
10	9.7	9.3	8.4	7.9	7.3	9.7	7.2	2.3
11	9.95	9.7	8.8	8.6	8.0	9.8	7.8	3.0
12	--	9.95	9.3	9.2	8.7	--	8.3	4.0
13	--	--	9.6	9.6	9.3	--	8.8	4.9
14	--	--	9.8	9.8	9.6	--	9.3	5.9
15	--	--	9.9	9.9	9.9	--	9.6	6.6
16	--	--	--	--	--	--	10.0	7.7
17	--	--	--	--	--	--	--	8.0
				Girls				
3	4.3	3.7	3.3	2.6	2.0	4.5	1.8	--
4	5.4	4.8	4.3	3.6	3.0	5.7	2.8	--
5	6.5	5.8	5.3	4.6	4.0	6.9	3.8	--
6	7.4	6.7	6.2	5.6	4.9	7.9	4.7	--
7	8.3	7.6	7.0	6.5	5.8	8.7	5.6	--
8	9.0	8.4	7.8	7.3	6.6	9.3	6.5	2.1
9	9.6	9.1	8.5	8.1	7.4	9.7	7.2	2.4
10	10.0	9.6	9.1	8.7	8.1	10.0	7.9	3.2
11	--	10.0	9.5	9.3	8.7	--	8.5	4.3
12	--	--	9.8	9.7	9.3	--	9.0	5.4
13	--	--	10.0	10.0	9.7	--	9.5	6.2
14	--	--	--	--	10.0	--	9.7	6.8
15	--	--	--	--	--	--	9.8	7.3
16	--	--	--	--	--	--	10.0	8.0
17	--	--	--	--	--	--	--	8.7

Band on 10 stages for each tooth.

(data from Nolla, 1960)

TABLE   1202

AGES (YEARS) OF REACHING DEVELOPMENTAL STAGES
FOR SELECTED PERMANENT TEETH
IN OHIO CHILDREN

| | Boys | | | | | Girls | | | | |
| | Percentiles | | | | | Percentiles | | | | |
Tooth	5th	15th	50th	85th	95th	5th	15th	50th	85th	95th
				Beginning Calcification						
P_1	1.9	2.1	2.4	2.9	3.1	1.6	1.7	2.1	2.4	2.8
P_2	2.8	3.0	3.4	3.9	4.4	2.6	2.9	3.4	4.3	4.8
M_1	0.09	0.10	0.16	0.23	0.24	0.02	0.06	0.14	0.21	0.23
M_2	3.0	3.1	3.6	4.3	5.2	2.7	3.0	3.5	4.3	4.6
M_3	7.4	8.1	9.1	10.4	11.0	7.6	8.3	9.0	10.4	11.4
				Beginning Root Formation						
P_1	6.1	6.6	7.1	7.9	8.2	6.0	6.4	6.9	7.6	7.9
P_2	6.9	7.3	7.9	8.5	9.4	6.6	7.0	7.7	8.5	9.1
M_1	3.2	3.5	4.0	4.3	4.9	2.9	3.1	3.7	4.4	4.7
M_2	7.6	7.9	8.5	9.6	10.4	7.1	7.6	8.4	9.4	10.2
M_3	12.4	13.2	14.2	15.5	16.6	11.8	12.9	14.9	16.9	17.3
				Apical Closure						
P_1	11.3	12.1	12.5	13.7	14.0	11.1	11.4	12.2	12.9	14.1
P_2	12.2	12.5	13.7	15.1	15.5	12.1	12.2	13.1	14.0	14.6
M_1	8.7	9.1	10.1	10.9	11.7	9.0	9.6	10.4	11.2	11.4
M_2	12.7	13.2	14.9	16.7	17.7	13.1	13.3	13.9	16.2	17.1

P_1 and P_2 = first and second premolars; M_1, M_2 and M_3 = first, second and third molars respectively

(data from Garn et al., 1959)

TABLE 1203

TIMING OF MATURATION
OF PERMANENT TEETH (years)
IN OHIO CHILDREN

Mandibular Tooth	Percentiles 5	95
Beginning Calcification		
P_1	1.6	3.0
P_2	2.7	4.7
M_1	0.04	0.24
M_2	2.8	4.8
M_3	7.5	10.9
Crown Completion		
P_1	6.0	8.1
P_2	6.7	9.3
M_1	3.1	4.9
M_2	7.3	10.2
M_3	12.0	17.1
Apical Closure		
P_1	11.2	14.0
P_2	12.1	15.4
M_1	8.8	11.6
M_2	12.8	17.6
M_3	Under 18 to over 26	

P_1 and P_2 = first and second premolars;
M_1, M_2 and M_3 = first, second and third
molars respectively.

(data from Lewis and Garn, 1960)

TABLE 1204

CHRONOLOGY OF CALCIFICATION OF THE PERMANENT MANDIBULAR FIRST MOLAR (months)
IN OHIO CHILDREN

Stage of calcification	Boys		Girls	
	Mean	S.D.	Mean	S.D.
Coalescence of at least 2 centers	7.0	1.4	7.0	1.4
Outline of cusps completed	20.7	3.0	19.5	2.6
Half of crown completed	28.4	4.5	25.9	2.8
2/3 of crown completed	35.1	4.0	32.3	3.0
Crown completed	41.5	5.6	39.3	4.2
Minimal root formation	45.0	4.9	42.3	3.7
1/4 of root completed	69.1	8.1	64.4	4.5
1/3 of root completed	74.1	9.2	69.0	5.3
1/2 of root completed	76.8	8.8	74.2	5.4
2/3 of root completed	84.3	8.4	80.7	5.9
3/4 of root completed	90.0	8.1	84.7	5.9
Root canal terminally divergent	100.4	7.7	94.1	8.1
Root canal terminally convergent	106.6	7.4	102.5	6.0

RELATIONSHIPS OF TIMING OF CALCIFICATION, ERUPTION AND EMERGENCE
OF FIRST PERMANENT MANDIBULAR MOLAR TOOTH (months)
IN OHIO CHILDREN

Stage	Boys		Girls	
	Mean	S.D.	Mean	S.D.
Crown completed	41.5	5.6	39.3	4.2
Minimal root formation	45.0	4.9	42.3	3.7
Alveolar emergence	64.6	8.0	61.2	7.7
1/4 of root completed	69.1	8.1	64.4	4.5
1/3 of root completed	74.1	9.2	69.0	5.3
Clinical emergence (Hurme, '48)	74.5	9.6	71.3	9.6
1/2 of root completed	76.8	8.8	74.2	5.4

(data from Gleiser and Hunt, 1955)

TABLE   1205

AGE AT STAGES OF DEVELOPMENT OF PERMANENT MANDIBULAR TEETH (years)
FOR OHIO CHILDREN

| | | Percentiles | | | | | |
| | | Boys | | | Girls | | |
Tooth	Developmental Stage	5	50	95	5	50	95
1st premolar	Beginning calcification of cusp	1.9	2.4	3.1	1.6	2.1	2.8
	Beginning formation of root	6.1	7.1	8.2	6.0	6.9	7.9
	Apical closure of root	11.3	12.5	14.0	11.1	12.2	14.1
2d premolar	Beginning calcification of cusp	2.8	3.4	4.4	2.6	3.4	4.8
	Beginning formation of root	6.9	7.9	9.4	6.6	7.7	9.1
	Apical closure of root	12.2	13.7	15.5	12.1	13.1	14.6
1st molar	Beginning calcification of cusp	0.1	0.2	0.2	0.0	0.1	0.2
	Beginning formation of root	3.2	4.0	4.9	2.9	3.7	4.7
	Apical closure of root	8.7	10.1	11.7	9.0	10.4	11.4
2d molar	Beginning calcification of cusp	3.0	3.6	5.2	2.7	3.5	4.6
	Beginning formation of root	7.6	8.5	10.4	7.1	8.4	10.2
	Apical closure of root	12.7	14.9	17.7	13.1	13.9	17.1

(data from MEDICAL RADIOGRAPHY AND PHOTOGRAPHY published by Health Sciences Markets
Division, Eastman Kodak Company. Courtesy of S. M. Garn, Ph.D., C. G. Rohmann, and F. N.
Silverman, M. D., 1967)

TABLE   1206

NUMBER OF CENTERS OF CALCIFICATION OBSERVED IN THE PERMANENT MANDIBULAR
FIRST MOLAR FROM RADIOGRAPHS AT THREE MONTHS OF AGE
IN OHIO CHILDREN

Sex	Mean	S.D.
Boys	2.78	1.59
Girls	2.40	1.12

(data from Gleiser and Hunt, 1955)

TABLE 1207

MEAN AGES (YEARS) OF REACHING DEVELOPMENTAL STAGES
FOR SELECTED PERMANENT TEETH
IN OHIO CHILDREN

		Mean	
Tooth and Stage		Boys	Girls
P_1	I	2.2	1.8
P_1	II	7.4	7.2
P_1	III	10.1	9.7
P_1	IV	10.9	10.3
P_1	V	13.0	12.5
P_2	I	3.3	3.3
P_2	II	8.2	8.0
P_2	III	11.1	10.3
P_2	IV	12.2	11.3
P_2	V	14.1	13.4
M_1	II	4.3	4.0
M_1	III	5.8	5.7
M_1	IV	6.9	6.9
M_1	V	10.3	10.6
M_2	I	3.5	3.2
M_2	II	8.9	8.7
M_2	III	11.2	10.7
M_2	IV	12.7	11.8
M_2	V	15.0	14.6
M_3	I	9.0	9.0
M_3	II	14.6	15.2

P_1 and P_2 = first and second premolars; M_1; M_2 and M_3 = first, second and third molars respectively.

(data from Garn et al., 1958)

TABLE 1208

MEAN TIME INTERVAL (years) BETWEEN VARIOUS STAGES OF ROOT FORMATION
IN BOSTON CHILDREN FOR PERMANENT TEETH

Stage interval	Sex	Canine	Pm$_1$	Pm$_2$
Root 1/4 - 1/2	M	2.32	1.70	1.64
	F	1.80	1.66	1.29
Root 1/2 - 3/4	M	1.59	1.36	1.36
	F	1.21	1.05	1.24
Root 3/4 - root full length open apex	M	0.56	0.51	0.86
	F	0.55	0.69	0.67

MEAN AGES (years) OF ATTAINING VARIOUS ROOT FORMATION STAGES
IN BOSTON CHILDREN

		Canine Age of attainment		Pm$_1$ Age of attainment		Pm$_2$ Age of attainment	
Stage	Sex	Mean	S.D.	Mean	S.D.	Mean	S.D.
Root 1/4	M	5.74	0.63	6.88	0.74	7.75	0.83
	F	5.30	0.59	6.49	0.71	7.50	0.80
Root 1/2	M	8.06	0.86	8.58	0.91	9.39	0.98
	F	7.10	0.76	8.15	0.87	8.79	0.93
Root 3/4	M	9.65	1.01	9.94	1.04	10.75	1.11
	F	8.31	0.88	9.20	0.96	10.03	1.04
Root full length; open apex	M	10.21	1.07	10.45	1.08	11.61	1.20
	F	8.86	0.93	9.89	1.03	10.70	1.11

Pm$_1$ and Pm$_2$ = first and second premolars

(data from Fanning, 1962 c)

TABLE 1209

MEAN TIME INTERVAL (years) BETWEEN QUARTER STAGES OF
ROOT FORMATION FOR PERMANENT MANDIBULAR CANINES AND PREMOLARS
IN BOSTON CHILDREN

Tooth	Sex	Root 1/4 - 1/2	Root 1/2 - 3/4	Root 3/4 - full length (open apex)
Canine	Male	2.3	1.6	0.6
	Female	1.8	1.2	0.6
First premolar	Male	1.7	1.4	0.5
	Female	1.7	1.1	0.7
Second premolar	Male	1.6	1.4	0.9
	Female	1.3	1.2	0.7

(data from Moorrees et al., 1962)

TABLE 1210

THIRD MOLAR DEVELOPMENT
IN OHIO CHILDREN

Stage of Development	Median Age (years) Boys	Girls	Combined-Sex Percentiles 15	50	85
Small Follicle	8.7	8.6	7.7	8.6	10.2
Full Follicle	9.2	9.1	8.1	9.1	10.2
Cusp Calcification	9.4	9.6	8.3	9.4	10.8
Crown Completion	13.6	14.2	12.4	13.9	16.0
Root Formation	14.4	15.0	13.3	14.8	16.7
Half Root	16.7	17.2	15.4	16.9	17.9
Alveolar Eruption	16.9	17.1	15.9	17.0	17.9
Cusp Level	17.9	17.8	17.2	17.8	--
Apical Completion	20.2	19.9	17.7	20.0	--

(data from Garn et al., 1962)

TABLE   1211

VARIABILITY OF AGES (YEARS) AT WHICH SELECTED PERMANENT MANDIBULAR
TEETH REACH PARTICULAR MATURITY STAGES AS GIVEN BY VARIOUS AUTHORS

Mandibular Tooth	Logan and Kronfeld (1933)	Schour and Massler (1940)	Spector (1956)	Lewis and Garn (1960)
		Beginning Calcification		
P_1	1.50---2.00	1.50---2.00	1.75	1.6---3.0
P_2	2.00---2.50	2.00---2.50	2.50	2.7---4.7
M_1	0.08---0.33	Birth	Birth	0.04---0.24
M_2	2.00---2.50	2.50---3.00	2.75	2.8---4.8
M_3	7.00---9.00	7.00--10.00	7.00--10.00	7.5--10.9
		Crown Completion		
P_1	--	5.00---6.00	--	6.0---8.1
P_2	--	6.00---7.00	--	6.7---9.3
M_1	--	2.50---3.00	--	3.1---4.9
M_2	--	7.00---8.00	--	7.3--10.2
M_3	--	12.00--16.00	--	12.0--17.1
		Apical Closure		
P_1	--	12.00--13.00	12.00--13.00	11.2--14.0
P_2	--	12.00--14.00	12.00--14.00	12.1--15.4
M_1	--	9.00--10.00	9.00--10.00	8.8--11.6
M_2	--	14.00--16.00	14.00--16.00	12.8--17.6
M_3	--	18.00--25.00	18.00--25.00	Under 18 to over 26

P_1 and P_2 = first and second premolar; M_1, M_2 and M_3 = first, second and third molars respectively.

(data from Lewis and Garn, 1960)

TABLE 1212

AGES (years) AT WHICH DENTAL STAGES ARE PRESENT
FOR WHITE CHILDREN IN NEW YORK CITY

Stages		Lower limit (Mean − 1 S.D.)	Mean (years)	Upper limit (Mean + 1 S.D.)
Edentulous	Male	0	0.18	0.38
	Female	0	0.08	0.24
Deciduous	Male	2.47	3.27	4.07
	Female	2.59	3.33	4.07
I_1M_1	Male	7.07	8.04	9.02
	Female	7.08	8.03	8.99
Pm	Male	10.54	11.73	12.93
	Female	9.87	11.22	12.57
M_2	Male	12.61	13.64	14.67
	Female	12.51	13.60	14.69
M_3 or $\geq$ 19 years	Male	19.81	21.40	22.98
	Female	19.75	21.54	23.34

I_1M_1 = at least three permanent central incisors and at least three
permanent first molars erupted; Pm = at least three premolars erupted;
M_2 = at least three permanent second molars erupted.

(data from Sillman, 1964, American Journal of Orthodontics 50:824-842)

TABLE   1213

AGES (months) AT WHICH DECIDUOUS TEETH ERUPT
IN OHIO CHILDREN

Tooth	Boys		Girls	
	Mean	S.D.	Mean	S.D.
Central Incisor				
Lower (L 1)	7.3	1.6	7.8	2.1
(U 1)	9.1	1.5	9.6	2.0
Lateral Incisors				
Lower (L 2)	13.0	2.8	13.8	3.6
Upper (U 2)	10.4	2.4	11.9	2.7
Canines				
Lower (L 3)	19.3	2.9	20.2	3.4
Upper (U 3)	18.9	2.7	20.1	3.2
First Molars				
Lower (L 4)	16.2	1.9	15.6	2.2
Upper (U 4)	16.0	2.3	15.7	2.3
Second Molars				
Lower (L 5)	25.9	3.8	27.1	4.2
Upper (U 5)	27.6	4.4	28.4	4.3

(data from Robinow et al., 1942)

TABLE   1214

AGE INTERVALS (years) BETWEEN ERUPTION OF CERTAIN DECIDUOUS TEETH
IN SWEDISH CHILDREN

	Boys				Girls			
	Mean	S.D.	Mean	S.D.	Mean	S.D.	Mean	S.D.
First – last tooth	0.38	0.03	--	--	0.41	0.03	--	--
First incisor – earliest	Upper jaw		Lower jaw		Upper jaw		Lower jaw	
first molar	0.55	0.04	0.58	0.04	0.54	0.04	0.47	0.04
First incisor – earliest canine	0.73	0.05	0.61	0.04	0.74	0.06	0.71	0.06
First canine – earliest first molar	0.65	0.05	0.69	0.05	0.64	0.05	0.67	0.05
First canine – earliest second molar	0.59	0.04	0.63	0.05	0.53	0.04	0.62	0.05

(data from Lysell et al., 1962)

TABLE 1215

AGES AT ERUPTION OF FIRST TOOTH AND
NUMBER OF TEETH ERUPTED AT ONE YEAR
FOR INFANTS IN WASHINGTON, D.C.

	Eruption of First Tooth (Age in Weeks)		Number of Teeth at One Year	
	Mean	S.D.	Mean	S.D.
Negro "clinic" infants				
Males	29.0	7.0	5.8	3.0
Females	29.0	8.6	6.1	2.6
Negro "private" infants				
Males	26.4	7.0	6.2	2.0
Females	27.7	6.7	5.6	2.2
White "private" infants				
Males	29.1	7.2	6.9	2.2
Females	31.6	7.5	6.4	2.4
Negro "private" and "clinic" infants				
Males	27.7	7.7	6.0	2.8
Females	28.9	7.8	5.9	2.6

(data from Ferguson, Scott, and Bakwin, 1957, The Journal of Pediatrics 50:327-331)

TABLE 1216

AGE (months) AT ERUPTION OF
FIRST AND LAST DECIDUOUS TOOTH
IN SWEDISH CHILDREN

Tooth	Boys		Girls	
	Mean	S.D.	Mean	S.D.
First tooth	7.63	0.18	8.00	0.25
Last tooth	29.82	0.44	30.08	0.42

(data from Lysell et al., 1962)

TABLE 1217

AGE (months) AT ERUPTION OF DECIDUOUS TEETH.
FOR EACH PAIR OF TEETH, THE FIRST TOOTH TO ERUPT AND THE LAST ARE SHOWN IN COLUMNS A
AND B AND THE MEAN FOR THE PAIR IN COLUMN C:
DATA FROM SWEDISH CHILDREN

| | A Age | | B Age | | C Age | |
Tooth	Mean	S.D.	Mean	S.D.	Mean	S.D.
			Boys			
Lower medial incisor	7.65	0.19	8.11	0.20	7.88	0.19
Upper medial incisor	9.71	0.17	10.17	0.18	10.01	0.17
Lower lateral incisor	12.67	0.28	13.78	0.32	13.23	0.29
Upper lateral incisor	10.83	0.23	11.57	0.25	11.20	0.23
Lower canine	19.58	0.33	20.28	0.34	19.92	0.34
Upper canine	19.02	0.30	19.57	0.31	19.30	0.31
Lower first molar	15.95	0.24	16.84	0.24	16.39	0.23
Upper first molar	15.73	0.25	16.44	0.26	16.08	0.25
Lower second molar	26.69	0.39	27.63	0.42	27.14	0.40
Upper second molar	28.27	0.42	29.50	0.44	28.89	0.42
			Girls			
Lower medial incisor	8.03	0.26	8.37	0.26	8.20	0.26
Upper medial incisor	10.25	0.20	10.69	0.23	10.47	0.21
Lower lateral incisor	12.59	0.36	13.64	0.38	13.11	0.37
Upper lateral incisor	11.04	0.25	12.08	0.30	11.55	0.27
Lower canine	19.20	0.35	19.73	0.34	19.47	0.35
Upper canine	18.84	0.33	19.52	0.34	19.18	0.33
Lower first molar	15.71	0.24	16.53	0.24	16.12	0.24
Upper first molar	15.57	0.21	16.28	0.24	15.93	0.22
Lower second molar	26.60	0.34	27.55	0.36	27.07	0.34
Upper second molar	28.73	0.40	29.97	0.43	29.35	0.41

(data from Lysell et al., 1962)

TABLE 1218

MEAN AGES OF TOOTH EMERGENCE IN U. S. CHILDREN
IN THE TEN-STATE NUTRITION SURVEY

| | Boys | | Girls | |
Tooth	Negro Mean	White Mean	Negro Mean	White Mean
Maxillary				
I1	6.8	7.3	6.8	7.1
I2	8.1	8.4	7.9	7.8
C	10.9	11.5	10.2	10.7
P1	10.2	10.9	9.7	10.4
P2	10.8	11.5	10.5	10.8
M1	6.5	6.2	6.2	6.4
M2	12.3	12.5	12.0	11.7
M3	20.6	24.5	22.2	25.6
Mandibular				
I1	6.0	6.2	5.8	6.4
I2	7.2	7.5	6.7	7.1
C	10.2	10.9	9.4	9.9
P1	10.3	11.0	9.7	10.5
P2	11.1	11.8	10.7	11.1
M1	6.2	6.2	5.9	6.2
M2	11.9	10.2	11.3	11.5
M3	19.1	25.1	19.2	24.4

I1 = first incisor; I2 = second incisor; C = canine; P1 =
first premolar; P2 = second premolar; M1 = first molar;
M2 = second molar; M3 = third molar.

(data from Garn et al., 1972)

TABLE 1219

PERCENTAGE OF INDIVIDUALS HAVING EMERGENCE OF MAXILLARY AND MANDIBULAR THIRD MOLARS
WITH AND WITHOUT EXTRACTION OF OTHER PERMANENT TEETH IN BOSTON CHILDREN

| | With extraction of other permanent teeth | | | | Without extraction of other permanent teeth | | | |
| | Maxilla percentage | | Mandible percentage | | Maxilla percentage | | Mandible percentage | |
Age (years)	Males	Females	Males	Females	Males	Females	Males	Females
13.0-13.9	0	5.3	0	4.6	0	0	0	0
14.0-14.9	1.5	4.8	4.3	1.4	0	0	0.7	1.5
15.0-15.9	5.7	10.7	12.6	5.6	1.7	3.4	2.3	7.6
16.0-16.9	16.5	10.0	32.2	40.0	6.1	2.9	8.1	8.8
17.0-17.9	29.7	52.4	38.0	56.4	14.1	10.5	15.9	14.0
18.0-18.9	43.3	46.9	58.5	44.4	15.1	19.6	23.1	25.0

(data from Fanning, 1962a)

TABLE 1220

AGES (years) AT ERUPTION FOR THE PERMANENT TEETH OF WHITE CHILDREN LIVING IN THE NORTHERN TEMPERATE ZONE

Order of emergence	Tooth		Mean ages of emergence		S.D. (both sexes)	Sex difference
	Max.	Mand.	Boys	Girls		
1	–	M_1	6.21	5.94	0.80	0.27
2	M_1	–	6.40	6.22	0.80	0.18
3	–	I_1	6.54	6.26	0.78	0.28
4	I_1	–	7.47	7.20	0.81	0.27
5	–	I_2	7.70	7.34	0.88	0.36
6	I_2	–	8.67	8.20	0.98	0.47
7, boys 8, girls	Pm_1	–	10.40	10.03	1.47	0.37
8, boys 7, girls	–	C	10.79	9.86	1.27	0.93
9	–	Pm_1	10.82	10.18	1.47	0.64
10	Pm_2	–	11.18	10.88	1.57	0.30
11	–	Pm_2	11.47	10.89	1.68	0.58
12	C	–	11.69	10.98	1.37	0.71
13	–	M_2	12.12	11.66	1.36	0.46
14	M_2	–	12.68	12.27	1.37	0.41

(data from Hurme, 1949)

TABLE 1221

AGE (years) OF GINGIVAL EMERGENCE OF PERMANENT TEETH
IN IOWA CHILDREN

| | Maxillary Arch | | | | | |
| | Means for Girls | | Means for Boys | | Standard Deviation* | |
Tooth	Right	Left	Right	Left	Girls	Boys
Central incisor	6.9	6.9	7.0	7.0	0.81	0.54
Lateral incisor	7.9	7.9	8.0	8.1	0.74	0.77
Canine	10.8	11.0	11.2	11.2	1.35	1.00
First premolar	10.3	10.3	10.5	10.5	1.23	0.97
Second premolar	11.1	11.2	11.4	11.4	1.21	0.99
First molar	6.2	6.2	6.1	6.2	0.61	0.51
Second molar	11.9	11.9	11.9	11.9	0.93	0.74

| | Mandibular Arch | | | | | |
| | Means for Girls | | Means for Boys | | Standard Deviation* | |
Tooth	Right	Left	Right	Left	Girls	Boys
Central incisor	6.0	6.0	6.0	6.0	0.60	0.65
Lateral incisor	7.1	7.1	7.2	7.2	0.76	0.66
Canine	9.8	9.8	10.5	10.5	1.10	0.92
First premolar	10.1	10.1	10.6	10.6	1.04	0.87
Second premolar	11.0	10.8	11.3	11.4	1.18	0.93
First molar	6.0	6.0	6.1	6.1	0.63	0.46
Second molar	11.2	11.2	11.6	11.5	0.96	0.87

*Composite S.D.s for each sex calculated from S.D.s for homologous teeth on the right and
left sides of the arch.

(data from Knott and Meredith, 1966)

TABLE 1222

AGE OF GINGIVAL EMERGENCE OF PERMANENT TEETH IN WHITE IOWA BOYS

Tooth	Maxillary Arch			Mandibular Arch		
	Right Mean	Left Mean	S.D.*	Right Mean	Left Mean	S.D.*
Central incisor	7.0	7.0	0.54	6.0	6.0	0.65
Lateral incisor	8.0	8.1	0.77	7.2	7.2	0.66
Canine	11.2	11.2	1.00	10.5	10.5	0.92
First premolar	10.5	10.5	0.97	10.6	10.6	0.87
Second premolar	11.4	11.4	0.99	11.3	11.4	0.93
First molar	6.1	6.2	0.51	6.1	6.1	0.46
Second molar	11.9	11.9	0.74	11.6	11.5	0.87

* Composite S.D.'s calculated from the S.D.'s for homologous teeth on the right and left sides of the arch.

(data from Sturdivant et al., 1962)

TABLE 1223

VARIABILITY (S.D.) IN PERMANENT
TOOTH EMERGENCE TIMING (years)
FOR CHILDREN IN THE TEN-STATE NUTRITION SURVEY

Tooth	White		Black	
	Boys	Girls	Boys	Girls
Maxilla				
I1	0.77	0.75	0.82	0.84
I2	1.01	0.91	0.99	1.14
C	1.39	1.40	1.63	1.59
P1	1.41	1.38	1.54	1.46
P2	1.48	1.56	1.55	1.54
M1	0.79	0.74	0.79	0.93
M2	1.34	1.22	1.36	1.32
Mandible				
I1	0.81	0.79	0.77	0.87
I2	0.78	0.82	1.05	0.88
C	1.14	1.26	1.52	1.61
P1	1.37	1.28	1.38	1.29
P2	1.61	1.50	1.44	1.45
M1	0.79	0.76	0.77	0.85
M2	1.38	1.23	1.47	1.33

I1 and I2 = central and lateral incisors; C = canine; P1 and P2 = first and second premolars and M1 and M2 = first and second molars.

(data from Garn, Nagy, Sandusky and Trowbridge, 1973)

TABLE 1224

RACIAL DIFFERENCE IN THE TIMING
OF PERMANENT TOOTH EMERGENCE (years)

| | BOYS | | GIRLS | |
Tooth	Caucasoid Mean	Negro Mean	Caucasoid Mean	Negro Mean
		Maxilla		
I1	7.34	6.93	6.98	6.74
I2	8.39	7.95	7.97	7.67
C	11.29	10.93	10.62	10.54
P1	10.64	10.45	10.17	10.00
P2	11.21	11.09	10.88	10.71
M1	6.40	6.30	6.35	6.00
M2	12.44	12.30	11.95	11.63
		Mandible		
I1	6.30	6.07	6.18	5.89
I2	7.47	6.97	7.13	6.58
C	10.52	10.34	9.78	9.71
P1	10.70	10.47	10.17	10.00
P2	11.43	11.10	10.97	10.74
M1	6.33	6.11	6.15	5.73
M2	12.00	11.90	11.49	11.20

I1 and I2 = central and lateral incisor; C = canine; P1 and P2 =
first and second premolar and M1 and M2 = first and second molar.

(Reprinted with permission from Archives of Oral Biology 18; Garn,
Sandusky, Nagy, et al., 1973, Pergamon Press, Ltd.)

TABLE 1225

MEDIAN AGES (years) OF ERUPTION OF
PERMANENT TEETH IN PIMA INDIANS

| | Maxillary Teeth | | | | | | |
	$_1I^1$	$_2I^2$	$_1C^1$	$_1Pm^1$	$_2Pm^2$	$_1M^1$	$_2M^2$
				Boys			
Median	7.83	8.74	11.66	10.08	11.33	5.98	11.67
S.D.	.71	.75	1.41	1.28	1.36	.77	1.21
				Girls			
Median	7.47	8.34	10.94	9.63	10.73	5.80	10.38
S.D.	.69	.98	1.59	1.23	1.34	.79	1.29

| | Mandibular Teeth | | | | | | |
	$_1I_1$	$_2I_2$	$_1C_1$	$_1Pm_1$	$_2Pm_2$	$_1M_1$	$_2M_2$
				Boys			
Median	6.26	7.65	10.78	10.43	11.39	5.89	11.29
S.D.	.89	1.21	1.28	1.29	1.45	.67	1.25
				Girls			
Median	6.15	7.32	9.66	9.87	10.73	5.43	10.80
S.D.	.75	1.02	1.25	1.30	1.46	.99	1.05

$_1I^1$ and $_2I^2$ = central and lateral incisors; $_1C^1$ = canine; $_1Pm^1$ and $_2Pm^2$ = first and second premolars and $_1M^1$ and $_2M^2$ = first and second molars.

(data from Dahlberg and Menegaz-Bock 1958)

TABLE 1226

AGE (years) OF CLINICAL ERUPTION OF PERMANENT
TEETH IN MONTREAL CHILDREN (complete crown emerged)

Tooth	BOYS		GIRLS	
	Mean	S.D.	Mean	S.D.
Left Maxilla				
I_1	7.66	1.11	7.32	1.11
I_2	8.68	1.12	8.25	1.11
C	11.75	1.12	11.59	1.17
PM_1	10.83	1.12	10.78	1.15
PM_2	11.50	1.13	11.42	1.16
M_1	6.83	1.11	6.79	1.11
M_2	12.50	1.10	12.42	1.09
Left Mandible				
I_1	6.90	1.11	6.57	1.10
I_2	7.88	1.11	7.52	1.11
C	11.16	1.12	10.40	1.11
PM_1	11.44	1.13	11.03	1.12
PM_2	12.12	1.15	11.61	1.12
M_1	6.98	1.19	6.66	1.12
M_2	12.36	1.11	11.93	1.08

I_1 and I_2 = central and lateral incisors; C = canine; PM_1 and PM_2 = first and second premolar; M_1 and M_2 = first and second molar.

(data from Perreault et al., 1975)

TABLE 1227

AGES (years) AT CLINICAL ERUPTION OF PERMANENT TEETH
ON THE LEFT SIDE OF FRENCH-CANADIAN BOYS

	Maxillary	
Tooth	Mean	S.D.
I_2	7.33	1.11
I_2	8.24	1.12
C	11.11	1.12
Pm_1	10.19	1.14
Pm_2	10.04	1.13
M_1	6.22	1.14
M_2	12.14	1.13

	Mandibular	
I_1	6.27	1.11
I_2	7.42	1.11
C	10.45	1.11
Pm_1	11.08	1.15
Pm_2	11.61	1.16
M_1	6.22	1.15
M_2	11.72	1.11

I_1 and I_2 = central and lateral incisor;
C = canine; Pm_1 and Pm_2 = first and second
premolars; M_1 and M_2 = first and second molar.

(data from Perreault at al., 1974)

TABLE   1228

AGES (years) AT CLINICAL ERUPTION OF PERMANENT TEETH
ON THE LEFT SIDE OF FRENCH-CANADIAN GIRLS

| Tooth | Maxilla | |
	Mean	S.D.
I_1	6.89	1.11
I_2	7.80	1.10
C	10.66	1.15
Pm_1	10.08	1.16
Pm_2	10.81	1.16
M_1	6.22	1.14
M_2	12.01	1.13
	Mandible	
I_1	5.98	1.12
I_2	7.07	1.12
C	9.62	1.11
Pm_1	10.44	1.15
Pm_2	11.18	1.17
M_1	5.95	1.16
M_2	11.32	1.11

I_1 and I_2 = medial and lateral central incisors;
C = canine; Pm1 and Pm2 = first and second molars;
M1 and M2 = first and second molars.

(data from Perreault et al., 1974)

TABLE   1229

AGE (years) OF EMERGENCE OF PERMANENT TEETH
FOR INDIAN CHILDREN IN NORTHERN
ONTARIO AND FRENCH-CANADIAN CHILDREN

	Indian		French-Canadian	
	Mean	S.D.	Mean	S.D.
BOYS				
Maxilla				
I^1	7.61	0.84	7.33	1.11
I^2	8.58	0.58	8.24	1.12
C	10.87	1.15	11.11	1.12
Pm^1	9.68	1.69	10.19	1.14
Pm^2	10.90	1.82	10.04	1.13
M^1	6.23	0.63	6.22	1.14
M^2	11.87	1.40	12.14	1.13
Mandible				
I_1	6.22	0.48	6.27	1.11
I_2	7.18	0.69	7.42	1.11
C	10.36	0.95	10.45	1.11
Pm_1	10.83	1.09	11.08	1.15
Pm_2	11.43	1.62	11.61	1.16
M_1	5.75	1.19	6.22	1.15
M_2	11.71	1.46	11.72	1.11
GIRLS				
Maxilla				
I^1	7.41	0.69	6.89	1.11
I^2	8.23	0.78	7.80	1.10
C	10.67	1.51	10.66	1.15
Pm^1	9.66	1.96	10.08	1.16
Pm^2	10.06	1.63	10.81	1.16
M^1	5.88	0.69	6.22	1.14
M^2	11.90	0.92	12.01	1.13
Mandible				
I_1	5.84	0.56	5.98	1.12
I_2	6.94	0.64	7.07	1.12
C	9.46	1.20	9.62	1.11
Pm_1	10.30	1.49	10.44	1.15
Pm_2	10.69	1.86	11.18	1.17
M_1	5.49	0.79	5.95	1.16
M_2	11.64	1.80	11.32	1.11

I^1 and I^2 = central and lateral incisors; C = canine; Pm^1 and
Pm^2 = first and second premolars and M_1 and M_2 = first and
second molars.

(data from Mayhall et al., 1977)

TABLE   1230

AGE (years) OF EMERGENCE OF PERMANENT TEETH IN CANADIAN ESKIMOS

Tooth	Foxe Basin Eskimo		French-Canadians	
	Mean	S.D.	Mean	S.D.
Males				
Maxilla				
I	7.43	0.76	7.33	1.11
I	8.47	0.57	8.24	1.12
C	11.10	1.52	11.11	1.12
Pm	9.57	1.62	10.19	1.14
Pm	10.70	1.66	10.04	1.13
M	5.61	0.64	6.22	1.14
M	11.39	1.36	12.14	1.13
Mandible				
I	6.32	0.67	6.27	1.11
I	7.20	0.85	7.42	1.11
C	9.52	1.27	10.45	1.11
Pm	10.32	1.84	11.08	1.15
Pm	11.38	2.46	11.61	1.16
M	5.40	0.87	6.22	1.15
M	10.78	1.43	11.72	1.11
Females				
Maxilla				
I	6.84	0.90	6.89	1.11
I	7.67	0.99	7.80	1.10
C	10.01	1.18	10.66	1.15
Pm	9.02	1.52	10.08	1.16
Pm	9.83	0.92	10.81	1.16
M	5.58	0.58	6.22	1.14
M	10.69	1.30	12.01	1.13
Mandible				
I	6.08	0.66	5.98	1.12
I	6.86	0.86	7.07	1.12
C	8.90	1.29	9.62	1.11
Pm	8.97	0.69	10.44	1.15
Pm	10.36	1.25	11.18	1.17
M	5.54	0.80	5.95	1.16
M	10.45	1.10	11.32	1.11

(data from Mayhall et al., 1978)

TABLE   1231

THE PERCENTAGES OF EACH PERMANENT TOOTH ERUPTED OR ERUPTING
INTO THE MOUTH AT SPECIFIED AGES
IN AUSTRALIAN CHILDREN

Upper Jaw

Age (years)	BOYS I_1	I_2	C	Pm_1	Pm_2	M_1	M_2	GIRLS I_1	I_2	C	Pm_1	Pm_2	M_1	M_2
4- 5	--	--	--	--	--	--	--	--	--	--	--	--	--	--
5- 6	--	--	--	--	--	9	--	--	--	--	--	--	14	--
6- 7	10	3	--	--	--	62	--	21	4	--	--	--	65	--
7- 8	63	21	--	--	--	100	--	81	47	--	8	1	96	--
8- 9	87	70	--	14	4	100	--	99	73	--	14	7	100	--
9-10	100	93	5	20	10	100	7	100	97	15	46	19	100	10
10-11	100	100	23	66	40	100	20	100	100	37	54	32	100	13
11-12	100	100	50	81	67	100	22	100	100	93	95	80	100	64
12-13	100	100	87	97	68	100	76	100	100	95	97	94	100	87
13-14	100	100	90	97	93	100	93	100	100	97	98	100	100	100
14-15	100	100	98	100	98	100	100	100	100	99	99	97	100	100
15-16	100	100	99	100	100	100	100	100	100	100	100	100	100	100

Lower Jaw

Age (years)	BOYS I_1	I_2	C	Pm_1	Pm_2	M_1	M_2	GIRLS I_1	I_2	C	Pm_1	Pm_2	M_1	M_2
4- 5	--	--	--	--	--	--	--	--	--	--	--	--	--	--
5- 6	18	--	--	--	--	12	--	15	2	--	--	--	15	--
6- 7	60	10	--	--	--	72	--	77	21	--	--	--	86	--
7- 8	100	57	--	--	--	100	--	100	68	--	3	1	100	--
8- 9	100	88	5	8	--	100	--	100	96	14	12	1	100	--
9-10	100	100	16	9	--	100	6	100	100	44	26	12	100	19
10-11	100	100	50	40	19	100	24	100	100	78	53	26	100	31
11-12	100	100	74	67	35	100	44	100	100	97	86	62	100	65
12-13	100	100	96	74	62	100	72	100	100	100	92	87	100	94
13-14	100	100	98	93	86	100	93	100	100	100	96	90	100	98
14-15	100	100	100	98	91	100	100	100	100	100	100	97	100	100
15-16	100	100	100	100	100	100	100	100	100	100	100	100	100	100

(data from Halikis, 1961b)

TABLE 1232

PERCENTAGE OF TEETH ERUPTED AT PARTICULAR AGES IN LONDON BOYS

Age (yrs.)	Lwr. 6	Upr. 6	Lwr. 1	Upr. 1	Lwr. 2	Upr. 2	Upr. 4	Lwr. 4	Lwr. 3	Upr. 5	Upr. 3	Lwr. 5	Lwr. 7	Upr. 7
5	2	6	4	--	--	--	--	--	--	--	--	--	--	--
5¼	5	3	2	--	--	--	--	--	--	--	--	--	--	--
5½	9	4	4	2	--	--	--	--	--	--	--	--	--	--
5¾	20	19	11	--	--	--	--	--	--	--	--	--	--	--
6	41	34	31	4	6	2	--	--	--	--	--	--	--	--
6¼	34	33	36	9	4	4	--	--	--	--	--	--	--	--
6½	60	53	42	13	6	6	--	--	--	--	--	--	--	--
6¾	68	70	73	35	13	4	--	--	--	--	--	--	--	--
7	76	74	80	36	31	7	--	--	--	--	--	--	--	--
7¼	85	85	90	43	24	--	--	--	--	--	--	--	--	--
7½	94	94	94	55	39	9	4	--	--	--	--	--	--	--
7¾	96	91	94	62	43	18	--	--	--	--	--	--	--	--
8	--	--	95	81	65	24	10	2	--	--	--	--	--	--
8¼	--	--	--	84	72	44	7	--	--	--	--	--	--	--
8½	--	--	--	93	90	50	9	--	--	2	--	6	--	--
8¾	--	--	--	96	94	68	10	7	--	1	--	6	--	--
9	--	--	--	--	94	77	13	4	--	2	--	2	--	--
9¼	--	--	--	--	97	81	12	1	--	2	2	--	--	--
9½	--	--	--	--	98	91	28	15	5	17	--	3	--	--
9¾	--	--	--	--	--	91	35	18	2	17	6	4	--	--
10	--	--	--	--	--	97	25	17	15	8	5	4	--	--
10¼	--	--	--	--	--	--	41	28	31	19	13	9	1	--
10½	--	--	--	--	--	--	37	27	21	15	12	8	2	2
10¾	--	--	--	--	--	--	61	42	45	31	21	16	7	5
11	--	--	--	--	--	--	61	62	53	45	33	34	27	26
11¼	--	--	--	--	--	--	68	49	58	49	40	25	22	20
11½	--	--	--	--	--	--	65	63	68	41	29	30	16	16
11¾	--	--	--	--	--	--	78	59	79	55	52	42	37	30
12	--	--	--	--	--	--	85	75	76	65	52	53	47	43
12¼	--	--	--	--	--	--	89	77	81	68	66	56	61	55
12½	--	--	--	--	--	--	84	83	94	67	67	55	54	50
12¾	--	--	--	--	--	--	92	77	90	75	68	58	57	53
13	--	--	--	--	--	--	94	86	93	78	71	60	79	74
13¼	--	--	--	--	--	--	98	93	97	86	93	83	72	62
13½	--	--	--	--	--	--	--	--	98	90	90	87	86	94
13¾	--	--	--	--	--	--	--	--	98	93	91	79	75	82
14	--	--	--	--	--	--	--	--	--	92	93	90	91	91
14¼	--	--	--	--	--	--	--	--	--	--	--	--	95	89

1 and 2 = central and lateral incisors; 3 = canine; 4 and 5 = first and second premolars;
6 and 7 = first and second molars.
(data from Parfitt, 1954)

TABLE   1233

PERCENTAGE OF TEETH ERUPTED AT PARTICULAR AGES IN LONDON GIRLS

Age (yrs.)	Lwr. 1	Lwr. 6	Upr. 6	Upr. 1	Lwr. 2	Upr. 2	Lwr. 3	Upr. 4	Lwr. 4	Upr. 5	Upr. 3	Lwr. 5	Lwr. 7	Upr. 7
5	12	--	--	--	--	--	--	--	--	--	--	--	--	--
5¼	5	1	1	--	--	--	--	--	--	--	--	--	--	--
5½	17	22	17	1	--	--	--	--	--	--	--	--	--	--
5¾	32	25	22	6	--	--	--	--	--	--	--	--	--	--
6	33	36	33	4	3	--	--	--	--	--	--	--	--	--
6¼	62	52	57	10	11	3	--	--	--	--	--	--	--	--
6½	63	62	59	21	6	3	--	--	--	--	--	--	--	--
6¾	79	70	69	27	20	3	--	--	--	--	--	--	--	--
7	94	95	83	55	42	5	--	--	--	--	--	--	--	--
7¼	91	95	95	50	35	7	--	--	--	--	--	--	--	--
7½	93	98	97	79	61	16	--	1	--	--	--	--	--	--
7¾	--	97	97	89	67	30	--	1	5	--	--	--	--	--
8	--	--	97	82	73	50	2	1	5	--	--	--	--	--
8¼	--	--	--	94	83	58	4	7	4	--	--	--	--	--
8½	--	--	--	94	94	63	11	9	6	5	--	5	--	--
8¾	--	--	--	97	94	91	26	13	15	6	2	6	--	--
9	--	--	--	--	99	84	19	17	9	5	3	4	--	--
9¼	--	--	--	--	97	91	21	29	23	13	6	6	--	--
9½	--	--	--	--	98	92	21	19	10	11	7	4	--	--
9¾	--	--	--	--	--	96	45	35	35	18	16	22	--	--
10	--	--	--	--	--	91	53	47	45	31	22	34	11	4
10¼	--	--	--	--	--	--	57	54	57	30	26	25	8	8
10½	--	--	--	--	--	--	65	53	59	35	27	24	19	8
10¾	--	--	--	--	--	--	60	56	60	30	32	36	18	13
11	--	--	--	--	--	--	74	75	68	53	47	44	30	23
11¼	--	--	--	--	--	--	90	74	65	57	54	48	35	35
11½	--	--	--	--	--	--	85	82	78	52	57	62	44	36
11¾	--	--	--	--	--	--	92	79	69	58	66	61	47	50
12	--	--	--	--	--	--	96	87	87	67	66	51	51	51
12¼	--	--	--	--	--	--	85	90	94	77	75	72	64	47
12½	--	--	--	--	--	--	--	95	95	75	86	79	82	78
12¾	--	--	--	--	--	--	--	96	96	93	82	85	91	82
13	--	--	--	--	--	--	98	--	81	93	75	77	81	93
13¼	--	--	--	--	--	--	--	--	--	91	95	93	87	85
13½	--	--	--	--	--	--	--	--	--	--	--	95	90	70
13¾	--	--	--	--	--	--	--	--	--	--	--	97	91	88
14	--	--	--	--	--	--	--	--	--	91	--	91	83	90
14¼	--	--	--	--	--	--	--	--	--	--	--	--	--	95

1 and 2 = central and lateral incisors; 3 = canine; 4 and 5 = first and second premolars;
6 and 7 = first and second molars.
(data from Parfitt, 1954)

TABLE   1234

ECONOMIC IMPACT ON TOOTH EMERGENCE TIMING (years) FOR CHILDREN IN THE TEN–STATE NUTRITION SURVEY

| | White children Income-needs group | | | | Black children Income-needs group | | | |
Tooth	Lower Mean	Higher Mean	d^1	Z^2	Lower Mean	Higher Mean	d^1	Z^2
BOYS								
Maxilla								
I1	7.40	7.12	0.28	0.36	6.96	6.79	0.17	0.21
I2	8.51	8.10	0.41	0.41	7.97	7.74	0.23	0.23
C	11.45	11.12	0.33	0.24	10.97	10.42	0.55	0.34
P1	10.63	10.67	-0.04	-0.03	10.45	10.20	0.25	0.16
P2	11.17	11.30	-0.13	-0.09	11.22	10.82	0.40	0.26
M1	6.47	6.24	0.23	0.29	6.25	6.12	0.13	0.17
M2	12.45	12.52	-0.07	-0.05	12.32	12.59	-0.27	-0.20
Mandible								
I1	6.37	6.23	0.14	0.17	6.11	5.56	0.55	0.71
I2	7.54	7.34	0.20	0.26	6.98	6.82	0.16	0.15
C	10.61	10.46	0.15	0.13	10.38	10.21	0.17	0.11
P1	10.79	10.57	0.22	0.16	10.40	10.43	-0.03	-0.02
P2	11.47	11.53	-0.06	-0.04	11.18	10.73	0.45	0.31
M1	6.36	6.35	0.01	0.01	6.10	5.89	0.21	0.27
M2	12.05	11.98	0.07	0.05	11.96	12.38	-0.42	-0.29
Mean difference	--	--	--	0.13	--	--	--	0.17
GIRLS								
Maxilla								
I1	7.02	6.77	0.25	0.33	6.75	6.77	-0.02	-0.02
I2	7.95	7.99	-0.04	-0.04	7.64	7.26	0.38	0.33
C	10.78	10.49	0.29	0.21	10.66	10.28	0.38	0.24
P1	10.18	10.23	-0.05	-0.04	10.06	10.05	0.01	0.01
P2	10.87	10.83	0.04	0.03	10.73	10.69	0.04	0.03
M1	6.35	6.30	0.05	0.07	5.95	6.61	-0.66	-0.71
M2	12.01	12.01	0.00	0.00	11.61	11.71	-0.10	-0.08
Mandible								
I1	6.29	5.92	0.37	0.47	5.87	5.66	0.21	0.24
I2	7.15	7.05	0.10	0.12	6.55	6.82	-0.27	-0.31
C	9.84	9.69	0.15	0.12	9.81	9.01	0.80	0.50
P1	10.15	10.15	0.00	0.00	10.09	9.41	0.68	0.53
P2	10.96	11.07	-0.11	-0.07	10.75	10.93	-0.18	-0.12
M1	6.15	6.13	0.02	0.03	5.67	6.57	-0.90	-1.06
M2	11.50	11.49	0.01	0.01	11.21	11.25	-0.04	-0.03
Mean difference	--	--	--	0.09	--	--	--	-0.03

TABLE 1234 (continued)

ECONOMIC IMPACT ON TOOTH EMERGENCE TIMING (years)
FOR CHILDREN IN THE TEN-STATE NUTRITION SURVEY

1 - d represents the difference in years and decimals.
2 - Z represents the difference in years expressed in standard deviation units.
I1 and I2 = central and lateral incisors; C = canine; P1 and P2 = first and second premolars and M1 and M2 = first and second molars.

(data from Garn, Nagy, Sandusky and Trowbridge, 1973)

TABLE 1235

PERCENTAGE OF BOSTON FEMALES WITH
FOUR THIRD MOLARS EMERGED

Age (years)	With extraction of other permanent teeth	Percentage	Without extraction of other permanent teeth
13.0-13.9	0		0
14.0-14.9	0		0
15.0-15.9	0		2
16.0-16.9	0		3
17.0-17.9	19		3
18.0-18.9	16		8

(data from Fanning, 1962a)

TABLE   1236

PERCENTAGE FREQUENCY WITH WHICH COMPARABLE TEETH ON THE TWO
SIDES OF THE DENTAL ARCHES DIFFER IN TIME OF GINGIVAL EMERGENCE
IN IOWA CHILDREN

Differences Categories (months)	Incisor Teeth				Canine Teeth		Premolar Teeth				Molar Teeth			
	Central		Lateral				First		Second		First		Second	
	G*	B*	G	B	G	B	G	B	G	B	G	B	G	B
Maxillary Arch														
Less than 1**	36	44	28	32	20	36	32	25	12	30	52	33	24	26
1 - 3	36	35	24	26	20	18	18	27	26	19	16	33	32	25
3 - 6	20	19	38	30	38	27	24	22	32	21	28	27	26	32
6 - 9	4	2	2	7	14	11	4	13	16	19	--	--	10	11
9 - 12	4	--	6	5	4	3	10	5	4	5	--	--	6	3
12 - 18	--	--	2	--	--	5	4	2	8	2	--	--	2	3
18 - 24	--	--	--	--	2	--	6	3	2	2	--	--	--	--
More than 24	--	--	--	--	2	--	2	3	--	2	--	--	--	--
Mandibular Arch														
Less than 1	58	44	30	35	26	30	14	33	24	26	44	63	32	29
1 - 3	28	37	32	32	20	28	26	26	16	18	30	26	24	34
3 - 6	14	15	32	24	30	31	36	25	24	39	20	11	24	23
6 - 9	--	2	4	7	18	9	10	12	22	9	6	--	12	9
9 - 12	--	2	2	2	--	--	6	2	8	3	--	--	4	5
12 - 18	--	--	--	--	4	2	6	2	2	--	--	--	4	--
18 - 24	--	--	--	--	--	--	--	--	--	3	--	--	--	--
More than 24	--	--	--	--	2	--	2	--	4	2	--	--	--	--

 * The letters G and B represent girls and boys, respectively.
** Difference = eruption age for a specified tooth on one side of a designated arch
   minus eruption age of the corresponding tooth on the other side of the arch.
   Differences tabulated without regard to sign.

(data from Knott and Meredith, 1966)

TABLE 1237

DIFFERENCE IN AGES AT ERUPTION (months)
BETWEEN THE TWO SIDES OF THE MOUTH
IN SWEDISH CHILDREN

Tooth	Boys		Girls	
	Mean	S.D.	Mean	S.D.
Lower medial incisor	+0.18	0.08	+0.11	0.07
Upper medial incisor	+0.13	0.06	+0.12	0.09
Lower lateral incisor	+0.36	0.17	+0.31	0.17
Upper lateral incisor	+0.14	0.11	+0.16	0.18
Lower canine	−0.08	0.10	−0.07	0.10
Upper canine	+0.21	0.09	+0.20	0.12
Lower first molar	+0.25	0.12	+0.16	0.14
Upper first molar	+0.17	0.11	+0.07	0.12
Lower second molar	+0.33	0.14	+0.33	0.14
Upper second molar	+0.42	0.17	+0.08	0.21

Positive values denote eruption of a tooth on the left before eruption
of the corresponding tooth on the right, negative values the contrary.

(data from Lysell et al., 1962)

TABLE   1238

PERCENTAGE FREQUENCIES WITH WHICH CORRESPONDING TEETH
OF THE RIGHT MAXILLARY AND MANDIBULAR ARCHES
DIFFER IN TIME OF GINGIVAL ERUPTION
IN IOWA CHILDREN

Difference Categories (months)	Incisor Teeth				Canine Teeth		Premolar Teeth				Molar Teeth			
	Central		Lateral				First		Second		First		Second	
	G*	B*	G	B	G	B	G	B	G	B	G	B	G	B
+36 to +42**	--	--	--	2	4	--	--	--	--	--	--	--	--	--
+30 to +36	--	--	--	--	--	2	2	--	--	--	--	--	2	--
+24 to +30	--	--	4	--	6	--	2	--	2	--	2	--	--	--
+18 to +24	4	7	4	4	14	10	2	--	6	--		--	--	4
+12 to +18	44	32	26	23	24	19	12	3	6	3	2	2	16	5
+ 6 to +12	40	46	36	47	40	32	22	21	24	23	16	7	44	34
0 to + 6	10	12	22	17	6	30	26	30	24	30	52	45	26	28
0 to - 6***	2	3	8	7	6	7	14	32	16	30	22	41	10	16
- 6 to -12	--	--	--	--	--	--	6	9	16	7	4	5	2	11
-12 to -18	--	--	--	--	--	--	12	3	2	5	2	--	--	2
-18 to -24	--	--	--	--	--	--	--	2	--	2	--	--	--	--
-24 to -30	--	--	--	--	--	--	2	--	2	--	--	--	--	--
-30 to -36	--	--	--	--	--	--	--	--	--	--	--	--	--	--
-36 to -42	--	--	--	--	--	--	--	--	--	--	--	--	--	--
-42 to -48	--	--	--	--	--	--	--	--	2	--	--	--	--	--
Mean****	0.9	1.0	0.8	0.8	1.1	0.7	0.2	-0.1	0.2	0.1	0.2	0.0	0.6	0.3

   * The letters G and B symbolize girls and boys respectively.
  ** Maxillary tooth erupts between 3.0 years and 3.5 years later than its mandibular antagonist.
 *** Gingival eruption of maxillary tooth precedes that of comparable mandibular tooth.
**** Mandibular central incisor erupts earlier than maxillary central incisor by 0.9 years for girls and 1.0 years for boys, mandibular lateral incisor erupts earlier than maxillary lateral incisor by 0.8 years for both sexes, and so forth.

(data from Knott and Meredith, 1966)

TABLE 1239

DIFFERENCES BETWEEN THE MEDIAN AGES OF EMERGENCE (years) AND STANDARD ERROR OF THE
DIFFERENCE FOR HOMOLOGOUS TEETH (maxillary minus mandibular)
IN PIMA INDIAN CHILDREN

	Right						
	1I	2I	C	1Pm	2Pm	1M	2M
Boys	$1.57 \pm .11$	$1.11 \pm .14$	$0.95 \pm .20$	$-.31 \pm .17$	$-.04 \pm .26$	$.03 \pm .14$	$.46 \pm .19$
Girls	$1.38 \pm .12$	$1.02 \pm .12$	$1.46 \pm .21$	$-.22 \pm .16$	$-.11 \pm .25$	$.25 \pm .16$	$.48 \pm .18$

	Left						
	I1	I2	C	Pm1	Pm2	M1	M2
Boys	$1.57 \pm .12$	$1.07 \pm .15$	$0.68 \pm .18$	$-.40 \pm .16$	$-.09 \pm .19$	$.14 \pm .12$	$.31 \pm .19$
Girls	$1.35 \pm .11$	$1.02 \pm .12$	$1.09 \pm .17$	$-.25 \pm .19$	$.11 \pm .18$	$0.59 \pm .21$	$0.68 \pm .17$

DIFFERENCES BETWEEN THE MEDIAN AGES OF EMERGENCE (years) AND STANDARD ERROR OF THE
DIFFERENCE OF TOOTH PAIRS (right minus left) IN PIMA INDIAN CHILDREN

	Maxillary Teeth						
	$_1I^1$	$_2I^2$	$_1C^1$	$_1Pm^1$	$_2Pm^2$	$_1M^1$	$_2M^2$
Boys	$-.02 \pm .10$	$+.07 \pm .13$	$-.11 \pm .20$	$+.15 \pm .17$	$-.01 \pm .19$	$-.09 \pm .13$	$.00 \pm .20$
Girls	$-.09 \pm .11$	$+.04 \pm .11$	$+.29 \pm .22$	$-.22 \pm .17$	$+.02 \pm .19$	$+.02 \pm .14$	$-.10 \pm .19$

	Mandibular Teeth						
	$_1I_1$	$_2I_2$	$_1C_1$	$_1Pm_1$	$_2Pm_2$	$_1M_1$	$_2M_2$
Boys	$-.02 \pm .15$	$+.03 \pm .16$	$-.16 \pm .18$	$+.06 \pm .16$	$-.06 \pm .26$	$+.02 \pm .13$	$-.15 \pm .19$
Girls	$-.02 \pm .13$	$+.04 \pm .13$	$-.08 \pm .16$	$.25 \pm .18$	$+.24 \pm .19$	$+.36 \pm .21$	$+.10 \pm .17$

I1 = central incisor; I2 or 21, lateral incisor; C = canine; Pm1 = first premolar,
Pm2 = second premolar, M1 = first molar and M2 = second molar.

(data from Dahlberg and Menegaz-Bock, 1958)

TABLE 1240

DIFFERENCE IN AGES AT ERUPTION (months)
BETWEEN UPPER AND LOWER JAW
IN SWEDISH CHILDREN

Tooth	Boys		Girls	
	Mean	S.D.	Mean	S.D.
Right medial incisor	-2.09	0.15	-2.28	0.18
Left medial incisor	-2.15	0.14	-2.27	0.17
Right lateral incisor	+2.15	0.24	+1.63	0.29
Left lateral incisor	+1.92	0.24	+1.45	0.29
Right canine	+0.47	0.15	+0.15	0.14
Left canine	+0.77	0.13	+0.41	0.14
Right first molar	+0.35	0.14	+0.24	0.15
Left first molar	+0.27	0.15	+0.15	0.14
Right second molar	-1.79	0.27	-2.15	0.25
Left second molar	-1.69	0.23	-2.41	0.25

Positive values denote eruption of a tooth in the upper jaw before
eruption of the corresponding tooth in the lower jaw, negative values
for contrary.

(data from Lysell et al., 1962)

TABLE 1241

TOTAL NUMBER OF DECIDUOUS TEETH ERUPTED IN
YPSILANTI (MI) CHILDREN BY HALF-YEAR AGE INTERVALS

Age range (months)	Mean Age	No. of teeth	
		Mean	S.D.
WHITE BOYS			
3.00 - 8.99	6.63	1.92	2.50
9.00 - 14.99	11.38	6.00	3.29
15.00 - 20.99	18.27	12.42	3.11
21.00 - 26.99	24.01	16.00	1.83
27.00 - 32.99	30.06	18.33	1.86
BLACK BOYS			
3.00 - 8.99	6.47	0.59	0.94
9.00 - 14.99	11.69	6.08	3.15
15.00 - 20.99	19.00	12.00	3.81
21.00 - 26.99	23.65	16.56	2.40
27.00 - 32.99	30.43	20.00	0.00
WHITE GIRLS			
3.00 - 8.99	6.99	1.89	2.17
9.00 - 14.99	11.77	5.74	2.78
15.00 - 20.99	17.95	13.11	3.19
21.00 - 26.99	23.27	16.33	2.10
27.00 - 32.99	29.60	19.46	1.04
BLACK GIRLS			
3.00 - 8.99	6.43	0.70	1.09
9.00 - 14.99	11.82	5.24	3.17
15.00 - 20.99	17.84	13.44	4.12
21.00 - 26.99	24.68	17.64	1.96
27.00 - 32.99	29.10	19.14	1.57

(data from Infante, 1975)

TABLE   1242

NUMBER OF FIRST PERMANENT MOLARS ERUPTED FOR GIRLS
IN LONDON

Age (years)	0	1	2	3	4
5	20	0	1	0	0
5¼	36	3	0	0	0
5½	30	3	6	3	3
5¾	13	5	3	0	6
6	18	6	6	5	7
6¼	14	1	4	0	17
6½	10	3	1	1	18
6¾	3	4	6	1	14
7	0	0	6	3	26
7¼	1	0	3	0	35
7½	0	1	0	1	39
7¾	0	0	0	1	34
8	0	0	0	2	36

(data from Parfitt, 1954)

TABLE 1243

AGES (years) AT WHICH SPECIFIC NUMBERS OF PERMANENT TEETH
ARE ERUPTED OR ERUPTING IN THE MOUTH AS A WHOLE
IN AUSTRALIAN CHILDREN

Number of Teeth	Age (years) Boys	Girls	Number of Teeth	Age (years) Boys	Girls
1	5.6	5.4	15	9.8	9.3
2	6.0	5.8	16	10.1	9.5
3	6.3	6.0	17	10.3	9.8
4	6.6	6.2	18	10.6	10.0
5	6.9	6.6	19	10.9	10.3
6	7.3	6.9	20	11.2	10.6
7	7.5	7.1	21	11.5	10.8
8	7.8	7.4	22	11.7	11.1
9	8.1	7.7	23	12.0	11.3
10	8.4	7.9	24	12.3	11.6
11	8.7	8.2	25	12.6	11.8
12	8.9	8.5	26	13.0	12.2
13	9.2	8.8	27	13.5	12.7
14	9.5	9.0	28	15.2	14.6

(data from Halikis, 1961b)

TABLE   1244

PROPORTIONS OF DECIDUOUS TEETH PRESENT IN BOSTON CHILDREN

Deciduous Teeth	7 Years *		8 Years		9 Years		10 Years	
	A**	B***	A	B	A	B	A	B
Males								
Maxillary second molars	0.96	0.89	0.94	0.69	0.89	0.44	0.76	0.29
Maxillary first molars	0.94	0.69	0.88	0.46	0.79	0.21	0.60	0.12
Maxillary canines	0.99	0.79	0.95	0.61	0.86	0.34	0.72	0.17
Mandibular second molars	0.94	0.82	0.87	0.64	0.85	0.38	0.63	0.23
Mandibular first molars	0.95	0.61	0.82	0.40	0.73	0.17	0.52	0.10
Mandibular canines	0.95	0.61	0.85	0.34	0.68	0.12	0.48	0.08
Females								
Maxillary second molars	0.98	0.84	0.95	0.68	0.85	0.36	0.62	0.13
Maxillary first molars	0.95	0.61	0.94	0.39	0.67	0.17	0.31	0.05
Maxillary canines	0.99	0.64	0.98	0.52	0.76	0.23	0.44	0.08
Mandibular second molars	0.90	0.77	0.90	0.65	0.74	0.33	0.49	0.10
Mandibular first molars	0.95	0.54	0.88	0.33	0.59	0.09	0.22	0.01
Mandibular canines	0.93	0.42	0.79	0.16	0.46	0.03	0.16	0.01

*   Age in years at last birthday at beginning of the study.
**   Initial examination.
*** Examination two years after the initial examination.

(data from Becker et al., 1972)

TABLE   1245

MEAN TOTAL NUMBER OF DECIDUOUS TEETH IN WHITE AND BLACK MICHIGAN CHILDREN

Age Group (months)	Mean Age (months)	Mean Number of Teeth	S.D.	Mean Age (months)	Mean Number of Teeth	S.D.
	White Boys			**White Girls**		
3.00 to 8.99	6.63	1.92	2.50	6.99	1.89	2.17
9.00 to 14.99	11.38	6.00	3.29	11.77	5.74	2.78
15.00 to 20.99	18.27	12.42	3.11	17.96	13.11	3.19
21.00 to 26.99	24.01	16.00	1.86	23.27	16.33	2.10
27.00 to 32.99	--	--	--	29.60	19.46	1.04
	Black Boys			**Black Girls**		
3.00 to 8.99	6.47	0.59	0.94	6.43	0.70	1.09
9.00 to 14.99	11.69	6.08	3.15	11.82	5.24	3.17
15.00 to 20.99	19.00	12.00	3.81	17.84	13.44	4.12
21.00 to 26.99	--	--	--	24.68	17.64	1.96
27.00 to 32.99	30.43	20.00	0.00	--	--	--

(data from Infante, 1974)

TABLE 1246

MEANS FOR INTERVALS (years) BETWEEN FIRST
PIERCING OF GINGIVA AND EMERGENCE OF WHOLE
CROWN OF PERMANENT TEETH IN MONTREAL CHILDREN

Tooth	Mean	
	Boys	Girls
Left Maxilla		
I_1	0.33	0.43
I_2	0.44	0.45
C	0.64	0.93
PM_1	0.64	0.70
PM_2	0.46	0.61
M_1	0.61	0.57
M_2	0.36	0.41
Left Mandible		
I_1	0.62	0.59
I_2	0.46	0.45
C	0.71	0.78
PM_1	0.36	0.59
PM_2	0.51	0.43
M_1	0.76	0.71
M_2	0.64	0.61

I_1 and I_2 = central and lateral incisor; C =
canine; PM_1 and PM_2 = first and second premolar;
M_1 and M_2 = first and second molar.

(data from Perreault et al., 1975)

TABLE 1247

PERCENTAGES OF MOST COMMON SUPERNUMERARY TEETH IN
PRIMARY AND PERMANENT DENTITION
IN ARKANSAS CHILDREN

| | Primary | | Permanent | | |
	Boys	Girls	Boys	Girls	Percentage
Mesiodens	--	--	9	4	36
Central	1	--	2	1	11
Lateral	4	6	4	4	50
Cuspid	--	--	--	--	--
Bicuspid	--	--	--	1	3
Molar	--	--	--	--	--
Total	5	6	15	10	

(data from Luten, 1967)

TABLE 1248

POSITION OF SUPERNUMERARY TEETH RELATED TO SEX IN
ARKANSAS CHILDREN

Sex	Upper	Lower	Right	Mid-line	Left	Erupt-ed	Unerupt-ed	Perma-nent	Primary
Male	20	--	8	2	10	7	13	15	5
Female	13	3	6	2	8	7	9	10	6
Totals	33	3	14	4	18	14	22	25	11

(data from Luten, 1967)

TABLE 1249

AGE (years) OF RESORPTION OF THE MESIAL ROOT OF THE DECIDUOUS FIRST MOLAR
IN BOSTON CHILDREN WITH AND WITHOUT TREATED OR UNTREATED CARIOUS LESIONS
OF THE DISTAL CROWN SURFACES

Stage of mesial root	Sex	Caries free		Untreated carious lesion		Treated carious lesion
		Mean (months)	S.D. (months)	Mean (months)	S.D. (months)	Mean (months)
Res_1	M	60.55	6.87	57.43	4.85	60.00
	F	54.55	4.94	57.00	5.56	--
$Res^{1/4}$	M	75.39	10.69	71.54	11.86	73.50
	F	68.40	12.68	68.31	10.73	--
$Res^{1/3}$	M	89.03	9.08	79.00	11.42	87.75
	F	81.71	11.34	84.89	8.26	87.00
$Res^{1/2}$	M	99.88	11.46	92.09	12.59	98.00
	F	95.71	8.64	94.80	10.96	96.00
$Res^{2/3}$	M	109.00	13.00	101.40	13.19	108.67
	F	102.82	9.83	103.00	12.91	108.00
$Res^{3/4}$	M	113.20	11.23	109.80	13.87	116.40
	F	111.67	10.08	110.00	11.90	112.80
Res_C	M	127.20	11.83	110.31	12.84	121.20
	F	112.00	7.82	113.29	9.97	105.00
Exf.	M	126.27	13.05	115.06	15.47	122.25
	F	118.88	8.75	116.29	10.07	120.00

Res = resorption; Exf = exfoliated.

(data from Fanning, 1962b. Reprinted with permission from Archives of Oral Biology 7:595-601. Copyright 1962, Pergamon Press, Ltd.)

TABLE 1250

AGE OF RESORPTION OF THE MESIAL ROOT OF THE DECIDUOUS SECOND MOLAR IN
BOSTON CHILDREN WITH AND WITHOUT TREATED OR UNTREATED CARIOUS LESIONS OF THE
MESIAL CROWN SURFACE

Stage of Mesial Root	Sex	Caries Free		Untreated Carious Lesion		Treated Carious Lesion
		Mean (months)	S.D. (months)	Mean (months)	S.D. (months)	Mean (months)
Res.$_i$	M	72.55	10.43	68.00	11.71	67.50
	F	68.18	7.74	68.67	7.94	--
Res.$_{1/4}$	M	85.80	9.98	79.85	13.48	93.00
	F	82.96	10.93	79.00	9.86	90.00
Res.$_{1/3}$	M	95.81	7.57	89.05	14.09	96.00
	F	94.43	8.77	89.14	12.88	94.80
Res.$_{1/2}$	M	110.55	10.35	94.94	16.06	112.15
	F	105.79	9.84	101.00	8.82	117.00
Res.$_{2/3}$	M	121.09	10.26	101.00	13.97	114.00
	F	115.29	10.26	109.50	11.94	124.00
Res.$_{3/4}$	M	129.43	13.03	102.75	12.60	138.00
	F	119.40	10.37	109.20	13.01	123.00
Res.$_c$	M	138.67	11.36	109.50	12.37	120.00
	F	138.00	6.00	123.00	4.25	--
Exf.	M	135.00	11.11	110.57	14.24	123.60
	F	131.25	11.76	102.00	8.49	126.00

Res. = resorbed; Exf. = exfoliated.

(data from Fanning, 1962b. Reprinted with permission from Archives of Oral Biology $\underline{7}$:
595-601. Copyright 1962, Pergamon Press, Ltd.)

TABLE 1251

AGE OF RESORPTION OF THE DISTAL ROOT OF THE DECIDUOUS FIRST MOLAR IN
BOSTON CHILDREN WITH AND WITHOUT TREATED OR UNTREATED CARIOUS LESIONS OF THE
DISTAL CROWN SURFACE

Stage of Distal Root	Sex	Caries Free		Untreated Carious Lesion		Treated Carious Lesion
		Mean (months)	S.D. (months)	Mean (months)	S.D. (months)	Mean (months)
Res.$_i$	M	67.26	10.14	61.88	7.77	70.00
	F	59.00	6.18	56.63	5.73	--
Res.$_{1/4}$	M	87.24	8.47	75.14	9.05	83.00
	F	72.22	13.47	67.14	9.43	102.00
Res.$_{1/3}$	M	98.44	11.77	86.40	15.59	96.00
	F	88.00	11.43	82.80	7.78	88.00
Res.$_{1/2}$	M	107.42	12.90	94.50	14.81	105.43
	F	99.25	9.52	92.77	12.49	100.50
Res.$_{2/3}$	M	112.50	12.07	94.29	11.83	115.20
	F	105.60	10.19	102.19	10.71	109.50
Res.$_{3/4}$	M	119.63	10.78	96.86	17.09	117.60
	F	115.20	11.93	106.91	13.94	110.00
Res.$_c$	M	129.75	14.68	105.20	16.30	102.00
	F	118.29	7.44	117.88	10.41	108.00
Exf.	M	128.00	12.85	112.20	16.30	122.25
	F	120.00	8.48	117.88	10.41	124.00

Res. = resorbed; Exf. = exfoliated.

(data from Fanning, 1962b. Reprinted with permission from Archives of Oral Biology 7: 595-601. Copyright 1962, Pergamon Press, Ltd.)

TABLE 1252

AGE (years) OF RESORPTION OF THE DISTAL ROOT OF THE DECIDUOUS SECOND MOLAR
IN BOSTON CHILDREN WITH AND WITHOUT TREATED OR UNTREATED CARIOUS LESIONS
OF THE DISTAL CROWN SURFACES

Stage of distal root	Sex	Caries free		Untreated carious lesion		Treated carious lesion
		Mean (months)	S.D. (months)	Mean (months)	S.D. (months)	Mean (months)
Res$_1$	M	83.00	9.31	77.14	11.73	78.00
	F	81.33	7.47	79.20	13.01	--
Res$^1/_4$	M	95.44	7.33	90.82	11.29	84.00
	F	92.50	9.52	91.38	12.80	90.00
Res$^1/_3$	M	110.81	10.91	100.09	13.02	100.50
	F	102.27	10.42	99.00	8.60	96.00
Res$^1/_2$	M	116.63	8.68	111.43	14.23	106.50
	F	109.09	10.43	110.67	14.39	114.00
Res$^2/_3$	M	125.08	8.38	121.50	15.63	102.00
	F	110.50	10.69	113.33	12.92	126.00
Res$^3/_4$	M	133.33	11.58	117.75	13.58	117.00
	F	119.40	8.70	118.00	14.03	144.00
Res$_C$	M	136.50	11.45	121.20	16.10	102.00
	F	128.25	10.11	132.00	9.17	--
Exf.	M	135.00	11.11	132.00	19.21	120.00
	F	133.71	10.29	141.00	4.25	150.00

Exf = exfoliated.

(data from Fanning, 1962b. Reprinted with permission from Archives of Oral Biology 7:
595-601. Copyright 1962, Pergamon Press, Ltd.)

TABLE 1253

AGE (years) FOR STAGES OF ROOT RESORPTION
OF THE DECIDUOUS MAXILLARY LEFT FIRST INCISOR
IN BOSTON CHILDREN

	Boys Percentiles					Girls Percentiles				
Stage	10	25	50	75	90	10	25	50	75	90
$Res^{1/3}$	--	--	--	--	--	3.75	4.20	4.75	5.36	5.97
$Res^{1/2}$	4.25	4.75	5.37	6.06	6.75	4.07	4.55	5.14	5.80	6.45
$Res^{2/3}$	4.60	5.14	5.80	6.55	7.28	4.56	5.09	5.74	6.46	7.18
$Res^{3/4}$	5.16	5.76	6.49	7.31	8.12	5.05	5.62	6.33	7.12	7.90
Exf	5.59	6.23	7.02	7.89	8.76	5.44	6.06	6.82	7.66	8.49

AGE (years) FOR STAGES OF ROOT RESORPTION
OF THE DECIDUOUS MAXILLARY LEFT SECOND INCISOR
IN BOSTON CHILDREN

	Boys Percentiles					Girls Percentiles				
Stage	10	25	50	75	90	10	25	50	75	90
Res_i	--	--	--	--	--	4.18	4.58	5.07	5.60	6.13
$Res^{1/4}$	4.76	5.31	5.98	6.73	7.48	4.69	5.14	5.67	6.26	6.84
$Res^{1/3}$	5.03	5.60	6.31	7.09	7.87	5.08	5.55	6.13	6.76	7.40
$Res^{1/2}$	5.50	6.12	6.89	7.74	8.58	5.46	5.97	6.58	7.26	7.91
$Res^{2/3}$	6.37	7.07	7.94	8.91	9.87	6.02	6.57	7.25	7.98	8.70
$Res^{3/4}$	--	--	--	--	--	6.34	6.92	7.63	8.40	9.15

AGES (years) FOR STAGES OF ROOT RESORPTION
OF THE DECIDUOUS MANDIBULAR LEFT CANINE
IN BOSTON CHILDREN

	Boys Percentiles					Girls Percentiles				
Stage	10	25	50	75	90	10	25	50	75	90
Res_i	5.39	5.76	6.19	6.65	7.10	4.12	4.59	5.17	5.81	6.45
$Res^{1/4}$	6.60	7.05	7.56	8.12	8.65	5.91	6.56	7.35	8.22	9.09
$Res^{1/3}$	7.80	8.32	8.92	9.57	10.19	7.06	7.82	8.74	9.77	10.79

Res_i = root shows blunting or rounding at apex. $Res^{1/4}$, $^{1/3}$ refer to the amount of the root that is resorbed.

TABLE 1253 (continued)

AGES (years) FOR STAGES OF ROOT RESORPTION OF THE
DECIDUOUS MANDIBULAR LEFT FIRST MOLAR
IN BOSTON CHILDREN

| | Boys | | | | | Girls | | | | |
| | Percentiles | | | | | | | | | |
Stage	10	25	50	75	90	10	25	50	75	90
				Mesial Root						
Res_i	--	--	--	--	--	2.96	3.31	3.73	4.20	4.66
$Res^{1/4}$	4.57	5.13	5.82	6.58	7.34	4.36	4.84	5.43	6.07	6.71
$Res^{1/3}$	5.74	6.42	7.25	8.18	9.12	5.48	6.06	6.78	7.56	8.34
$Res^{1/2}$	7.10	7.92	8.92	10.05	11.18	6.22	6.87	7.66	8.54	9.42
				Distal Root						
Res_i	--	--	--	--	--	3.33	3.72	4.21	4.76	5.31
$Res^{1/4}$	3.86	4.73	5.89	7.30	8.81	4.58	5.10	5.74	6.45	7.16
$Res^{1/3}$	4.59	5.60	6.93	8.57	10.32	5.28	5.87	6.59	7.40	8.19
$Res^{1/2}$	5.01	6.10	7.54	9.26	11.19	5.88	6.53	7.32	8.20	9.08
$Res^{2/3}$	--	--	--	--	--	7.32	8.11	9.08	10.15	11.21

Res_i = root shows blunting or rounding at apex. $Res^{1/4}$, $^{1/3}$, etc., refer to the amount of
the root that is resorbed.

AGE (years) FOR STAGES OF ROOT RESORPTION
OF THE DECIDUOUS MANDIBULAR LEFT SECOND MOLAR
IN BOSTON CHILDREN

| | Boys | | | | | Girls | | | | |
| | Percentiles | | | | | Percentiles | | | | |
Stage	10	25	50	75	90	10	25	50	75	90
				Mesial Root						
$Res^{1/4}$	4.79	5.64	6.72	8.00	9.26	4.20	4.78	5.50	6.32	7.14
$Res^{1/3}$	5.48	6.43	7.65	9.08	10.58	5.42	6.14	7.04	8.06	9.08
$Res^{1/2}$	6.19	7.25	8.61	10.20	11.87	7.09	9.01	9.15	10.44	11.75
				Distal Root						
$Res^{1/4}$	5.36	6.13	7.09	8.21	9.26	4.99	5.71	6.63	7.68	8.75
$Res^{1/3}$	6.63	7.56	8.73	9.33	10.96	5.98	6.83	7.91	9.15	10.40

(data from Fanning, 1961)

TABLE 1254

AGE (years) FOR THREE STAGES OF PRIMARY CANINE ROOT RESORPTION
AND FOR GINGIVAL EMERGENCE OF PERMANENT SUCCESSORS
IN IOWA CHILDREN

	Percentiles					
	10	50	90	10	50	90
Tooth Stage		Boys	.		Girls	

Mandibular Canine

Root Resorption:						
Initial	6.4	8.5	9.9	6.0	7.4	9.1
One-half	7.9	9.6	10.6	7.3	8.7	10.2
Three-fourths	8.3	9.9	10.9	7.7	9.1	10.5
Successor Emergence	9.3	10.5	11.6	8.5	9.7	11.2

Maxillary Canine

Root Resorption:						
Initial	7.0	8.8	10.3	6.6	7.9	9.4
One-half	8.6	10.1	11.1	8.2	9.5	11.4
Three-fourths	9.0	10.4	11.6	8.6	9.8	11.9
Successor Emergence	10.0	11.2	12.6	9.4	10.8	12.7

TIME (months) FROM THREE STAGES OF PRIMARY CANINE ROOT RESORPTION
AND FOR GINGIVAL EMERGENCE OF PERMANENT SUCCESSORS

Interval	Canine Tooth	Percentiles				
		10	25	50	75	90
Initial Root Resorption	Mandibular	11	15	26	36	42
to Successor Emergence	Maxillary	18	23	32	43	50
One-half Root Resorption	Mandibular	7	8	10	14	20
to Successor Emergence	Maxillary	9	11	14	18	24
Three-fourths Root Resorption	Mandibular	3	5	7	9	14
to Successor Emergence	Maxillary	5	7	9	13	17

(data from O'Meara and Knott, 1967)

TABLE 1255

MEAN AGES OF EXFOLIATION OF DECIDUOUS MOLAR
AND CANINE TEETH DETERMINED FROM FOUR STUDIES

Deciduous Teeth	Hellman (1923)	Clements et al., (1957)	Stone et al., (1951)	Present Study
Males				
Maxillary canine	11.6	11.1	11.6	10.5
Maxillary first molar	10.3	8.8	11.0	10.0
Maxillary second molar	10.5	9.6	11.7	10.9
Mandibular canine	10.8	10.2	10.7	9.5
Mandibular first molar	9.7	8.2	11.3	9.8
Mandibular second molar	10.6	8.9	12.0	10.6
Females				
Maxillary canine	10.8	10.3	10.8	10.5
Maxillary first molar	9.6	8.5	10.5	10.1
Maxillary second molar	10.8	9.4	10.9	10.9
Mandibular canine	9.7	9.2	9.8	9.6
Mandibular first molar	9.3	8.0	10.6	9.8
Mandibular second molar	10.4	8.5	11.6	10.6

(data from Becker et al., 1972)

TABLE 1256

AGES (years) OF LOSS OF DECIDUOUS TEETH
IN AUSTRALIAN CHILDREN AGED 2 TO 15 YEARS

Tooth	Boys		Girls	
	Mean	S.D.	Mean	S.D.
Maxilla				
i_1	6.8	1.63	6.3	2.09
i_2	7.5	1.5	7.1	2.04
c	11.3	1.61	10.4	1.71
m_1	8.5	2.5	8.5	2.12
m_2	8.9	2.64	8.9	2.48
Mandible				
i_1	6.4	0.61	6.1	0.77
i_2	7.4	0.89	7.0	0.82
c	10.5	1.47	9.5	1.25
m_1	7.4	3.19	7.3	2.71
m_2	7.4	2.89	6.7	3.24

(data from Halikis, 1961a)

TABLE 1257

RATES OF EXTRACTION AND EXFOLIATION OF DECIDUOUS TEETH IN BOSTON CHILDREN

Deciduous Teeth	7-9 Years		8-10 Years		9-11 Years		10-12 Years	
	Extrac-tion	Exfolia-tion	Extrac-tion	Exfolia-tion	Extrac-tion	Exfolia-tion	Extrac-tion	Exfolia-tion
Males								
Maxillary second molars	0.05	0.03	0.07	0.20	0.06	0.45	0.04	0.58
Maxillary first molars	0.09	0.17	0.07	0.41	0.07	0.67	0.07	0.72
Maxillary canines	0.01	0.19	0.00	0.36	0.00	0.61	0.02	0.74
Mandibular second molars	0.07	0.06	0.06	0.20	0.06	0.49	0.04	0.59
Mandibular first molars	0.14	0.22	0.09	0.43	0.06	0.71	0.06	0.73
Mandibular canines	0.00	0.31	0.01	0.60	0.01	0.82	0.01	0.83
Females								
Maxillary second molars	0.07	0.07	0.06	0.22	0.04	0.54	0.07	0.72
Maxillary first molars	0.07	0.29	0.08	0.51	0.06	0.69	0.02	0.82
Maxillary canines	0.00	0.35	0.01	0.46	0.00	0.70	0.00	0.82
Maxillary second molars	0.04	0.10	0.05	0.23	0.08	0.48	0.10	0.69
Maxillary first molars	0.07	0.37	0.07	0.56	0.01	0.83	0.09	0.86
Maxillary canines	0.00	0.55	0.01	0.79	0.01	0.92	0.00	0.92

(data from Becker et al., 1972)

TABLE 1258

PERCENTAGE OF DECIDUOUS TEETH SHED
IN LONDON SCHOOLCHILDREN

Age (years)	Tooth									
	Lwr. a	Upr. a	Lwr. b	Upr. b	Lwr. c	Lwr. d	Upr. d	Upr. c	Upr. e	Lwr. e
2	--	2	--	--	--	--	--	--	--	--
3	--	1	--	--	--	--	--	--	--	1
4	--	2	--	1	--	3	2	--	1	2
5	14	8	3	3	--	14	9	--	5	11
6	58	28	16	8	2	22	13	2	6	17
7	97	76	62	42	4	29	20	2	10	23
8	100	98	92	87	16	38	34	11	20	31
9	--	99	99	96	40	52	51	29	29	36
10	--	100	100	100	61	69	67	37	46	52
11	--	--	--	--	85	82	83	63	60	57
12	--	--	--	--	96	94	93	81	80	76
13	--	--	--	--	99	99	99	95	92	89
14	--	--	--	--	100	100	100	98	97	97
15	--	--	--	--	--	--	--	100	98	98
16	--	--	--	--	--	--	--	--	100	100

a = central incisor; b = lateral incisor; C = canine; d = first deciduous molar
and e = second deciduous molar.

(data from Parfitt, 1954)

TABLE 1259

DIFFERENCE IN TIME BETWEEN THE PERMANENT TOOTH PIERCING
THE GUM AND ERUPTION OF 1/3 OF THE CROWN
FOR CHILDREN IN LONDON (England)

Teeth	Months
U 1	2.7
L 1	2.5
U 2	3.6
L 2	3.6
U 3	6.3
L 3	5.2
U 4	3.7
L 4	4.2
U 5	1.0
L 5	1.7
U 6	2.7
L 6	1.8
U 7	2.8
L 7	4.9

U = upper; L = lower; 1 - 7 are the
central, and lateral incisors, canine,
first and second premolars and first
second molars respectively.

(data from Parfitt, 1954)

TABLE 1260

TIME TAKEN (mos.) ERUPTING DIFFERENT RELATIVE AMOUNTS OF
MAXIMUM INTRAORAL HEIGHT ATTAINED BY PERMANENT INCISOR TEETH
IN THE FIRST FOUR YEARS AFTER GINGIVAL EMERGENCE
IN IOWA CHILDREN

Relative Intraoral Height %		Percentiles			
	Median	10	30	70	90

Maxillary Left Central Incisor

50	3.0	2.5	2.7	3.4	4.5
70	5.9	3.6	4.9	7.0	8.2
90	18.4	10.4	15.2	22.3	27.9
100	43.8	27.2	38.8	48.0	48.0

Mandibular Left Central Incisor

50	3.2	2.2	2.7	3.8	5.5
70	7.8	3.5	5.8	9.1	11.7
90	19.6	12.8	16.5	23.3	28.4
100	45.0	28.1	38.9	48.0	48.0

Maxillary Left Canine

50	3.1	2.0	2.7	3.7	4.4
70	5.7	4.4	5.0	6.6	8.9
90	12.5	7.9	10.4	16.0	26.6
100	48.0	19.8	36.7	48.0	48.0

Mandibular Left Canine

50	3.5	2.5	2.9	4.0	5.6
70	7.3	4.8	6.5	8.0	11.3
90	19.4	11.3	16.1	22.5	35.2
100	48.0	26.9	47.1	48.0	48.0

Relative Intraoral Height %		Percentiles			
	Median	10	30	70	90

Maxillary Left Second Premolar

50	2.4	1.7	2.0	2.6	3.1
70	3.6	2.8	3.4	4.3	5.6
90	13.8	5.3	8.2	15.5	24.5
100	48.0	22.6	40.0	48.0	48.0

Mandibular Left Second Premolar

50	2.5	2.0	2.3	2.8	3.4
70	3.6	2.9	3.2	5.2	6.1
90	13.9	6.2	8.8	17.0	22.2
100	48.0	21.5	34.0	48.0	48.0

(data from Giles et al., 1963)

TABLE 1261

ORDER OF ERUPTION OF DECIDUOUS TEETH
IN SWEDISH CHILDREN

Tooth	Order									
	1	2	3	4	5	6	7	8	9	10
Boys										
Lower medial incisor	95	1	--	--	--	--	--	--	--	--
Upper medial incisor	9	78	8	1	--	--	--	--	--	--
Lower lateral incisor	2	6	18	63	4	3	--	--	--	--
Upper lateral incisor	5	32	48	10	1	--	--	--	--	--
Lower canine	--	--	--	1	2	8	36	49	--	--
Upper canine	--	--	--	--	1	4	78	12	--	1
Lower first molar	--	--	1	6	46	42	1	--	--	--
Upper first molar	--	--	1	13	57	21	2	2	--	--
Lower second molar	--	--	--	--	--	--	--	2	76	18
Upper second molar	--	--	--	--	--	--	--	3	27	66
Girls										
Lower medial incisor	73	2	--	--	--	--	--	--	--	--
Upper medial incisor	8	52	12	3	--	--	--	--	--	--
Lower lateral incisor	2	11	17	35	3	6	--	1	--	--
Upper lateral incisor	3	30	32	9	--	1	--	--	--	--
Lower canine	--	--	--	--	2	5	36	32	--	--
Upper canine	--	--	--	--	3	5	53	14	--	--
Lower first molar	--	--	1	7	44	21	1	1	--	--
Upper first molar	--	--	1	12	40	21	--	1	--	--
Lower second molar	--	--	--	--	--	--	--	--	69	6
Upper second molar	--	--	--	--	--	--	--	--	15	60

Note: The data relate to the first tooth in each pair. The horizontal rows show in how many cases a given tooth erupted first, second, third etc. The vertical columns show how often different teeth erupted first, second, third etc. If two teeth erupted simultaneously, they are marked in the same column: e.g., if first teeth, in column I. The next tooth to erupt is in such cases shown in column 3.

(data from Lysell et al., 1962)

STATURE PREDICTION

TABLE 1262

PERCENT OF MATURE STATURE FOR EARLY, AVERAGE AND
LATE-MATURING CHILDREN IN CALIFORNIA GROUPED ACCORDING TO C.A. AT
REACHING COMPLETE SKELETAL MATURITY

BOYS

C.A. (years)	Early-Maturing		Average-Maturing		Late-Maturing	
	Mean	S.D.	Mean	S.D.	Mean	S.D.
10.7	83.2	0.9	80.0	1.3	79.2	1.6
11.2	83.5	2.0	81.6	1.5	80.0	1.8
11.7	85.3	2.5	82.8	1.4	81.3	1.3
12.2	87.4	2.6	84.4	1.4	82.5	1.5
12.7	89.6	3.0	86.2	1.5	83.8	1.5
13.2	92.3	2.8	88.4	1.7	85.4	1.6
13.7	94.8	2.4	90.7	1.9	87.0	1.6
14.2	96.9	1.8	93.2	1.9	88.9	2.0
14.7	98.1	1.4	95.3	1.6	91.2	2.0
15.2	99.0	0.9	97.0	1.2	93.6	2.0
15.7	99.5	0.5	98.4	0.7	95.9	1.5
16.2	100.1	0.4	99.2	0.5	97.8	1.0
16.7	100.2	0.4	99.8	0.3	98.6	0.7
17.2	100.6	0.5	100.1	0.2	99.3	0.5
17.7	100.6	0.3	100.3	0.2	99.9	0.4

TABLE 1262   (continued)

PERCENT OF MATURE STATURE FOR EARLY, AVERAGE AND
LATE-MATURING CHILDREN IN CALIFORNIA GROUPED ACCORDING C.A. AT
REACHING COMPLETE SKELETAL MATURITY

GIRLS

C.A. (years)	Early-Maturing		Average-Maturing		Late-Maturing	
	Mean	S.D.	Mean	S.D.	Mean	S.D.
10.7	89.6	2.4	86.5	2.3	84.6	1.8
11.2	91.8	2.3	88.8	2.5	85.2	1.4
11.7	93.9	2.5	91.0	2.5	87.3	1.9
12.2	95.6	2.3	92.8	2.3	90.0	2.1
12.7	97.2	1.6	95.1	1.6	91.8	2.0
13.2	98.2	0.9	96.9	1.1	94.3	2.3
13.7	98.9	0.9	98.0	0.7	95.6	2.3
14.2	99.4	0.5	98.7	0.5	96.8	1.8
14.7	99.8	0.3	99.0	0.8	97.9	1.2
15.2	100.0	0.2	99.5	0.4	98.7	0.9
15.7	100.2	0.3	99.8	0.2	99.2	0.6
16.2	100.2	0.3	99.8	0.2	99.4	0.2
16.7	100.3	0.3	100.1	0.2	99.7	0.2
17.2	100.4	0.2	100.2	0.2	99.9	0.2
17.7	100.5	0.2	100.2	0.2	100.1	0.2

C.A. = chronological age

(data from Bayley, 1943a)

TABLE 1263

PERCENTAGES OF MATURE STATURE ACHIEVED FOR CHILDREN
WITH SKELETAL AGES (Greulich-Pyle, in years-months) WITHIN ONE YEAR
OF THEIR CHRONOLOGICAL AGES

BOYS

Skeletal Age	7-0	7-3	7-6	7-9	8-0	8-3	8-6	8-9	9-0	9-3	9-6	9-9
% of Mature	69.5	70.2	70.9	71.6	72.3	73.1	73.9	74.6	75.2	76.1	76.9	77.7

Skeletal Age	10-0	10-3	10-6	10-9	11-0	11-3	11-6	11-9	12-0	12-3	12-6	12-9
% of Mature	78.4	79.1	79.5	80.0	80.4	81.2	81.8	82.7	83.4	84.3	85.3	86.3

Skeletal Age	13-0	13-3	13-6	13-9	14-0	14-3	14-6	14-9	15-0	15-3	15-6	15-9
% of Mature	87.6	89.0	90.2	91.4	92.7	93.8	94.8	95.8	96.8	97.3	97.6	98.0

Skeletal Age	16-0	16-3	16-6	16-9	17-0	17-3	17-6	17-9	18-0	18-3	18-6	--
% of Mature	98.2	98.5	98.7	98.9	99.1	99.3	99.4	99.5	99.6	99.8	100.0	--

GIRLS

Skeletal Age	6-0	6-3	6-6	6-10	7-0	7-3	7-6	7-10	8-0	8-3	8-6	8-10
% of Mature	72.0	72.9	73.8	75.1	75.7	76.5	77.2	78.2	79.0	80.1	81.0	82.1

Skeletal Age	9-0	9-3	9-6	9-9	10-0	10-3	10-6	10-9	11-0	11-3	11-6	11-9
% of Mature	82.7	83.6	84.4	85.3	86.2	87.4	88.4	89.6	90.6	91.0	91.4	91.8

Skeletal Age	12-0	12-3	12-6	12-9	13-0	13-3	13-6	13-9	14-0	14-3	14-6	14-9
% of Mature	92.2	93.2	94.1	95.0	95.8	96.7	97.4	97.8	98.0	98.3	98.6	98.8

Skeletal Age	15-0	15-3	15-6	15-9	16-0	16-3	16-6	16-9	17-0	17-6	18-0	--
% of Mature	99.0	99.1	99.3	99.4	99.6	99.6	99.7	99.8	99.9	99.95	100.0	--

(data from Bayley and Pinneau, 1952, The Journal of Pediatrics 40:423-441)

TABLE 1264

PERCENTAGES OF ADULT STATURE ACHIEVED IN ACCELERATED CALIFORNIA CHILDREN;
SKELETAL AGE (GREULICH-PYLE, IN YEARS-MONTHS) ONE YEAR OR MORE ADVANCED
OVER CHRONOLOGICAL AGE

BOYS

Skeletal Age	7-0	7-3	7-6	7-9	8-0	8-3	8-6	8-9	9-0	9-3	9-6	9-9
% of Mature	67.0	67.6	68.3	68.9	69.6	70.3	70.9	71.5	72.0	72.8	73.4	74.1

Skeletal Age	10-0	10-3	10-6	10-9	11-0	11-3	11-6	11-9	12-0	12-3	12-6	12-9
% of Mature	74.7	75.3	75.8	76.3	76.7	77.6	78.6	80.0	80.9	81.8	82.8	83.9

Skeletal Age	13-0	13-3	13-6	13-9	14-0	14-3	14-6	14-9	15-0	15-3	15-6	15-9
% of Mature	85.0	86.3	87.5	89.0	90.5	91.8	93.0	94.3	95.8	96.7	97.1	97.6

Skeletal Age	16-0	16-3	16-6	16-9	17-0	--	--	--	--	--	--	--
% of Mature	98.0	98.3	98.5	98.8	99.0	--	--	--	--	--	--	--

GIRLS

Skeletal Age	7-0	7-3	7-6	7-10	8-0	8-3	8-6	8-10	9-0	9-3	9-6	9-9
% of Mature	71.2	72.2	73.2	74.2	75.0	76.0	77.1	78.4	79.0	80.0	80.9	81.9

Skeletal Age	10-0	10-3	10-6	10-9	11-0	11-3	11-6	11-9	12-0	12-3	12-6	12-9
% of Mature	82.8	84.1	85.6	87.0	88.3	88.7	89.1	89.7	90.1	91.3	92.4	93.5

Skeletal Age	13-0	13-3	13-6	13-9	14-0	14-3	14-6	14-9	15-0	15-3	15-6	15-9
% of Mature	94.5	95.5	96.3	96.8	97.2	97.7	98.0	98.3	98.6	98.8	99.0	99.2

Skeletal Age	16-0	16-3	16-6	16-9	17-0	17-6	--	--	--	--	--	--
% of Mature	99.3	99.4	99.5	99.7	99.8	99.95	--	--	--	--	--	--

(data from Bayley and Pinneau, 1952: The Journal of Pediatrics 40:423-441)

TABLE 1265

PERCENTAGES OF ADULT STATURE ACHIEVED IN RETARDED CALIFORNIA CHILDREN;
SKELETAL AGE (GREULICH-PYLE, IN YEARS-MONTHS)
ONE YEAR OR MORE RETARDED FOR CHRONOLOGICAL AGE

BOYS

Skeletal Age	6-0	6-3	6-6	6-9	7-0	7-3	7-6	7-9	8-0	8-3	8-6	8-9
% of Mature	68.0	69.0	70.0	70.9	71.8	72.8	73.8	74.7	75.6	76.5	77.3	77.9

Skeletal Age	9-0	9-3	9-6	9-9	10-0	10-3	10-6	10-9	11-0	11-3	11-6	11-9
% of Mature	78.6	79.4	80.0	80.7	81.2	81.6	81.9	82.1	82.3	82.7	83.2	83.9

Skeletal Age	12-0	12-3	12-6	12-9	13-0	--	--	--	--	--	--	--
% of Mature	84.5	85.2	86.0	86.9	88.0	--	--	--	--	--	--	--

GIRLS

Skeletal Age	6-0	6-3	6-6	6-10	7-0	7-3	7-6	7-10	8-0	8-3	8-6	8-10
% of Mature	73.3	74.2	75.1	76.3	77.0	77.9	78.8	79.7	80.4	81.3	82.3	83.6

Skeletal Age	9-0	9-3	9-6	9-9	10-0	10-3	10-6	10-9	11-0	11-3	11-6	11-9
% of Mature	84.1	85.1	85.8	86.6	87.4	88.4	89.6	90.7	91.8	92.2	92.6	92.9

Skeletal Age	12-0	12-3	12-6	12-9	13-0	13-3	13-6	13-9	14-0	14-3	14-6	14-9
% of Mature	93.2	94.2	94.9	95.7	96.4	97.1	97.7	98.1	98.3	98.6	98.9	99.2

Skeletal Age	15-0	15-3	15-6	15-9	16-0	16-3	16-6	16-9	17-0	--	--	--
% of Mature	99.4	99.5	99.6	99.7	99.8	99.9	99.9	99.95	100.0	--	--	--

(data from Bayley and Pinneau, 1952: The Journal of Pediatrics 40:423-441)

TABLE 1266

ERRORS OF ADULT STATURE PREDICTION (in) IN CALIFORNIA
CHILDREN WITH THE METHOD OF BAYLEY AND PINNEAU
(1952)

Age	BOYS		GIRLS	
(years and months)	Mean	S.D.	Mean	S.D.
8 - 0	-.86	1.73	-.13	1.47
8 - 6	--	--	--	--
9 - 0	-.65	1.46	+.02	1.27
9 - 6	-.58	1.33	-.41	1.13
10 - 0	-.49	1.37	+.002	1.33
10 - 6	-.37	1.20	+.06	1.11
11 - 0	-.36	1.15	-.04	1.14
11 - 6	-.22	0.94	-.12	1.15
12 - 0	-.13	1.06	+.16	1.09
12 - 6	-.09	0.78	+.23	1.09
13 - 0	-.21	0.62	+.01	1.21
13 - 6	-.15	0.55	+.05	1.32
14 - 0	+.10	0.42	+.13	1.21
14 - 6	+.001	0.40	-.23	0.85
15 - 0	+.08	0.38	+.03	0.88
15 - 6	+.04	0.32	+.07	0.65
16 - 0	-.02	0.26	+.01	0.49
16 - 6	-.07	0.25	+.12	0.35
17 - 0	+.02	0.20	+.19	0.41
17 - 6	-.08	0.22	+.14	0.30
18 - 0	-.05	0.11	+.14	0.38

(data from Bayley and Pinneau, 1952. The Journal of Pediatrics 40:
423-441)

TABLE 1267

PERCENT OF MATURE STATURE BY AGE
FOR BOSTON CHILDREN

		Boys					Girls		
Age (years)	Total Mean	S.D.	Accel. Mean	Ret. Mean	Total Mean	S.D.	Accel. Mean	Ret. Mean
1	42.66	1.08	42.5	43.1	45.24	1.42	46.4	44.7
2	49.62	1.16	50.1	49.7	52.58	1.67	54.3	51.4
3	54.47	1.14	55.2	54.5	58.41	1.59	60.3	56.8
4	58.58	1.33	59.5	58.3	63.19	1.65	65.2	61.3
5	62.36	1.44	63.5	61.8	67.35	2.01	70.0	65.3
6	65.94	1.66	67.4	65.3	71.17	2.34	73.8	69.3
7	68.67	1.81	70.6	68.0	74.22	1.85	76.5	72.0
8	71.97	1.96	74.0	71.4	77.60	2.13	80.6	75.3
9	75.18	2.09	77.6	74.5	81.17	2.28	84.3	78.3
10	78.17	2.25	80.6	77.2	84.64	2.77	88.6	81.6
11	80.88	2.56	83.9	79.7	88.50	3.32	93.2	84.5
12	84.13	3.05	87.5	82.2	92.50	3.27	96.6	87.8
13	87.94	3.96	91.8	84.8	95.91	2.49	98.4	91.9
14	92.07	4.12	95.3	88.7	98.03	1.36	99.2	95.9
15	95.41	3.32	97.5	90.1	99.10	.67	99.4	98.2
16	97.64	2.15	98.9	95.0	99.53	.48	99.7	99.3
17	98.89	1.31	99.6	97.9	99.71	1.18	99.8	99.7
18	99.59	.72	99.9	99.2	100.00	0.00	100.0	100.0

Accel. = accelerated; Ret. = retarded. These children have skeletal ages that differ by
more than one year from their chronological ages.

(data from Bayley, 1962)

TABLE 1268

PERCENT OF MATURE STATURE ACHIEVED AT SUCCESSIVE
CHRONOLOGICAL AGES IN CALIFORNIA CHILDREN

AGE		BOYS		GIRLS	
		Mean	S.D.	Mean	S.D.
Months	1	30.18	0.77	32.40	1.44
	2	32.40	0.93	34.51	1.56
	3	33.93	1.00	35.96	1.31
	4	35.21	0.95	37.50	1.08
	5	36.50	0.99	38.78	1.08
	6	37.67	0.93	39.84	1.20
	7	38.44	0.95	40.69	1.20
	8	39.22	1.10	41.79	1.37
	9	40.08	1.07	42.20	1.22
	10	40.80	1.14	43.09	1.37
	11	41.53	1.16	44.10	1.24
	12	42.23	1.04	44.67	1.42
	15	44.02	1.19	46.90	1.18
	18	45.64	1.34	48.76	1.37
	24	48.57	1.44	52.15	1.34
	30	51.14	1.40	54.75	1.22
Years	3.0	53.53	1.34	57.16	1.20
	4.0	57.72	1.38	61.84	1.45
	5.0	61.60	1.49	66.24	1.45
	6.0	65.31	1.58	70.29	1.60
	7.0	69.08	1.60	74.28	1.61
	8.0	72.40	1.68	77.57	1.87
	9.0	75.61	1.68	81.19	2.00
	9.5	77.21	1.66	83.03	2.13
	10.0	78.40	1.76	84.76	2.42
	10.5	79.82	1.77	86.85	2.71
	11.0	81.30	1.94	88.65	2.88
	11.5	82.54	2.00	90.81	3.06
	12.0	84.00	2.23	92.61	3.27
	12.5	85.43	2.49	94.72	2.61
	13.0	87.32	3.02	95.96	2.15
	13.5	89.22	3.57	97.17	1.70
	14.0	91.00	3.96	98.27	1.24
	14.5	92.60	3.85	98.74	0.93
	15.0	94.60	3.74	99.31	0.68
	15.5	96.00	3.31	99.54	0.48
	16.0	97.09	2.71	99.62	0.35
	16.5	97.95	2.12	99.75	0.34
	17.0	98.79	1.43	99.95	0.25
	17.5	99.28	1.01	99.91	0.25
	18.0	99.55	0.58	99.96	0.11

(data from Bayley and Pinneau, 1952, The Journal of Pediatrics 40:423-441)

TABLE   1269

MEAN PERCENTAGES OF ADULT STATURE ACHIEVED BY
WHITE BOYS IN CLEVELAND (OH) IN
RELATION TO SOMATOTYPE

Chronological Age (years)	Ectomorphs	Mesomorphs
2	48.7	49.5
3	53.0	54.4
4	56.8	58.6
5	65.0	62.4
6	61.0	66.2
7	68.6	69.8
8	71.9	73.0
9	74.9	76.2
10	77.9	79.5
11	80.6	82.1
12	83.2	85.6
13	86.6	89.6
14	90.4	93.9
15	94.5	97.2
16	96.4	98.4
17	98.2	99.3

(data from Dupertuis and Michael, 1953)

TABLE 1270

βWEIGHTS FOR THE PREDICTION OF ADULT STATURE IN BOYS

| Age | | βRL | βW | βMPS | βSA | βO |
yr.	mo.					
1	0	0.966	0.199	0.606	−0.673	1.632
1	3	1.032	0.086	0.580	−0.417	− 1.841
1	6	1.086	−0.016	0.559	−0.205	− 4.892
1	9	1.130	−0.106	0.540	−0.033	− 7.528
2	0	1.163	−0.186	0.523	0.104	− 9.764
2	3	1.189	−0.256	0.509	0.211	−11.618
2	6	1.207	−0.316	0.496	0.291	−13.114
2	9	1.219	−0.369	0.485	0.349	−14.278
3	0	1.227	−0.413	0.475	0.388	−15.139
3	3	1.230	−0.450	0.466	0.410	−15.729
3	6	1.229	−0.481	0.458	0.419	−16.081
3	9	1.226	−0.505	0.451	0.417	−16.228
4	0	1.221	−0.523	0.444	0.405	−16.201
4	3	1.214	−0.537	0.437	0.387	−16.034
4	6	1.206	−0.546	0.431	0.363	−15.758
4	9	1.197	−0.550	0.424	0.335	−15.400
5	0	1.188	−0.551	0.418	0.303	−14.990
5	3	1.179	−0.548	0.412	0.269	−14.551
5	6	1.169	−0.543	0.406	0.234	−14.106
5	9	1.160	−0.535	0.400	0.198	−13.672
6	0	1.152	−0.524	0.394	0.161	−13.267
6	3	1.143	−0.512	0.389	0.123	−12.901
6	6	1.135	−0.499	0.383	0.085	−12.583
6	9	1.127	−0.484	0.378	0.046	−12.318
7	0	1.120	−0.468	0.373	0.006	−12.107
7	3	1.113	−0.451	0.369	−0.034	−11.948
7	6	1.106	−0.434	0.365	−0.077	−11.834
7	9	1.100	−0.417	0.361	−0.121	−11.756
8	0	1.093	−0.400	0.358	−0.167	−11.701
8	3	1.086	−0.382	0.356	−0.217	−11.652
8	6	1.079	−0.365	0.354	−0.270	−11.592
8	9	1.071	−0.349	0.353	−0.327	−11.498
9	0	1.063	−0.333	0.353	−0.389	−11.349
9	3	1.054	−0.317	0.353	−0.455	−11.118
9	6	1.044	−0.303	0.355	−0.527	−10.779
9	9	1.033	−0.289	0.357	−0.605	−10.306
10	0	1.021	−0.276	0.360	−0.690	− 9.671
10	3	1.008	−0.263	0.363	−0.781	− 8.848

TABLE 1270 (continued)

βWEIGHTS FOR THE PREDICTION OF ADULT STATURE IN BOYS

Age		βRL	βW	βMPS	βSA	β0
yr.	mo.					
10	6	0.993	−0.252	0.368	−0.878	−7.812
10	9	0.977	−0.241	0.373	−0.983	−6.540
11	0	0.960	−0.231	0.378	−1.094	−5.010
11	3	0.942	−0.222	0.384	−1.211	−3.206
11	6	0.923	−0.213	0.390	−1.335	−1.113
11	9	0.902	−0.206	0.397	−1.464	1.273
12	0	0.881	−0.198	0.403	−1.597	3.958
12	3	0.859	−0.191	0.409	−1.735	6.931
12	6	0.837	−0.184	0.414	−1.875	10.181
12	9	0.815	−0.177	0.418	−2.015	13.684
13	0	0.794	−0.170	0.421	−2.156	17.405
13	3	0.773	−0.163	0.422	−2.294	21.297
13	6	0.755	−0.155	0.422	−2.427	25.304
13	9	0.738	−0.146	0.418	−2.553	29.349
14	0	0.724	−0.136	0.412	−2.668	33.345
14	3	0.714	−0.125	0.401	−2.771	37.183
14	6	0.709	−0.112	0.387	−2.856	40.738
14	9	0.709	−0.098	0.367	−2.922	43.869
15	0	0.717	−0.081	0.342	−2.962	46.403
15	3	0.732	−0.062	0.310	−2.973	48.154
15	6	0.756	−0.040	0.271	−2.949	48.898
15	9	0.792	−0.015	0.223	−2.885	48.402
16	0	0.839	−0.014	0.167	−2.776	46.391

β values are regression weights for recumbent length (RL), weight (W), midpoint stature (MPS) and skeletal age (SA). β0 is the intercept of the regression equation.

(data from Roche, Wainer, and Thissen, 1975b, Pediatrics 56:1026-1033, Copyright American Academy of Pediatrics 1975; and 1975c, Monographs in Paediatrics 3:1-114, S. Karger, Basel)

TABLE   1271

β WEIGHTS FOR THE PREDICTION OF ADULT STATURE IN GIRLS

Age						
yr.	mo.	βRL	βW	βMPS	βSA	β0
1	0	1.087	−0.271	0.386	0.434	21.729
1	3	1.112	−0.369	0.367	0.094	20.684
1	6	1.134	−0.455	0.349	−0.172	19.957
1	9	1.153	−0.530	0.332	−0.374	19.463
2	0	1.170	−0.594	0.316	−0.523	19.131
2	3	1.183	−0.648	0.301	−0.625	18.905
2	6	1.195	−0.693	0.287	−0.690	18.740
2	9	1.204	−0.729	0.274	−0.724	18.604
3	0	1.210	−0.757	0.262	−0.736	18.474
3	3	1.215	−0.777	0.251	−0.729	18.337
3	6	1.217	−0.791	0.241	−0.711	18.187
3	9	1.217	−0.798	0.232	−0.684	18.024
4	0	1.215	−0.800	0.224	−0.655	17.855
4	3	1.212	−0.797	0.217	−0.626	17.691
4	6	1.206	−0.789	0.210	−0.600	17.548
4	9	1.199	−0.777	0.205	−0.582	17.444
5	0	1.190	−0.761	0.200	−0.571	17.398
5	3	1.180	−0.742	0.197	−0.572	17.431
5	6	1.168	−0.721	0.193	−0.584	17.567
5	9	1.155	−0.697	0.191	−0.609	17.826
6	0	1.140	−0.671	0.190	−0.647	18.229
6	3	1.124	−0.644	0.189	−0.700	18.796
6	6	1.107	−0.616	0.188	−0.766	19.544
6	9	1.089	−0.587	0.189	−0.845	20.489
7	0	1.069	−0.557	0.189	−0.938	21.642
7	3	1.049	−0.527	0.191	−1.043	23.011
7	6	1.028	−0.498	0.192	−1.158	24.602
7	9	1.006	−0.468	0.194	−1.284	26.416
8	0	0.983	−0.439	0.196	−1.418	28.448
8	3	0.960	−0.411	0.199	−1.558	30.690
8	6	0.937	−0.384	0.202	−1.704	33.129
8	9	0.914	−0.369	0.204	−1.853	35.747
9	0	0.891	−0.334	0.207	−2.003	38.520
9	3	0.868	−0.311	0.210	−2.154	41.421
9	6	0.845	−0.389	0.212	−2.301	44.415
9	9	0.824	−0.269	0.214	−2.444	47.464
10	0	0.803	−0.250	0.216	−2.581	50.525
10	3	0.783	−0.233	0.217	−2.710	53.548

TABLE 1271 (continued)

β WEIGHTS FOR THE PREDICTION OF ADULT STATURE IN GIRLS

Age						
yr.	mo.	βRL	βW	βMPS	βSA	β0
10	6	0.766	−0.217	0.217	−2.829	56.481
10	9	0.749	−0.203	0.217	−2.936	59.267
11	0	0.736	−0.190	0.216	−3.029	61.841
11	3	0.724	−0.179	0.214	−3.108	64.136
11	6	0.716	−0.169	0.211	−3.170	66.093
11	9	0.711	−0.159	0.206	−3.217	67.627
12	0	0.710	−0.151	0.201	−3.245	68.670
12	3	0.713	−0.143	0.193	−3.254	69.140
12	6	0.720	−0.136	0.184	−3.244	68.966
12	9	0.733	−0.129	0.173	−3.214	68.061
13	0	0.752	−0.121	0.160	−3.166	66.339
13	3	0.777	−0.113	0.144	−3.100	63.728
13	6	0.810	−0.105	0.127	−3.015	60.150
13	9	0.850	−0.085	0.106	−2.915	55.522
14	0	0.898	−0.083	0.083	−2.800	49.781

β values are regression weights for recumbent length (RL), weight (W), midpoint stature (MPS) and skeletal age (SA). β0 is the intercept of the regression equation.

(data from Roche, Wainer, and Thissen, 1975 b, Pediatrics 56:1026-1033. Copyright American Academy of Pediatrics 1975)

TABLE   1272

AGE OF REACHING 90 PERCENT
MATURE STATURE (years) IN RELATION
TO SOMATOTYPE IN CALIFORNIA BOYS

	Mean	S.D.
Extreme Groups		
Mesomorphs	13.73	.91
Ectomorphs	13.95	.55
Remainder of Sample		
Not Mesomorphs	13.74	.85
Not Ectomorphs	13.72	.89
Total Sample	13.74	.87

(data from Livson and McNeill, 1962)

TABLE   1273

PERCENT OF EIGHTEEN-YEAR SIZE ATTAINED AT SPECIFIC AGES
IN BOSTON CHILDREN

Age (years)	GIRLS			BOYS		
	Stature	Femur and Tibia length	Foot length	Stature	Femur and Tibia length	Foot length
1	46.1	34.8	49.9	42.6	31.2	45.2
2	53.2	41.7	57.8	49.7	39.2	53.0
4	63.6	55.1	67.8	58.7	49.8	62.0
8	77.4	73.9	82.1	72.4	68.7	75.8
12	91.3	91.6	97.0	84.5	85.4	91.3
16	99.5	100.0	100.0	98.6	99.6	100.0

(data from Blais et al., 1956)

TABLE 1274

PERCENTILES FOR AGES AT WHICH ADULT STATURE IS REACHED
IN OHIO CHILDREN

Ages	Boys Percentiles			Girls Percentiles		
	10	50	90	10	50	90
Chronological age (years)	18.4	21.2	23.5	15.8	17.3	21.1
Years after PHV	4.4	7.8	10.3	4.6	6.0	9.8
Years after menarche	--	--	--	3.8	4.8	6.7
Years after SA 13 years	4.8	8.1	10.9	--	--	--
Years after SA 11 years	--	--	--	4.7	6.3	9.7

PHV = age at peak height velocity
SA = skeletal age

(data from Roche and Davila, 1972. Pediatrics 50:874-880. Copyright American
Academy of Pediatrics 1972)

SUPPLEMENTAL INFORMATION

Published sources of raw data

Relevant sources of data not included

Published Sources of Raw Data

(The names of authors supplying
serial data are underlined)

STATURE AND RECUMBENT LENGTH

Stature: Gray, 1941; Palmer, 1944; Garn and Moorrees, 1951; Tuddenham and Snyder, 1954; Cheek, 1968; Heiber, 1975; Parízková, 1976

Recumbent length: Deming, 1957; Babson and Bramhall, 1969; Thompson, 1951; Fomon et al., 1970, 1971

Recumbent length in low birth weight infants: Seckel and Rolfes, 1959

Recumbent length in relation to diet: Paiva, 1953; Davidson et al., 1967; Fomon et al., 1970, 1975

PHYSIQUE

Stature/$\sqrt[3]{\text{weight}}$: Garn and Moorrees, 1951

Wetzel Grid: Garn and Moorrees, 1951

Differences between recumbent length and head circumference: Philip, 1975

WEIGHT

Weight: Gray, 1941; Palmer, 1944; Thompson, 1951; Friis-Hanson, 1961; Garn and Moorrees, 1951; Tuddenham and Snyder, 1954; Beal et al., 1962; Berg, 1968; Young et al., 1968; Cheek, 1968; Babson and Bramhall, 1969; Fomon et al., 1970, 1971; Fomon et al., 1975; Heiber, 1975; Parízková, 1976

Weight in low birth weight infants: Seckel and Rolfes, 1959

Weight in relation to diet: Davidson et al., 1967; Fomon et al., 1970, 1975

Weight in relation to stature: Pryor, 1966

Weight in relation to biiliac diameter: Pryor, 1966

HEAD

Head length: Gray, 1941

Head width: Gray, 1941

Head height: Gray, 1941; Kelly and Reynolds, 1947

Head circumference: Thompson, 1951; Bray et al., 1969

Head circumference in low birth weight infants: Seckel and Rolfes, 1959

TRUNK

Trunk length: Tyler, 1966; Thompson, 1951

Sitting height, Gray, 1941; Palmer, 1944

Stem length: Tuddenham and Snyder, 1954

TRUNK (continued)

Chest width: Gray, 1941; Palmer, 1944; Schwarz, 1946; Thompson, 1951

Biacromial diameter: Thompson, 1951; Tuddenham and Snyder, 1954

Bicristal (biiliac) diameter: Gray, 1941; Palmer, 1944; Meredith and Carl, 1946; Thompson, 1951; Tuddenham and Snyder, 1954

Chest depth: Gray, 1941; Palmer, 1944

Pelvic depth: Palmer, 1944

Chest circumference: Gray, 1941; Thompson, 1951

Chest circumference in low birth weight infants: Seckel and Rolfes, 1959

Bust circumference: Tyler, 1966

Waist circumference: Tyler, 1966

Hip circumference: Tyler, 1966

LOWER EXTREMITY

Leg length: Palmer, 1944

Trochanteric height: Berg, 1968

Pubic height: Palmer, 1944

Symphysis height: Thompson, 1951

Tibiale height: Berg, 1968

Foot length: Berg, 1968

Calf circumference: Tuddenham and Snyder, 1954

UPPER EXTREMITY

Arm length: Palmer, 1944

CRANIUM

Cranial capacity: Bray et al., 1969

Nasion-sella: Nanda, 1956

Nasion-basion: Brodie, 1955

Basion-sphenoccipitale as percentage of nasion-basion: Brodie, 1955

Sphenoccipitale-sphenoethmoidale as percentage of nasion-basion: Brodie, 1955

Sphenoethmoidale-nasion as percentage of nasion-basion: Brodie, 1955

Nasion-sella-basion: Brodie, 1955

Fontanelle size: Philip, 1975

FACE

Sella-gnathion increments: Bambha and Van Natta, 1963

Face height: Gray, 1941

Anterior face height: Prakash and Margolis, 1952

Perpendicular face height: Prakash and Margolis, 1952

Upper incisal height: Prakash and Margolis, 1952

Upper molar height: Prakash and Margolis, 1952

Lower incisal height: Prakash and Margolis, 1952

Lower molar height: Prakash and Margolis, 1952

Nasion to incisal edge of upper central incisor: Prakash and Margolis, 1952

Nasion to occlusal surface of upper first molar: Prakash and Margolis, 1952

Nasion-gnathion: Nanda, 1956

Sella-gonion: Nanda, 1956

Sella-gnathion: Nanda, 1956

Nose height: Gray, 1941

Face breadth: Gray, 1941

Nose breadth: Gray, 1941

Overbite: Prakash and Margolis, 1952

MANDIBLE

Mandibular size: Nanda, 1956; Gilda, 1974

Mandibular size in relation to age at ossification of adductor sesamoid: Heiber, 1975

Bigonial diameter: Newman and Meredith, 1956

Direction of mandibular growth: Björk and Skieller, 1972

Position of mandibular foramen: Benham, 1976

DENTAL ARCH

Width: Sillman, 1965

Length: Sillman, 1965

Distance from last molar tooth to distal border of ramus: Brodie, 1941, 1942

Increments in widths and depths: Knott, 1961

PELVIS

    Widths and heights: Coleman, 1969

    Iliac index: Armendares et al., 1967

LONG BONES

    Ulnar length: Valk, 1974

    Tibial length: Babson and Bramhall, 1969

    Tibia, total and distal increments: Garn et al., 1968

SKELETON: MATURATION

    Onset of ossification: Pyle et al., 1961

    Skeletal age: Pařízková, 1976

    Epiphyseal fusion in shoulder, elbow, hand, knee, and foot: Pyle et al., 1961

    Epiphyseal fusion - distal phalanx II (hand): Buehl and Pyle, 1942

    Carpal areas: Kelly and Reynolds, 1947

    Width of knee epiphyses: Philip, 1975

OTHER

    Age of pubescent spurts: Bambha, 1961

    Skeletal and chronological ages at pubertal events: Hansman and Maresh, 1961

HEIGHTS

    Suprasternal height: Thompson, 1951

    Vertex-suprasternal notch: Thompson, 1951

    Vertex-symphysis: Thompson, 1951

RELIABILITY

    Anthropometric measurements: Knott, 1941

Relevant Sources of Reference Data That Could Not
Be Included Because of Copyright Restrictions

Chung, C. S., Niswander, J. D., Runck, D. W., Bilben, S. E., and Kau, M. C. W., 1971, Genetic and epidemiologic studies of oral characteristics in Hawaii's school children. II. Malocclusion, Am. J. Hum. Genet., 23:471.

Eppright, E. S., and Sidwell, V. D., 1954, Physical measurements of Iowa school children, J. Nutr., 54:543.

Ferris, A. G., Laus, M. J., Hosmer, D. W., and Beal, V. A., 1980, The effect of diet on weight gain in infancy, Am. J. Clin. Nutr., 33:2635.

Foster, T. A., Voors, A. W., Webber, L. S., Frerichs, R. R., and Berenson, G. S., 1977, Anthropometric and maturation measurements of children, ages 5 to 14 years, in a bi-racial community--The Bogalusa Heart Study, Am. J. Clin. Nutr., 30:582.

Frisancho, A. R., 1974, Triceps skinfold and upper arm muscle size norms for assessment of nutritional status, Am. J. Clin. Nutr., 27:1052.

Frisch, R. E., and Revelle, R., 1969, The height and weight of adolescent boys and girls at the time of peak velocity of growth in height and weight: longitudinal data, Hum. Biol., 41:536.

Frisch, R. E., and Revelle, R., 1970, Height and weight at menarche and a hypothesis of critical body weights and adolescent events, Science, 169:397.

Frisch, R. E., and Revelle, R., 1971, Height and weight at menarche and a hypothesis of menarche, Arch. Dis. Childh., 46:695.

Frisch, R. E., and Revelle, R., 1971, The height and weight of girls and boys at the time of initiation of the adolescent growth spurt in height and weight and the relationship to menarche, Hum. Biol., 43:140.

Frisch, R. E., Revelle, R., and Cook, S., 1971, Height, weight and age at menarche and the "critical weight" hypothesis, Science, 174:1148.

Garn, S. M., 1966, Body size and its implications, in: "Review of Child Development Research," Vol. 2, L. W. Hoffman and M. L. Hoffman, eds., Russell Sage Foundation, New York.

Hansman, C. F., 1962, Appearance and fusion of ossification centers in the human skeleton, Am. J. Roentgenol., 88:476.

Huenemann, R. L., Shapiro, L. R., Hampton, M. C., and Mitchell, B. W., 1966, A longitudinal study of gross body composition and body conformation and their association with food and activity in a teen-age population. Views of teen-age subjects on body conformation, food and activity, Am. J. Clin. Nutr., 18:325.

Johnston, F. E., and Beller, A., 1976, Anthropometric evaluation of the body composition of black, white, and Puerto Rican newborns, Am. J. Clin. Nutr., 29:61.

Johnston, F. E., McKigney, J. I., Hopwood, S., and Smelker, J., 1978, Physical growth and development of urban native Americans: A study in urbanization and its implications for nutritional status, Am. J. Clin. Nutr., 31:1017.

Kelly, H. J., and Reynolds, L., 1947, Appearance and growth of ossification centers and increases in the body dimensions of white and Negro infants, Am. J. Roentgenol. Rad. Ther., 57:477.

Penchaszadeh, V. B., Hardy, J. B., Mellitts, E. D., Cohen, B. H., and McKusick, V. A., 1972, Growth and development in an "inner city" population: An assessment of possible biological and environmental influences. I. Intra-uterine growth, Johns Hopkins Med. J., 130:384.

Pomerance, H. H., 1979, "Growth Standards in Children," Harper and Row, Hagerstown, MD.

Reed, T. E., 1967, Research on blood groups and selection from the Child Health and Development Studies, Oakland, California. 1. Infant birth measurements, Am. J. Hum. Genet., 19:732.

Young, C. M., Sipin, S. S., and Roe, D. A., 1968, Body composition of pre-adolescent and adolescent girls. II. Anthropometric measurements, J. Am. Diet. Assoc., 53:357.

Zavaleta, A., and Malina, R. M., 1980, Growth, fatness, and leanness in Mexican-American children, Am. J. Clin. Nutr., 33:2008.

Zuk, G. H., 1958, The plasticity of the physique from early adolescence through adulthood, J. Genet. Psychol., 92:205.

ANNOTATED BIBLIOGRAPHY

ANNOTATED BIBLIOGRAPHY

The following citations are presented in two styles. One style provides a brief sample description and lists the variables included in the tables of this volume. The other style provides bibliographic data only for publications listed in the footnotes to some tables, usually in relation to methodology, and for publications included in the table of raw data sources.

Abraham, S., Lowenstein, F. W., and O'Connell, D. E., 1975, "Preliminary Findings of the First Health and Nutrition Examination Survey, United States, 1971-1972. Anthropometric and Clinical Findings," DHEW Publ. No. (HRA) 75-1229. U.S. Govt. Print. Off., Washington. Data from the first half of a projected nationally representative sample (N = 10,126) of those aged 1 to 74 years. (Tables of stature and weight in relation to income level)

Acheson, R. M., 1957, The Oxford method of assessing skeletal maturity, Clin. Orthop., 10:19.

Acheson, R. M., and Dupertuis, C. W., 1957, The relationship between physique and rate of skeletal maturation in boys, Hum. Biol., 29:167. Data from 128 white Cleveland (OH) males, from an upper socioeconomic group, who were enrolled in the Brush Foundation Study. (Table of maturity scores for the hip and pelvis)

Adams, M. S., and Niswander, J. D., 1968, Birth weight of North American Indians, Hum. Biol., 40:226. Data from 14,376 American Indian natives born in United States Public Health Service medical facilities. Only tribes with more than 50 births of each sex in a 2-year period were considered. Multiple births, stillbirths, children with major malformations, neonatal deaths, and children of unknown tribe or birth weight were excluded. The final sample included infants from 18 tribes. (Table of birth weight)

Adams, M. S., and Niswander, J. D., 1973, Birth weight of North American Indians. A correction and amplification, Hum. Biol., 45:351. Data from 37,475 infants. (Tables of birth weight)

Adams, M. S., Brown, K. S., Iba, B. Y., and Niswander, J. D., 1970, Health of Papago Indian children, Public Health Rep., 85:1047. Data from 920 Papago Indians aged 5 to 15 years living in Southwestern Arizona along the Mexican border. (Tables of birth weight, stature, and weight)

Adeloye, A., Kattan, K. R., and Silverman, F. N., 1975, Thickness of the normal skull in the American blacks and whites, Am. J. Phys. Anthropol., 43:23. Data from 211 children aged less than 19 years. (Tables of cranial thickness)

Ahlgren, J. G. A., Ingervall, B. F., and Thilander, B. L., 1973, Muscle activity in normal and postnormal occlusion, Am. J. Orthod., 64:445. Data from 15 boys aged 9 to 11 years with normal occlusion examined in a school dental service in Gothenburg (Sweden). (Table of voltage amplitude in chewing, activity in rest position and during swallowing)

Aisenson, M. R., 1950, Closing of the anterior fontanelle, Pediatrics, 6:223. Data from 1677 infants in New York City. They were from lower income groups but not indigent. Almost all were white and largely of Italian extraction. Breast-fed infants and black infants were excluded. (Table of percentages with fontanelle closed)

Altemus, L. A., 1955, Horizontal and vertical dentofacial relationships in normal and Class II, Division I malocclusion in girls 11-15 years, Angle Orthod., 25:120. Serial data from 20 girls with normal occlusion enrolled in the Philadelphia Center for Research in Child Growth. (Table of mandibular length, anterior and posterior ramal length, anterior and posterior alveolar length, body length, symphyseal height, and ramal, incisal, occlusal and mandibular angles)

Anderson, D. L., Thompson, G. W., and Popovich, F., 1975, Adolescent variation in weight, height, and mandibular length in 111 females, Hum. Biol., 47:309. Data from girls aged 8 to 16 years enrolled in the Burlington Growth Centre (Ontario). (Tables of stature, weight and mandibular length, and ratios of weight to stature, weight to mandibular length, and mandibular length to stature in relation to age at peak height velocity)

Anderson, M., and Green, W. T., 1948, Lengths of the femur and the tibia. Norms derived from orthoroentgenograms of children from five years of age until epiphysial closure, Am. J. Dis. Child., 75:279. Data from 255 children most of whom had suffered from poliomyelitis on the opposite side and were examined at the Children's Hospital in Boston (MA). The remainder (N = 42) were enrolled in the Harvard School of Public

Health Growth Study.   (Table of lengths of femur and tibia in relation to skeletal age)

Anderson, M., Blais, M., and Green, W. T., 1956, Growth of the normal foot during childhood and adolescence. Length of the foot and interrelations of foot, stature, and lower extremity as seen in serial records of children between 1-18 years of age, Am. J. Phys. Anthropol., 14:287. Serial data from 512 patients with normal muscle power in the neck and trunk and in one entire lower extremity. These children were seen in the Children's Medical Center, Boston (MA). (Tables of foot length, ratios of foot length to stature, tibial length and sum of tibial and femoral lengths, foot proportions)

Anderson, M., Green, W. T., and Messner, M. B., 1963, Growth and predictions of growth in the lower extremities, J. Bone Joint Surg., 45A:1. Data from 100 children (51 normal from the Harvard School of Public Health Study; 49 with poliomyelitis on the opposite side) studied annually for at least 8 years. (Tables of increments in femoral and tibial lengths; increments in stature; total increase in stature after particular ages; growth remaining in femur and tibia in relation to chronological and skeletal ages)

Anderson, M., Messner, M. B., and Green, W. T., 1964, Distribution of lengths of the normal femur and tibia in children from one to eighteen years of age, J. Bone Joint Surg., 46A:1197. Data from 134 Boston (MA) children examined serially. (Table of lengths of femur and tibia)

Anderson, M., Hwang, S.-C., and Green, W. T., 1965, Growth of the normal trunk in boys and girls during the second decade of life related to age, maturity, and ossification of the iliac epiphyses, J. Bone Joint Surg., 47A:1554. Data from 134 children enrolled in the Harvard School of Public Health Growth Study. (Tables of sitting height, potentials for growth in sitting height at particular skeletal ages, and age of onset of ossification in iliac epiphysis)

Armendares, S., Urrusti-Sanz, J., and Diaz-del-Castillo, E., 1967, Iliac index in newborns. Comparative values at term, in prematurity, and in Down's syndrome, Am. J. Dis. Child., 113:229.

Arya, B. S., Savara, B. S., Thomas, D., and Clarkson, Q., 1974, Relation of sex and occlusion to mesiodistal tooth size, Am. J. Orthod., 66:479. Serial data from 48 boys and 47 girls aged 4.5 to 14 years examined at the University of Oregon Dental School. None had orthodontic treatment; 25 boys and 20 girls had normal occlusions. (Tables of occlusion)

Austin, J. H. M., and Gooding, C. A., 1971, Roentgenographic measurement of skull size in children, Radiology, 99:641. Data from radiographs of 262 California children aged less than 6 years. The normality of the children was not described. (Table of cranial indices, length, width, height and volume in relation to intercondylar width)

Baber, W. E., and Meredith, H. V., 1965, Childhood change in depth and height of the upper face, with special reference to Downs' A point, Am. J. Orthod., 51:913. Data from 20 white girls examined serially who were living in or near Iowa City (IA). (Tables of facial depth and height and increments in these)

Babson, S. G., 1970, Growth of low-birth-weight infants, J. Pediatr., 77:11. Data from low-birth-weight infants in Portland (OR) who were single-born whites without obvious congenital abnormalities or fetal infection. Each had daily measurements of weight and weekly measurements of recumbent length and head circumference from the first week of age until discharge from hospital and then at least 3 further sets of measurements during the first year. None had a gross neurological deficit or retardation through 1.5 years. Infants separated to A) gestational age 27 to 29 weeks, weight appropriate; B) gestational age 31 to 33 weeks, weight appropriate; C) full term but birth weight <2000 gm. Data recorded from birth to 1 year after "term." (Table of weight, recumbent length and head circumference)

Babson, S. G., and Bramhall, J. L., 1969, Diet and growth in the premature infant. The effect of different dietary intakes of ash-electrolyte and protein on weight gain and linear growth, J. Pediatr., 74:890.

Babson, S. G., Behrman, R. E., and Lessel, R., 1970, Liveborn birth weights for gestational age of white middle class infants, Pediatrics, 45:937. Data from 40,000 singleton newly born infants with gestational ages between 27 and 44 weeks. These infants were born in Portland (OR) from 1959 through 1966. The mothers were primarily of North European descent who received private prenatal care; 80% were delivered by obstetricians and almost all lived near sea level. (Table of birth weight)

Baer, M. J., and Durkatz, J., 1957, Bilateral asymmetry in skeletal maturation of the hand

and wrist: A roentgenographic analysis, Am. J. Phys. Anthropol., 15:181. Data from 474 roentgenograms of 245 children (239 Caucasian, 5 Negro, 1 Mongolian). These healthy Detroit (MI) children were from families of middle to upper socioeconomic status. (Table of frequency of bilateral asymmetry)

Bakwin, H., and Patrick, T. W., Jr., 1944, The weight of Negro infants, J. Pediatr., 24:405. Data from 677 observations of 114 New York City infants. About 40% were breastfed for the first 3 to 4 months. (Table of weight)

Bambha, J. K., 1961, Longitudinal cephalometric roentgenographic study of face and cranium in relation to body height, J. Am. Dent. Assoc., 63:776. Serial data from 49 white children who were participants in the Child Research Council Growth Study (Denver, CO). (Tables of age at maximum rates of growth in face, cranium and stature and their timing in relation to each other, duration of period of acceleration, and gain during spurt)

Bambha, J. K., and Van Natta, P., 1963, Longitudinal study of facial growth in relation to skeletal maturation during adolescence, Am. J. Orthod., 49:481.

Baughan, B., and Demirjian, A., 1978, Sexual dimorphism in the growth of the cranium, Am. J. Phys. Anthropol., 49:383. Serial data from about 200 Montreal children of French-Canadian ancestry. The schools they attended were representative of different socio-economic areas of Montreal and there was a wide range of paternal occupations. (Table of cranial height)

Baughan, B., Demirjian, A., and Levesque, G.-Y., 1979, Skeletal maturity standards for French-Canadian children of school-age with a discussion of the reliability and validity of such measures, Hum. Biol., 51:353. Data from 4,084 Montreal school children aged 6 to 17 years. (Table of Tanner-Whitehouse skeletal ages)

Baum, A. T., 1951, A cephalometric evaluation of the normal skeletal and dental pattern of children with excellent occlusions, Angle Orthod., 21:96. Data from 62 children in the Seattle (WA) public schools aged 11 to 13 years. (Tables of Downs' analysis data)

Bayley, N., 1943a, Skeletal maturing in adolescence as a basis for determining percentage of completed growth, Child Dev., 14:1. Data from participants in the Oakland Growth Study. (Table of percentage of adult height attained in relation to rate of maturing)

Bayley, N., 1943b, Size and body build of adolescents in relation to rate of skeletal maturing, Child Dev., 14:47. Data from participants in the Oakland Growth Study. (Tables of the percentage of adult size achieved by skeletal age groupings for stem length, biiliac diameter, biacromial diameter, stem length/stature, biiliac diameter/stature, and biiliac diameter/biacromial diameter in relation to rate of maturation)

Bayley, N., 1962, The accurate prediction of growth and adult height, Mod. Probl. Paediatr., 7:234. Data from 185 children examined serially in the Harvard School of Public Health Growth Study and the Berkeley Growth Study. (Table of percentage of adult stature achieved by chronological age and by skeletal age)

Bayley, N., and Pinneau, S. R., 1952, Tables for predicting adult height from skeletal age revised for use with the Greulich-Pyle hand standards, J. Pediatr., 40:423. Data from 192 normal white children examined serially in the Berkeley Growth Study. (Tables of percentage mature stature achieved by chronological age, rate of maturing, and errors in prediction)

Beal, V. A., Meyers, A. J., and McCammon, R. W., 1962, Iron intake, hemoglobin and physical growth during the first two years of life, Pediatrics, 30:518.

Beaudry, P. H., and Sutherland, J. M., 1960, Birth weights of infants of toxemic mothers, J. Pediatr., 56:505. Data from babies born in Cincinnati (OH) in 1956; 4068 controls and 418 born to toxemic mothers. All were liveborn; infants weighing less than 500 gm were excluded. (Table of birth weight in relation to racial group)

Beck, G. J., and van den Berg, B. J., 1975, The relationship of the rate of intrauterine growth of low-birth-weight infants to later growth, J. Pediatr., 86:504. Data from 488 children (295 white, 193 black) in Oakland (CA) whose parents form a broadly-based group typical of an employed population. All infants were single-born (birth weights 1501 to 2500 gm) and without severe congenital anomalies. (Table of stature and weight)

Becker, H. M., Glass, R. L., and Shiere, F. R., 1972, Exfoliation of the deciduous teeth during the ages of the mixed dentition, J. Dent. Res., 51:498. Serial data from 705 white Massachusetts children aged 7 to 10 years who did not need orthodontic treatment. (Tables of proportions of deciduous teeth present, rates of extraction and exfoliation, and ages of exfoliation)

Bench, R. W., 1963, Growth of cervical vertebrae as related to tongue, face, and denture behavior, Am. J. Orthod., 49:183. Serial data from 165 California children aged 2 to 19 years who were patients in an orthodontic practice. (Tables of vertical distances and horizontal distances relating to pterygoid root, posterior nasal spine, tongue, mandible, hyoid, third cervical vertebra, orbitale, menton, Frankfort horizontal plane, basion, and posterior aspect of symphysis)

Benham, N. R., 1976, The cephalometric position of the mandibular foramen with age, J. Dent. Child., 43:233.

Berg, S. J., 1968, "Relationship Between Selected Body Measurements and Success in the Standing Broad Jump," M.S. Thesis, Department of Physical Education, Washington State University, Pullman.

Biggerstaff, R. H., Allen, R. C., Tuncay, O. C., and Berkowitz, J., 1977, A vertical cephalometric analysis of the human craniofacial complex, Am. J. Orthod., 72:397. Data from 83 children aged 12 and 16 years examined serially in Ann Arbor (MI). These children were attending the University Elementary School in Ann Arbor. None were treated orthodontically during or previous to the cephalometric film series. (Table of skeletal and dentoalveolar ratios)

Björk, A., 1955, Cranial base development, Am. J. Orthod., 41:198. Data from 243 normal Swedish boys examined at 12 and 20 years. (Table of increments in cranial base angles, forehead and foramen angles and in cranial base size)

Björk, A., and Helm, S., 1967, Prediction of the age of maximum puberal growth in body height, Angle Orthod., 37:134. Data from 52 healthy children examined serially at the Royal Dental College, Copenhagen. (Tables of ages at peak height velocity, menarche, adductor sesamoid ossification, eruption of all canines and premolars and eruption of all second molars)

Björk, A., and Palling, M., 1954, Adolescent age changes in sagittal jaw relation, alveolar prognathy, and incisal inclination, Acta Odontol. Scand., 12:201. Data from a random sample of 243 Swedish boys examined at 12 and 20 years. (Tables of prognathism, incisor inclination and changes in these)

Björk, A., and Skieller, V., 1972, Facial development and tooth eruption. An implant study at the age of puberty, Am. J. Orthod., 62:339.

Blair, E. S., 1954, A cephalometric roentgenographic appraisal of the skeletal morphology of Class I, Class II, Div. 1 and Class II, Div. 2 (Angle) malocclusion, Angle Orthod., 24:106. Data from 40 children, aged 10 to 14 years, studied in the Department of Orthodontics, University of Illinois. (Table of mandibular length, first molar position, cranial base length and shape, and facial angles)

Blais, M. M., Green, W. T., and Anderson, M., 1956, Lengths of the growing foot, J. Bone Joint Surg., 38A:998. Data from 512 children attending the Children's Medical Center, Boston (MA). The measurements were made on the normal feet of children with unilateral paralysis. (Tables of foot length, percentage adult size attained for stature, and the lengths of the femur, tibia, and foot)

Bowden, B. D., 1971, Sesamoid bone appearance as an indicator of adolescence, Aust. Orthod. J., 2:242. Serial data from 112 Australian children of British ancestry enrolled in the University of Melbourne Growth Study. (Tables of ages at appearance, in relation to chronological and skeletal age, and peak height velocity)

Bowker, W. D., and Meredith, H. V., 1959, A metric analysis of the facial profile, Angle Orthod., 29:149. Data from 48 white children examined serially who were living in or near Iowa City (IA). (Table of distances of profile points to N-Pg)

Boyd, J. D., 1945, Clinical appraisal of infant's head size, Am. J. Dis. Child., 69:71. Data from 100 white infants in Iowa with a mean of 6.5 serial assessments each. Most were first-born children of unmarried mothers and most of the infants were healthy and active. (Tables of head circumference in relation to age and recumbent length and head circumference in relation to birth weight)

Brader, A. C., 1957, A cephalometric x-ray appraisal of morphological variations in cranial base and associated pharyngeal structures: implications in cleft palate therapy, Angle Orthod., 27:179. Data from 60 normal Pennsylvania children with a preponderance of Class I occlusions. (Table of cranial base angles, position of atlas, hyoid and maxillary tuberosity, size and position of adenoidal tissue, and length of soft palate)

Brandner, M. E., 1970, Normal values of the vertebral body and intervertebral disc index during growth, Am. J. Roentgenol., 110:618. Data from 187 Swiss children aged from birth to 17 years. (Tables of vertebral body height/sagittal diameter of vertebral body; intervertebral disc thickness/vertebral body height)

Bray, P. F., Shields, W. D., Wolcott, G. J., and Madsen, J. A., 1969, Occipitofrontal head circumference--an accurate measure of intracranial volume, J. Pediatr., 75:303.

Brodie, A. G., 1941, On the growth pattern of the human head from the third month to the eighth year of life, Am. J. Anat., 68:209.

Brodie, A. G., 1942, On the growth of the jaws and the eruption of teeth, Angle Orthod., 12:109.

Brodie, A. G., Jr., 1955, The behaviour of the cranial base and its components as revealed by serial cephalometric roentgenograms, Angle Orthod., 25:148.

Buehl, C. C., and Pyle, S. I., 1942, The use of age at first appearance of three ossification centers in determining the skeletal status of children, J. Pediatr., 21:335. Serial data from 60 white participants in the Brush Foundation Study (Cleveland, OH). (Table of age at onset of ossification in distal ulnar epiphysis, adductor sesamoid, crest of ilium, epiphyseo-diaphyseal fusion in distal phalanx II, and menarche)

Burmeister, L. F., Flatt, A. E., and Weiss, M. W., 1974, "Size and Strength Development of the Hand in Elementary School Children," Iowa State Services for Crippled Children, Iowa City. Data from 1741 children in grades K to 6. Approximately 81% were white, 5% were black, and 7% were Spanish. (Tables of hand width, palm length, hand length; second, third, fourth, and fifth finger lengths)

Burson, C. E., 1952, A study of individual variation in mandibular bicanine dimension during growth, Am. J. Orthod., 38:848. Serial data from 24 children enrolled in the Child Research Council Study (Denver, CO). (Table of bicanine width)

Burstone, C. J., 1959, Integumental contour and extension patterns, Angle Orthod., 29:93. Data from 37 Caucasian children, aged 13.4 to 15.6 years, chosen as having "good" faces. (Table of integumental extension values)

Caffey, J., Ames, R., Silverman, W. A., Ryder, C. T., and Hough, G., 1956, Contradiction of the congenital dysplasia--predislocation hypothesis of congenital dislocation of the hip through a study of the normal variation in acetabular angles at successive periods in infancy, Pediatrics, 17:632. Data from 627 newborn infants in New York City examined during the first week of life, of whom 551 and 527 were examined also at 6 and 9 months respectively. (Tables of acetabular angles)

CAHPER, 1968, "The Physical Work Capacity of Canadian Children Aged 7 to 17," Canadian Association for Health, Physical Education and Recreation, Ottawa, Canada. Data from 2017 children selected randomly within 175 randomly selected schools throughout Canada. (Tables of weight)

Cannon, J., 1970, Craniofacial height and depth increments in normal children, Angle Orthod., 40:202. Serial data from 52 Australian children of British ancestry enrolled in the University of Melbourne Growth Study. (Tables of status and increments in craniofacial depth and height, facial profile, mandibular size, and incisal inclination)

Carroll, J. L., Knott, V. B., and Meredith, H. V., 1966, Change in several calvariofacial distances and angles during the decade of childhood following age 5 years, Growth, 30:47. Data from 69 children examined biennially from 5 to 15 years of age. These were North American white children, predominantly of Northwest European ancestry. They were of above average socioeconomic status and were born and reared in Iowa. (Tables of distances from the occipital condyles to bregma, nasion, ANS, and pogonion; bregma-nasion, nasion-ANS, ANS-pogonion, and increments in these measurements; angles N-bregma-condyle, bregma-condyle-N, bregma-N-condyle, N-condyle-ANS, condyle-N-ANS, N-ANS-condyle, condyle-ANS-pogonion, ANS-condyle-pogonion, ANS-pogonion-condyle, and increments in these angles)

Chaconas, S. J., and Bartroff, J. D., 1975, Prediction of normal soft tissue facial changes, Angle Orthod., 45:12. Serial data from 20 Caucasian children enrolled in the Brush Foundation Study (Cleveland, OH) who had Class I occlusions. (Table of profile measurements)

Cheek, D. B., 1968, "Human Growth: Body Composition, Cell Growth, Energy and Intelligence," Lea and Febiger, Philadelphia.

Chen, H., Ceres, E., Casquejo, C., Boriboon, K., and Wooley, P., Jr., 1974, Inter-nipple distance in normal children from birth to 14 years, and in children with Turner's,

Noonan's, Down's, and other aneuploides, Growth, 38:421. Data from 472 normal children in Detroit (MI) aged from birth to 14 years. (Tables of inter-nipple distance, ratios of inter-nipple distance to chest circumference, chest width, and chest depth)

Cherry, F. F., 1968, Growth from birth to five years of New Orleans underprivileged Negro children, Bulletin, Tulane U. Med. Fac., 27:233. Serial data from 99 boys and 101 girls from families with a monthly income less than $160 plus $15 additional for each child. (Tables of recumbent length, weight, chest circumference and head circumference)

Cherry, F. F., Bancroft, H., and Newsom, W. T., 1959, Growth of Negro premature infants, Pediatrics, 24:13. Data from 817 low birth weight infants from birth to 12 weeks of age born in New Orleans (LA) in 1955 to 1957. (Tables of weight and recumbent length in relation to birth weight, head circumference, differences between head circumference and chest circumference, ages of regaining of birth weight and of reaching multiples of birth weight)

Cheyne, V. D., and Oba, J. T., 1943, Average weights of the permanent teeth, including the relative amounts of enamel to dentin and cementum, J. Dent. Res., 22:181. Data from 759 sound teeth extracted from individuals in Indiana. (Tables of weights of permanent teeth, relative weights of dentin and cementum)

Choovivathanavanich, P., and Kanthavichitra, N., 1970, Arm circumference in children, Lancet, i:44. Data from 316 well-nourished children (168 Negro, 140 Puerto Rican) in New York City. (Table of arm circumference and weight)

Christie, A., 1949, Prevalence and distribution of ossification centers in the newborn infant, Am. J. Dis. Child., 77:355. Data from 1112 newly born infants (298 white boys, 267 white girls, 271 Negro boys and 276 Negro girls). These infants were born at Johns Hopkins Hospital (Baltimore, MD). Multiple births and syphilitic infants were excluded. The families were of low socioeconomic status. The radiographs were taken within 72 hours of birth. (Table of prevalence of centers for calcaneum, talus, distal epiphysis of femur, proximal epiphysis of tibia, cuboid, head of humerus, capitate, hamate, lateral cuneiform, and head of femur by weight groups)

Clarke, H. H., and Degutis, E. W., 1962, Comparison of skeletal age and various physical and motor factors with the pubescent development of 10, 13, and 16 year old boys, Res. Quart., 33:356. Data from 237 Caucasian boys in the Medford (OR) public schools. (Table of skeletal age in relation to puberty status)

Clarke, H. H., and Hayman, N. R., 1962, Reduction of bone assessments necessary for the skeletal age determination of boys, Res. Quart., 33:202.

Clements, E. M. B., Davies-Thomas, E., and Picket, K. G., 1957, Age at which the deciduous teeth are shed, Br. Med. J., 5034:1508.

Coben, S. E., 1955, The integration of facial skeletal variants. A serial cephalometric roentgenographic analysis of craniofacial form and growth, Am. J. Orthod., 41:407. Data from 25 boys and 22 girls each examined at both 8 and 16 years. (Tables of facial depth and height, facial profile and incisal inclination, cranial base length, and increments in these)

Cohen, A. M., and Vig, P. S., 1976, A serial growth study of the tongue and intermaxillary space, Angle Orthod., 46:332. Serial data from 50 children (25 boys, 25 girls) aged 4 to 19 years who had been born in King's Hospital, London. This retrospective study used the first 50 individuals in B. C. Leighton's 1960 series. (Tables of intermaxillary space area)

Coleman, W. H., 1969, Sex differences in the growth of the human bony pelvis, Am. J. Phys. Anthropol., 31:125. Serial data from 14 boys and 16 girls. These white children from southwestern Ohio were enrolled in the Fels Longitudinal Study. (Tables of mean movement of points from 9 to 18 years)

Collins, E., 1973, The illusion of widely spaced nipples in the Noonan and Turner syndromes, J. Pediatr., 83:557. Data from 247 healthy Australian children aged 2 to 10 years. (Table of chest width, internipple distance, nipple areolar width, sternal length, internipple width/chest width, and stature)

Corlett, E. L., 1947, Mandibular incisor position relative to basal bone, Am. J. Orthod., 33:21. Data from 363 Iowa children of upper socioeconomic status aged 4 to 15 years. (Table of incisor inclination)

Cotton, W. N., Takano, W. S., and Wong, W. M. M., 1951, The Downs analysis applied to three other ethnic groups, Angle Orthod., 21:213. Data from 20 American-born Chinese children aged 11 to 16 years. They had normal arch relationships and a good facial pattern. (Table of profile analysis)

Craig, C. E., 1951, The skeletal patterns characteristic of Class I and Class II Division 1 malocclusions in Norma Lateralis, Angle Orthod., 21:44. Data from 34 children aged 12 years with Class I malocclusions. The records were obtained from the Department of Orthodontia, University of Illinois. (Table of face depth, face height, cranial base lengths, cranial vault size)

Cronqvist, S., 1968, Roentgenologic evaluation of cranial size in children. A new index, Acta Radiol. (Diag.), 7:97.

Cruise, M. O., 1973, A longitudinal study of the growth of low birth weight infants. I. Velocity and distance growth, birth to 3 years, Pediatrics, 51:620. Data obtained in Buffalo (NY) from 202 low birth weight infants and 113 full-term single birth Caucasian controls; all mothers had prenatal care and all infants were free of congenital anomalies, severe respiratory distress and sepsis. (Tables of increments in weight, recumbent length and head circumference)

Crump, E. P., Horton, C. P., Masuoka, J., and Ryan, D., 1957, Growth and development. I. Relation of birth weight in Negro infants to sex, maternal age, parity, prenatal care, and socioeconomic status, J. Pediatr., 51:678. Data from 2081 deliveries in Nashville (TN). (Table of birth weight in relation to age and parity of mother)

Cullen, R. L., and Vidić, B., 1972, The dimensions and shape of the human maxillary sinus in the perinatal period, Acta Anat., 83:411. Data from 17 cadavers from St. Louis (MO) aged from 7 prenatal months to 2 postnatal days. (Table of height, depth and width of sinus)

Cunningham, A. S., 1979, Morbidity in breast-fed and artificially fed infants. II., J. Pediatr., 95:685. Data from 724 infants born at a rural hospital in New York State. Those with birth weights below the 5th percentile or serious neonatal problems were excluded. Of the infants, 505 were seen regularly in a clinic for the first year of life. (Tables of weight in relation to feeding)

Cureton, T. K., and Barry, A. J., 1964, Improving the physical fitness of youth. A report of research in the Sports-Fitness School of the University of Illinois, Monogr. Soc. Res. Child Dev., 29: Serial 95, No. 4. Data from 707 boys aged 7 to 15 years. (Tables of stature, weight, chest expansion, difference between chest and abdominal circumferences)

Cycle II of the Health Examination Survey, National Center for Health Statistics. Unpublished results. This survey was conducted in 1963-1965 and included 7,119 children aged 6 to 11 years who were representative of the noninstitutionalized population of the U.S. (Tables of weight/stature[3], sitting height/stature, biacromial diameter, bicristal diameter, chest circumference, waist circumference, hip circumference, foot length, calf circumference, upper arm circumference, forearm circumference, and upper arm "muscle" circumference)

Cycle III of the Health Examination Survey, National Center for Health Statistics. Unpublished results. This survey was conducted in 1966-1970 and included 6,768 youths aged 12 through 17.9 years who were representative of the noninstitutionalized youths in the U.S. (Tables of weight/stature[3], sitting height/stature, iliocristal height, bizygomatic diameter, bigonial diameter, cervicale height, biacromial diameter, bicristal diameter, bitrochanteric diameter, chest circumference, waist circumference, hip circumference, subischial length, thigh length, calf length, foot length, calf circumference, knee breadth, ankle breadth, upper arm length, forearm length, upper arm circumference, forearm circumference, upper arm "muscle" circumference, elbow breadth, and wrist breadth)

Dahlberg, A. A., and Menegaz-Bock, R. M., 1958, Emergence of the permanent teeth in Pima Indian children, J. Dent. Res., 37:1123. Data from 957 children aged 3.25 to 14.75 years. (Tables of median ages of eruption, lateral differences in these ages, and maxillary-mandibular differences in these ages)

Davidson, M., Levine, S. Z., Bauer, C. H., and Dann, M., 1967, Feeding studies in low-birth-weight infants. I. Relationships of dietary protein, fat, and electrolyte to rates of weight gain, clinical courses, and serum chemical concentrations, J. Pediatr., 70:695.

Dedick, A. P., and Caffey, J., 1953, Roentgen findings in the skull and chest in 1,030 newborn infants, Radiology, 61:13. Data from lateral head radiographs of infants in New York City born 1945-1950. Of the group, 493 were male (304 white, 189 Negro) and 537 female (343 white, 194 Negro). (Tables of ossification centers present)

DeKock, W. H., 1972, Dental arch depth and width studied longitudinally from 12 years of age to adulthood, Am. J. Orthod., 62:56. Serial data from 16 boys and 10 girls in Iowa City (IA) with acceptable occlusion. They were primarily of Northwest European ancestry and above average socioeconomic status. (Tables of dental arch width and depth and increments in these)

DeKock, W. H., Knott, V. B., and Meredith, H. V., 1968, Changes during childhood and youth in facial depths from integumental profile points to a line through bregma and sellion, Am. J. Orthod., 54:111. Data from 40 white boys and 20 white girls living in or near Iowa City (IA). They were examined serially from 5 to 17 years with some missed visits. (Tables of face depths and increments in these)

Deming, J., 1957, Application of the Gompertz curve to the observed pattern of growth in length of 48 individual boys and girls during the adolescent cycle of growth, Hum. Biol., 29:83.

Deming, J., and Washburn, A. H., 1963, Application of the Jenss curve to the observed pattern of growth during the first eight years of life in forty boys and forty girls, Hum. Biol., 35:484. Data from white children in Denver (CO) measured serially from birth to 8 years; statures at older ages were corrected to be approximately equivalent to recumbent lengths. Curves fitted of the form $y = c + dx - e^{a+bx}$. (Tables of increments of recumbent length and weight)

Demirjian, A., 1980, "Anthropometry Report: Height, Weight and Body Dimensions," Minister of National Health and Welfare, Ottawa, Canada. Data from 5,515 children and adolescents included in a national survey made in 1970-1972. (Tables of stature, weight, weight for stature, sitting height, relative sitting height, relative leg length, anterior superior iliac spine height, anterior superior iliac spine height/stature, and head circumference)

Demirjian, A., and Jeniček, M., 1972, Latéralité corporelle des enfants Canadiens français à Montréal, Kinanthropologie, 4:158. Data from 5,055 children aged 6 to 17 years attending Catholic schools. (Tables of biacromial and bicristal diameters for stature and age, biacromial diameter/stature, and bicristal diameter/stature)

Demirjian, A., and Marcus, R., 1971, Effects of tooth extraction on arch dimensions of French Canadian children, J. Can. Dent. Assoc., 37:230. Data from 1,352 dental casts of children aged 6, 7, 10 and 11 years living in or near Montreal. (Tables of dental arch widths)

Demirjian, A., Jeniček, M., and Dubuc, M. B., 1972, Les normes staturo-pondérales de l'enfant urbain Canadien français d'âge scolaire, Can. J. Public Health, 63:14. Data from 2,722 boys aged 6 to 17 years and 2,352 girls aged 6 to 16 years. They were all of French-Canadian descent for three generations and from three socioeconomically different parts of Montreal. (Tables of stature, weight, and weight for stature)

Demirjian, A., Bailey, D. A., De Pena, J., Auger, F., and Jeniček, M., 1976, Somatic growth of Canadian children of various ethnic origins, Can. J. Public Health, 67:209. Data from French-Canadian children in Montreal (1,054 boys, 1,050 girls), children of "anglophone" origin in Saskatoon (1,611 boys, 1,591 girls) and Eskimos of North Foxe Basin, N.W.T. (Igloolik; 335 boys, 329 girls). All the children were aged 7 to 15 years. (Tables of stature and weight)

Dodge, W. F., and West, E. F., 1970, Arm circumference in school children, Lancet, i:417. Data from 10,051 children in Galveston County (TX) of whom 62% were Anglo, 27% Negro, and 11% Latin. The Negro and Latin children tended to be from lower socioeconomic groups. (Table of weight, stature, and arm circumference)

Downs, W. B., 1948, Variations in facial relationships: their significance in treatment and prognosis, Am. J. Orthod., 34:812.

Downs, W. B., 1952, The role of cephalometrics in orthodontic case analysis and diagnosis, Am. J. Orthod., 38:162.

Drummond, R. A., 1968, A determination of cephalometric norms for the Negro race, Am. J. Orthod., 54:670. Data from 40 Waco (TX) children aged 8 to 23 years. (Tables of craniofacial size and shape and incisor inclination)

Dupertuis, C. W., and Michael, N. B., 1953, Comparison of growth in height and weight between ectomorphic and mesomorphic boys, Child Dev., 24:203. Data from 125 boys included among upper socioeconomic white Cleveland (OH) children examined serially between 1928 and 1942. (Table of percentages of adult stature attained in relation to somatotype)

Duval-Beaupère, G., and Combes, J., 1971, Segments supérieur et inférieur an cours de la croissance physiologique des filles. Etude longitudinale de la croissance de 54 filles, Arch Fr. Pédiatr., 28:1057. Data from girls in Paris. (Table of ages at peak growth in recumbent length and upper and lower segments; lengths of these segments, rates of growth in recumbent length, and upper and lower segments)

Duval-Beaupère, G., and Sayet, A., 1979, Contribution respective des segments supérieurs et inférieurs à la croissance du garçon, Arch Fr. Pédiatr., 36:369. Serial data from 64 boys in Paris examined each 6 months from birth to maturity. (Table of mean rates of growth of stature and upper and lower segments)

Edholm, O. G., Adam, J. M., and Best, T. W., 1974, Day-to-day weight changes in young men, Ann. Hum. Biol., 1:3. Data from 122 British soldiers in Northern Ireland aged 17.5 to 23.5 years. Weights were obtained nude, early in the morning after urinating and defecating. (Table of weight change)

Edlin, J. C., Whitehouse, R. H., and Tanner, J. M., 1976, Relationship of radial metaphyseal band width to stature velocity, Am. J. Dis. Child., 130:160. Serial data from 35 boys and 32 girls included in the Harpenden Longitudinal Growth Study near London (England). (Tables of metaphyseal band width in relation to age and stature velocity)

Eichorn, D. H., and Bayley, N., 1962, Growth in head circumference from birth through young adulthood, Child Dev., 33:257. Serial data from 31 boys and girls enrolled in the Berkeley (CA) Growth Study. They had white English-speaking parents who were somewhat above average socioeconomically. They were predominantly of Northern European ancestry; slightly more than 20% reported French, Spanish or Italian ancestry in small proportions. (Tables of head circumference and increments in this by chronological age and in relation to peak height velocity)

Ellis, J. D., Carron, A. V., and Bailey, D. A., 1975, Physical performance in boys from 10 through 16 years, Hum. Biol., 47:263. Serial data from 106 boys enrolled in the Saskatchewan Child Growth and Development Study. (Tables of stature and weight)

Elsasser, W. A., 1953, Studies on dentofacial morphology. II. Orthometric analysis of facial pattern, Am. J. Orthod., 39:193. Data from 909 native-born white children living in Idaho, aged 6 to 12 years. Age- and sex-specific differences were small. (Table of soft tissue profile depths)

Elsasser, W. A., and Wylie, W. L., 1948, The craniofacial morphology of mandibular retrusion, Am. J. Phys. Anthropol., 6:461. Data from 93 white California children with Class I occlusion and 90 with Class II Division 1 occlusion. All were aged about 11 years. (Table of facial depth and height, mandibular length and height, cranial base length, cranial base shape, head breadth, length and shape)

Engelman, J. A., 1965, Measurement of perioral pressures during playing of musical wind instruments, Am. J. Orthod., 51:856. Data from 20 subjects aged 10 to 17 years who were patients in the Department of Orthodontics at Washington University School of Dentistry, St. Louis (MO). They had not begun orthodontic treatment or were not in a period of active treatment. Each had at least one year's experience with the instrument used. (Table of perioral pressures)

Erhardt, C. L., Joshi, G. B., Nelson, F. G., Kroll, B. H., and Weiner, L., 1964, Influence of weight and gestation on perinatal and neonatal mortality by ethnic group, Am. J. Public Health, 54:1841. Data from 647,871 infants (513,197 white; 134,664 non-white) born in New York City whose gestational ages and birth weights were reported. (Tables of birth weight for gestational age and race and duration of pregnancy)

Fanning, E. A., 1961, A longitudinal study of tooth formation and root resorption, New Zealand Dent. J., 57:202. Data from serial radiographs of 48 white boys and 51 white girls examined in the Harvard School of Public Health Growth Study. (Tables of chronology of formation of deciduous mandibular left canine, first molar and second molar and resorption of the deciduous maxillary left first incisor, second incisor, and the deciduous mandibular left first incisor, second incisor, left canine, first molar and second molar)

Fanning, E. A., 1962a, Third molar emergence in Bostonians, Am. J. Phys. Anthropol., 20:339. Data from 2,549 high school students aged 13 to 17.9 years. (Tables of percentage of individuals with emergence of third molars and percentage with four third molars emerged)

Fanning, E. A., 1962b, The relationship of dental caries and root resorption of deciduous molars, Arch. Oral Biol., 7:595. Data from serial radiographs of 207 healthy white

Boston (MA) children primarily of middle socioeconomic status. (Tables of chronology of resorption of mesial and distal roots of deciduous first and second molars)

Fanning, E. A., 1962c, Effect of extraction of deciduous molars on the formation and eruption of their successors, Angle Orthod., 32:44. Serial data from 134 children enrolled in the Harvard School of Public Health Growth Study. These were white children living in or near Boston (MA). (Tables of ages at root formation stages and time intervals between these)

Faust, M. S., 1977, Somatic development of adolescent girls, Monogr. Soc. Res. Child Dev., 42:Serial 169, No. 1. Serial data from 97 white girls enrolled in the Guidance Study or the Berkeley (CA) Growth Study. (Tables of stem length/stature, biacromial/bicristal diameter in relation to chronological age and skeletal age; timing of maximum rates of growth in stature, stem length, leg length, shoulder width, and hip width and in subcutaneous fat thickness, weight and strength in relation to peak height velocity)

Ferguson, A. D., Scott, R. B., and Bakwin, H., 1957, Growth and development of Negro infants. VII. Comparison of the deciduous dentition in Negro and white infants (A preliminary study), J. Pediatr., 50:327. Data from 808 Negro and 175 white normal healthy infants from families of low to middle socioeconomic levels in Washington (DC). (Table of age at eruption of first tooth and number of teeth at one year)

Ferris, A. G., Beal, V. A., Laus, M. J., and Hosmer, D. W., 1979, The effect of feeding on fat deposition in early infancy, Pediatrics, 64:397. Serial data from 92 female infants in Western Massachusetts. Race not stated. (Tables of weight in relation to age)

Fleming, H. B., 1961, An investigation of the vertical overbite during the eruption of the permanent dentition, Angle Orthod., 31:53. Data from serial casts at the University of Michigan Elementary and High Schools Growth Study. The children had normal occlusion and were aged 9 to 16 years. Race not stated. The number at each age varied from 48 to 79. (Table of vertical overbite)

Fomon, S. J., Filer, L. J., Jr., Thomas, L. N., and Rogers, R. R., 1970, Growth and serum chemical values of normal breastfed infants, Acta Paediatr. Scand., Suppl. 202. Serial data from 149 infants in Iowa City (IA). (Tables of daily increments in recumbent length and weight)

Fomon, S. J., Thomas, L. N., Filer, L. J., Jr., Ziegler, E. E., and Leonard, M. T., 1971, Food consumption and growth of normal infants fed milk-based formulas, Acta Paediatr. Scand., Suppl. 223:1. Data from 142 full-term white infants in Iowa City (IA). (Tables of weight and recumbent length in relation to type of feeding, and increments in recumbent length)

Fomon, S. J., Filer, L. T., Jr., Thomas, L. N., Anderson, T. A., and Nelson, S. E., 1975, Influence of formula concentration on caloric intake and growth of normal infants, Acta Paediatr. Scand., 64:172.

Fomon, S. J., Ziegler, E. E., Filer, L. J., Jr., Anderson, T. A., Edwards, B. B., and Nelson, S. E., 1978, Growth and serum chemical values of normal breastfed infants, Acta Paediatr. Scand., Suppl. 273. Serial data from 233 normal breastfed infants in Iowa City (IA). (Tables of weight and recumbent length)

Francis, C. C., 1948, Growth of the human pituitary fossa, Hum. Biol., 20:1. Serial data from 400 white children and 391 Negro children included in the Brush and Bolton studies (Cleveland, OH). (Tables of pituitary fossa size)

Freeman, M. G., Graves, W. L., and Thompson, R. L., 1970, Indigent Negro and Caucasian birth weight-gestational age tables, Pediatrics, 46:9. Data from 7,547 Caucasian and 9,800 Negro infants in Atlanta (GA), all from indigent families (annual income less than $2,080 for an individual or $3,120 for a family of four after deducting taxes and alimony). (Tables of birth weight in relation to gestational age)

Frerichs, R. R., Srinivasan, S. R., Webber, L. S., Rieth, M. C., and Berenson, G. S., 1978, Serum lipids and lipoproteins at birth in a biracial population: The Bogalusa Heart Study, Pediatr. Res., 12:856. Data from 419 infants (256 white, 163 black) examined immediately after delivery. (Table of birth weight, total cholesterol, and lipoproteins)

Friis-Hansen, B., 1961, Body water compartments in children: changes during growth and related changes in body composition, Pediatrics, 28:169.

Fuchs, M., Iosub, S., Bingol, N., and Gromisch, D. S., 1980, Palpebral fissure size revisited, J. Pediatr., 96:77. Data from 120 Hispanic, 101 black, and 200 white infants in New York City. (Table of palpebral fissure size)

Fujioka, M., and Young, L. W., 1978, The sphenoidal sinuses: radiographic patterns of normal development and abnormal findings in infants and children, Radiology, 129:133. Data from 1,401 normal infants and children in Pittsburgh (PA). Race not stated. (Table of prevalence of radiographic patterns)

Gamm, S. H., and Gianelly, A. A., 1970, Polygonic interpretation of the Steiner analysis, Am. J. Orthod., 58:479. Data from 137 Boston (MA) children considered to have normal dental occlusions. An "ideal" sample was formed from 77 of these children. They were aged 8 to 13 years (42 boys, 35 girls). (Table of craniofacial angles)

Garcia, C. J., 1975, Cephalometric evaluation of Mexican Americans using the Downs and Steiner analyses, Am. J. Orthod., 68:67. Data from 25 boys and 34 girls of Mexican American descent residing in the East Los Angeles (CA) area. They were aged 14.4 to 17.2 years. All had complete permanent dentitions (except for third molars) and excellent occlusions. (Tables of profile analysis including incisor position and inclination)

Garn, S. M., 1966, Fels parent-specific standards for height. Privately printed. Data from children living in southwestern Ohio enrolled in the Fels Longitudinal Study; sample size not given. (Tables of mean statures in relation to mid-parent stature)

Garn, S. M., 1973, Stature norms and nutritional surveys, Ecol. Food Nutr., 2:79. Data from 827 Negro and 878 white girls aged 1 through 13 years examined in the Ten-State Nutrition Survey, 1968-1970. (Table of stature within racial groups in relation to income)

Garn, S. M., and McCreery, L. D., 1970, Variability of postnatal ossification timing and evidence for a "dosage" effect, Am. J. Phys. Anthropol., 32:139. Data from children examined serially in the Fels Longitudinal Study. These white children were living in southwestern Ohio. (Table of standard deviations and coefficients of variation for ossification centers)

Garn, S. M., and Moorrees, C. F. A., 1951, Stature, body-build, and tooth emergence in Aleutian Aleut children, Child Dev., 22:261.

Garn, S. M., Lewis, A. B., Koski, K., and Polacheck, D. L., 1958, The sex difference in tooth calcification, J. Dent. Res., 37:561. Data from 255 white children in southwestern Ohio, enrolled in the Fels Longitudinal Study, for whom there are serial lateral radiographs. (Table of ages of reaching stages of formation of permanent teeth)

Garn, S. M., Lewis, A. B., and Polacheck, D. L., 1959, Variability of tooth formation, J. Dent. Res., 38:135. Data from serial lateral radiographs of 255 white children in southwestern Ohio enrolled in the Fels Longitudinal Study. They were of northwestern European ancestry and of middle socioeconomic class. (Table of ages at various stages of permanent tooth formation)

Garn, S. M., Rohmann, C. G., and Apfelbaum, B., 1961, Complete epiphyseal union of the hand, Am. J. Phys. Anthropol., 19:365. Data from 100 white children in southwestern Ohio examined serially in the Fels Longitudinal Study. (Table of age at fusion in metacarpals and phalanges)

Garn, S. M., Rohmann, C. G., and Robinow, M., 1961, Increments in hand-wrist ossification, Am. J. Phys. Anthropol., 19:45. Data from 154 normal children examined serially in the Fels Longitudinal Study. These white children were living in southwestern Ohio. (Table of annual increments in number of hand-wrist centers)

Garn, S. M., Lewis, A. B., and Bonné, B., 1962, Third molar formation and its developmental course, Angle Orthod., 32:270. Serial data from 140 white children living in southwestern Ohio and enrolled in the Fels Longitudinal Study. (Table of age of calcification and root formation)

Garn, S. M., Rohmann, C. G., and Silverman, F. N., 1965, Missing secondary ossification centers of the foot, Inheritance and developmental meaning, Ann. Radiol. (Paris), 8:631. Data from 420 children examined serially in the Fels Longitudinal Study. These white children were living in southwestern Ohio. (Table of frequency of missing foot centers)

Garn, S. M., Rohmann, C. G., and Blumenthal, T., 1966, Ossification sequence polymorphism and sexual dimorphism in skeletal development, Am. J. Phys. Anthropol., 24:101. Data from 174 children examined serially in the Fels Longitudinal Study. These white children were living in southwestern Ohio. (Table of frequency of carpal and tarsal ossification sequences and sequence polymorphisms in hand-wrist, elbow, knee, and shoulder)

Garn, S. M., Lewis, A. B., and Kerewsky, R. S., 1966a, Sexual dimorphism in the buccolingual tooth diameter, J. Dent. Res., 45:1819. Data from 117 adolescents from southwestern Ohio representing 75 families enrolled in the Fels Longitudinal Growth Study. (Table of sex differences in mesiodistal and buccolingual diameters)

Garn, S. M., Lewis, A. B., and Kerewsky, R. S., 1966b, The meaning of bilateral asymmetry in the permanent dentition, Angle Orthod., 36:55. Data from 239 white adolescents in southwestern Ohio who were enrolled in the Fels Longitudinal Study. (Table of symmetry)

Garn, S. M., Rohmann, C. G., and Silverman, F. N., 1967, Radiographic standards for postnatal ossification and tooth calcification, Med. Radiogr. Photogr., 43:45. Data from white children living in southwestern Ohio enrolled in the Fels Longitudinal Study. The total number of children was about 600; the number of radiographs varied widely by age and site. (Tables of age at onset of ossification centers, cortical thickness of second metacarpal and ages for selected developmental stages of permanent teeth)

Garn, S. M., Lewis, A. B., and Kerewsky, R. S., 1967a, Buccolingual size asymmetry and its developmental meaning, Angle Orthod., 37:186. Data from 118 white adolescents in southwestern Ohio who were enrolled in the Fels Longitudinal Study. (Table of buccolingual asymmetry)

Garn, S. M., Lewis, A. B., and Kerewsky, R. S., 1967b, Sex differences in tooth shape, J. Dent. Res., 46:1470. Data from 55 males and 48 females. These were white individuals living in southwestern Ohio and enrolled in the Fels Longitudinal Study. (Table of buccolingual/mesiodistal diameter)

Garn, S. M., Hempy, H. O., III, and Schwager, P. M., 1968, Measurement of localized bone growth employing natural markers, Am. J. Phys. Anthropol., 28:105.

Garn, S. M., Lewis, A. B., and Walenga, A. J., 1968, Maximum-confidence values for the human mesiodistal crown dimension of human teeth, Arch. Oral Biol., 13:841. Data from 658 children in southwestern Ohio; the majority were of northwestern European ancestry. These children were enrolled in the Fels Longitudinal Study. (Table of mesiodistal diameters of teeth)

Garn, S. M., Hertzog, K. P., Poznanski, A. K., and Nagy, J. M., 1972, Metacarpophalangeal length in the evaluation of skeletal malformation, Radiology, 105:375. Serial data from white children in Southwestern Ohio enrolled in the Fels Longitudinal Study. Sample size not given. (Table of metacarpal and phalangeal lengths)

Garn, S. M., Sandusky, S. T., Nagy, J. M., and McCann, M. B., 1972, Advanced skeletal development in low-income Negro children, J. Pediatr., 80:965. Data from 4,988 participants in the Ten-State Nutrition Survey of 1968–1970. (Table of differences in onset of ossification within the hand-wrist by race)

Garn, S. M., Wertheimer, F., Sandusky, S. T., and McCann, M. B., 1972, Advanced tooth emergence in Negro individuals, J. Dent. Res., 51:1506. Data from 1,951 low-income Negro and white individuals included in the Michigan sub-set of the Ten-State Nutrition Survey, 1968–1970. (Table of ages of eruption of permanent teeth)

Garn, S. M., Clark, D. C., and Trowbridge, F. L., 1973, Tendency toward greater stature in American black children, Am. J. Dis. Child., 126:164. Data from more than 10,000 low-income children aged 1 to 15 years included in the Ten-State Nutrition Survey 1968–1970. (Table of statures in groups matched for income and fatness)

Garn, S. M., Nagy, J. M., Sandusky, S. T., and Trowbridge, F. L., 1973, Economic impact on tooth emergence, Am. J. Phys. Anthropol., 39:233. Data from about 10,000 Negro and white children aged 4.5 to 16.5 years included in the Ten-State Nutrition Survey. (Tables of variability in permanent tooth emergence timing in relation to age and income)

Garn, S. M., Sandusky, S. T., Nagy, J. M., and Trowbridge, F. L., 1973, Negro-caucasoid differences in permanent tooth emergence at a constant income level, Arch. Oral Biol., 18:609. Data from 9,656 generally lower income children included in the Ten-State Nutrition Survey of 1968–1970. (Table of tooth emergence by race)

Garn, S. M., Sandusky, S. T., Rosen, N. N., and Trowbridge, F. L., 1973, Economic impact on postnatal ossification, Am. J. Phys. Anthropol., 38:1. Data from 1,790 children included in the 1968–1970 Ten-State Nutrition Survey. Age range not given. (Table of age of onset of selected carpals and ulna in relation to income)

Garn, S. M., Poznanski, A. K., and Larson, K. E., 1975, Magnitude of sex differences in dichotomous ossification sequences of the hand and wrist, Am. J. Phys. Anthropol., 42:85. Data from 3,029 children of European descent. Age not stated. (Table of sex differences in dichotomous ossification sequences)

Garn, S. M., Poznanski, A. K., and Larson, K., 1976, Metacarpal lengths, cortical diameters and areas from the 10-State Nutrition Survey including: estimated skeletal weights, weight, and stature for whites, blacks, and Mexican-Americans, in: "Proceedings of First Workshop on Bone Morphometry," Z. F. G. Jaworski, ed., University of Ottawa Press, Ottawa. Data from 9,053 children aged 1 to 17 years. (Tables of stature and weight)

Garn, S. M., Shaw, H. A., and McCabe, K. D., 1977, Birth size and growth appraisal, J. Pediatr., 90:1049. Data from 10,390 white children in the Collaborative Perinatal Project of the National Institute of Neurological and Communicative Disorders and Stroke. (Table of weight, length and head circumference at 7 years in relation to birth weight)

Garn, S. M., Shaw, H. A., and McCabe, K. D., 1978, Effect of socioeconomic status on early growth as measured by three different indicators, Ecol. Food Nutr., 7:51. Data from 16,850 white children and 17,405 black children aged from birth to 7 years. They were examined in the Collaborative Perinatal Project of the National Institute of Neurological and Communicative Disorders and Stroke. (Table of stature, weight and head circumference in relation to socioeconomic status)

George, S. L., 1978, A longitudinal and cross-sectional analysis of the growth of the postnatal cranial base angle, Am. J. Phys. Anthropol., 49:171. Serial data from 32 white children aged one month to 5.75 years. These white children were studied at the Child Research Council (Denver, CO). (Tables of means and increments for N-S-Ba, A-S-Ba, and the clival angle. The clival angle is A-Sp-Ba where A = frontale and Sp = sphenoidale)

Gerber, F. R., Taylor, F. H., De Levie, M., Drash, A. L., and Kenny, F. M., 1972, Normal standards for exophthalometry in children 10 to 14 years of age: relation to age, height, weight and sexual maturation, J. Pediatr., 81:327. Data from 482 public school students in Pittsburgh (PA); all were Caucasian except one Oriental. (Table of exophthalometric values in relation to age and pubertal status)

Gilda, J. E., 1974, Analysis of linear facial growth, Angle Orthod., 44:1.

Giles, N. B., Knott, V. B., and Meredith, H. V., 1963, Increase in intraoral height of selected permanent teeth during the quadrennium following gingival emergence, Angle Orthod., 33:195. Data from 75 North American white children residing in or near Iowa City (IA). (Tables of intraoral height and time taken to erupt relative amounts)

Gindhart, P. S., 1971, "Growth of the Tibia and Radius Studied Longitudinally in Normal Children in Relation to Childhood Disorders," Dissertation, University of Texas at Austin. Data from about 330 white children in southwestern Ohio examined serially from birth to 9 years in the Fels Longitudinal Study. (Tables of tibia and radius lengths, and increments in these)

Gleiser, I., and Hunt, E. E., Jr., 1955, The permanent mandibular first molar: its calcification, eruption and decay, Am. J. Phys. Anthropol., 13:253. Data from white children (25 boys, 25 girls) examined serially at the Forsyth Dental Infirmary for Children, Boston (MA). (Tables of timing of stages, number of centers of calcification, rate of elongation, and amount of root formation at emergence)

Goldman, H. I., Freudenthal, R., Holland, B., and Karelitz, S., 1969, Clinical effects of two different levels of protein intake on low-birth-weight infants, J. Pediatr., 74: 881. Data from infants of low birth weight excluding those with major congenital malformations, intestinal obstruction, or Rh disease. (Table of maximum weight loss in newly born infants and number of days to regain birth weight)

Goodman, J. L., 1942, Changes in size and contour of thorax during first postnatal week, Am. J. Dis. Child., 64:674. Data from 68 newly born white infants (36 boys, 32 girls) born in the University Hospitals, Iowa City (IA). The mothers were of lower than average socioeconomic status. (Table of chest width and depth and thoracic index)

Gray, H., 1941, Individual growth-rates from birth to maturity for 15 physical traits, Hum. Biol., 13:306.

Green, L. J., 1968, Associations: Occlusion and osseous face, Angle Orthod., 38:40. Data from 90 white children aged 9 years in Iowa City (IA). (Tables of jaw depth difference and face depth difference)

Greulich, W. W., 1957, A comparison of the physical growth and development of American-born and native Japanese children, Am. J. Phys. Anthropol., 15:489. Data from 898 children aged 4.5 to 18.5 years living in the San Francisco Bay (CA) area. (Tables of stature, sitting height, weight, sitting height/stature, and skeletal age)

Greulich, W. W., and Pyle, S. I., 1959, "Radiographic Atlas of Skeletal Development of the Hand and Wrist," Stanford University Press, California.

Grewe, J. M., 1970, Intercanine width variability in American Indian children, Angle Orthod., 40:353. Data from 1,233 Navajo and 606 Chippewa children in New Mexico and Minnesota, respectively. (Tables of intercanine widths)

Gross, R. T., and Moses, L. E., 1956, Weight gains in the first four weeks of infancy: A comparison of three diets, Pediatrics, 18:362. Data from 282 Negro and 125 white infants born at the Stanford University Hospital (CA). They received breast milk, evaporated milk diluted 1:2, or a modified evaporated milk in equicaloric concentration containing less protein but more fat and carbohydrate. (Tables of increments in weight)

Haas, L. L., 1952, Roentgenological skull measurements and their diagnostic applications, Am. J. Roentgenol., 67:197. Data from 1,300 racially mixed patients of the University of Illinois Hospitals. The racial distribution was not reported. (Tables of antero-posterior diameter, cranial height and width, cephalic modulus and index)

Haas, L. L., 1954, The size of the sella turcica by age and sex, Am. J. Roentgenol., 72:754. Data from children without obvious pathology examined in the University of Illinois Hospitals. About 20% of the group are black. (Table of sellar area)

Halikis, S. E., 1961a, The variability of eruption of permanent teeth and loss of deciduous teeth in Western Australian children. II. Times of loss of the deciduous teeth, Aust. Dent. J., 6:141. Data from 862 children attending a dental clinic. (Table of ages at loss of deciduous teeth)

Halikis, S. E., 1961b, The variability of eruption of permanent teeth and loss of deciduous teeth in Western Australian children. III. The post-eruptive age of the permanent teeth, Aust. Dent. J., 6:312. Data from 862 children aged 5 to 15 years attending a dental clinic. (Tables of ages at which particular numbers are erupted, and cumulative percentages of teeth erupted)

Hamill, P. V. V., Johnston, F. E., and Grams, W., 1970, "Height and Weight of Children," DHEW Publ. (HSA) No. 1000, Series 11, No. 104, U.S. Govt. Print. Off., Washington. Data from 7,119 U.S. children aged 6 to 11 years who constituted a nationally representative sample. (Tables of stature, weight, and weight for stature by region)

Hamill, P. V. V., Johnston, F. E., and Lemeshow, S., 1972, "Height and Weight of Children: Socioeconomic Status. United States," DHEW Publ. No. (HSM) 73-1601, U.S. Govt. Print. Off., Washington. Data from 7,119 children aged 6 to 11 years who constituted a nationally representative sample. (Tables of stature and weight in relation to family income, education and place of residence)

Hamill, P. V. V., Johnston, F. E., and Lemeshow, S., 1973, "Body Weight, Stature, and Sitting Height: White and Negro Youths 12-17 Years. United States," DHEW Publ. No. (HRA) 74-1608, U.S. Govt. Print. Off., Washington. Data from 6,727 youths aged 12-17 years who constituted a nationally representative sample. (Tables of stature, weight, and sitting height by race)

Hamill, P. V. V., Drizd, T. A., Johnson, C. L., Reed, R. B., and Roche, A. F., 1977, "NCHS Growth Curves for Children, Birth - 18 Years, United States," DHEW Publ. No. (PHS) 78-1650, U.S. Govt. Print. Off., Washington. Data from 867 white children studied serially at The Fels Research Institute (0-3 years) and from nationally representative samples (2-18 years). (Tables of recumbent length, weight, stature, head circumference, weight for recumbent length, weight for stature, and recumbent length-stature differences)

Hansman, C. F., 1962, Appearance and fusion of ossification centers in the human skeleton, Am. J. Roentgenol., 88:476.

Hansman, C. F., 1966, Growth of interorbital distance and skull thickness as observed in roentgenographic measurements, Radiology, 86:87. Serial data from healthy white children of above average socioeconomic status enrolled in the Child Research Council (Denver, CO). The number at each age ranges from 49 to 102 over the age range from 6 months to 18 years. (Tables of interorbital distance and cranial thickness at lambda)

Hansman, C. F., and Maresh, M. M., 1961, A longitudinal study of skeletal maturation, Am. J. Dis. Child., 101:305.

Hanson, M. L., and Cohen, M. S., 1973, Effects of form and function on swallowing and the developing dentition, Am. J. Orthod., 64:63. Serial data from 225 children aged 4.5 to 5 years who were selected at random after excluding those with severe mental or physical impairments. Race not stated. (Table of open bite, overjet, overbite, crossbite, and diastemas)

Hanson, M. R., 1965, "Motor Performance Testing of Elementary School Age Children," Ph.D. Thesis, University of Washington, Pullman. Data from 22 schools in Minnesota with physical education programs at least 3.5 days/week. Included were 2,870 children in grades 1 through 6. Race not stated. (Tables of stature and weight)

Harding, V. S. V., 1952, A method of evaluating osseous development from birth to 14 years, Child Dev., 23:247. Serial data from 323 white children in Boston (MA) with "middle-class" families. (Tables of ages at onset of ossification of selected bones in the hand and foot, and ages at fusion of selected bones)

Harland, W. R., Grillo, G. P., Cornoni-Huntley, J., and Leaverton, P. E., 1979, Secondary sex characteristics of boys 12 to 17 years of age. The U.S. Health Examination Survey, J. Pediatr., 95:293. Data from a nationally representative sample of 6,768 boys. (Tables of prevalence of genital and pubic hair stages in relation to chronological and skeletal age, and uric acid concentration)

Harper, H. A. S., Poznanski, A. K., and Garn, S. M., 1974, The carpal angle in American populations, Invest. Radiol., 9:217. Data from 572 individuals (273 white, 299 black), aged 4 to 24 years, randomly selected from those examined in the Ten-State Nutrition Survey, 1968-1970. (Table of carpal angles)

Harris, J. E., 1962, A cephalometric analysis of mandibular growth rate, Am. J. Orthod., 48:161. Serial data from 40 white children enrolled in the University of Michigan Elementary School Growth Study. The sample was random except that those receiving orthodontic treatment were excluded. (Table of length of mandible)

Harsha, D. W., Frerichs, R. R., and Berenson, G. S., 1978, Densitometry and anthropometry of black and white children, Hum. Biol., 50:261. Data from 242 Louisiana youths aged 6 to 16 years (143 white, 99 black). (Table of weight, stature, cervicale height, iliac crest height, gluteal furrow height, sitting height, upper arm length, bizygomatic diameter, bigonial diameter, wrist diameter, biacromial diameter, chest breadth and depth, biiliac and bitrochanteric diameters, knee and ankle diameters, skinfolds at triceps, biceps, subscapular, suprailiac, subcostal, femoral, and calf sites; circumferences of upper arm, chest, waist and hip; body density and percent body fat)

Heiber, R. G., 1975, "The Relationship of the Ulnar Sesamoid Bone in Males to Circumpuberal Growth Rates of the Mandible and Body Height and Weight," Thesis for Master of Science, Ohio State University, Columbus. Serial data from 17 white boys enrolled in the Fels Longitudinal Study. (Tables of increments in stature, weight, S-Gn, Ar-Gn and Ba-Gn in relation to age at ossification of the adductor sesamoid)

Heimer, C. B., and Freedman, A. M., 1965, The physical development of prematurely born Negro infants. Growth up to 2 1/2 years of age, Am. J. Dis. Child., 109:500. Data from 348 low birth weight infants and 50 with birth weights >2500 gm born in New York City between 1956 and 1959. Measurements were made at birth (weight only), 6, 17, 32, 56, 80, 104, and 134 weeks. (Tables of weight, recumbent length in relation to birth weight, head circumference, head length, head width, and increments of these)

Heller, C. A., Scott, E. M., and Hammes, L. M., 1967, Height, weight and growth of Alaskan Eskimos, Am. J. Dis. Child., 113:338. Serial data from 643 infants and cross-sectional data from 561 children (280 boys, 275 girls) aged 3 to 16 years. (Tables of stature and weight)

Hellman, M., 1923, The process of dentition and its effect on occlusion, Dent. Cosmos., 65:1329.

Hellman, M., 1927, Changes in the human face brought about by development, Int. J. Orthod., 13:475.

Hellman, M., 1933, Growth of the face and occlusion of teeth in relation to orthodontic treatment, Int. J. Orthod., 19:1116.

Henderson, S. G., and Sherman, L. S., 1946, The roentgen anatomy of the skull in the newborn infant, Radiology, 46:107. Data from 100 healthy infants varying in birth weight from 5 lb. 1 oz for a 4-week premature infant to 9 lb. 4 oz. for a full-term child. Of these infants, 61 were black and the remainder were white. Almost all were radiographed within 10 days of birth. (Tables of length and depth of pituitary fossa; anteroposterior length, height and width of cranium; cranial volume, separation of cranial bones, and saddle angle)

Henriques, A. C., 1953, The growth of the palate and the growth of the face during the period of the changing dentition, Am. J. Orthod., 39:836. Data from 600 children, aged 7 to 12 years, enrolled in the Philadelphia Center for Child Growth. Age was expressed in the summarized data as the nearest whole year. Race was not specified. (Tables of dental arch width, palate length, and dental arch length in relation to age and dental stage)

Herness, L. E., Rule, J. T., and Williams, B. H., 1973, A longitudinal cephalometric study of incisor overbite from ages five to eleven, Angle Orthod., 43:279. Data from 20 children enrolled in a growth study at The Ohio State University College of Dentistry, Columbus. All had Class I occlusions and none received orthodontic treatment. They were from middle class and professional families and predominantly of middle and northern European background. (Table of increments in overbite)

Higley, L. B., 1954, Cephalometric standards for children 4 to 8 years of age, Am. J. Orthod., 40:51. Serial data from 175 white children living in or near Iowa City (IA) who were above average socioeconomically. (Tables of various lengths and angles within the facial area)

Hinck, V. C., and Hopkins, C. E., 1965, Concerning growth of the sphenoid sinus, Arch. Otolaryngol., 82:62. Serial data from 76 white children aged 3-18 years. (Table of clivus-sinus distance)

Hinck, V. C., Hopkins, C. E., and Savara, B. S., 1962, Sagittal diameter of the cervical spinal canal in children, Radiology, 79:97. Data from 333 radiographs of white children examined serially from 3 to 18 years in the Child Development Study at the University of Oregon Dental School. (Table of sagittal diameters of spinal canal)

Hinck, V. C., Hopkins, C. E., and Clark, W. M., 1965, Sagittal diameter of the lumbar spinal canal in children and adults, Radiology, 85:929. Data from white patients at the University of Oregon Medical School. Patients with conditions likely to affect growth and development were excluded. In the age range 3 to 18 years, 67 radiographs were measured. (Table of sagittal diameter of spinal canal)

Hinck, V. C., Clark, W. M., Jr., and Hopkins, C. E., 1966, Normal interpediculate distances (minimum and maximum) in children and adults, Am. J. Roentgenol., 97:141. Data from 353 white children 3 through 18 years old with records in the roentgenographic files of the University of Oregon Medical School. (Tables of interpediculate distances)

Hirsch, N., Hall, S. R., and Bachand, R., 1969, A cephalometric evaluation of 8-year-old Caucasians, Am. J. Orthod., 56:128. Data from 18 boys and 12 girls aged 8 years. These children were a random sample of those attending the Eastman Dental Clinic in Rochester (NY). (Table of facial depth, angles and profile)

Hitchcock, H. P., 1969, A cephalometric description of Class I malocclusion, Am. J. Orthod., 55:124. Data from 153 Caucasian children in Alabama with a mean age of 12.2 years and with Class I occlusions. (Table of facial shape)

Hoerr, N. L., Pyle, S. I., and Francis, C. C., 1962, "Radiographic Atlas of Skeletal Development of the Foot and Ankle: A Standard of Reference," Charles C Thomas, Springfield (IL). Data from children enrolled in the Brush Foundation Study. These white Cleveland (OH) children were above average socioeconomically. (Table of ages at onset of ossification)

Holcomb, A. E., and Meredith, H. V., 1956, Width of the dental arches at the deciduous canines in white children 4 to 8 years of age, Growth, 20:159. Data from 150 white children living in or near Iowa City (IA). (Tables of arch widths and increments in these widths)

Hopkin, G. B., 1967, Neonatal and adult tongue dimensions, Angle Orthod., 37:132. Data from 32 Scottish neonatal infants. (Table of tongue size)

Hopkins, J. W., 1947, Height and weight of Ottawa elementary school children of two socio-economic strata, Hum. Biol., 19:68. Data from 2,565 boys and 2,668 girls aged 6 to 12 years attending six elementary schools between 1933 and 1945. Three schools were in better class areas and three in less prosperous areas. (Tables of stature and weight within each socioeconomic group)

Huenemann, R. L., Hampton, M. C., Behnke, A. R., Shapiro, L. R., and Mitchell, B. W., 1974, "Teenage Nutrition and Physique," Charles C Thomas, Springfield (IL). Data from high school students in Berkeley (CA) measured between 1961 and 1965. Those included numbered about 1000 who were studied each year at successive grade levels (9th through 12th). (Tables of abdomen, ankle, biceps, buttocks, calf, chest, forearm, knee, shoulder, thigh and wrist circumferences; ankle, biacromial, biiliac, bitrochanteric, chest and wrist diameters; stature, weight, lean body weight, and percent body fat)

Hunter, C. J., 1966, The correlation of facial growth with body height and skeletal maturation at adolescence, Angle Orthod., 36:44. Serial data from 61 white participants in the growth study conducted by the Child Research Council (Denver, CO). (Tables of timing of pubertal spurt in stature; stature, skeletal age and facial size at this time; mandibular growth in relation to the spurt, and percentages and amounts of facial growth completed at the spurt)

Hurme, V. O., 1949, Ranges of normalcy in the eruption of permanent teeth, J. Dent. Child., 16:11. Composite data from literature. (Table of ages at eruption and order of eruption of permanent teeth)

Infante, P. F., 1974, Sex differences in the chronology of deciduous tooth emergence in white and black children, J. Dent. Res., 53:418. Data from 376 children aged 4 to 33 months. (Table of number of deciduous teeth erupted)

Infante, P. F., 1975, An epidemiological study of deciduous tooth emergence and growth in white and black children of Southeastern Michigan, Ecol. Food Nutr., 4:117. Data from 376 lower socioeconomic children aged 4 to 33 months. (Tables of number of erupted teeth, stature, weight, head and arm circumferences, subscapular skinfold, and birth weight)

Ingervall, B., 1974, Relation between height of the articular tubercle of the temporomandibular joint and facial morphology, Angle Orthod., 44:15. Data from 38 Swedish children with normal occlusions. (Table of inclination of condylar path)

Ingervall, B., Seeman, L., and Thilander, B., 1972, Relation between combined widths of incisors and dimensions of dental arches, Scand. J. Dent. Res., 80:181. Data from 324 Swedish children aged 8 years 10 months to 11 years 10 months. Eighty-two of the children had normal occlusions. The occlusal status of the remainder is described in Ingervall, B., Seeman, L., and Thilander, B., 1972, Frequency of malocclusion and need of orthodontic treatment in 10-year-old children in Gothenburg, Svensk tandelak.-T., 65:7. (Table of dental arch widths and length)

Jamison, P. L., 1978, Anthropometric variation, in: "Eskimos of Northwestern Alaska: A Biological Perspective," P. L. Jamison, S. L. Segura, and F. A. Milan, eds., Dowden, Hutchinson and Ross, Inc., Stroudsburg (PA). Data from 470 children in Alaska aged 1 to 18 years. (Tables of stature and weight)

Jenss, R. M., and Bayley, N., 1937, A mathematical method for studying the growth of a child, Hum. Biol., 9:556.

Johnston, F. E., 1962, "A Longitudinal Study of Skeletal Maturation and its Relationship to Growth in Philadelphia White Children," Ph.D. Thesis, University of Pennsylvania. Data from 120 children aged 5 to 18 years who were examined serially. (Tables of relative skeletal age, standard deviations in relation to skeletal and chronological age for weight, stature, sitting height; biacromial, bicristal and bitrochanteric diameters; total, upper, and lower face heights; bizygomatic and bigonial diameters; porion-nasion, porion-subnasale, porion-gnathion, condylion-gonion, condylion-gnathion, gonion-gnathion, and nasal height)

Johnston, F. E., Roche, A. F., Schell, L. M., and Wettenhall, H. N. B., 1975, Critical weight at menarche: critique of a hypothesis, Am. J. Dis. Child., 129:19. Data from 652 girls. (Table of weight and age at menarche)

Jones, B. H., and Meredith, H. V., 1966, Vertical change in osseous and odontic portions of human face height between the ages of 5 and 15 years, Am. J. Orthod., 52:902. Data from 40 white children living in or near Iowa City (IA) who were examined serially from 5 to 15 years of age. (Tables of nasal height, maxillary subnasal height, dental height, mandibular height, total face height and increments in these)

Jones, J. D., 1966, The eruption of the lower incisor and the accompanying development of the symphysis and point B, Angle Orthod., 36:358. Serial data from 22 pairs of identical twins in California. Race not stated. (Table of increments in point B development and status of this)

Joseph, M. C., and Meadow, S. R., 1969, The metacarpal index of infants, Arch. Dis. Childh., 44:515. Data from 50 normal English children. (Table of metacarpal index)

Kano, K., and Chung, C. S., 1975, Do American born Japanese children still grow faster than native Japanese?, Am. J. Phys. Anthropol., 43:187. Data from unmixed Japanese children (2,954 boys, 3,213 girls) in Hawaii aged 11 to 17 years. (Tables of stature, weight, and weight/stature^2)

Kasius, R. V., Randall, A., Tompkins, W. T., and Wiehl, D. G., 1955, Maternal and newborn
    nutrition studies at Philadelphia Lying-In Hospital. Newborn studies. III. Size and
    growth of babies during the first three months of life, Milbank Mem. Fund Q., 33:341.
    Data from 1,335 infants (988 white, 347 Negro; about 75% of the white infants had at
    least one parent of Italian ancestry). Those weighing less than 5.5 lbs. at birth
    were excluded, as were twins and infants of mothers with syphilis or a serious chronic
    disease. Measurements were made at birth and at 1, 2 and 3 months. (Tables of recum-
    bent length, weight, chest circumference at xiphoid and increments of these in first,
    second and third months, weight in relation to birth weight and mother's pregravid
    weight, age and race)
Kasius, R. V., Randall, A., Tompkins, W. T., and Wiehl, D. G., 1957, Maternal and newborn
    nutrition studies at Philadelphia Lying-In Hospital. Newborn studies. V. Size and
    growth of babies during the first year of life, Milbank Mem. Fund Q., 35:323. Data
    from 1,391 singleton infants (354 Negro, 1,037 white; about 75% of the white infants
    had at least one parent of Italian ancestry) weighing more than 5.5 lbs. at birth.
    Babies of mothers with syphilis or a serious chronic disease were excluded. (Tables
    of weight, chest circumference at xiphoid, recumbent length, crown-rump length, rump-
    sole length, bicristal diameter, head and calf circumferences; increments and percentage
    gains from birth to specified ages in weight, chest circumference at xiphoid, crown-
    sole length, bicristal diameter, crown-rump length, rump-sole length, head circumference
    and calf circumference)
Kasius, R. V., Randall, A., Tompkins, W. T., and Wiehl, D.G., 1958, Maternal and newborn
    nutrition studies at Philadelphia Lying-In Hospital. Newborn studies. VI. Infant size
    at birth and parity, length of gestation, maternal age, height and weight status,
    Milbank Mem. Fund Q., 36:335. Data from the same infants as those studied in earlier
    parts of this report. (Tables of birth weight, chest circumference at xiphoid, crown-
    sole length, crown-rump length, rump-sole length, bicristal diameter and head circum-
    ference in relation to gestational age and race and to the mother's pregravid weight
    status, parity and age)
Keene, H. J., 1971, Epidemiologic study of tooth size variability in caries free naval
    recruits, J. Dent. Res., 50:1331. Data from 387 white individuals. (Tables of mesio-
    distal diameters of permanent teeth in birth weight groups, with and without excluding
    those with hypodontia)
Keene, H. J., 1979, Mesiodistal crown diameters of permanent teeth in male American Negroes,
    Am. J. Orthod., 76:95. Data from 56 Negro males with intact dentitions and no evidence
    of caries experience. (Table of mesiodistal diameters)
Kelly, H. J., and Reynolds, L., 1947, Appearance and growth of ossification centers and
    increases in the body dimensions of white and Negro infants, Am. J. Roentgenol. Rad.
    Ther., 57:477.
Kelly, H. J., Souders, H. J., Johnston, A. T., Bound, L. E., Humscher, H. A., and Macy, I.
    G., 1943, Daily decreases in the body total and stem lengths of normal children,
    Hum. Biol., 15:65. Serial data from 10 children in Detroit (MI) aged 4 to 8 years;
    race not stated. (Table of diurnal decreases in recumbent length and in stem length)
Kessler, A., and Scott, R. B., 1950, Growth and development of Negro infants. II. Relation
    of birth weight, body length and epiphysial maturation to economic status, Am. J. Dis.
    Child., 80:370. Data from 300 normal newly born singletons weighing at least 5 pounds
    at birth. If the mothers had complicated pregnancies, the infants were excluded.
    (Table of weight, length, and epiphysial size in relation to economic status and milk
    intake)
Kincaid, R. M., 1951, The frequency of deglutition in man: its relationship to overbite,
    Angle Orthod., 21:34. Data from 50 individuals aged 9 to 14 years being treated for
    malocclusion in the Orthodontia Clinic at the University of Illinois. Race not stated.
    (Table of posterior dental height)
King, E. W., 1952, A roentgenographic study of pharyngeal growth, Angle Orthod., 22:23.
    Serial data from about 30 boys and 30 girls aged 3 months to 15 years in Cleveland (OH).
    Race not stated. (Tables of atlas to pterygomaxillary fissure, hyoid bone to mandibu-
    lar symphysis and anterior cranial base, posterior nasal spine to anterior cranial base,
    anterior arch of atlas to cranial base, atlas to sella turcica, and sella turcica to
    pterygomaxillary fissure)

Knott, V. B., 1941, Physical measurement of young children: A study of anthropometric reli-
     abilities for children three to six years of age, University of Iowa Studies in Child
     Welfare, 18 (No. 3):1. Data from 131 white children in Iowa City (IA). (Table of
     differences between recumbent length and stature; crown-rump length and sitting height;
     erect and recumbent leg length; and recumbent and erect chest circumference)
Knott, V. B., 1961, Size and form of the dental arches in children with good occlusion
     studied longitudinally from age 9 years to late adolescence, Am. J. Phys. Anthropol.,
     19:263. Data from 29 white children living in or near Iowa City (IA). (Tables of
     dental arch widths, depths, and depth/width)
Knott, V. B., 1963, Stature, leg girth, and body weight of Puerto Rican school children
     measured in 1962, Growth, 27:157. Data from 1,800 school children aged 7 to 17 years.
     (Tables of stature, calf circumference, and weight)
Knott, V. B., 1969, Ontogenetic changes of four cranial base segments in girls, Growth, 33:
     123. Data from 37 white girls living in or near Iowa City (IA) who were studied
     serially from 5.5 to 17 years. (Table of increments in cranial base lengths)
Knott, V. B., 1971, Change in cranial base measures of human males and females from age 6
     years to early adulthood, Growth, 35:145. Data from 36 males and 30 females examined
     serially from 6 to 25 years. These white individuals were living in or near Iowa City
     (IA). (Tables of cranial base lengths and angles)
Knott, V. B., 1972, Longitudinal study of dental arch widths at four stages of dentition,
     Angle Orthod., 42:387. Data from about 35 healthy Caucasoid individuals of above
     average socioeconomic status and primarily of northwest European ancestry, living in or
     near Iowa City (IA). (Tables of dental arch widths and increments in these)
Knott, V. B., 1973, Growth of the mandible relative to a cranial base line, Angle Orthod.,
     43:305. Data from 40 Caucasoid children living in or near Iowa City (IA). The children
     were largely of northwest European descent and above average socioeconomic level.
     (Tables of cranial base length and shape, mandibular depth and shape, anterior and
     posterior facial height and shape, and increments in these measures)
Knott, V. B., and Johnston, R., 1970, Height and shape of the palate in girls: A longitudi-
     nal study, Arch. Oral Biol., 15:849. Data from 50 girls at 5, 9 and 12 years of age,
     and at 15 years for 25 of them and at 17 years for 12 of them. These were white
     children living in Iowa City (IA). (Tables of palatal height, width and depth and of
     height/width, height/depth, and depth/width; increments in height and height/width)
Knott, V. B., and Meredith, H. V., 1963, Body size of United States schoolboys at ages from
     11 years to 15 years, Hum. Biol., 35:507. Data from 520 white boys in Iowa City (IA)
     measured within 3 months of a birthday. (Tables of biparietal diameter, bizygomatic
     diameter, stature and weight)
Knott, V. B., and Meredith, H. V., 1966, Statistics on eruption of the permanent dentition
     from serial data for North American white children, Angle Orthod., 36:68. Data from 50
     girls predominantly of northwest European ancestry. They were above average in socio-
     economic status, residents of Iowa City (IA) and unselected by use of odontic criteria.
     (Tables of ages at eruption of permanent teeth, lateral differences in eruption, and
     arch differences)
Knott, V. B., and O'Meara, W. F., 1967, Serial data on primary incisor root resorption and
     gingival emergence of permanent successors, Angle Orthod., 37:212. Data from 70 boys
     and 75 girls living in or near Iowa City (IA). They were healthy white children pre-
     dominantly of northwest European ancestry and above average in socioeconomic status.
     (Tables of ages at resorption and eruption of successors and time from resorption to
     eruption)
Kondo, S., and Eto, M., 1975, Physical growth studies on Japanese-American children in com-
     parison with native Japanese, in: "Comparative Studies on Human Adaptability of
     Japanese, Caucasians and Japanese Americans," S. M. Horvath, S. Kondo, H. Matsui, and
     H. Yoshimura, eds., University of Tokyo Press, Tokyo. Data from 956 Japanese-American
     children in Los Angeles (CA). (Tables of stature, weight, weight/stature^3, sitting
     height, relative sitting height, chest circumference and skeletal age)
Kornfeld, W., 1953, Chest development in early childhood, J. Pediatr., 42:715. Data from
     2,473 white children from birth to 6 years, seen in a private practice in New Jersey
     during routine examinations. The measurements were made with the infant supine to 12
     months and at midrespiration. (Tables of chest circumference superior to nipples and
     at xiphoid and differences between these circumferences)

Koski, K., 1961, Growth changes in the relationships between some basicranial planes and the palatal plane, Suomen Hammaslääkäriseuran Toimituksia [Proc. Finnish Dent. Soc., I. Rytoma, ed.], 57:15. Data from 57 Finnish children examined serially. (Table of angles between craniofacial planes)

Kowalski, C. J., and Walker, G. F., 1971a, The Tweed triangle in a large sample of normal individuals, J. Dent. Res., 50:1690. Data from 438 white individuals, with normal occlusion. (Tables of Frankfort horizontal plane-mandibular plane angle)

Kowalski, C. J., and Walker, G. F., 1971b, Distribution of the mandibular incisor-mandibular plane angle in "normal" individuals, J. Dent. Res., 50:984. Data from 438 white individuals with normal occlusion. (Table of mandibular incisor-mandibular plane angle)

Kramer, W. S., and Ireland, R. L., 1959, Measurements of the primary teeth, J. Dent. Child., 26:252. Data from 1600 molar teeth, 419 cuspids and 736 incisors. Race and place not specified. (Tables of deciduous tooth size)

Kraus, B. S., 1961, The Western Apache: Some anthropometric observations, Am. J. Phys. Anthropol., 19:227. Data from 208 boys and 165 girls aged 1 to 18 years. (Tables of stature, weight, head length, head width, and bizygomatic width)

Krogman, W. M., 1970, Growth of head, face, trunk, and limbs in Philadelphia white and Negro children of elementary and high school age, Monogr. Soc. Res. Child Dev., 35: Serial 136, No. 3. Data from 942 white and 988 Negro children in Philadelphia (PA) who were examined annually. (Tables of stature, weight, sitting height, sitting height/stature, suprasternal height, symphyseal height, anterior trunk length, biacromial diameter, chest diameter, chest depth, bicristal diameter, bitrochanteric diameter, acromiale height, radiale height, stylion height, dactylion III height, total arm length, upper arm length, forearm length, anterior superior iliac spine height, tibiale height, sphyrion-fibulare height, leg length, thigh length, lower leg length, head length, head breadth, minimum frontal breadth, total face height, upper face height, lower face height, dental height, bicondylar diameter (closed and open), bizygomatic diameter, bigonial diameter, nasal breadth, porion-nasion depth, porion-subnasale depth, porion-prosthion depth, porion-intradentale superius labiale depth, porion-incision depth, porion-intradentale inferius labiale depth, porion-gnathion depth, porion-gonion, gonion-gnathion, condylion-nasion depth, condylion-gnathion depth, mandibular ramus height, interpalpebral diameter, mouth width, total lip height, and lower lip height)

Krogman, W. M., and Chung, D., 1965, The craniofacial skeleton at the age of one month, Angle Orthod., 35:305. Data from 150 white and Negro children in Philadelphia (PA). (Table of craniofacial size and shape and incisor inclination)

Kuhns, L. R., and Finnström, O., 1976, New standards of ossification in the newborn, Radiology, 119:655. Data from a Michigan hospital population aged less than 3 days. All were within 2 s.d. of appropriate values for gestational age in weight, recumbent length, head circumference, thoracic spine length, and tooth mineralization. All were from pregnancies not affected by viral or chronic illnesses and the neonates were free of known illness except the respiratory distress syndrome. (Tables of ages at onset of ossification in proximal humerus, distal femur, proximal tibia, and some tarsal bones)

Laestadius, N. D., Aase, J. M., and Smith, D. W., 1969, Normal inner canthal and outer orbital dimensions, J. Pediatr., 74:465. Data from 472 healthy Caucasian children in the State of Washington, grouped by age to the nearest year except before 3 years when they were grouped by 6-month intervals. (Tables of inner canthal and outer orbital diameters)

Lavine, L. S., Moss, M. L., and Noback, C. R., 1962, Digital epiphyseal fusion in adolescence. A longitudinal study, J. Pediatr., 61:571. Serial data from 157 normal children in New York City of whom 11 were Negro. They were examined each 3 months. (Table of order of fusion)

Lebret, L., 1962, Growth changes of the palate, J. Dent. Res., 41:1391. Data from serial dental casts of 30 children. Some were enrolled in the Longitudinal Growth Study of the Harvard School of Public Health and some were obtained from children attending a private school in Delaware. (Table of palatal height and breadth)

Leighton, B. C., 1960, "Investigation into the Changes that Take Place in the Form and Relationship of the Dental Arches During the First Three and a Half Years of Life, and the Part Played in these Changes by Environmental Factors," M.D.S. Thesis, University of London (England).

Lewis, A. B., and Garn, S. M., 1960, The relationship between tooth formation and other maturational factors, Angle Orthod., 30:70. Serial data from 255 white children of middle socioeconomic status living in southwestern Ohio. (Table of ages of calcification, crown completion and apical closure in selected permanent teeth)

Lewis, A. B., and Roche, A. F., Unpublished serial craniofacial measurements of children (34 boys, 33 girls) enrolled in the Fels Longitudinal Study. They are all white, and were living in southwestern Ohio. (Tables of cranial base angles and lengths, basion-articulare, cranial vault height and length)

Lewis, A. B., and Roche, A. F., 1974, Cranial base elongation in boys during pubescence, Angle Orthod., 44:83. Serial data from 58 white boys enrolled in the Fels Longitudinal Study. These boys were of middle socioeconomic status and living in southwestern Ohio. (Tables of the sizes of spurts in cranial base segments)

Lewis, A. B., and Roche, A. F., 1977, The saddle angle: constancy or change?, Angle Orthod., 47:46. Serial data from 76 boys and 53 girls with a Class I occlusion. These white children were of middle socioeconomic status. They were living in southwestern Ohio and were enrolled in the Fels Longitudinal Study. (Tables of size of N-S-Ba and increments in N-S-Ba)

Lichty, J. A., Ting, R. Y., Bruns, P. D., and Dyar, E., 1957, Studies of babies born at high altitude. 1. Relation of altitude to birth weight, Am. J. Dis. Child., 93:666. Data from 180 Colorado mothers (race not stated) with 615 newborns living at an altitude of 10,000 to 11,000 feet. (Table of birth weight)

Livson, N., and McNeill, D., 1962, Physique and maturation rate in male adolescents, Child Dev., 33:145. Serial data from 177 California boys in the Berkeley, Guidance, and Oakland Growth Studies. (Table of age at reaching 90% adult stature in relation to somatotype)

Lloyd-Jones, O., 1941, Race and stature; A study of Los Angeles school children, Res. Quart. Am. Assoc. Health Phys. Educ., 12:83. Data from 163,008 children in Los Angeles (CA) aged 5 to 18 years (121,820 white; 22,354 Mexican; 5,142 Negro; 3,692 Japanese). (Tables of weight, weight for stature, and stature)

Logan, W. H. G., and Kronfeld, R., 1933, Development of the human jaws and surrounding structures from birth to fifteen years, J. Am. Dent. Assoc., 20:379.

Lubchenco, L. O., Hansman, C., Dressler, M., and Boyd, E., 1963, Intrauterine growth as estimated from liveborn birthweight data at 24 to 42 weeks of gestation, Pediatrics, 32:793. Data from 5,635 Caucasian infants in Denver (CO) in whom gestational age was considered certain, and who did not have gross pathological conditions; infants with weights far above the 90th percentile were excluded. (Table of weight in relation to gestational age)

Lubchenco, L. O., Hansman, C., and Boyd, E., 1966, Intrauterine growth in length and head circumference as estimated from live births at gestational ages from 26 to 42 weeks, Pediatrics, 37:403. Data from 4,720 Caucasian infants in Denver (CO) in whom gestational age was considered certain, and who did not have gross pathological conditions. Infants with weights far above the 90th percentile were excluded. (Tables of recumbent length, head circumference, and weight/recumbent length[3])

Luten, J. R., Jr., 1967, The prevalence of supernumerary teeth in primary and mixed dentitions, J. Dent. Child., 34:346. Data from 798 children in a private pedodontic practice. Place and race not stated. (Table of prevalence of supernumerary teeth and their position)

Lysell, L., Magnusson, B., and Thilander, B., 1962, Time and order of eruption of the primary teeth. A longitudinal study, Odontologisk Revy, 13:217. Serial data from 366 Swedish children. (Tables of age of eruption of first and last deciduous tooth, each deciduous tooth, intervals between teeth and lateral differences in age at eruption, maxillo-mandibular differences in age at eruption, and order of eruption)

Malina, R. M., Unpublished data from 1,448 black and 1,396 white Philadelphia (PA) school children aged 6 to 13 years. The white sample was from a middle- to upper-middle socioeconomic level whereas the Negro sample was from a lower socioeconomic level. (Tables of weight, stature, sitting height, biacromial width, bicristal width, bitrochanteric width, biepicondylar width, bicondylar width, circumferences of upper arm, forearm and calf and increments in these)

Malina, R. M., 1969, Skeletal maturation rate in North American Negro and white children, Nature, 223:1075. Data from 486 Negro and 320 white children in Philadelphia (PA) aged 6 to 12 years. The white children were at higher socioeconomic levels than the Negro children. (Table of annual increments in Tanner-Whitehouse scores)

Malina, R. M., 1970, Skeletal maturation studied longitudinally over one year in American whites and Negroes six through thirteen years of age, Hum. Biol., 42:377. Data from 806 children (486 Negro, 320 white) living in Philadelphia (PA). (Tables of skeletal age and increments)

Malina, R. M., 1972, Weight, height and limb circumferences in American white and Negro children: Longitudinal observations over a one year period, J. Trop. Pediatr., 13:280. Data from 830 children aged 6 through 13 years. The white sample was from a middle to upper middle socioeconomic level whereas the Negro sample was from a lower socioeconomic level; all the children lived in Philadelphia (PA). (Tables of weight increments, stature increments, and calf circumference)

Malina, R. M., Hamill, P. V. V., and Lemeshow, S., 1973, "Selected Body Measurements of Children 6-11 Years. United States," DHEW Publ. No. (HSM) 73-1605, U.S. Govt. Print. Off., Washington. Data from 7,119 children aged 6 to 11 years who constituted a nationally representative sample. (Tables of sitting height, acromion-olecranon length, elbow-wrist length, foot length, foot breadth, hand length and width, chest width and depth, elbow-elbow breadth, waist circumference, and hip circumference)

Malina, R. M., Mueller, W. H., and Holman, J. D., 1976, Parent-child correlations and hereditability of stature in Philadelphia black and white children 6 to 12 years of age, Hum. Biol., 48:475. Data from 806 children. The white children were of higher socioeconomic status than the black children. (Table of stature)

Margolis, H. I., 1947, A basic facial pattern and its application in clinical orthodontics. I. The maxillofacial triangle, Am. J. Orthod. Oral Surg., 33:631.

Marshall, W. A., 1971, Evaluation of growth rate in height over periods of less than one year, Arch. Dis. Childh., 46:414. Serial data from 260 London (U.K.) children aged 7 to 10 years. (Tables of rate of growth in stature per year over intervals of 3 and 6 months)

Marshall, W. A., and Ahmed, L., 1976, Variation in upper arm length and forearm length in normal British girls: photogrammetric standards, Ann. Hum. Biol., 3:61. Serial data from 282 healthy girls aged 5 to 18 years participating in the Harpenden and London (U.K.) Longitudinal Growth Studies. (Tables of upper arm and forearm lengths in relation to age and sitting height)

Martin, W. E., 1955, "Children's Body Measurements for Planning and Equipping Schools," DHEW, Office of Education, Special Publ. No. 4, U.S. Govt. Print. Off., Washington. Data from 3,318 children aged 5 to 21 years from 10 schools in southern Michigan in the area between Ann Arbor and Detroit. Blacks formed 11.7% of the sample; 0.7% were other non-white children. (Tables of ankle height, foot length, heel to ankle length, foot width, elbow to wrist length, wrist to knuckle length, wrist to finger tip length, hand width, head circumference, vertex to shoulder length, shoulder to trochanter length, trochanter to knee length, shoulder width, and upper arm length [arc])

Matheny, W. D., and Meredith, H. V., 1947, Mean body size of Minnesota schoolboys of Finnish and Italian ancestry, Am. J. Phys. Anthropol., 5:343. Data from 1,986 boys (1,102 Finnish lineage, 884 Italian lineage) living in communities near Hibbing (MN). Those with gross physical pathology were excluded. (Tables of stature; head and neck length, i.e., suprasternale to vertex; trunk length, i.e., suprasternale height when sitting; leg length, biacromial diameter, chest circumference at xiphoid in midrespiration, bicristal diameter, arm length, head length, biparietal diameter, nasion-menton height, bizygomatic diameter, circumferences of upper arm, upper thigh and maximum calf, and weight)

Mayhall, J. T., Belier, P. L., and Mayhall, M. F., 1977, Permanent tooth emergence timing of Northern Ontario Indians, Ontario Dentist, 54:8. Data from 599 school children aged 4 to 20 years of the Cree or Ojibway tribes. (Table of age of emergence of permanent teeth)

Mayhall, J. T., Belier, P. L., and Mayhall, M. F., 1978, Canadian Eskimo permanent tooth emergence timing, Am. J. Phys. Anthropol., 49:211. Data from 368 Inuit children and adolescents. (Table of age of emergence)

Mazess, R. B., and Cameron, J. R., 1972, Growth of bone in school children: comparison of radiographic morphometry and photon absorptiometry, Growth, 36:77. Data from 322 white children aged 6 to 14 years from Middleton (WI). (Tables of width for distal radius, medullary cavity for distal radius, cortical thickness and cortical area and cortical area/total area for distal radius)

Mazess, R. B., and Cameron, J. R., 1974, Bone mineral content in normal U.S. Whites, in: "Proceedings International Conference on Bone Mineral Measurement," U.S. Govt. Print. Off., Washington.

Mazess, R. B., and Mather, W. E., 1974, Bone mineral content of North Alaskan Eskimos, Am. J. Clin. Nutr., 27:916.

Mazess, R. B., and Mather, W. E., 1975, Bone mineral content in Canadian Eskimos, Hum. Biol., 47:45. Data from 177 children measured using direct photon absorptiometry. (Tables of stature and weight)

McCammon, R. W., Unpublished data from the Child Research Council (Denver, CO). Serial data recorded from 334 white children of middle socioeconomic status living in or near Denver (CO). These unpublished data are more complete than those in McCammon, 1970. (Tables of stature, recumbent length, weight, weight/stature^2, weight/recumbent length^2, upper arm circumference, and relative skeletal age)

McCammon, R. W., 1970, "Human Growth and Development," Charles C Thomas, Springfield (IL). Data from 334 white children studied serially at the Child Research Council, Denver (CO). (Tables of sitting height, crown-rump length, chest circumference, chest width, chest depth, iliac circumference, biiliac diameter, biacromial diameter, circumferences of upper arm, forearm, hand, wrist, foot, ankle, calf and knee; internal chest width; lengths of humerus, radius, ulna, femur, tibia, and fibula; ratios with recumbent length or stature of sitting height, biiliac diameter, biacromial diameter, chest width, and chest depth)

McGowan, A., Jordan, M., and MacGregor, J., 1975, Skinfold thickness in neonates, Biol. Neonate, 23:66. Data from 48 male and 39 female Scottish neonates who were apparently normal. (Table of recumbent length, birth weight and skinfold thicknesses)

Méhes, K., and Kitzvéger, E., 1974, Inner canthal and intermammillary indices in the newborn infant, J. Pediatr., 85:90. Data from 506 infants in Hungary examined within the first 5 days of life. They included infants with major abnormalities and those in whom gestational age could not be determined; race not stated. (Tables of inner canthal and intermammillary indices; the relationship of these to birth weight and age)

Meredith, H. V., 1953, Growth in head width during the first twelve years of life, Pediatrics, 12:411. Composite data from reported studies. (Tables of head width and increments in head width)

Meredith, H. V., 1954, Growth in bizygomatic face breadth during childhood, Growth, 18:111. Data from 289 white children in Iowa City (IA). (Tables of bizygomatic diameter and increments in this)

Meredith, H. V., 1955, A longitudinal study of change in size and form of the lower limbs on North American white school boys, Growth, 19:89. Data from 70 white boys measured annually from 5 to 11 years. The boys were American-born and living in or near Iowa City (IA). They were of upper socioeconomic status and predominantly of northwest European ancestry. (Tables of lower limb length, calf circumference, calf circumference/lower limb length, and increments in these)

Meredith, H. V., 1957, Change in the profile of the osseous chin during childhood, Am. J. Phys. Anthropol., 15:247. Data from 34 white children living in or near Iowa City (IA), of above average socioeconomic status, who were examined serially from 4 to 14 years. (Table of the depth of the anterior concavity of the mandibular symphysis)

Meredith, H. V., 1958, A time-series analysis of growth in nose height during childhood, Child Dev., 29:19. Serial data obtained annually from 80 children aged 5 to 12 years. They were predominantly of northwest European ancestry, of above average socioeconomic status, and living in or near Iowa City (IA). (Tables of skeletal nose height [nasion to anterior nasal spine] and increments in this height)

Meredith, H. V., 1959a, Change in a dimension of the frontal bone during childhood and adolescence, Anat. Rec., 134:769. Data from 30 white girls living in or near Iowa City (IA). (Table of N-bregma and increments in this)

Meredith, H. V., 1959b, A longitudinal study of growth in face depth during childhood, Am. J. Phys. Anthropol., 17:125. Data from 125 white children living in or near Iowa City (IA) who were examined serially from 5 to 11 years. (Table of increments in ANS-occipital condyles distance)

Meredith, H. V., 1960, Changes in form of the head and face during childhood, Growth, 24:215.
    Serial data from 5 to 11 years for 60 American-born white girls who were singleton
    births and living in or near Iowa City (IA). They were from above average socioeconomic
    families. (Tables of head width/length, face width/head width, face depth/bizygomatic
    diameter, face depth/face height, face height/face width, bigonial diameter/bizygomatic
    diameter, nose height/face depth, nose height/face height, and nose width/face width)

Meredith, H. V., 1961, Serial study of change in a mandibular dimension during childhood and
    adolescence, Growth, 25:229. Data recorded annually from 132 children aged 5 to 17
    years. These children were living in or near Iowa City (IA) and were predominantly of
    northwest European ancestry. They were above average socioeconomically. (Table of
    mandibular body length and increments in this length)

Meredith, H. V., 1962, Childhood interrelations of anatomic growth rates, Growth, 26:23.
    Serial data from 67 white boys in Iowa City (IA) from families of above average socio-
    economic status. (Tables of increments from 5 to 10 years in head length, biparietal
    diameter, anterior nasal spine to line connecting occipital condyles, bizygomatic
    diameter, pogonion to gonion depth, bigonial diameter, nasion to anterior nasal spine,
    bialar width, stem length, biacromial diameter, bicristal diameter, chest circumference
    at xiphoid, upper arm circumference, calf circumference, leg length, and weight)

Meredith, H. V., and Carl, L. J., 1946, Individual growth in hip width: A study covering
    the age period from 5 to 9 years based upon seriative data for 55 nonpathologic white
    children, Child Dev., 17:157.

Meredith, H. V., and Chadha, J. M., 1962, A roentgenographic study of change in head height
    during childhood and adolescence, Hum. Biol., 34:299. Serial data from 130 white
    children living in or near Iowa City (IA). (Table of bregma-sella height and increments
    in this)

Meredith, H. V., and Cox, G. C., 1954, Widths of the dental arches at the permanent first
    molars in children 9 years of age, Am. J. Orthod., 40:134. Data from 94 white children
    living in or near Iowa City (IA). (Table of dental arch width)

Meredith, H. V., and Culp, S. S., 1951, Body form in childhood: ratios quantitatively
    describing four slender-to-stocky continua on boys four to eight years of age, Child
    Dev., 22:3. Data from white children living in or near Iowa City (IA). These children
    were aged 4 to 8 years, were of predominantly northwest European ancestry and above
    average socioeconomically. (Table of arm girth/upper limb length, calf girth/lower
    limb length, chest girth/stem length, and $\sqrt[3]{W}$/stature)

Meredith, H. V., and Higley, L. B., 1951, Relationships between dental arch widths and widths
    of the face and head, Am. J. Orthod., 37:193. Data from 82 white children examined at
    5 years of whom 64 were examined again at 7 years. They were living in or near Iowa
    City (IA). They were predominantly of northwest European ancestry and were of above
    average socioeconomic status. (Tables of maxillary arch width/upper face width)

Meredith, H. V., and Hixon, E. H., 1954, Frequency, size, and bilateralism of Carabelli's
    tubercle, J. Dent. Res., 33:435. Data from casts of 100 white children aged 9 years
    living in or near Iowa City (IA). (Table of prevalence of Carabelli's tubercle)

Meredith, H. V., and Hopp, W. M., 1956, A longitudinal study of dental arch width at the
    deciduous second molars on children 4 to 8 years of age, J. Dent. Res., 35:879. Data
    from 77 white children living in or near Iowa City (IA). (Table of dental arch widths)

Meredith, H. V., and Knott, V. B., 1962, Descriptive and comparative study of body size on
    United States schoolgirls, Growth, 26:283. Data from 300 white school girls in Iowa
    City (IA). (Tables of stature, stem length, biparietal width, head length, bizygomatic
    diameter, maximum calf circumference, foot length and weight)

Meredith, H. V., and Knott, V. B., 1968, Coronal breadth of human primary anterior teeth,
    Am. J. Phys. Anthropol., 28:49. Data from 90 white children of each sex who were living
    in or near Iowa City (IA). These children were predominantly of northwest European
    ancestry and of above average socioeconomic status. (Tables of mesiodistal diameters
    of teeth)

Meredith, H. V., and Meredith, E. M., 1950, Annual increment norms for ten measures of
    physical growth on children four to eight years of age, Child Dev., 21:141. Data from
    white children in Iowa City (IA) who were physically normal and were measured within 3
    days of each birthday from 4 to 8 years. The sample size varies from 160 to 200 for
    annual increments. (Tables of weight, stature, stem length, arm and leg lengths, bia-
    cromial and bicristal diameters, chest circumference at xiphoid in midrespiration, upper
    arm circumference and maximum calf circumference)

Meredith, H. V., and Meredith, E. M., 1953, The body size and form of present-day white elementary school children residing in west-central Oregon, Child Dev., 24:83. Data from 941 white children aged 7 to 11 years attending public elementary schools in the Eugene School District (OR). (Tables of stature, stem length, biacromial and bicristal diameters, chest circumference at xiphoid in midrespiration, abdominal circumference at umbilicus, arm length, upper arm circumference, leg length, maximum calf circumference and weight and some ratios x 100: upper arm circumference/arm length, calf circumference/leg length, chest circumference/stem length, biacromial diameter/bicristal diameter, and chest circumference/abdominal circumference)

Meredith, H. V., and Spurgeon, J. H., 1976a, Comparative findings on the skelic index of black and white children and youths residing in South Carolina, Growth, 40:75. Data from 1,210 children aged 9 to 15 years (634 white, 576 black) attending 27 schools in Columbia (SC). (Tables of sitting height, leg length, leg length/sitting height)

Meredith, H. V., and Spurgeon, J. H., 1976b, Body size and form of black and white female youths measured during 1974-1975 at Columbia, South Carolina, Child Dev., 47:360. Data from 386 children (186 white, 200 black) aged 13 years ± 2 months. (Tables of stature, sitting height, leg length, bicristal diameter, upper arm circumference, maximum calf circumference, weight, and some ratios x 100: leg length/sitting height, calf circumference/leg length, and bicristal diameter/stature)

Meredith, H. V., Knott, V. B., and Hixon, E. H., 1958, Relations of the nasal and subnasal components of facial height in childhood, Am. J. Orthod., 44:285. Data from 55 children examined annually from 4 to 12 years. They were American-born white children, predominantly of northwest European ancestry and above average socioeconomic status. These children were living in or near Iowa City (IA). (Table of upper facial height/lower facial height)

Michelson, N., 1945, Studies in the physical development. V. The ossification time of the os pisiforme, Hum. Biol., 17:143. Data from 180 underprivileged white girls in New York City. (Table of percentage prevalence)

Miller, H. C., and Hassanein, K., 1971, Diagnosis of impaired fetal growth in newborn infants, Pediatrics, 48:511. Data from 1,724 infants in Kansas City (KS) which is about 800 feet above sea level. About 60% of the infants are white; about 80% of the mothers were service patients or on welfare. About 25% of the infants were born to mothers who were single, divorced or separated. (Tables of recumbent length, head circumference, and birth weight in relation to gestational age)

Mills, L. F., 1966, Changes in dimension of the dental arches with age, J. Dent. Res., 45:890. Data from 1,253 white children in Maryland with neutrocclusion. (Table of dental arch length and breadth)

Montoye, H. S., 1978, "An Introduction to Measurement in Physical Education," Allyn & Bacon, Inc., Boston. Originally published by Phi Epsilon Kappa Fraternity, Indianapolis, 1970. (Tables of relative weight)

Moorrees, C. F. A., 1959, "The Dentition of the Growing Child. A Longitudinal Study of Dental Development Between 3 and 18 Years of Age," Harvard University Press, Cambridge. Serial data from casts of 132 white children enrolled in the Harvard School of Public Health Growth Study and from serial casts of 100 public school children in Wilmington (DE). (Tables of tooth size, dental arch length, overbite and increments in overbite, and dental arch widths and lengths)

Moorrees, C. F. A., and Chadha, J. M., 1962, Crown diameters of corresponding tooth groups in the deciduous and permanent dentition, J. Dent. Res., 41:466. Data from 132 white children examined serially in the Harvard School of Public Health Growth Study and 52 children attending a public school in Wilmington (DE). (Tables of tooth size)

Moorrees, C. F. A., and Chadha, J. M., 1965, Available space for the incisors during dental development--a growth study based on physiologic age, Angle Orthod., 35:12. Data from 132 white children examined serially in the Harvard School of Public Health Growth Study and 52 children attending a public school in Wilmington (DE). (Tables of incisor space and increments in them)

Moorrees, C. F. A., and Kean, M. R., 1958, Natural head position, a basic consideration in the interpretation of cephalometric radiographs, Am. J. Phys. Anthropol., 16:213. Data from Boston (MA) girls aged 18 to 20 years; race not stated. (Table of angles between reference planes and vertical)

Moorrees, C. F. A., and Reed, R. B., 1965, Changes in dental arch dimensions expressed on the basis of tooth eruption as a measure of biologic age, J. Dent. Res., 44:129. Data from 132 white children included in the longitudinal study of the Harvard School of Public Health and from 52 children attending a public school in Wilmington (DE). (Tables of dental arch widths and lengths and increments in these)

Moorrees, C. F. A., Thomsen, S. Ø., Jensen, E., and Yen, P. K.-J., 1957, Mesiodistal crown diameters of the deciduous and permanent teeth in individuals, J. Dent. Res., 36:39. Data from 184 children of European stock; 132 were enrolled in the Harvard School of Public Health Growth Study, the remainder were from a public school in Wilmington (DE). (Tables of mesiodistal crown diameters and comparison with successors)

Moorrees, C. F. A., Fanning, E. A., Grøn, A-M, and Lebret, L., 1962, The timing of orthodontic treatment in relation to tooth formation. Transactions of the European Orthodontic Society. Data from 246 white children of middle socioeconomic status in southwestern Ohio enrolled in the Fels Longitudinal Study. (Tables of intervals between stages of root formation and root formation at emergence)

Moyers, R. E., van der Linden, F. P. G. M., Riolo, M. L., and McNamara, J. A., Jr., 1976, "Standards of Human Occlusal Development," Monograph No. 5, Cranofacial Growth Series, Center for Human Growth and Development, University of Michigan, Ann Arbor. Data from 208 upper socioeconomic white Michigan children studied from 3 to 16 years with many incomplete sets of data but all had at least three annual records. (Tables of tooth size and shape, dental arch width, dental arch depth, and palatal height)

Naeye, R. L., Benirschke, K., Hagstrom, J. W. C., and Marcus, C. C., 1966, Intrauterine growth of twins as estimated from liveborn birth-weight data, Pediatrics, 37:409. Data from all twins (701 monochorionic and 1,748 dichorionic) born 1932-1963 in selected hospitals in Vermont, Massachusetts and New York; those with congenital malformations and hemolytic disease of the newborn were excluded. About 10% were non-Caucasian. (Tables of weight for gestational age and intrapair weight differences for gestational age)

Nakamura, S., Savara, B. S., and Thomas, D. R., 1972, Norms of size and annual increments of the sphenoid bone from four to sixteen years, Angle Orthod., 42:35. Serial data from 79 children studied in the Child Study Clinic at the University of Oregon Dental School. The children were predominantly of northwest European ancestry. (Tables of sphenoid size and increments in sphenoid size)

Nanda, R. S., 1956, Cephalometric study of the human face from serial roentgenograms, Ergebn. d. Anat. u. Entwicklungsgesch., 35:358.

Neumann, C. G., and Alpaugh, M., 1976, Birthweight doubling time: A fresh look. Pediatrics, 57:469. Data from 357 infants in Los Angeles (CA): 165 white, 66 Mexican-American, 41 Oriental, 33 black, and 52 miscellaneous. (Table of age of doubling time)

Newcomer, E. O., and Meredith, H. V., 1951, Eleven measures of body size on a 1950 sample of 15-year-old white schoolboys at Eugene, Oregon, Hum. Biol., 23:24. Data from 102 white boys. (Table of weight, stature, stem length, biacromial and bicristal diameters, chest circumference at xiphoid in midrespiration, abdominal circumference at umbilicus, arm and leg lengths, upper arm circumference, and calf circumference)

Newman, K. J., and Meredith, H. V., 1956, Individual growth in skeletal bigonial diameter during the childhood period from 5 to 11 years of age, Am. J. Anat., 99:157. Serial data from 72 white children living in or near Iowa City (IA). (Tables of bigonial diameter)

Nolla, C. M., 1960, The development of the permanent teeth, J. Dent. Child., 27:254. Serial data from 50 children aged 25 to 121 months examined at the Child Development Laboratories of the University of Michigan School. Race not stated. (Tables of maturation stages in relation to age)

Norval, M., Kennedy, R. L. J., and Berkson, J., 1951, Biometric studies of the growth of children of Rochester, Minnesota, Hum. Biol., 23:273. Data from 4,111 children aged between birth and 12 months examined in well-baby clinics or The Mayo Clinic, 1934-1945. They were predominantly of north European ancestry and grew up in a rural farming community. They were of average, or above average, economic status. (Tables of recumbent length, weight, weight for recumbent length, and monthly gain in weight)

Novak, L. P., Hamamoto, K., Orvis, A. L., and Burke, E. C., 1970, Total body potassium in infants. Determination by whole-body counting of radioactive potassium (^{40}K), Am. J. Dis. Child., 119:419. Data from 64 healthy white full-term Minnesota infants examined at an average age of 32 days. Tables of total body potassium, weight, and recumbent length)

Nutrition Survey of Lower Greasewood Chapter Navaho Tribe, 1968-1969 (no author), conducted by the Department of Community Medicine, University of Pittsburgh, The National Nutrition Survey and the United States Public Health Service, Indian Health Service. Data from 168 boys and 161 girls aged from birth to 14 years. (Tables of stature, weight, and head circumference)

Nyhan, W. L., and Wessel, M. A., 1954, Neonatal growth in weight of normal infants on four different feeding regimens, Pediatrics, 14:442. Data from 400 Connecticut infants observed from birth for 6 to 8 weeks: 100 breast-fed on a flexible schedule, 100 breast-fed on a four-hour schedule, 100 bottle-fed on a flexible schedule, and 100 bottle-fed on a four-hour schedule. Weights recorded before first morning feeding. (Table of weight gain and loss after birth in relation to feeding method)

Odland, L. M., Warnick, K. P., and Esselbaugh, N. C., 1958, "Cooperative Nutritional Status Studies in the Western Region. II. Bone Density," Bulletin 534, Montana Agricultural Experiment Station, Montana State College. Data from 1,277 children (613 boys, 664 girls) living in Oregon, Colorado, Arizona, New Mexico, Utah, Idaho, Montana, or Washington who were examined between 1947 and 1952. (Tables of bone density of calcaneus and middle phalanx V of the hand)

O'Meara, W. F., and Knott, V. B., 1967, Serial data on primary canine root resorption and gingival emergence of permanent successors, Angle Orthod., 37:261. Data from 56 boys and 60 girls living in or near Iowa City (IA). These were healthy white children predominantly of northwest European ancestry and above average socioeconomic status. (Tables of ages at resorption and emergence, and time from resorption to emergence)

O'Rahilly, T. X., 1951, Deciduous dental arch widths and widths of the face in early childhood, Am. J. Orthod., 37:698. Serial data from 130 children aged 4 to 7 years enrolled in the Facial Growth Study of the State University of Iowa. About 95% of the children were of northwest European ancestry. They were of above average socioeconomic status. (Tables of face widths, arch widths and increments in these)

Owen, G. M., Kram, K. M., Garry, P. J., Lowe, J. E., and Lubin, A. H., 1974, A study of nutritional status of preschool children in the United States, 1968-1970, Pediatrics, 53:597. Data from 3,441 children aged 1-6 years; approximately 80% white, 15% Negro, and 4% Spanish-American. These were 65% of randomly selected children in 74 communities. (Tables of weight [by race and for the whole group], head circumference, stature; stature and weight in relation to socioeconomic status)

Paiva, S. L., 1953, Pattern of growth of selected groups of breast-fed infants in Iowa City, Pediatrics, 11:38. Serial data from 45 healthy infants (race not stated) from families of above average socioeconomic status. The mothers were well-nourished and some supplementary feeding was given. (Tables of recumbent length and weight)

Palmer, C. E., 1944, Studies of the center of gravity in the human body, Child Dev., 15:99.

Parfitt, G. J., 1954, Variations in the age of shedding of deciduous and eruption of permanent teeth, Dent. Rec., 74:279. Data from 3,150 school children in London (U.K.). (Tables of percentage of teeth erupted, time taken to eruption of 1/3 of crown, number of first permanent molars erupted at particular ages, and percentage of deciduous teeth shed at particular ages)

Pařízková, J., 1976, Growth and growth velocity of lean body mass and fat in adolescent boys, Pediatr. Res., 10:647.

Partington, M. W., and Roberts, N., 1969, The heights and weights of Indian and Eskimo school children on James Bay and Hudson Bay, Can. Med. Assoc. J., 11:502. Data from three groups of children aged 5 to 15 years: 263 Eskimos on the east coast of Hudson Bay, 754 Cree Indians in the region of James Bay, and 119 Mohawk children of the Tyendinaga Reserve. (Tables of stature, weight, and birth weight)

Pelton, W. J., and Elsasser, W. A., 1955, Studies of dentofacial morphology. IV. Profile changes among 6,829 white individuals according to age and sex, Angle Orthod., 25:199. Data from 5,571 children aged 5 to 19 years in 10 geographic areas of the U.S. (Tables of distance from a vertical plane to subnasion, upper incisor, lower incisor, pogonion; vertical distances from nasion-subnasale and nasion-lowest chin point; upper and total face height)

Perreault, J. G., Demirjian, A., and Jeniček, M., 1974, Emergence des dents permanentes chez les enfants Canadiens-français, J. Can. Dent. Assoc., 40:306. Serial data from 1,535

Québec children aged 6 to 10 years.   (Tables of age at emergence)

Perreault, J. G., Chaumont, A., Jeníček, M., and Demirjian, A., 1975, Emergence des dents permanentes chez les enfants Canadiens d'origine française. 2 ième partie: Passage de la percée de la gencive à la couronne clinique complète, J. Can. Dent. Assoc., 41:572. Data from 1,535 semilongitudinal examinations of Québec children aged 6 to 13 years. (Tables of age at clinical eruption and time from first eruption to complete eruption)

Pett, L. B., and Ogilvie, G. F., 1957, The report on Canadian average weights, heights and skinfolds, Can. Bull. Nutr., 5:1. Data from a nationally representative sample. (Tables of stature, weight, and weight for stature)

Philip, A. G. S., 1975, Fontanel size and epiphyseal ossification in neonatal twins discordant by weight, J. Pediatr., 86:417.

Pileski, R. C. A., Woodside, D. G., and James, G. A., 1973, Relationship of the ulnar sesamoid bone and maximum mandibular growth velocity, Angle Orthod., 43:162. Data from 199 participants in the Burlington Orthodontic Research Center. This group is predominantly Caucasian and considered to be representative of the province of Ontario. (Table of age of mandibular peak growth velocity and age of ulnar sesamoid ossification)

Plets, J. H., Isaacson, R. J., Speidel, T. M., and Worms, F. W., 1974, Maxillary central incisor root length in orthodontically treated and untreated patients, Angle Orthod., 44:43. Data from 50 patients aged 12 to 20 years at the University of Minnesota School of Dentistry and 45 preorthodontic patients who had a mean age of 12 years 8 months and 45 post-orthodontic patients who were the preorthodontic group examined later. Race not stated. (Table of incisor tooth length)

Poznanski, A. K., 1974, "The Hand in Radiologic Diagnosis," W. B. Saunders Company, Philadelphia. Data from white children living in southwestern Ohio and enrolled in the Fels Longitudinal Study. (Table of ratios between lengths of the bones of the hand)

Prakash, P., and Margolis, H. I., 1952, Dento-craniofacial relations in varying degrees of overbite, Am. J. Orthod., 38:657. Data from 120 standardized radiographs of 120 white Boston (MA) individuals aged 12 to 30 years with all permanent teeth in occlusion (including third molars) and within the range of normal vertical craniofacial growth. There were 36 with normal dentitions, 44 with Angle Class I malocclusions, and 40 with Angle Class II malocclusions. (Table of overbite, anterior face height, perpendicular face height, lower incisal height, lower molar height, upper incisal height, upper molar height, nasion to incisal edge of upper central incisor, and nasion to occlusal surface of upper first molar)

Prosterman, L. L., 1962, "Craniofacial, Profile, and Denture Relations in Children with Normal Occlusion," M.S. Thesis, McGill University, Montreal, Québec, Canada. Data from 70 white Pennsylvania children aged 11.8 to 13.9 years. All had normal occlusion. (Tables of cranial base length and shape, facial depth, height, and profile, and inclination and position of incisor teeth)

Pryor, H. B., 1966, Charts of normal body measurements and revised width-weight tables in graphic form, J. Pediatr., 68:615.

Pryor, H. B., 1969, Objective measurement of interpupillary distance, Pediatrics, 44:973. Data from 12,853 California children aged between birth and 15 years. Some of the data are presented separately for Mexican-Americans and Japanese-Americans. (Tables of inter-inner canthus, inter-outer canthus and interpupillary distances)

Pyle, S. I., and Hoerr, N. L., 1955, "Radiographic Atlas of Skeletal Development of the Knee. A Standard of Reference," Charles C Thomas, Springfield (IL).

Pyle, S. I., Reed, R. B., and Stuart, H. C., 1959, Patterns of skeletal development in the hand, Pediatrics, 24:886. Data from 133 children examined serially in the Harvard School of Public Health Growth Study. These white children were living in or near Boston (MA). (Table of skeletal age)

Pyle, S. I., Stuart, H. C., Cornoni, J., and Reed, R. R., 1961, Onsets, completions, and spans of the osseous stage of development in representative bone growth centers of the extremities, Monogr. Soc. Res. Child Dev., 26:Serial 79, No. 1. Data from 78 white Boston (MA) children examined serially from birth to 18 years. (Tables of ages at onset and completion of ossification in selected centers)

Rajic, M. K., Lavallée, H., Shephard, R. J., Jéquier, J. C., Cabarre, R., and Beaucage, C., 1978, Height-weight comparison of Canadian school children, in: "Physical Fitness Assessment: Principles, Practice and Application," R. J. Shephard and H. Lavallée, eds., Charles C Thomas, Springfield (IL). Data from 546 children aged 6 to 12 years in the

Trois-Rivières region of Québec. (Tables of stature, weight, and log weight)

Rauh, J. L., Schumsky, D. A., and Witt, M. T., 1967, Heights, weights, and obesity in urban school children, Child Dev., 38:515. Data from 11,000 school children in Cincinnati (OH) who were a representative sample of all the children in these schools from kindergarten through the eleventh grade (71% white; almost all the remainder Negroes). Age was recorded to the last completed year. (Tables of stature, weight, and Wetzel physiques)

Redman, R. S., Shapiro, B. L., and Gorlin, R. J., 1966, Measurement of normal and reportedly malformed palatal vaults. II. Normal juvenile measurements, J. Dent. Res., 45:266. Data from 1,098 white children in Minnesota. (Table of palatal height, width and length and height/width)

Reed, R. D., and Stuart, H. C., 1959, Patterns of growth in height and weight from birth to eighteen years of age, Pediatrics, 24:904. Complete serial data (with some interpolations) from 134 white children in Boston (MA). (Table of stature and weight)

Reynolds, E. L., 1945, The bony pelvic girdle in early infancy. A roentgenometric study, Am. J. Phys. Anthropol., 3:321. Serial data from 95 white children in southwestern Ohio who had 467 sets of radiographs during the first year after birth. These children were of middle socioeconomic status. (Tables of pelvis height, interpubic breadth, ilium length and breadth, ischium length, pubis length, breadth of greater sciatic notch, iliac index, pelvis breadth, inlet depth and breadth, interiliac breadth, bi-ischial breadth, pelvic index, inlet index, sacral index, relative inlet breadth and anterior segment index)

Reynolds, E. L., 1946, Sexual maturation and the growth of fat, muscle and bone in girls, Child Dev., 17:121. Data from 48 normal white girls enrolled in the Fels Longitudinal Study. These girls from southwestern Ohio are of middle socioeconomic status. (Tables of calf breadth; muscle, fat and bone widths; and rates of change in these in early and late maturing girls)

Reynolds, E. L., 1947, The bony pelvis in prepuberal childhood, Am. J. Phys. Anthropol., 5:165. Data from 193 children who had 640 serial radiographs between the ages of 15 months and 9.5 years. These white children were of middle socioeconomic status and were living in southwestern Ohio. (Tables of pelvis height and breadth, inlet breadth, interiliac breadth, interpubic breadth, inter-tuberal breadth, ilium length, pubis length, sagittal diameter of inlet, ischium length, breadth of iliac notch, inter-obturator breadth, bitrochanteric breadth, length of femoral neck, pubic angle, pelvic angle, femoral angle, femoral-pelvic angle, pelvic index, inlet index, relative sacrum breadth and relative inlet breadth)

Reynolds, E. L., and Asakawa, T., 1951, Skeletal development in infancy. Standards for clinical use, Am. J. Roentgenol., 65:403. Data from 119 normal infants examined serially in the Fels Longitudinal Study. These white infants were living in southwestern Ohio and came from families of middle socioeconomic status. (Table of number of centers present in hand and foot)

Reynolds, E. L., and Sontag, L. W., 1944, Seasonal variations in weight, height, and appearance of ossification centers, J. Pediatr., 24:524. Serial data from 133 white children in southwestern Ohio. These children were of middle socioeconomic status. (Table of increments in stature, weight, and number of ossification centers in relation to season)

Rhoads, T. F., Rapaport, M., Kennedy, R., and Stokes, J., Jr., 1945, Studies on the growth and development of male children receiving evaporated milk. II. Physical growth, dentition, and intelligence of white and Negro children through the first four years as influenced by vitamin supplements, J. Pediatr., 26:415. Serial data from 233 children seen in the Outpatient Department of the Children's Hospital of Philadelphia (PA). These boys had birth weights of 5 lb. or more. They were all followed from about 42 days of age to at least 2 years. At 2 years of age, 58% were white and 42% Negro. The whites were predominantly of Irish, British, or German origin. (Tables of weight, length, head circumference, crown-rump length, chest breadth, chest circumference, pelvic breadth, and age at closure of anterior fontanelle)

Richardson, E. R., and Malhatra, S. K., 1975, Mesiodistal crown dimension of the permanent dentition of American Negroes, Am. J. Orthod., 68:157. Data from 162 individuals (81 male, 81 female). These individuals were either from an orthodontic practice or were included in the craniofacial growth study at Meharry Medical College or the dental clinic at that school. (Tables of tooth size)

Richardson, M. E., 1970, The early developmental position of the lower third molar relative to certain jaw dimensions, Angle Orthod., 40:226. Data from 162 Irish children aged 8.0 to 13.7 years. (Table of molar position)

Riedel, R. A., 1948, "A Cephalometric Roentgenographic Study of the Relation of the Maxilla and Associated Parts to the Cranial Base in Normal and Malocclusion of the Teeth," M.S.D. Thesis, Northwestern School of Dentistry.

Riedel, R. A., 1952, The relation of maxillary structures to cranium in malocclusion and in normal occlusion, Angle Orthod., 22:142. Data from 24 children aged 7 to 11 years with excellent occlusions. (Table of profile, incisor inclination, cranial base shape, and mandibular plane inclination)

Riolo, M. L., Moyers, R. E., McNamara, J. A., Jr., and Hunter, W. S., 1974, "An Atlas of Craniofacial Growth: Cephalometric Standards from the University School Growth Study," Monograph No. 2, Craniofacial Growth Series, Center for Human Growth and Development, University of Michigan, Ann Arbor. Data from 83 white individuals studied serially from 6 to 16 years of age. (Tables of S-N-A, B-N-S, A-N-B, N-S to ANS-PNS, N-S-Ba, inclination of upper incisor to lower incisor (central), inclination of lower central incisor to mandibular plane, Co-Pg, Co-Go, Ar-Pg, Go-Gn, S-N, S-SOS, and S-Ba)

Robinow, M., 1942, The variability of weight and height increments from birth to six years, Child Dev., 13:159. Serial data from 170 normal white children in southwestern Ohio, mostly from middle class homes. These children were enrolled in the Fels Longitudinal Study. (Tables of increments of nude weight and recumbent length)

Robinow, M., Richards, T. W., and Anderson, M., 1942, The eruption of deciduous teeth, Growth, 6:127. Serial data from white southwestern Ohio children enrolled in the Fels Longitudinal Study. The sample size was not reported. These children were from families of middle socioeconomic status. (Tables of ages at eruption)

Robinow, M., Johnston, M., and Anderson, M., 1943, Feet of normal children. A study of lateral x-rays of the weight-bearing foot, J. Pediatr., 23:141. Data from 800 serial radiographs of children aged 2.5 to 13 years. These were white children living in southwestern Ohio and enrolled in the Fels Longitudinal Study. They were from families of middle socioeconomic status. (Table of foot angles)

Robson, J. R. K., Larkin, F. A., Bursick, J. H., and Perri, K. P., 1975, Growth standards for infants and children: A cross-sectional study, Pediatrics, 56:1014. Data from 1,233 infants and children attending a clinic in Michigan (Washentaw County), of whom 522 were black and the remainder were white. (Tables of weight in blacks and whites)

Roche, A. F., Unpublished a, Serial data from white children living in southwestern Ohio. These children are enrolled in the Fels Longitudinal Study and are from families of middle socioeconomic status. (Tables of stature, recumbent length, weight, weight/stature^2, weight/recumbent length^2 and skeletal age)

Roche, A. F., Unpublished b, Measurements of the cervical vertebral column. Serial data from 32 Cleveland (OH) white children enrolled in the Bolton Study. (Tables of vertebral body height and intervertebral disc height)

Roche, A. F., 1953, Increase in cranial thickness during growth, Hum. Biol., 25:81. Serial data from 32 white Cleveland (OH) children enrolled in the Brush Foundation Study. (Tables of cranial thickness at nasion, between bregma and nasion, bregma, vertex, lambda and euryon)

Roche, A. F., 1968, Sex-associated differences in skeletal maturity, Acta Anat., 71:321. Serial data from Melbourne (Australia) children of British ancestry whose families were approximately representative of the Melbourne community in socioeconomic status. (Table of sex-associated differences in skeletal maturity)

Roche, A. F., and Davila, G. H., 1972, Late adolescent growth in stature, Pediatrics, 50:874. Serial data from 103 boys and 91 girls in the Fels Longitudinal Study. These were white children of middle socioeconomic status living in southwestern Ohio. (Tables of ages at which adult stature is reached and total increments in stature after selected ages or landmarks)

Roche, A. F., and French, N. Y., 1970, Differences in skeletal maturity levels between the knee and hand, Am. J. Roentgenol., 109:307. Data from 245 normal Melbourne (Australia) children of British ancestry who were aged 12 to 15 years. These children, who were enrolled in the University of Melbourne Growth Study, were from families that were approximately representative of the Melbourne community socioeconomically. (Table of differences between knee and hand skeletal ages)

Roche, A. F., and Lewis, A. B., 1974, Sex differences in the elongation of the cranial base during pubescence, Angle Orthod., 44:279. Serial data from 58 boys and 41 girls from southwestern Ohio. These white children were enrolled in the Fels Longitudinal Study. They were of middle socioeconomic status. (Tables of the prevalence and size of pubertal spurts, and age of occurrence and size of pubertal spurts in cranial base lengths and stature)

Roche, A. F., and Sunderland, S., 1959, Multiple ossification centers in the epiphyses of the long bones of the human hand and foot, J. Bone Joint Surg., 41B:375. Data from 120 normal children examined serially in the University of Melbourne Child Growth Study. These children were all of British ancestry and socioeconomically approximately representative of the Melbourne community. (Tables of prevalence of multiple centers)

Roche, A. F., Roberts, J., and Hamill, P. V. V., 1974, "Skeletal Maturity of Children 6-11 Years, United States," DHEW Publ. No. (HRA) 75-1622, U.S. Govt. Print. Off., Washington. Data from a nationally representative sample of 7,119 children. (Tables of skeletal age by chronological age for hand-wrist; number of bones maturing or adult)

Roche, A. F., Davila, G. H., and Mellits, E. D., 1975, Late adolescent changes in weight, in: "Biosocial Interrelations in Population Adaptation," E. S. Watts, F. E. Johnston, and G. W. Lasker, eds., Mouton, The Hague. Serial data from 118 boys and 111 girls. These white children, who were of middle socioeconomic status, were living in southwestern Ohio and were enrolled in the Fels Longitudinal Study. (Table of increments in weight after chronological ages and growth and developmental landmarks)

Roche, A. F., Roberts, J., and Hamill, P. V. V., 1975, "Skeletal Maturity of Children 6-11 Years: Racial, Geographic Area, and Socioeconomic Differentials. United States," DHEW Publ. No. (HRA) 76-1631, U.S. Govt. Print. Off., Washington. Data from a nationally representative sample of 7,119 children. (Tables of skeletal age by region, income, urban-rural factors, parental education, population size of place of residence; age at onset of ossification in relation to family income, race and region)

Roche, A. F., Wainer, H., and Thissen, D., 1975a, "Skeletal Maturity: The Knee Joint as a Biological Indicator," Plenum Medical Book Company, New York. Data from 7,897 anteroposterior radiographs of the left knee. These radiographs were taken serially of 552 white children of middle socioeconomic status who were living in southwestern Ohio. These children were enrolled in the Fels Longitudinal Study. (Tables of RWT knee skeletal ages, and standard errors of the estimates)

Roche, A. F., Wainer, H., and Thissen, D., 1975b, The RWT method for the prediction of adult stature, Pediatrics, 56:1026. Serial data from 200 white children in southwestern Ohio who were enrolled in the Fels Longitudinal Study. These children were from families of middle socioeconomic status. (Tables of coefficients for prediction equations)

Roche, A. F., Roberts, J., and Hamill, P. V. V., 1976, "Skeletal Maturity of Youths 12-17 Years. United States," DHEW Publ. No. (HRA) 77-1642. Data from a nationally representative sample of 6,768 youths. (Tables of distributions of skeletal age against chronological age, and range of bone-specific skeletal ages)

Roche, A. F., Lewis, A. B., Wainer, H., and McCartin, R., 1977, Late elongation of the cranial base, J. Dent. Res., 56:802. Serial data from 33 boys and 26 girls. These white children were of middle socioeconomic status. They were living in Southwestern Ohio and were enrolled in the Fels Longitudinal Study. (Table of ages at which 95% of the mature lengths of cranial base segments were reached)

Roche, A. F., Roberts, J., and Hamill, P. V. V., 1978, "Skeletal Maturity of Youths 12-17 Years. Racial, Geographic Area, and Socioeconomic Differentials, United States, 1966-1970," DHEW Publ. No. (PHS) 79-1654, U.S. Govt. Print. Off., Washington. Data from a nationally representative sample of 6,768 youths. (Tables of ages at epiphyseal fusion, distributions of skeletal age by region, race, income, and parental education)

Rosenstein, S. W., 1964, A longitudinal study of anteroposterior growth of the mandibular symphysis, Angle Orthod., 34:155. Serial data from 31 untreated Cleveland (OH) children enrolled in the Bolton Study. (Table of increments in A-P symphyseal thickness)

Rueda-Williamson, R., and Rose, H. E., 1962, Growth and nutrition of infants. The influence of diet and other factors on growth, Pediatrics, 30:639. Data from 67 full-term healthy infants in Boston (MA) measured serially from birth to 15 months. They were mostly from low socioeconomic groups; 88% were of northern European stock. (Table of weight and recumbent length)

Salzmann, J. A., and Ast, D. B., 1955, The Newburgh-Kingston Caries Fluoride Study in dento-facial growth and development cephalometric study, Am. J. Orthod., 41:674. Data from 759 children aged 6 to 10 years. Race not stated. (Tables of Margolis Triangle data, craniofacial shape, gonial angle, and incisor inclination)

Savara, B. S., and Singh, I. J., 1968, Norms of size and annual increments of seven anatomical measures of maxillae in boys from three to sixteen years of age, Angle Orthod., 38:104. Serial data from 52 boys of northwest European ancestry, predominantly from middle socioeconomic status families. (Tables of maxillary height, width and length and increments in these measures together with increments in relation to the pubertal spurt and age)

Schaefer, O., 1970, Pre- and post-natal growth acceleration and increased sugar consumption in Canadian Eskimos, Can. Med. Assoc. J., 103:1059. Data from 214 infants. (Table of birth weight)

Schour, I., and Massler, M., 1940, Studies in tooth development, J. Am. Dent. Assoc., 27:1918.

Schraer, H., 1958, Variation in the roentgenographic density of the os calcis and phalanx with sex and age, J. Pediatr., 52:416. Data from 1,484 children from Washington, Montana, New Mexico, Pennsylvania, and "miscellaneous U.S." (Table of density of calcaneus and middle phalanx V of the hand)

Schutte, J. E., 1979, "Growth and Body Composition of Lower and Middle Income Adolescent Black Males," Ph.D. Thesis, Southern Methodist University, Dallas. Data from 203 boys in Dallas County (TX) aged 10 to 18 years. (Tables of stature; weight; biacromial, biiliac, elbow, wrist, knee, and ankle diameters; maximum and minimum chest circumferences; buttocks, upper arm, forearm, thigh, and lower leg circumferences)

Schwarz, G. S., 1946, Determination of frontal plane area from the product of the long and short diameters of the cardiac silhouette, Radiology, 47:360.

Scott, J., 1975, Age, sex and contralateral differences in the volumes of human submandibular salivary glands, Arch. Oral. Biol., 20:885. Data from 19 glands of individuals aged 16 to 35 years. (Tables of volume and lateral differences in volume)

Scott, R. B., Cardozo, W. W., Smith, A. de G., and de Lilley, M. R., 1950, Growth and development of Negro infants. III. Growth during the first year of life as observed in private pediatric practice, J. Pediatr., 37:885. Data from 654 infants from families in the lower middle income level in Washington (DC). There are 248 to 320 observations in each sex at each month of age. Low birth weight infants (<2500 gm) and those suffering from severe illnesses or major congenital deformities were excluded. (Tables of birth weight and recumbent length)

Scott, R. B., Jenkins, M. E., and Crawford, R. P., 1950, Growth and development of Negro infants. I. Analysis of birth weights of 11,818 newly born infants, Pediatrics, 6:425. (Tables of birthweights)

Scott, R. B., Hiatt, H. H., Clark, B. G., Kessler, A. D., and Ferguson, A. D., 1962, Growth and development of Negro infants. IX. Studies on weight, height, pelvic breadth, and head and chest circumferences during the first year of life, Pediatrics, 29:65. Data from 111 normal, healthy infants from lower middle class families. None had congenital defects or serious illnesses and none was premature. (Tables of weight, recumbent length, bicristal diameter, head circumference, and chest circumference)

Seckel, H. P. G., and Rolfes, L. J., 1959, Long-range studies on statural growth of smallest surviving prematures, Ann. Pediatr. Fenn., 2:75.

Selby, S., 1961, Separate centers of ossification of the tip of the internal malleolus, Am. J. Roentgenol., 86:496. Data from 151 children examined serially in the Fels Longitudinal Study. These white children were living in southwestern Ohio and were of middle socioeconomic status. (Table of age of appearance and age of fusion of center for the tip of the internal malleolus)

Selby, S., Garn, S. M., and Kanareff, V., 1955, The incidence and familial nature of a bony bridge on the first cervical vertebra, Am. J. Phys. Anthropol., 13:129. Data from 1,414 serial radiographs of 447 individuals from southwestern Ohio. Of these, 106 were participants in the Fels Longitudinal Study; the remainder were near relatives. All were aged 14 years or older at the time of the most recent radiograph. (Table of incidence of bridging)

Sheldon, W. H., Lewis, N. D. C., and Tenney, A., 1969, Psychotic patterns and physical constitution, in: "Schizophrenia: Current Concepts and Research," D. V. Siva Sankar, ed., PJD Publications, Hicksville, NY.

Sillman, J. H., 1964, Dimensional changes of the dental arches: longitudinal study from birth to 25 years, Am. J. Orthod., 50:824. Serial data from 65 normal white individuals born in New York City. (Tables of dental stages in years, and increments in rugal length and width in relation to dental stages)

Sillman, J. H., 1965, Some aspects of individual dental development: longitudinal study from birth to 25 years, Am. J. Orthod., 51:1.

Silverman, F. N., 1957, Roentgen standards for size of the pituitary fossa from infancy through adolescence, Am. J. Roentgenol., 78:451. Serial data from 320 white children from southwestern Ohio enrolled in the Fels Longitudinal Study. These children were from families of middle socioeconomic status. (Tables of mean area, and areas for particular lengths and depths)

Simmons, K., 1944, The Brush Foundation Study of Child Growth and Development. II. Physical growth and development, Monogr. Soc. Res. Child Dev., 9:Serial 37, No. 1. Data from serial examinations of 999 children aged 3 months to 19 years. These children were selected from the Cleveland (OH) population to represent a group free of gross physical and mental defects. All were white, most were of north European ancestry, and the families were above average both economically and educationally. (Tables of weight, stature, recumbent length, sitting height, stem length, cristal height, symphyseal height, anterior iliac spinous height, tibial length, suprasternal height; biacromial, bicristal, and bitrochanteric diameters; chest width, total arm length, lengths of upper arm, forearm, and hand; chest circumference at inspiration and at expiration; head length, breadth, and height; and skeletal age)

Singh, I. J., Savara, B. S., and Newman, M. T., 1967, Growth in the skeletal and non-skeletal components of head width from 9 to 14 years of age, Hum. Biol., 39:182. Serial data from 18 boys and 20 girls of northwest European ancestry. (Tables of total head width, skeletal head width, inner cranial width, soft tissue thickness over the cranium, and cranial bone thickness; and annual increments of these)

Singh, R., 1976, A longitudinal study of the growth of trunk surface area measured by planimeter on standard somatotype photographs, Ann. Hum. Biol., 3:181. Serial data from 15 English boys enrolled in the Harpenden Growth Study. (Table of thorax area, abdomen area, and trunk index)

Slater, S., 1970, An evaluation of the metacarpal sign (short fourth metacarpal), Pediatrics, 46:468. Data from 232 children (consecutive patients) aged less than 13 years; those referred for an endocrine evaluation were excluded. (Table of prevalence of metacarpal sign)

Sloan, R. F., Bench, R. W., Mulick, J. F., Ricketts, R. M., Brummett, S. W., and Westover, J. L., 1967, The application of cephalometrics to cinefluorography: Comparative analysis of hyoid movement patterns during deglutition in Class I and Class II orthodontic patients, Angle Orthod., 37:26. Data from 15 subjects averaging 12 years in age and having Class I occlusions. (Table of hyoid movement during deglutition, craniofacial size and shape, and hyoid position)

Snyder, R. G., Spencer, M. L., Owings, C. L., and Schneider, L. W., 1975, "Anthropometry of U.S. Infants and Children," Paper No. 750423, 1975, SAE Automotive Engineering Congress and Exposition, Detroit, MI. Also published by Consumer Protection Safety Commission, Report. No. UM-HSRI-B1-75-5. Data from 4,027 infants and children in schools, day-care centers, nurseries and clinics in eight states. (Tables of stature, sitting height, knee-sole length, rump-sole length, shoulder-elbow length, lower arm length, hand length and width, foot length and breadth, fifth finger length and breadth, third finger length and breadth, head breadth, maximum shoulder breadth, chest breadth, waist breadth, lower torso breadth, head length, chest depth; circumferences of neck, chest, forearm, upper arm, mid-thigh, and maximum calf)

Spector, W. S., 1956, "Handbook of Biological Data," W. B. Saunders Co., Philadelphia.

Spurgeon, J. H., and Meredith, H. V., 1979, Body size and form of black and white male youths: South Carolina youths compared with youths measured at earlier times and other places, Hum. Biol., 51:187. Data from 405 boys in Richland County aged 15 years. (Table of stature, sitting height, lower limb length, biiliac width, arm and calf circumferences, weight, lower limb length/sitting height, biiliac width/lower limb length and calf circumference/lower limb length)

Spurgeon, J. H., Young, N. D., and Meredith, H. V., 1959, Body size and form of American-born boys of Dutch ancestry residing in Michigan, Growth, 23:55. Data from 246 boys in Holland (MI) aged 7, 11, or 15 years. About 60% attended public schools. They were all born in North America and had at least three grandparents of Dutch lineage. (Tables of stature, stem length, lower limb length, thoracic circumference, abdominal circumference, calf circumference, and weight)

Spurgeon, J. H., Meredith, E. M., and Meredith, H. V., 1978, Body size and form of children of predominantly black ancestry living in West and Central Africa, North and South America, and the West Indies, Ann. Hum. Biol., 5:229. Data from black children in South Carolina aged 6, 9, or 11 years. (Tables of stature, sitting height, leg length, biiliac diameter, arm and calf circumferences, weight, leg length/sitting height, hip circumference/leg length, and calf circumference/leg length)

Stennett, R. G., and Camp, D. M., 1969, Cross-sectional, percentile height and weight norms for a representative sample of urban, school-aged, Ontario children, Can. J. Public Health, 60:465. Data from 32,399 children in London (Ontario) aged 4.9 to 18 years. (Table of stature and weight)

Stone, H. H., Lawton, F. E., Bransby, E. R., and Hartley, H. O., 1951, Time of eruption of permanent teeth and time of shedding of deciduous teeth, Brit. Dent. J., 90:1.

Strickland, A. L., and Shearin, R. B., 1972, Diurnal height variation in children, J. Pediatr., 80:1023. Data from 100 children aged 3 to 14 years. The measurements were made on rising from bed in the morning and between 4 and 5 p.m. (Table of diurnal variations in stature)

Stuart, H. C., and Dwinell, P. H., 1942, The growth of bone, muscle and overlying tissues in children six to ten years of age as revealed by studies of roentgenographs of the leg area, Child Dev., 13:196. Data from about 130 white children studied serially from 6 to 10 years of age. These children were enrolled in the Harvard School of Public Health Growth Study. (Tables of tibial length, calf width, and tibial shaft width)

Stuart, H., and Meredith, H. V., 1946, Use of body measurements in the school health program. Part II. Methods to be followed in taking and interpreting measurements and norms to be used, Am. J. Public Health, 36:1373. Data from white children in Iowa City (IA). These data included 3,771 sets of serial measurements of several hundred children. (Tables of stature, weight, bicristal diameter, chest circumference at xiphoid, and calf circumference)

Stuart, H. C., Pyle, S. I., Cornoni, J., and Reed, R. B., 1962, Onsets, completions and spans of ossification in the 29 bone-growth centers of the hand and wrist, Pediatrics, 29:237. Data from 133 white children examined serially in the Harvard School of Public Health Growth Study. These children were living in or near Boston (MA). (Table of ages at onset and completion of ossification)

Sturdivant, J. E., Knott, V. B., and Meredith, H. V., 1962, Interrelations from serial data for eruption of the permanent dentition, Angle Orthod., 32:1. Data from 57 white boys living in or near Iowa City (IA). These boys were predominantly of northwest European ancestry and above average in socioeconomic status. (Tables of ages at eruption)

Swearingen, J. J., and Young, J. W., 1965, "Determination of centers of gravity of children, sitting and standing," Report AM-65-23, Office of Aviation Medicine, Civil Aeronautics Research Institute, Oklahoma City (OK). Data from 1,200 Oklahoma children aged 5 to 18 years. (Tables of upper arm length, forearm length, chest depth, and waist depth)

Swearingen, J. J., Badgley, J. M., Braden, G. E., and Wallace, T. F., 1969, "Determination of centers of gravity in infants," Aerospace Medical Association Annual Meeting, San Francisco (CA). Data from 135 infants aged up to 36 months. (Table of stature, weight, vertex to crotch, vertex to shoulder, and position of center of gravity)

Tanner, J. M., and Whitehouse, R. H., 1959, "Standards for Skeletal Maturity. Part I., International Children's Centre, Paris.

Tanner, J. M., Whitehouse, R. H., and Healy, M. J. R., 1962, "A New System for Estimating Skeletal Maturity from the Hand and Wrist, with Standards Derived from a Study of 2,600 Healthy British Children. Part II, The Scoring System," International Children's Centre, Paris.

Tanner, J. M., Whitehouse, R. H., Marshall, W. A., Healy, M. J. R., and Goldstein, H., 1975, "Assessment of Skeletal Maturity and Prediction of Adult Height (TW2 method)," Academic Press, London.

Taylor, W. H., and Hitchcock, H. P., 1966, The Alabama analysis, Am. J. Orthod., 52:245. Data from 40 white children with normal occlusions who were not treated orthodontically and had pleasing or acceptable facial development. These were school children in grades 3 through 6. (Table of craniofacial size and shape, incisor inclination and position)

Ten-State Nutrition Survey 1968-1970, 1972, DHEW Publ. No. (HSM) 72-8130. Data from 10 states plus New York City. The sample was drawn from low income areas but those examined are not all from low income groups; it includes 7,337 white, 7,779 black, and 1,431 Mexican-American children aged 1-18 years. (Tables of stature and weight by race and income)

Thompson, G. W., and Popovich, F., 1974, Relationship of mandibular measurements to stature and weight in humans, Growth, 38:187. Data from 111 girls examined serially from 4 to 18 years of age. When selected, they were a representative sample of 3-year-olds in Burlington (Ontario). (Table of mandibular body length, mandibular length, stature and weight)

Thompson, G. W., Popovich, F., and Anderson, D. L., 1976, Maximum growth changes in mandibular length, stature and weight, Hum. Biol., 48:285. Data from 111 girls enrolled in the Burlington Growth Centre (Ontario). (Tables of mandibular length, stature and weight; increments of these measures in relation to chronological and skeletal ages)

Thompson, H., 1951, Data on the growth of children during the first year after birth, Hum. Biol., 23:75. Data from 68 white American-born New Haven (CT) infants from middle class families. All were full-term healthy babies who were developing normally. All measurements were made with the infant recumbent. (Tables of increments in recumbent length, suprasternal height, symphysis height, vertex to suprasternal notch, vertex to symphysis, suprasternal notch to symphysis, biacromial diameter, chest breadth and chest circumference at level of nipple in mid-respiration, bicristal diameter, head circumference, and weight)

Todd, T. W., 1937, "Atlas of Skeletal Maturation," C. V. Mosby, St. Louis.

Tracy, W. E., Savara, B. S., and Brant, J. W. A., 1965, Relation of height, width and depth of the mandible, Angle Orthod., 35:269. Serial data from 27 girls enrolled in the Child Study Clinic, University of Oregon Dental School. They were of northwest European ancestry and of middle socioeconomic status. (Table of mandibular height, length and width)

Trim, P. T., and Meredith, H. V., 1952, Body form in Homo Sapiens: A study of five anthropometric ratios on white boys fifteen years of age, Growth, 16:1. Data from 102 boys of northwest European ancestry living in Oregon. (Tables of arm circumference/upper limb length, chest circumference/stem length, calf circumference/lower limb length, biacromial diameter/biiliac diameter, and chest circumference/abdominal circumference)

Trotter, M., 1971, The density of bones in the young skeleton, Growth, 35:221. Data from 143 skeletons from birth to 22 years representing both sexes and American white and Negro groups. (Tables of density of parts of bones, and mean densities of selected bones)

Trotter, M., 1973, Percentage ash weight of young human skeletons, Growth, 37:153. Data from 66 skeletons of American whites and Negroes of each sex, aged from birth to 23 years. (Tables of percentage ash weight of whole skeletons and of selected bones and parts of some bones)

Trotter, M., and Hixon, B. B., 1974, Sequential changes in weight, density and percentage ash weight of human skeletons from an early fetal period through old age, Anat. Rec., 179:1. Data from 144 skeletons from birth to 22 years. These represented each sex and white and Negro groups. (Tables of ash weight for total skeletons, for divisions of the skeleton, and percentages for each bone and division)

Tuddenham, R. D., and Snyder, M. M., 1954, "Physical Growth of California Boys and Girls from Birth to Eighteen Years," University of California Publications in Child Development, 1 (No. 2), University of California Press, Berkeley. Serial data from 137 white children enrolled in the Guidance Study in Berkeley (CA). (Tables of stem length, biacromial and biiliac diameter, and calf circumference)

Tyler, A., 1966, "The Sizing and Design for Physical Education Uniforms," Unpublished thesis for M.S. in Physical Education, University of North Carolina at Greensboro.

Uesato, G., Kinoshita, Z., Kawamoto, T., Koyama, I., and Nakanishi, Y., 1978, Steiner cephalometric norms for Japanese and Japanese-Americans, Am. J. Orthod., 73:321. Data from post-treatment lateral cephalograms from 50 Hawaiian American children aged 11 to 18 years (25 boys, 25 girls). Table of facial angles and dimensions and incisor inclinations and position)

Usher, R., and McLean, F., 1969, Intrauterine growth of liveborn Caucasian infants at sea level: standards obtained from measurements in 7 dimensions of infants born between 25 and 44 weeks of gestation, J. Pediatr., 74:901. Data from 300 Caucasian infants with varying socioeconomic backgrounds and national origins who were born alive in Montreal and measured within 36 hours of birth. Infants were excluded who had major congenital abnormalities, eyrthroblastosis or marked fetal malnutrition or who were born to diabetic mothers. (Tables of birth weight, recumbent length, foot length, circumferences of head, chest, abdomen, and thigh in relation to gestational age and to birth weight; smoothed values for birth weight; recumbent length in relation to gestational age; head circumference, recumbent length, and the difference between head circumference and chest circumference in relation to birth weight; head circumference in relation to recumbent length)

Valk, I. M., 1974, Ulnar growth in boys around the age of puberty, Growth, 38:437.

Verghese, K. P., Scott, R. B., Teixeira, G., and Ferguson, A. D., 1969, Studies in growth and development. XII. Physical growth of North American Negro children, Pediatrics, 44:243. Data from 2,632 children examined in well-baby clinics and a Police Boys' Club in the District of Columbia; all were normal and healthy and from low income families. (Tables of weight, stature, stem length, head circumference, chest circumference, biacromial diameter, bicristal diameter, upper arm length, forearm and hand lengths)

Vickers, V. S., and Stuart, H. C., 1943, Anthropometry in the pediatrician's office. Norms for selected body measurements based on studies of children of North European stock, J. Pediatr., 22:155. Mixed longitudinal data for children from birth to 10 years of age principally of north European stock (a large proportion Irish) living in or near Boston (MA) in low to middle economic circumstances. The number at each 6 months of age decreases with age from 136 to about 26 in each sex. (Tables of weight, head circumference, chest circumference at level of xiphoid, recumbent length, stature, bicristal diameter, chest breadth at level of xiphoid, crown-rump length and sitting height)

Vig, P. S., and Cohen, A. M., 1974, The size of the tongue and the intermaxillary space, Angle Orthod., 44:25. Data from 75 English children referred for orthodontic consultation. (Tables of intermaxillary space size)

Vig, P. S., and Cohen, A. M., 1979, Vertical growth of the lips: A serial cephalometric study, Am. J. Orthod., 75:405. Serial data from 50 untreated children in London (UK). (Tables of lip height, anterior lower face height, lower lip to upper incisors and lip separation)

Walker, R. N., 1974a, Standards for somatotyping children: I. The prediction of young adult height from children's growth data, Ann. Hum. Biol., 1:149. Mixed longitudinal data from normal white children in New Haven (CT) and in Berkeley (CA). There were 124 New Haven children and 136 Berkeley children in the study samples. (Table of stature)

Walker, R. N., 1974b, Standards for somatotyping children: II. The prediction of somatotyping ponderal index from children's growth data, Ann. Hum. Biol., 1:289. Data from 223 white children in New Haven (CT) examined serially at ages from 2 to 21 years with many incomplete series. Data from the Berkeley Growth Study were used for replication. (Table of stature/weight[3])

Walker, R. N., 1978, Pre-school physique and late-adolescent somatotype, Ann. Hum. Biol., 5:113. Data from 319 white pre-school children in New Haven (CT). (Table of somatotype)

Walker, R. N., 1979, Sheldon's trunk index and the growth of the thoracic and lumbar trunk, Ann. Hum. Biol., 6:315. Serial data from 82 English boys aged 5 to 18 years. These boys were enrolled in the Harpenden Growth Study. They were living in a children's home where they were well cared for and well nourished. (Tables of trunk index, thoracic and lumbar heights and areas)

Walker, R. N., and Tanner, J. M., 1980, Prediction of adult Sheldon somatotypes I and II from ratings and measurements at childhood ages, Ann. Hum. Biol., 7:213. Serial data from 82 boys of the Harpenden Growth Study (UK). (Table of somatotype)

Warner, W. L., Meeker, M., and Eells, K., 1949, "Social Class in America: A Manual of Procedure for the Measurement of Social Status," Science Research Associate, Chicago.

Weinberg, W. A., Dietz, S. G., Penick, E. C., and McAlister, W. H., 1974, Intelligence, reading achievement, physical size and social class. A study of St. Louis Caucasian boys aged 8-0 to 9-6 years, attending regular schools, J. Pediatr., 85:482. Data from

360 boys. (Tables of skeletal age, stature, weight, head circumference, head length and width)

Welch, J. P., Winsor, E. J., and Mackintosh, S. M., 1971, The distribution of height and weight, and the influence of socio-economic factors, in a sample of eastern Canadian urban school children, Can. J. Public Health, 62:373. Data from 10,901 children aged 6.5 to 13.5 years. (Table of stature and weight)

Wetzel, N. C., 1941, Physical fitness in terms of physique, development, and basal metabolism, J. Am. Med. Assoc., 116:1187.

Whitehouse, R. H., Tanner, J. M., and Healy, M. J. R., 1974, Diurnal variation in stature and sitting height in 12-14-year-old boys, Ann. Hum. Biol., 1:103. Data from 19 English boys measured at 9 a.m. and 2 p.m. and from 11 English boys measured at 10 a.m. and 5 p.m. (Table of mean differences)

Wilson, R. S., 1974, Growth standards for twins from birth to four years, Ann. Hum. Biol., 1:175. Data from 292 pairs of white twins of all socioeconomic groups in Louisville (KY). (Tables of weight increments, head circumference, and increments in head circumference)

Wilson, R. S., 1976, Concordance in physical growth for monozygotic and dizygotic twins, Ann. Hum. Biol., 3:1. Data from repeated measurements of 636 white twins in Louisville (KY). (Tables of recumbent length, stature, and within-pair differences in weight)

Wilson, R. S., 1979, Twin growth: initial deficit, recovery, and trends in concordance from birth to nine years, Ann. Hum. Biol., 6:205. Serial data from 900 white twins in the metropolitan Louisville (KY) area from all socioeconomic groups. (Tables of weight, recumbent length, stature, and intrapair differences for these variables)

Wingerd, J., Schoen, E. J., and Solomon, I. L., 1971, Growth standards in the first two years of life based on measurements of white and black children in a prepaid health care program, Pediatrics, 47:818. Data from children born between 1959 and 1969 in northern California (66% white, 23% black, 4% oriental, and 7% mixed or other). Children with serious congenital anomalies or the products of multiple births were excluded. Children who could stand were measured upright; these data were not distinguished from recumbent length. There were about six sets of measurements for each of the 15,031 children. (Tables of recumbent length, weight, and head circumference)

Wingerd, J., Solomon, I. L., and Schoen, E. J., 1973, Parent-specific height standards for preadolescent children of three racial groups, with method for rapid determination, Pediatrics, 52:555. Data from single-born children without severe congenital anomalies. There were available 59,000 measurements of 11,233 children aged 16.9 years. (Tables of stature in relation to race)

Wingerd, J., Peritz, E., and Sproul, A., 1974, Race and stature differences in the skeletal maturation of the hand and wrist, Ann. Hum. Biol., 1:201. Data from 637 children aged 5 to 9 years in San Francisco (CA) who were white, black or oriental. The white children and the black children represented the upper and lower 2.5% and the central 5% of the appropriate age, sex and race distributions for stature. (Tables of differences of bone-specific Greulich-Pyle skeletal ages for groups of children in comparison with means for white children of medium stature; variance of differences between bone-specific skeletal ages and means for white children of medium stature within each hand-wrist)

Wise, F. C., and Meredith, H. V., 1942, The physical growth of Alabama white girls attending WPA schools, Child Dev., 13:165. Data from 92 girls aged 1.75 to 5.25 years enrolled in preschools or day nurseries in or near Birmingham. (Tables of stature, sitting height, lower limb length [stature less sitting height], biacromial and bicristal diameters, chest circumference at xiphoid, abdominal circumference at umbilicus, chest width and chest depth at xiphoid, upper arm and calf circumferences, weight, subcutaneous fat thickness at abdomen, subscapular and triceps; and leg length/sitting height, chest circumference/abdominal circumference, bicristal diameter/sitting height, and chest width/chest depth)

Wolff, G., 1942, A study of height in white school children from 1937 to 1940 and a comparison of different height-weight indices, Child Dev., 13:65. Data from children in Hagerstown (MD) aged 6 through 16 years; 19,949 sets of data were recorded during a 4-year period. It is not clear to what extent the same children were measured in successive years. (Tables of weight/stature, weight/stature^2, and weight/stature^3 x 100)

Wood, B. F., 1971, Malocclusion in the modern Alaskan Eskimo, Am. J. Orthod., 60:344. Data from 100 children (58 boys, 42 girls) aged 11 to 20 years. They had all been exposed to a refined diet since birth and had received dental care. (Tables of intercanine and intermolar widths)

Wylie, W. L., 1946, The relationship between ramus height, dental height and overbite, Am. J. Orthod., 32:57. Data from 90 children (29 boys, 61 girls) in whom all deciduous teeth had been lost. These children were clinic patients in the Division of Orthodontics at the University of California, San Francisco. (Tables of face depth and height, ramus height and molar height in relation to age)

Yarbrough, C., Habicht, J-P., Klein, R. E., and Roche, A. F., 1973, Determining the biological age of the preschool child from a hand-wrist radiograph, Invest. Radiol., 8:233. Serial data from 156 white southwestern Ohio boys of middle socioeconomic status who were enrolled in the Fels Longitudinal Study. (Table of number of centers ossified in the hand-wrist)

Young, C. M., Sipin, S. S., and Roe, D. A., 1968, Body composition of pre-adolescent and adolescent girls. II. Anthropometric measurements, J. Am. Diet. Assoc., 53:357.

Young, J. W., 1966, "Selected Facial Measurements of Children for Oxygen-mask Design," Report No. AM 66-9, Federal Aviation Agency, Office of Aviation Medicine, Washington. Data from 978 white children aged 1 month to 17 years. (Tables of head length and breadth, bizygomatic and bigonial diameters, nasion-subnasale, nasion-stomion, nasion-supramentale, nasion-menton, interocular distance, nasal root breadth, nasal bridge breadth, nasal alar breadth, mouth width [bichelion diameter], nasal root projection, nasal tip projection in relation to posterior limits of the alae and to the facial plane)

Zavaleta, A. N., 1976, "Densitometric Estimates of Body Composition in Mexican American Boys," Ph.D. dissertation, University of Texas at Austin. Data from 95 boys aged 9 to 14.9 years living in a lower socioeconomic neighborhood. (Tables of stature, weight, sitting height, subischial length, sitting height ratio, biacromial and bicristal diameters, bicondylar diameter, and biepicondylar diameter)

INDEX

Abdomen area 994
Abdomen circumference 27, 35-7, 45, 984-91,
        1097-8, 1120
    and gestational age 475-6
Acetabular angle 1149-51, 1162-4
Acromiale height 430
Adductor sesamoid, age when adult 1217-8
    ossification onset 693, 1217-8, 1220-2,
        1225, 1240
Age, maternal, and birth weight (see also
        Maternal age) 183
Agenesis, third molar 861
Alar breadth 616-7
Altitude, and birth weight 356
Ankle circumference 984-91, 1036-40
Ankle diameter 213, 984-91, 1044, 1046-7
Ankle height 1022-3
Ankle, sex-associated differences in
        Greulich-Pyle skeletal age 1257
Anterior arch of atlas, position 553
Anterior fontanelle, age of closure 516
    size 1380
    and type of artificial feeding 517
Anterior iliac spine height 432-5
Anterior trunk height 879
Anthropometric measurements, reliability 1382
Apical base, sagittal relation 696
Arch, dental see Mandible and Maxilla
Arm length see Upper limb length
Arm, upper see Upper arm
Articulare see also Cranial base measurements
Articulare-sella-basion angle 551
    increments 554
Artificial feeding, and pelvis breadth 1167
    and recumbent length 138, 155-6
    and weight 155-6, 200-1, 302
    and weight loss after birth 200
Ash weights of bones as percent of total
        skeleton 1127-9
Atlas, anterior arch position 553, 649
    bridging 1152
    ossification centers 1213
    position 646-7

B (Downs)-nasion-sella angle 550
B (Downs)-sella distance 544
Basal angle see Saddle angle

Basion see also Cranial base measurements
Basion-articulare 534-5, 520-1, 541-5
    increments 538-9, 540, 545-6
Basion-opisthion plane see Foramen magnum plane
Basion, position 741
Basion-sphenoccipitale/nasion-basion 1380
Biacromial diameter 27, 35-7, 46, 48, 161-4,
        213, 349, 351-2, 354-5, 880-901, 984-
        91, 1380
    and family income 901
    and Greulich-Pyle skeletal age 890
    increments 161-4, 306-7, 497
    and maturation rate 891-2
    and pubertal spurt 117, 901
    and recumbent length 943, 945
    and stature 896-9, 943-6
Biacromial diameter/biiliac diameter 1097-8
Biacromial diameter/stature 898-9
Bialar nose width/bizygomatic diameter 610
Bialar projection 686
Biceps circumference 984-91
Biceps skinfold thickness 153, 213
Bicondylar diameter, femur 901, 1043
    mandible 602-3, 736
Bicristal diameter see Biiliac diameter
Biepicondylar diameter 901
Bigonial diameter 213, 608, 612-4, 739-40, 1379
    and Greulich-Pyle skeletal age 615
Bigonial diameter/bizygomatic diameter 610
Biiliac diameter 27, 35-8, 40-3, 46, 48, 161-4,
        204-13, 221, 223, 227, 349, 351-2,
        354-5, 372, 468, 893, 900-1, 921-48,
        983-9, 1097-8, 1380
    and family income 900
    and gestational age 221
    and Greulich-Pyle skeletal age 931
    increments 161-4, 306-7, 498
    and maternal age and parity 227, 372
    and maternal weight 223, 225, 227
    and maturation rate 938-41
    and recumbent length 943, 945
    and stature 41, 933-6, 941-6
Biiliac diameter/biacromial diameter, and
        maturation rate 947-8
Biiliac diameter/lower limb length 38
Biiliac diameter/sitting height 1120

Biparietal diameter 31-2, 46-9, 486-9, 492,
        495-6, 513, 535, 1296, 1379
  and bicondylar diameter 602-3, 736
  and birth weight 492
  increments 308, 491, 497-8
Biparietal diameter/head length 499
Biparietal skeletal head width *see* Cranial
        width
Birth weight (*see also* Low birth weight
        infants) 153-5, 171-203, 217-9, 222,
        225, 229, 234, 238, 240-1, 246, 249,
        356, 360-2, 369-74, 486, 492, 682
  age of doubling 200
  age of multiplying 203
  age of regaining 201, 203
  and altitude 356
  and biparietal diameter 492
  and chest circumference 476-7
  and foot length 1024
  and gestational age 172-82, 184-8, 196-8,
        218-21
  and head circumference 157, 450-1, 458,
        470, 477
  and head length 485
  and increments of head size and recumbent
        length 308
  and increments of weight 298, 308
  and inner canthal index 683
  and intermammillary index 683
  and maternal age, parity and weight 183,
        222, 224-7, 370-2
  and recumbent length 136-7, 140, 145-6, 157
  twins 194-5, 363-5, 367-8
  weight loss after 199-200
Bitrochanteric diameter 213, 948-52, 984-91,
        1159
  and Greulich-Pyle skeletal age 953
Bizygomatic diameter 31-2, 46-9, 213, 495,
        604-611, 641-2, 1380
  and Greulich-Pyle skeletal age 607
  increments 498, 611
Bizygomatic diameter/biparietal diameter 609
Body density 213
Bolton plane angle 528, 702-3
Bone density *see* specific bones
Breast feeding, and recumbent length 138,
        155-6, 373
  and weight 155-6, 200, 373-4
  and weight loss after birth 202
Bregma-nasion, increments 515
Bust circumference 1380
Buttock circumference 970, 984-91

Calcaneus, bone density 1207
  ossification onset 1217-8, 1223, 1226,
        1228, 1230
Calcified calcaneus/foot length 1025
Calf bone thickness, and maturation rate
        1944-5

Calf circumference 27, 32, 35-8, 40, 42-3,
        45-9, 204-12, 349, 468, 984-92, 1002,
        1028-34, 1037-40, 1380
  and family income 1040
  increments 306-7, 310-2, 498, 1003, 1035
Calf circumference/lower limb length 39, 41,
        1002, 1007, 1097-8, 1122
  increments 1003
Calf diameter and maturation rate 1044-5
Calf fat thickness and maturation rate 1044-5
Calf length 1001, 1011-2
Calf muscle thickness and maturation rate
        1044-5
Calf skinfold thickness 213
Capitate, ossification onset 1221, 1225
Carabelli's tubercle 868
Caries and age at exfoliation 1345-8
Carpal angle 1189
Carpal areas 1376
Carpus, ossification onset 1239-40
Cementum weight 865
Center of gravity, distances to top of head,
        shoulder, crotch, and foot 214
Centers of ossification *see* Ossification
        centers
Cephalic index 500, 535
Cephalic modulus *see* Cranial capacity
Cephalometric planes 718, 721-2
Cervical spinal canal, sagittal diameter 1152
Cervical vertebrae, body heights 1131-44
  horizontal position and levels 741-2
  ossification onset 1212
Cervicale height 213, 875-6
Chest area 992-4
Chest circumference 27, 35-7, 45-6, 48, 161-4,
        204-13, 219, 222, 225, 249, 349-51,
        353-4, 371, 467, 469, 473-6, 919,
        961-74, 984-91, 1380
  and artificial feeding 671
  difference between erect and recumbent
        measurements 167
  difference between supramammillary and
        xiphoid levels 974
  and gestational age 219, 474-5
  and head circumference 470, 477
  increments 161-4, 306-7, 309, 311-4, 498
  increments, in low birth weight infants 1380
  and maternal age, parity, and weight 222,
        225, 371
  and recumbent length 478, 480
  and socioeconomic status 970
  and stature 478-81
Chest circumference/abdomen circumference
        1097-8, 1120
Chest circumference/stem length 1097-8, 1121
Chest depth 27, 214, 914, 919, 954-8, 1120,
        1380
Chest depth/width 914, 1120
Chest depth/recumbent length 911, 913

Chest depth/stature 911-4
Chest height 992
Chest width 27, 161-4, 213, 902-10, 914-5,
        919, 984-91, 1380
  and artificial feeding 917
  increments 161-4
  internal 908-10
Chest width/depth 1120
Chest width/recumbent length 913
Chest width/stature 911-4
Chewing, muscle activity during 747
Chin position 688
Chin projection 684-5
Cholesterol, plasma 217
Circumference see specific site
Clavicle, ossification onset 1225, 1240
Cleft palate, cranio-facial angles in 553
Clival plane angle 551, 721
  increments 554
Condylar path 655
Condylion to pregonial notch 732
Convexity angle 539-42, 654, 694, 710, 713,
        715-7, 728
  increments 540, 556
Cranial base, ossification onset 1212-3
Cranial base angles 536-9, 548-53, 700-4,
        707-8, 712, 718, 721-2, 728, 1380
  increments 554-8
Cranial base lengths 518-27, 529-30, 533-42,
        559-62, 597-8, 656-7, 660-1, 669-70,
        730, 1380
  age at reaching 95% mature length 530
  and facial depth and height 536-40
  increments 515, 531-4, 540, 545-7, 558,
        563-6, 601, 662-4
  and occlusion 544
  and puberty 534, 670
Cranial base shape see Clival plane angle
        and Cranial base angles
Cranial base width 559-62
  increments 497
Cranial capacity 500, 514, 571, 1380
Cranial chords, increments 558
Cranial height 512-3, 522-3, 571, 639
  increments 515, 601, 663-4
  and occlusion 544
Cranial index 514
Cranial length 510-2, 571, 642
  increments 558, 663-4
  and occlusion 544
Cranial modulus see Cephalic modulus
Cranial shape 535
Cranial size 515
Cranial vault, ossification onset 1212-3
Cranial vault shape 723-4
  increments 555
Cranial vault thickness 496, 501-9
  increment 497
Cranial width 496, 512, 571
  increments 497, 558, 563-6

Cranial width/mandible width 514
Craniometric planes see specific measurement
Cristal height 213, 430-1, 436
Crossbite 746
Crotch to center of gravity, distance 214
Crotch to top of head, distance 214
Crown-rump length (see also Stem length) 215,
        222-3, 226, 371, 426
  and gestational age 220
  increments 309
  and maternal age, parity and weight 222-3,
        226, 371
  and sitting height, differences 167
Crown-sole length see Recumbent length
Cuboid, ossification 1223, 1226
Cuneiform, lateral and medial, age when adult
        1217-8
  ossification onset 1217-8, 1226, 1228, 1230

Dactylion III height 428
Deciduous teeth (see also Dental development)
        835-44
  ages at emergence 1214-6, 1342
  buccolingual diameters 838, 841-2, 866
  crown size 839
  emergence order 1857
  exfoliation 1352-3
  exfoliation, age at, influence of caries
        1345-8
  extraction 1353
  labiolingual diameters 839
  maturation 1297-9
  mesiodistal/buccolingual ratios 866
  mesiodistal diameters 835-41, 843-4, 850-3
  number erupted at specific ages 1315
  overall length 839
  resorption 1298-300, 1345-51
  root length 839-40, 842
  root size 840-2
  root spreading 840-2
Deglutition see swallowing
Dens height 728
Dental arch measurements 1372
Dental development (see also Deciduous and
        Permanent teeth, Mandible, and
        Maxilla) 1313
  intervals between eruption stages 1301, 1323,
        1351, 1355
  stages, arch and jaw measurements 770, 777,
        783-9, 791, 806-7, 810-3, 824-6,
        829-32, 1372
  supernumerary teeth 1344
Dental diastema 745
Dental eruption see Deciduous and Permanent
        teeth
Dental extraction see Deciduous teeth
Dental height 581, 598-600
  increments 587
Dental maturation stages 110, 1313
Dental resorption see Deciduous teeth

Dentine weight 865
Diastema, dental 745
Diurnal variation, in stature 106, 119
  in weight 292, 295, 302
Downs analysis 654

Education of parents, influence see Parental
    education
Elbow, diameter 213, 901, 1099-2001
  and family income 1114
  epiphyseal fusion 1382
Elbow to elbow breadth 1117
Emergence of teeth see Deciduous and
    Permanent teeth, Dental Development
Enamel weight 865
Epiphyseal fusion (see also specific bones)
    1217-8, 1289-92, 1382
Erect measurements, recumbent measurements,
    differences 165-7
Exophthalmos and puberty 683

Face breadth and width see specific diameter
Face depth 535-7, 540-2, 621-5, 627, 631,
    633-6, 638-43, 658, 660-1, 697, 699,
    719-20
  increments 498, 515, 532-3, 545-7, 644,
    662-4
  and occlusion 544
  percent complete at puberty 670
  and pubescent spurt 115, 669
  and skeletal age 532-3, 620, 626, 630, 632
Face depth/bizygomatic diameter 641
Face depth/cranial length 642
Face depth/face height 641
Face height 527, 535-8, 559-62, 581, 641
  anterior 530, 743, 1381
  anterior, increments 547, 644-5
  and bizygomatic diameter 609
  increments 563-6
  lower 537-40, 546, 557, 577, 588-90,
    597-8, 651, 685
  lower, and Greulich-Pyle skeletal age 591
  lower, increments 515, 545-6, 587, 601
  perpendicular 1382
  posterior 530, 647, 650, 728, 745
  posterior, increments 547, 644-5
  relative to cranial base length 536-8
  and skeletal maturity 532
  total 46-9, 537-8, 540, 546, 575, 581,
    599-600, 655-6, 729
  total, increments 545-6, 587, 601
  total, and Greulich-Pyle skeletal age 576
  total, and occlusion 544, 655
  total, percent complete at puberty 670
  upper 537-40, 577, 581-5, 599-600, 635,
    651, 658-9, 685, 688, 1381
  upper, and Greulich-Pyle skeletal age 579
  upper, increments515, 545-6, 585, 601
Face shape, upper and lower 723-8
Face width see specific diameters

Facial angles 540-2
Facial plane angles 654, 694, 701-5, 710, 713,
    715-8, 728
Facial profile 550, 784-9, 694-5, 697, 699-
    700, 707-8, 710, 712-6, 719-20, 730
  convexity angle 539-41, 654, 694, 710, 713,
    715-7, 729
  increments 540, 555-6
  and occlusion 697
Facial size, timing of pubescent spurt in
    relation to spurts in stature 670
Family income, influence 91-101, 357-9, 383-7,
    900, 1040, 1093, 1104, 1246, 1267-70,
    1272, 1332
Feeding, type of, influences 138, 155-6, 200-2,
    229, 302, 373-4, 426, 517, 917, 971
Femoral angle 1161
Femoral-pelvic angle 1161
Femoral skinfold thickness 213
Femur, bicondylar diameter 992
  density 1129
  distal end, increments, and Greulich-Pyle
    skeletal age 1202-3
  distal, epiphysis size 229
  ossification onset 1223, 1226, 1238-9
Femur length 1159, 1177-8, 1190, 1192-4, 1374
  increments 1200-1
  and tibial length/stature 1026
Fibula, epiphyseal fusion 1217-8
  ossification onset 1217-8, 1227, 1229, 1239
Fibular length 1205-6
Finger lengths 1074-5, 1108-11
Finger widths 1106-7
Flank skinfold thickness 153, 213
Fluoride content, influence on gonial angle
    706
Foot, distance to crotch 214
  epiphyseal fusion 1382
  longitudinal arch, angulation of 1049
  ossification onset 1237
  sex-associated differences in Greulich-Pyle
    skeletal age 1257
Foot circumference 1037-40
Foot height 1022-3
Foot length 1014-25, 1380
  and birth weight 1024
  and gestational age 1019
  ratios 1025-6
Foot width 1021-2, 1048
Foramen magnum plane angle
  increments 558
Forearm and hand length 351-2, 354-5
Forearm circumference 984-91, 1087-96
Forearm length 1061-8, 1072-3
Forearm length/sitting height 1062
Forehead angle, increments 558
Frankfort horizontal plane angle 528, 722
Frontal breadth, minimum 490
Fronto-occipital circumference see Head
    circumference

Geographic regions, and Greulich-Pyle
        skeletal age 1274-8
Gestational age, and abdomen circumference
        475-6
  and biiliac diameter 222
  and chest circumference 219, 474-5
  and crown-rump length 220
  and foot length 1021
  and head circumference 221, 447, 462, 474-5
  and inner canthal index 676
  and intermammillary index 676
  and recumbent length 26, 135, 139, 141,
        145-6, 218-20
  and rump-sole length 220
  and stature 26
  and thigh circumference 475
  and twins, weight and stature 367-8
  and weight 26, 172-88, 196-8, 218-9
  and weight increments 298
  and weight/recumbent length $\sqrt{3}$ 286
Girth see specific circumference
Gluteal furrow height 213
Gnathion see Cranial base measurements
Gonial angle 535-6, 539-41, 700, 730
  and fluoride content 706
  increments 540, 543, 545-7
Greater sciatic notch breadth 1156
Greulich-Pyle, see Skeletal age,
        Greulich-Pyle
Growth velocity 158

Hamate, ossification onset 1221, 1225
Hand, epiphyseal fusion 1382
Hand circumference 1094-4
Hand length 1069-73
Hand phalanges, epiphyseal fusion 1217-8,
        1289-92
  length 1179-86
Hand width 1072-3, 1105, 1108-11
Hand-wrist bones, epiphyseal fusion
        (see also Wrist) 1289-92
  ossification onset 1211, 1237-8
  sex-associated differences 1257
  skeletal age 1252-87, 1295-6
Head balance axis, increments 558
Head circumference 34, 154, 157, 161-4, 221,
        223-4, 225-6, 350-1, 353-4, 441-60,
        465-81, 1296
  and artificial feeding 465
  and birth weight 157, 369, 450-1, 458,
        470, 477
  and chest circumference 470, 477
  and gestational age 221, 447, 449, 462,
        474-5
  increments 161-4, 308, 447, 461-4, 475-6
  in low birth weight infants 1369
  and maternal age 227-8, 372
  and maternal weight 223-4, 227-8
  and parity 372
  and recumbent length 471-3, 478, 480, 1369

Head circumference (continued)
  and socioeconomic status 390-1, 469
  and stature 478-81
Head depth 32
Head height 493-4, 1369
Head length 46-9, 482-5, 495, 513, 535,
        1296, 1369
  and biparietal diameter 499
  and birth weight 485
  increments 308, 498
  and socioeconomic status 469
Head plus neck length 46, 48
Head position, natural 722
Head, top of, distances to crotch and to
        shoulder 214
Head shape see Cephalic index
Head width see Biparietal diameter
Heel-ankle length 1921-2
Heel to midpoint between calcaneus and
        cuboid/foot length 1025
Height see Stature and specific height
Hip see Acetabular angle and Femoral pelvic
        angle
Hip circumference 213, 979-81, 1380
Hip circumference/lower limb length 1007
Hip, ossification onset 1237-8, 1240
  skeletal age 1289
Hip width see Biiliac diameter and
        Bitrochanteric diameter
His plane angle 721
Humerus, density 1129-30
  epiphyseal fusion 1217-8, 1289
  length 1168-9
  ossification onset 1212-3, 1217-8, 1223,
        1225, 1227, 1229, 1237, 1240
Hyoid, movement during swallowing 728
  ossification centers 1212-3
  position 553, 725-8, 741-2
Hypodontia, agenesis of third molar 859
  and mesiodistal diameters 854

Iliac circumference 982-3
  and recumbent length 415, 418
  and stature 415-8
Iliac crest height 213, 430-1, 436
Iliac crest epiphysis, ossification onset 409
Iliac index 1156, 1180
Ilium breadth 1155
Ilium length 1154
Incisal height 655
Incisal inclination (see also Dental develop-
        ment) 654, 690-1, 695, 696-9, 702,
        709-10, 712-5, 717-9, 729, 743
Incisor position (see also Dental development)
        527, 535, 537-40, 654-5, 675, 688,
        696, 698, 710, 712-5, 719-20
Income see Family income and Socioeconomic
        status
Increments see specific measurement
Inferior segment see Lower limb length

Inner canthal distance 678
Inner canthal index 676, 682
Inter-inner canthal distance 677, 680-1
Intermammillary index 676, 682
Intermaxillary measurements 833-4
Internipple distance 915, 918-9
Interocular diameter 673
Interorbital diameter 671-2
Inter-outer canthal distance 679-80
Interpalpebral width see Inter-inner canthal
        distance
Interpedicular distance 1148-50
Inter-pubic breadth 1154
Interpupillary distance 675, 680
Intervertebral disc height 1139-44
Intervertebral disc height/vertebral body
        height 1147
Intraoral height of permanent teeth,
        relative 1356
Ischium length 1140

Jaw, sagittal relation 796

Knee, ankle, and hand-wrist skeletal age:
        1257, 1280
Knee, epiphyseal fusion 1382
    epiphyses, width 1382
    ossification onset 1237-8
    sex-associated differences in Greulich-Pyle
        skeletal age, 1257
Knee circumference 984-91, 1037-40
Knee diameter 213
Knee-sole length 1013
Knee width 213, 1041-4
    femur bicondylar diameter 901

Lean body mass 984-91
Leg length see Lower limb length
Lip height 596-7
Lip length, normal and extended 619
Lip projection 684-5, 688, 695, 719
Lip separation 593
Lip thickness, upper 718
Lipoproteins, plasma 217
Log weight 55
Low birth weight infants 136-7, 145-6, 298,
        308, 362, 450, 462, 485, 492, 1379
Lower incisor see Incisor
Lower jaw see Mandibular measurements
Lower limb length 27, 35-8, 40, 42-3, 45-6,
        48, 214, 349, 419, 998, 1000,
        1002-7, 1121, 1380
    and calf circumference 39, 41, 1002, 1007,
        1097-8, 1121
    differences between erect and recumbent
        measurements 167
    and growth velocity 158
    and hip circumference 1007
    increments 114, 306-7, 498, 1003
    and pubescent spurt 117, 158

Lower limb length (continued)
    and sitting height 38, 40, 1006-7, 1120
    and stature 1004-5
Lower limb skeleton weight 1124
Lower limb skeleton weight/total skeleton
        weight 1123
Lower lip-upper incisor distance 594
Lower segment length (see also Lower limb
        length) 419, 879, 1008
Lumbar area and height 992
Lumbar spinal canal, sagittal diameter 1151
Lunate, age when adult 1217-8
    ossification onset 1217-8, 1221-2, 1225,
        1227, 1229, 1231

Mandible alveolar length 729
Mandible body length 535-42, 636, 638, 641,
        657, 729-30, 736
    and Greulich-Pyle skeletal age 628
    increments 540, 543, 545-6, 638
    and occlusion 544
    percent complete at puberty 670
Mandible depth 530
    increments 498, 547, 644-5
Mandible elongation rate, age at peak 693
Mandible growth direction 1381
Mandible height, increments 587
Mandible length 216, 252, 276, 528, 535, 637
    and pubescent spurt 116, 252, 276
    and weight 276
    total 629, 697, 729-30, 736, 742
    total, increments 737
Mandible length/stature 276
Mandible shape see Gonial angle, Ramus
        inclination angle
Mandible width 514
    increments 498
Mandibular arch depth 766-9, 828
    increments 808-9
Mandibular arch depth/mandibular arch width
        828
Mandibular arch length 808, 827
    and dental stage 811, 813
Mandibular arch width 777-9, 791-805, 808, 828
    and dental stage 777, 791, 807
    increments 807-9, 815, 817, 819, 821-2,
        824-6
Mandibular first molar position 528, 535
Mandibular foramen, position 1381
Mandibular incisor position see Incisor
Mandibular incisor segment space and dental
        stage 830, 832
Mandibular plane angle 536-42, 654, 660-1, 694,
        699-700, 704, 709-10, 713-8, 728-9
    and occlusion 698
    increments 540, 543, 545-7, 662-4
Mandibular position, increments 556
    relative 541-2
Mandibular ramus height 539-46, 599-600, 656,
        729, 732-6

Mandibular ramus height (continued)
  and Greulich-Pyle skeletal age 592
  increments 543, 545-7
Mandibular size 1381
Mandibular symphysis, concavity 693
  depth 692
  depth, increments 692, 738
  height 729
  position 714
Mastication, muscle activity during 747
Maternal age influence 183, 225-8, 371-2
Maternal parity influence 183, 225-8, 371-2
Maternal weight influence 222-8, 370
Maturation duration (see also specific bone)
      1217-8
Maturation rate, influence of 1001-2, 1044-5
Mature stature prediction see Stature
      prediction
Maxilla size 652-3
  increments 665-8
Maxillary arch depth 763-5, 769, 828
  increments 808-9
Maxillary arch depth/maxillary arch width 828
Maxillary arch length 808, 827
  and dental stage 810, 812
  increments 827
Maxillary arch width 771-9, 799-805, 808,
      818, 823, 828
  and dental stage 770, 777, 783-9, 806
  increments 808-9, 814, 816, 818, 820, 822-6
Maxillary arch width/upper face width 823
Maxillary incisor see Incisor
Maxillary incisor segment space, and dental
      stage 829 831
Maxillary position 527, 535, 648
  increments 543, 545-6
Maxillary sinus size and position 572-3
Maxillary subnasal height increments 587
Maxillary tuberosity, position 553
Maximum shoulder width 883
Menarche 376, 1220
Metacarpal index 1187
Metacarpal lengths 1179-82
  hand phalangeal lengths, ratios 1183-6
Matacarpal, second, cortical thickness 1188
Metacarpal sign 1187
Metacarpals, epiphyseal fusion 1217-8
  ossification onset 1221-3, 1225, 1227,
      1229, 1231, 1239, 1247-8
Metatarsals, epiphyseal fusion 1217-8, 1289
  ossification onset 1212-3, 1226, 1228,
      1230, 1239-40, 1249-50
Metatarsal/foot length 1025
Middle phalanx fifth, bone density
      coefficient 1189, 1209
Mid-thigh circumference 1027
Milk intake influence 229
Molar height 546, 655, 1381
Molar inclination, first 743
Molar, last, distance to distal border of
      ramus 1381

Molar position 655
Molar, third, agenesis 859
Mouth width 618

Nasion see also Cranial base measurements
Nasion-articulare, increments 558
Nasion-basion distance 1380
Nasion-bregma 509, 651
Nasion-gnathion 1381
Nasion-incial edge of upper incisor 1381
Nasion-menton 46-9, 686
Nasion-menton/bizygomatic diameter 609
Nasion-occlusal surface of upper first molar
      1381
Nasion-sella 558, 1380
Nasion-sella-articulare, increments 57
Nasion-sella-basion angle see Saddle angle
Nasion-sella-gnathion angle 535
Nasion-sella-palatal plane angle 550
Nasion-sella plane angle 536-7, 722, 730
Nasion-supramentale 586
Navicular, age when adult 1217-8
  ossification onset 1217-8, 1226, 1228, 1230
Neck circumference 975
Newborn anthropometry, acetabular angle 1164-5
  ankle circumference 1036-7, 1039
  biacromial diameter 885
  biacromial diameter/recumbent length 943,
      945
  biceps skinfold thickness 153
  biiliac diameter 924, 937
  biiliac diameter/recumbent length 943, 945
  biparietal diameter 488, 492
  calf circumference 1037, 1039
  chest circumference 249, 964, 967, 969,
      973-4
  chest depth 914, 956
  chest depth/chest width 914
  chest depth/recumbent length 913
  chest diameter, internal 908
  chest width 904, 906, 914
  chest width/recumbent length 905, 911, 913
  flank skinfold thickness 153
  foot circumference 1037, 1039
  foot length 1021, 1024
  forearm circumference 1094-5
  greater sciatic notch breadth 1156
  hand circumference 1094-5
  head circumference 154, 441, 444-447, 449-
      51, 456, 458-60, 462, 468-70, 473-81
  head length 585
  iliac circumference 982
  iliac index 1156
  ilium breadth 1155
  ilium length 1154
  inner canthal index 676, 678, 682
  inter-inner canthal distance 677
  inter-outer canthal distance 679
  intermammillary index 676, 682
  inter-pubic breadth 1154

Newborn anthropometry
        (continued)
  interpupillary distance 675
  ischium length 1155
  knee circumference 1037, 1039
  lower limb circumference 1037, 1039
  onset of ossification 1212-6, 1223
  onset of ossification, in relation to
        weight groups 1237-8
  outer-orbital diameter 674
  pelvis measurements 1153-4
  plasma cholesterol, lipoprotein, and
        triglycerides 217
  pubis length 1155
  quadriceps skinfold thickness 153
  recumbent length 126, 128, 133-43, 145-7,
        154, 249
  saddle angle 548
  sitting height 401
  sitting height/recumbent length 411, 417
  subscapular skinfold thickness 153
  tongue size 574
  triceps skinfold thickness 153
  upper arm circumference 1094-5
  weight (see also Birthweight) 378
  weight/recumbent length $\sqrt{3}$ 286
  wrist circumference 1094-5
Nipple width 915
Nipples, distance between see Internipple
        distance
Nose breadth, alar 616-7
  bridge 673
  root 616
Nose height 578, 582, 586, 1381
  increments 587
Nose height/bizygomatic diameter 642
Nose height/face depth 643
Nose height/subnasal height 580
Nose length 566
  increments 498
Nose projection 684-7
Nose width 1381
  increments 498

Occlusal plane angle 654, 694, 698-700, 710,
        713, 715-7, 719-20, 729
Occlusion 544, 655, 696-8
Onset of ossification see Ossification onset
Open bite 745
Os coxae, ossification onset 1226
Ossification 1382
Ossification centers 1212-6, 1219-20,
        1232-3, 1237-8, 1247-8
  missing 1251
  multiple 1257-60
  number, in foot and hand-wrist 1232-4
  number, seasonal differences 304-5, 1245
Ossification onset 1211-6, 1217-31, 1237-40,
        1247-50
  adductor sesamoid 110, 1217-8, 1220-2,
        1225, 1240

Ossification onset (continued)
  atlas 1213
  calcaneus 1217-8, 1223, 1226, 1228, 1230
  capitate 1221, 1225
  carpus 1239-40
  cervical vertebrae 1212
  clavicle 1225, 1240
  cranial base 1212-3
  cranial vault 1212-3
  crest of ilium 1217-8, 1220
  cuboid 1217-8, 1223, 1226
  cuneiform, intermediate 1226, 1228, 1230
  cuneiform, lateral 1217-8, 1226
  cuneiform, medial 1217-8, 1226, 1228, 1230
  family income, influence 1246
  femur 1223, 1226-7, 1229, 1238
  fibula 1217-8, 1227, 1229, 1239
  foot 1237-8
  hamate 1221, 1225
  hand-wrist 1211, 1221-2
  hip 1237-8, 1240
  humerus 1212-3, 1217-8, 1223, 1225, 1227,
        1229, 1238-40
  hyoid 1212-3
  iliac crest epiphysis 409
  and infant's weight 1213, 1237-8
  knee 1237-8
  lunate 1217-8, 1221-2, 1225, 1227, 1229, 1231
  and maternal age 1213
  metacarpals 1217-9, 1221-3, 1225, 1227, 1229,
        1231, 1239, 1247-8
  metatarsals 1217-9, 1226, 1228, 1230, 1239-
        40, 1249-50
  navicular 1217-8, 1226, 1228, 1230
  in newborn, 1212-6, 1223
  os coxae 1226
  patella 1217-8, 1226-7, 1229, 1230
  phalanges, finger 1217-22, 1227, 1229, 1231,
        1239, 1247-8
  phalanges, toe 1217-20, 1226, 1228, 1230,
        1239-40, 1249-50
  pisiform 1224
  radius 1217-8, 1221, 1225, 1227, 1229, 1239-40
  scaphoid 1217-8, 1221-2, 1225, 1227, 1229,
        1231, 1240
  scapula 1212-3, 1225, 1239-40
  seasonal influences 1235
  sequence 1231-4
  sesamoid bones of the hand 1223
  talus 1223
  tibia 1217-8, 1225, 1228, 1230, 1238-40
  trapezium 1221-2, 1225, 1227, 1229, 1231
  trapezoid 1221-2, 1225, 1227, 1229, 1231
  triquetral 1217-8, 1221-2, 1225, 1227, 1229,
        1231
  ulna 1217-8, 1220-2, 1225, 1227, 1229, 1231,
        1240
  variability of age 1239-40
  by weight groups 1237-8
Outer-orbital diameter 674

Overbite 655, 744-6, 869-70, 1381
Overjet 746
Oxford method, skeletal age 1288

Palatal plane angle 718, 721
Palate depth 760-1
Palate height 748-55, 758, 760-2
  increments 762
Palate length 756, 759, 762
Palate shape 760-2
Palate width 757-62
Palpebral fissure width 682
Para-umbilical skinfold 27
Parental education, influence 104-5, 388-9,
  1270-1
Parity, influence 183, 225-8, 371-2
Patella, ossification onset 1226-7, 1229,
  1240
Pelvic angle 1161
Pelvic depth 1380
Pelvic growth direction 1163
Pelvic index 1153, 1160
Pelvis, anterior segment index 1153
  biischial breadth 1153
  breadth 1153-54, 1157-60, 1167, 1382
  height 1154, 1157, 1382
  iliac index 1156
  iliac notch breadth 1159
  ilium breadth 1155
  ilium length 1154, 1158
  inlet breadth 1153, 1157
  inlet index 1153, 1160
  inlet, sagittal diameter 1153, 1158
  ischim length 1155, 1158
  newborn 1153-6
  pubic angle 1161
  pubic height 1382
  pubic length 1155, 1158
  skeletal age 1288
Percent body fat 213, 984-91
Perioral pressure 746
Permanent teeth (see also Dental development)
  843-869
  age at emergence 1300, 1302, 1307, 1309,
    1311, 1317-8, 1351
  agenesis of third molar see Hypodontia
  arch differences at age of emergence
    1236-8
  buccolingual diameters 856-8, 867-8
  buccolingual diameters, lateral differences
    860
  calcification centers, number 1208
  Carabelli's tubercle 868
  cementum weight 865
  dentine weight 865
  enamel weight 865
  enamel weight/dentine and cementum weight
    865
  intervals between maturation stages 1310-1
  intra-oral height at intervals after
    emergence 862-3

Permanent teeth (continued)
  lateral differences at age of emergence 1357
  lower central incisor position 868
  lower third molar position 869
  mandibular first molar, rate of maturation
    861
  maturation 1302-8, 1310-12
  mesiodistal diameters 843-58
  mesiodistal diameters, and hypodontia 854,
    859
  mesiodistal diameters, lateral differences
    859
  mesiodistal/buccolingual diameters 866-7
  molar calcification 861
  molar maturation 861
  number emerged at specific ages 1240-1
  relative intra-oral height, rate of change
    1356
  root length 853
  root length/total tooth length 853
  total tooth length 853
  weight 864
Phalanges, hand, epiphyseal fusion 1217-8,
    1289-92
  finger, ossification onset 1217-22, 1227,
    1229, 1231, 1239, 1247-8
  foot, epiphyseal fusion 1217-8
  toe, ossification onset 1217-20, 1226, 1228,
    1230, 1239-40, 1249-50
Pharynx depth 553
Physique 1375
Physique, Sheldon 1119
Physique, Wetzel 1115-8
Pisiform, ossification onset 1224
Pituitary fossa, size 568-71
Plasma cholesterol, lipoproteins and
    triglycerides 217
Post-cranial skeleton weight 1126
Post-cranial skeleton weight/total skeleton
    weight 1125
Potassium, total body 154
Profile see Facial profile
Prognathism 696-7
Pterygoid root position 741
Pterygoid width 559, 561
  increments 563-6
Pubescence, and adult stature 1375
  and exophthalmos 683
  and facial size 670
  and Greulich-Pyle skeletal age 1258-9
  and hand-wrist ossification 1211
  and hand-wrist skeletal age 1295
Pubescent events, chronological and skeletal
    ages 1382
Pubescent spurt (see also specific variable)
    115-8, 158, 216, 251-2, 276, 421, 425,
    534, 669-70, 737, 900, 1295-6, 1377,
    1382
Pubic or pubis measurements see Pelvic or
    pelvis

Quadriceps skinfold thickness 153

Radiale height 428
Radius, density 1129-30
    compact bone area 1173
    epiphyseal fusion 1217-8
    epiphysis 1130
    length 1169-71
    length, increments 1172, 1204
    metaphyseal band width 1174-5
    medullary cavity width 1173
    ossification onset 1217-8, 1221, 1225,
        1227, 1229, 1239-40
    total cross-sectional area 1173
Ramus inclination angle 536-42
    increments 540, 543, 545-6
Recumbent length 28, 120-54, 159-64, 216,
        218-20, 222, 225-6, 229, 247-9, 251-
        3, 911, 913, 943, 945, 1379
    and birth weight 136-7, 140, 153-4, 157,
        218
    and gestational age 135, 139, 141, 145-6,
        218-20, 286
    growth velocity 158
    and head circumference 154, 157, 471-3,
        478, 480, 1379
    and iliac circumference 415, 418
    increments 144-52, 161-4, 308-9, 311-4
    low birth weight infants 145-6, 1379
    and maternal age and parity 371
    and pubescent spurt 158
    and stature 165-7
    and twins, differences 142-4, 159-60, 364
    and type of feeding 138, 155-6, 373, 1379
    and weight, ratios 278-81, 285
Recumbent measurements, erect measurements,
        differences 165-7
Regional differences, U.S. anthropometry
        106, 356, 1274-8
Relative sitting height 405
Relative weight 290-1
Roche-Wainer-Thissen method see Skeletal age
Root length, permanent teeth 855
Rugal length, increments 759
Rugal width, increments 759
Rump-sole length (see also Lower Limb length)
        215, 220, 223, 226-7, 371-2, 999
    and gestational age 220
    increments 309

Sacral index 1153, 1160
Saddle angle 548-9, 551-3, 712, 728, 1380
    increments 554, 557-8
Scalp thickness 496-7
Scaphoid, age when adult 1217-8
    ossification onset 1217-8, 1221-2, 1225,
        1227, 1229, 1231, 1240
Scapula, ossification onset 1212-3, 1225,
        1239-40
Seasonal effects, on ossification 304-5, 1245

Seasonal effects (continued)
    on stature 111-2
    on stature increments 111, 304-5
    on weight increments 304-5, 375
Segment lengths, lower and upper 879
Sella see also Cranial base measuremnts and
            Pituitary fossa
Sella-articulare, increments 558-601
Sella-basion, increments 558
Sella-gnathion 1381
Sella-gonion 1381
Sella-nasion-gnathion angle 535
Sequence of ossification onset 1241-4
Sesamoid bones of the hand, ossification
        onset 1223
Sheldon physique 1119
Shoulder circumference 984-90
Shoulder, epiphyseal fusion 1382
    ossification onset 1237-8
Shoulder-center of gravity, distance 214
Shoulder-top of head, distance 214
Shoulder-trochanter, length 873-4
Shoulder-vertex, distance 214
Shoulder width see Biacromial diameter
Sinus see specific name
Sitting height 27, 38, 40, 42-3, 51, 213-4,
        395-408, 1006-7, 1062, 1120, 1379
    and biiliac diameter 1120
    crown-rump length differences 167
    and Greulich-Pyle skeletal age 408
    and lower limb length 38, 40, 1006-7, 1120
    potential remaining at skeletal ages 409
    and recumbent length 415
    and stature 51, 410-9
Skeletal age 1382
    and adult stature, percent achieved 1361-5
    and family income 1267-70, 1272
    and geographic region 1274-8
    Greulich-Pyle (see also specific measurement)
        63, 1253-7, 1260-79
    hand-wrist 1252-87, 1281-4
    hip and pelvis 1288
    knee 1280
    lateral differences 1293-4
    Oxford method 1288
    parental education 1270-1
    pubescence 115-6, 1258-9, 1295-6
    Roch-Wainer-Thissen method 1281-2
    Sex-associated differences in the ankle,
        hand-wrist, and knee 1257
    and socioeconomic status 469
    Tanner-Whitehouse method 1283-7
    urban/rural differences 1273-4
Skeleton weight (see also specific bones)
        1125-6
Skinfold thickness (see also specific site) 27
Skull weight 1126
Skull weight/total skeleton weight 1125
Socioeconomic status (see also Family income),
        and birth weight 229

Socioeconomic status (continued)
  and chest circumference 970
  and distal femoral epiphysis size 229
  and head circumference 390-1, 469
  and head length 469
  and proximal tibial epiphysis size 229
  and recumbent length 134, 229
  and skeletal age 469
  and stature 54, 102-3, 250, 390-1, 469
  and weight 54, 250, 357-9, 377-82, 390-1,
      469
Soft palate thickness 553
Somatotype and percent adult stature
      achieved 1371
Specific gravity 213
Sphenoccipital-sphenoethmoidale/nasion-
      basion angle 1380
Sphenoethmoidale-nasion/nasion-basion angle
      1380
Sphenoidal plane angle 721
Sphenoid sinus, shape 567
  size 559-62
Sphyrion-fibulare height 438
Spine height, anterior iliac 432-5
Squamous occipital, inclination 571
Standing height see Stature
Stature 9-24, 26-32, 34-8, 40, 42-6, 48,
      50-8, 62-106, 204-14, 216, 250-2,
      276, 349-51, 353-4, 410, 415-9, 736,
      984-91, 1014-5, 1026, 1296, 1379
  adult, age of achievement 1374-5
  adult, percent achieved 1367-9
  adult, percent achieved relative to foot
      length, femur and tibia length 1374
  adult, percent achieved relative to
      maturation rate and skeletal age
      1361-5
  adult, percent achieved relative to
      somatotype 1369
  adult, prediction 1366, 1368-71
  and biiliac diameter 41, 933-6, 943-6,
      953-6
  and chest measurements 478-81, 911-5
  diurnal variation 106-119
  erect and recumbent length measurements
      165-7
  and head circumference 478-81
  and iliac circumference 415-8
  increments 54, 107, 113-4, 304-7
  for midparent stature 59-62
  and pubescent spurt 115, 118, 251-2, 669
  ratios 25, 276-7, 282-5, 287-9, 421, 425,
      478-81, 1121, 1379
  and skeletal age 63
  and skinfold thickness 100
  and socioeconomic status 54, 102-3, 250,
      390-1, 469
  and stem length 421, 425
  twins 22-3, 142, 159-60, 364, 367

Stature (continued)
  velocity 110-2, 251-2, 1180-1
  and weight 25, 52, 256-77, 282-5, 287-9,
      1121, 1379
Stem length 32, 35-7, 45, 146, 349-51, 353-4,
      420-1, 1379
  and chest circumference 1097-8, 1121
  increments 306-7, 498
  and maturation rate 421-4
  and pubescence 117, 421, 425
Stem length/stature 421-2, 426
Sternal length 915
Strength, influence of pubescent spurt 117
Stylion height 429
Subcostal skinfold thickness 153, 213
Subischial length 419, 998
Submandibular glands, volume 574
Subnasal height 580-1
Subscapular skinfold thickness 27, 34, 153, 213
Superior segment, growth velocity and pubescent
      spurt 158
Supernumerary teeth 1344
Suprailiac skinfold thickness 213
Supramammillary and xiphoid levels, chest
      circumference 973-4
Suprasternal height 161-4, 877-8, 1382
Suprasternal notch-symphysis 161-4, 1382
Suture width 571
Swallowing, hyoid and muscle movement during
      728, 747
Symphyseal angle 718
Symphyseal height 161-4, 437, 1000, 1380

Talus, age when adult 1217-8, 1289
  ossification onset 1223
Teeth see Deciduous and Permanent, Dental
      development, Incisor, Molar
Thigh circumference 46-9, 984-91
  and family income 1040
  and gestational age 473-4
Thigh length 1008-11
Thigh skinfold 213
Thorax see Chest
Thumb length 1108-1111
Tibia, density 1129-30
  epiphyseal fusion 1217-8, 1289
  epiphysis 1130
  length 1026, 1190-8, 1382
  length, and foot length 1025
  length, and Greulich-Pyle skeletal age
      1192-1201
  length, increments 1199-1200, 1204, 1382
  length, and percent adult stature achieved
      1376
  medial malleolus, separate centers 1251
  ossification onset 1217-8, 1223, 1228, 1230,
      1238-40
  proximal end, increments, and Greulich-Pyle
      skeletal age 1202-3

Tibia (continued)
  proximal, epiphysis, size 229
  width 1205
Tibiale height 438, 1380
Toe phalanges, ossification onset 1217-20,
        1226, 1228, 1230, 1239-40, 1249-50
Tongue position 553-741
Tongue size 574
Torso width, lower 916
Total body potassium 154
Trapezium, ossification onset 1221-2, 1225,
        1227, 1229, 1231
Trapezoid, ossification onset 1221-2, 1225,
        1227, 1229, 1231
Triceps skinfold 27, 153, 213
Triglycerides, plasma 217
Triquetral, age when adult 1217-8
  ossification onset 1217-8, 1221-2, 1225,
        1227, 1229, 1231
Trochanter-knee length (see also Thigh
        length) 873-4
Trochanteric height 1380
Trunk area index 993-4
Trunk height, anterior 879
Trunk length (see also Shoulder-trochanter
        length) 46, 48, 220, 1002, 1379
Twins, birth weight 197, 243-4, 363-9
  head circumference 460
  head circumference, increments 464
  recumbent length 142-4, 159-60, 364
  stature 22-3, 142, 159-60, 364, 367
  weight 243-4, 363-9
  weight, increments 303

Ulna, epiphyseal fusion 1217-8, 1289-91
  length 1176-7, 1382
  ossification onset 1217-8, 1220-2, 1225,
        1227, 1229, 1231, 1240
Ulnar sesamoid ossification 532-3
Upper arm circumference 27, 34-8, 40, 42-4,
        46-9, 213, 349, 1076-82, 1094-6
  and family income 1093
  biceps circumference 984-91
  increments 306-7, 498, 1083
Upper arm circumference/upper limb length
        1097-8, 1121
Upper arm length 213, 351-2, 354-5,
        1054-61, 1072-3
Upper arm/sitting height 1062
Upper arm muscle circumference 1084-6
Upper incisor see Incisor
Upper limb length 35-7, 46-8, 349, 1055-6,
        1097-8, 1121, 1380
  increments 306-7
Upper limb skeleton weight 1126
Upper limb skeleton weight/total skeleton
        weight 1125
Upper segment length 114, 879
Urban/rural differences 97-9
  and Greulich-Pyle skeletal age 1273-4
  and weight 357-9

Vertebral body height ratios 1145-7
Vertex-center of gravity, distance 214
Vertex-crotch, distance 214
Vertex-shoulder, distance 214, 466-7
Vertex-suprasternal notch, distance 161-4, 1382
Vertex-symphysis, distance 161-4, 1382

Waist circumference 137, 213
Waist width 920, 960, 976-8
Weight (see also Birth weight) 25-38, 40,
        42-58, 80, 82, 154-6, 204-16, 23-51,
        315-91, 736, 984-91, 1296, 1379
  and biiliac diameter 1379
  dietary influences 1378
  differences between twins 363-8
  diurnal variation 292, 295, 302
  and family income 384-7
  and feeding, type of 155-6, 200-2, 302,
        373-4
  and gestational age 26, 172-88, 196-8, 218-9
  groups, and ossification onset 1237-8
  increments 54, 116, 161-4, 292-314, 374, 498
  log of 55
  loss after birth, and artificial feeding 202
  in low-birth-weight infants 1379
  and mandibular length 276
  and maternal age, parity, and weight 222,
        225, 370
  maternal see Maternal weight
  at menarche 376
  and pubescent spurt 116-7, 251
  and recumbent length 253-5, 278-81, 286
  and skeletal age 242
  and socioeconomic status 54, 250, 357-9,
        377-82, 390-1, 469
  and stature 25, 52, 256-77, 282-5, 287-9,
        1121, 1379
  in twins 194-5, 243-4, 303, 363-9
  and urban/rural differences 357-9
  and wrist circumference
Wetzel grid 1379
Wetzel physique 1115-8
Wrist circumference 984-91, 1094-6
Wrist diameter 213, 984-91, 1102-4
Wrist-finger tip length 1072-3
Wrist-knuckle length 1072-3

Xiphisternal circumference see Chest
        circumference
Xiphoid and supramammillary levels, chest
        circumference 973-4

Y-axis 554, 694, 698, 703, 710, 713, 716-8,
        728
  and fluoride content 706
  and occlusion 698

Data for U.S. Children -- By Race

Alaskan Eskimos 56-7, 90, 348, 818
American-Dutch 45
American-Finnish 48-9
American Indians 50, 82-3, 190-5, 340, 443,
         495, 800-1, 1323, 1337
American-Italian 46-7
American-Japanese 51, 75, 81, 264-7, 289,
         330, 341, 396, 680, 694, 720, 972,
         1266, 1286
American-Oriental 694, 984-91, 1264-5
Blacks 38-44, 65-9, 71-5, 77-9, 93-4, 100-2,
         109, 130-4, 137, 178-81, 183, 213,
         215, 217, 219-24, 232, 249-50, 264-7,
         288, 296, 301, 308-15, 317-21, 327,
         330, 332, 334-9, 350-5, 360-2, 370-2,
         377-9, 385-6, 396, 403-4, 411, 427,
         430-1, 433, 438, 450-1, 454-7, 465,
         468-70, 484-5, 489-90, 492, 509,
         517, 575, 578, 582-4, 588-9, 595,
         597-8, 608, 613-4, 617-8, 621-5, 627,
         633-4, 636, 681-2, 694, 713-4, 718
         732-4, 847, 854-5, 876, 878-9, 882,
         888-9, 900, 910, 917, 923-9, 949,
         951-2, 959, 969-71, 978, 981, 984-91,
         998, 1000-1, 1006-8, 1010, 1012,
         1017, 1029, 1033, 1035, 1040, 1042-4,
         1047, 1054, 1057, 1064, 1068, 1073,
         1078, 1082-3, 1086, 1089, 1091-3,
         1100-4, 1125-6, 1129-30, 1166-7,
         1189, 1211-2, 1217-8, 1231, 1237-8,
         1262-5, 1268-9, 1271-2, 1275, 1277-9,
         1283-5, 1291, 1315, 1321-2, 1332-3,
         1339, 1342

General 9-15, 64-9, 74, 120-1, 171, 184-7,
         230-3, 253-4, 156-63, 287, 316,
         318-21, 323, 325, 357-9, 369-77,
         381-4, 387-8, 395-7, 410-1, 430-1,
         441-2, 482, 486, 581, 585, 602-1,
         605, 612-3, 654, 689, 713, 823,
         843-7, 875, 880-1, 883, 902-3, 916,
         920-2, 949, 955-6, 961-3, 975-7,
         979-80, 997-9, 1008-9, 1011-6, 1018,
         1023, 1027-9, 1031, 1036, 1041-2,
         1046-8, 1055-6, 1058, 1063, 1065,
         1071-2, 1074-7, 1079, 1084-5, 1087-8,
         1090, 1099-100, 1102-3, 1105-7,
         1125-9, 1189, 1207, 1211-4, 1223,
         1231-4, 1235-8, 1244, 1246, 1253-6,
         1262-3, 1267-71, 1273-9, 1290-2,
         1295-6, 1302, 1312, 1317, 1321-2,
         1332-3, 1351-2, 1365-6, 1372-5

Hawaiian-Japanese 52
Latin-Americans 44, 66-8, 75, 80, 84, 264-7,
         319-20, 330, 337, 379, 419, 654,
         675, 682, 719, 900, 1034, 1089

Data for Canadian Children -- By Race

Eskimos 57-8, 200, 1328
French-Canadian 88-9, 275, 346-7, 801, 804-5,
         1325-7
General 53, 85-7, 251-2, 268-74, 344-7, 406-7,
         413-4, 434-5, 459, 513, 1001, 1005
Igloolik 24, 245
Indians 190-4, 199, 1327

Data for U.S. Children --
     By Place of Residence

Alabama 27, 376, 678, 715, 719, 1720
Arkansas 1344
Berkeley, CA 25, 331
Boston, MA 16, 28, 107-8, 122, 234, 246-8,
         376, 398-9, 409, 426, 444-5, 699, 722,
         758, 770-1, 790-4, 801-2, 804-7,
         810-21, 827, 829-32, 835, 843-4, 848,
         851-3, 870, 905, 925-6, 965, 1019-20,
         1025-6, 1045, 1190-4, 1200-3, 1205,
         1217-22, 1255, 1289, 1297-9, 1302,
         1310-1, 1317, 1333, 1342, 1345-50,
         1353, 1369, 1376
Buffalo, NY 145-6, 462
California 26, 29, 74, 117, 129-32, 296,
         331-4, 421-5, 446, 452-5, 463, 514,
         535, 546, 675, 677, 679-80, 692, 728,
         741-2, 891-3, 900, 939-42, 984-91,
         1032, 1264-5, 1272, 1361-2, 1365-6,
         1368, 1374
Cincinnati, OH 70-3, 171, 322-5, 509, 1115-8
Cleveland, OH 17, 123, 235, 400, 418, 430,
         436-7, 483, 487, 493, 501-6, 646-8,
         695-6, 725-7, 738, 877, 883, 905,
         926, 950, 965-6, 1053, 1059, 1066,
         1069, 1195, 1220, 1237-40, 1252,
         1286, 1369
Colorado 358, 671-2, 911-4, 937-8, 943-6,
         956-7, 967-8, 1037-40, 1080-1, 1094-6,
         1260
Connecticut 25, 76, 161-4, 993, 1118
Delaware 758
Denver, CO 18-9, 115, 124-5, 139, 144, 172-3,
         236-7, 278-9, 282-3, 286, 299, 356,
         376, 401-2, 415-8, 447, 477-81, 507-8,
         551, 554, 663, 669-70, 737, 798,
         885-9, 906-9, 981-2, 1168-9, 1176-8,
         1196-7, 1205-6
Georgia 174-81
Idaho 685
Illinois 30, 500, 512, 544, 570, 598, 649-50,
         730, 968
Indiana 685, 864-5
Iowa 138, 148-52, 157, 167, 292-5, 374-5, 448,
         488, 491, 494, 498-9, 509, 515, 527-
         31, 547, 552, 555-6, 580-1, 587, 606,
         608-11, 635, 639-45, 651, 656, 658-65,

Iowa (continued)
        684, 693, 700-4, 723-4, 739, 760-2,
        1003, 1108-11, 1121, 1300-1, 1319-20,
        1334, 1336, 1351, 1356
Iowa City, IA 31-3, 155-6, 204-12, 306-7
Kansas 141, 182, 449
Los Angeles, CA 75, 200, 264-7, 330
Louisiana 213
Louisville, KY 22-3, 142-4, 159-60, 243-4,
        303, 363, 368, 460, 464
Maryland 277, 799, 808
Massachusetts 196-7, 201
Michigan 34, 45, 146, 290-1, 326-7, 466-7,
        524-6, 549-50, 629, 637, 689-91,
        731, 735, 743-4, 748-9, 763-8, 780-2,
        795-7, 844, 855, 857, 866, 873-4,
        887, 918-9, 1022-3, 1072-3, 1223,
        1238, 1293-4, 1339, 1342
Minnesota 21, 128, 240-1, 255, 297, 762, 855
Missouri 469, 572, 1296
Nebraska 839-42
New England 368
New Jersey 374, 973-4
New Orleans, LA 134, 136, 203, 360, 469
New York 196-7, 201, 705-6, 716, 1187
New York City 137, 184-7, 308, 315, 362, 485,
        492, 516, 559, 1164-6, 1212-6,
        1224, 1313
Ohio 20-1, 113, 118, 126-7, 147, 238-9, 280-1,
        284-5, 299-300, 304-5, 328-9, 375-6,
        510-1, 518-23, 530, 532-4, 548, 557,
        568-9, 573, 707-8, 849, 858-61, 867,
        869, 1044-5, 1049, 1131-44, 1152-63,
        1171-2, 1179-86, 1188, 1198-9, 1204,
        1225-6, 1232-4, 1239-43, 1245, 1251,
        1276, 1281-2, 1305, 1311, 1314, 1375
Oklahoma 214, 513, 577, 586, 604, 616, 619,
        673, 686-7, 740, 958, 960, 1060, 1067
Oregon 35-7, 154, 188, 349, 496-7, 559-66,
        652-3, 665-8, 736, 850, 1097-8,
        1148-52, 1258-9
Pennsylvania 536-40, 543, 545-6, 553, 567,
        571, 601, 683, 717
Philadelphia, PA 63, 77-9, 109, 138, 215,
        219-28, 242, 249, 303-4, 309-14,
        335-6, 361, 370-2, 376, 406-7, 408,
        412, 427-30, 433, 439, 465, 468, 484,
        490, 517, 575-8, 579-80, 582-4, 588-9,
        591-2, 595, 597-8, 602-3, 605, 607-8,
        614-5, 617-8, 620-8, 630-4, 636, 681,
        709, 717, 729, 732-4, 746, 783-9,
        878-9, 888-90, 910, 917, 926-7, 929,
        950-2, 959, 972, 1000-1, 1010, 1033,
        1035, 1043, 1052, 1066, 1071, 1082-3,
        1091-2, 1101, 1167, 1261, 1283-5
Seattle 674, 678, 710-2
South Carolina 38, 43, 402, 1006
Tennessee 183, 855
Texas 44, 66, 250, 337
Utah 746-7

Washington, DC 133-4, 189, 229, 338-9, 350-5,
        929, 969, 1315
Wisconsin 1173

                            Canada

Montreal 24, 88-9, 135, 140, 198, 218, 245,
        275, 346-7, 458, 473-7, 894-8, 930-5,
        1021, 1024, 1287, 1324, 1343
Ontario 116, 216, 276, 693, 736-7
Ottawa 54, 103, 380
Saskatchewan 54
Saskatoon 24, 245
Trois-Rivières, Québec 55

                            Other

Aleut Islands 190, 192
Australia 542-3, 599-601, 915, 918-9, 1211,
        1247-50, 1257, 1280, 1329, 1341, 1352
Denmark 110, 558, 696-7, 757
England 111-2, 119, 302, 574, 590, 593-4,
        596, 833-4, 992, 994, 1059-62, 1174-5,
        1187, 1330-1, 1340, 1354-5
Finland 721
France 114, 118, 158
Ireland 869
Scotland 153, 574
Sweden 657, 745-6, 1314-6, 1335, 1338, 1357